DISCARDED

Geothermal Systems:
Principles and Case Histories

Geothermal Systems: Principles and Case Histories

Edited by

L. Rybach
Institute of Geophysics, ETH Zurich, Switzerland

and

L. J. P. Muffler
U.S. Geological Survey, Menlo Park, California, U.S.A.

A Wiley–Interscience Publication

JOHN WILEY & SONS
Chichester · New York · Brisbane · Toronto

Copyright © 1981 by John Wiley & Sons Ltd.

All rights reserved.

No part of this book may be reproduced by any means, nor transmitted, nor translated into a machine language without the written permission of the publisher.

British Library Cataloguing in Publication Data:

Geothermal systems.
  1. Earth temperature
  I. Rybach, L.   II. Muffler, L. J. P.
  551   QE509   80-40290

ISBN 0 471 27811 4

Photoset in Malta by Interprint Limited and printed in Great Britain by The Pitman Press, Bath, Avon.

List of Contributors

M. Choussy	Comision Ejecutiva Hidroelectrica del Rio Lempa, El Salvador.
M. D. Crittenden, Jr	U.S. Geological Survey, Menlo Park, California 94025, U.S.A.
G. Cuéllar	Comision Ejecutiva Hidroelectrica del Rio Lempa, El Salvador.
I. G. Donaldson	Physics and Engineering Laboratory, Department of Scientific and Industrial Research, Lower Hutt, New Zealand.
D. Escobar	Comision Ejecutiva Hidroelectrica del Rio Lempa, El Salvador.
R. O. Fournier	U.S. Geological Survey, Menlo Park, California 94025, U.S.A.
J. Gálfi	Research Institute for Water Management, 1095 Budapest, Kvassay, J.u.1., Hungary.
S. K. Garg	Systems, Sciences and Software, La Jolla, California 92038, U.S.A.
M. A. Grant	Applied Mathematics Division, Department of Scientific and Industrial Research, Wellington, New Zealand.
F. Horváth	Eötvös University, 1083 Budapest, Kun Béla Tér 2, Hungary.
D. R. Kassoy	Department of Mechanical Engineering, University of Colorado, Boulder, Colorado 80309, U.S.A.
K. Korim	Water Exploration Enterprise, 1051 Budapest, Zriniyi u.1., Hungary.
A. W. Laughlin	Geosciences Division, Los Alamos Scientific Laboratory, Los Alamos, New Mexico 87545, U.S.A.
J. T. Lumb	Department of Scientific and Industrial Research, Wellington, New Zealand.
L. J. P. Muffler	U.S. Geological Survey, Menlo Park, CA 94025, U.S.A.
H. Nakamura	Japan Metals and Chemicals Co., Ltd., 8-4 Koami-cho, Nihonbashi, Chuo-ku, Tokyo 103, Japan.
P. Ottlik	Hungarian Water Authority, 1011 Budapest, Fö u. 44, Hungary.
L. Rybach	Institute of Geophysics, ETH Zürich, CH-8093 Zürich Switzerland.
V. Stefánsson	National Energy Authority, Geothermal Division, Reykjavík, Iceland.
L. Stegena	Eötvös University, 1083 Budapest, Kun Béla Tér 2, Hungary.
K. Sumi	Geological Survey of Japan, Higashi 1-chome 1-3, Yatabe-machi, Tsukuba-gun, Ibaraki-ken 305, Japan.

Contents

Preface   **PRINCIPLES**   xiii

1. **Geothermal Systems, Conductive Heat Flow, Geothermal Anomalies**

    L. Rybach 3

    1.1. Geothermal Systems: Definition and Classification 3
    1.2. Geophysical and Geochemical Signatures of Geothermal Systems 5
        1.2.1. Convective Hydrothermal Systems Related to Shallow Intrusions . 5
        1.2.2. Circulation Systems with Meteoric Water in Areas of High to Normal Regional Heat Flow 6
        1.2.3. Low Temperature ($<150°C$) Aquifers in High-porosity Environments 7
        1.2.4. Hot Dry Rock (HDR) 8
    1.3. Conductive Heat Transfer and Regional Heat Flow 9
        1.3.1. Fundamentals 9
        1.3.2. Heat Sources. 13
        1.3.3. Conductive Heat Flow 16
        1.3.4. Heat Content of Rocks, Geothermal Resource Base 25
    1.4. Geothermal Anomalies and Their Plate Tectonic Framework 25
        1.4.1. Spreading Ridges 26
        1.4.2. Continental Rifts 29
        1.4.3. Convergent Margins (Subduction zones) 30
        1.4.4. Intraplate Thermal Anomalies. 31

2. **Convective Heat and Mass Transfer in Hydrothermal Systems**

    S. K. Garg and D. R. Kassoy. 37

    2.1. Mathematical Modelling of Hydrothermal Systems. 37
        2.1.1. Mass and Energy Transport in Fractured Media 38
        2.1.2. Balance Laws for Fluid Flow Through Porous Media 39
        2.1.3. Boussinesq Approximation. 42
    2.2. Idealized Convective Heat and Mass Transport 43

	2.2.1. Nondimensional Equations and Parameters	43
	2.2.2. Idealized Modelling	45
2.3.	Plausible Models of Geothermal Reservoirs	52
	2.3.1. Pipe Models	52
	2.3.2. Large-scale Convection Cell Models	54
	2.3.3. Hot Springs and Fault Zones	57
	2.3.4. Fault Zone-controlled Charging of a Geothermal Reservoir	59
2.4.	Pre-production Models of Hydrothermal Systems	60
	2.4.1. Long Valley Caldera	62
	2.4.2. Salton Sea Geothermal Reservoir	64
	2.4.3. Wairakei Geothermal Field	68

3. Prospecting for Geothermal Resources

J. T. Lumb 77

3.1.	Introduction	77
3.2.	Prospecting Strategies	78
3.3.	Hydrological Aspects	79
3.4.	Exploration Techniques	80
	3.4.1. Literature Survey and Preliminary Investigations	81
	3.4.2. Geological, Hydrological and Mineralogical Studies	82
	3.4.3. Geochemistry (including Isotope Chemistry)	83
	3.4.4. Geophysical Methods	87
	3.4.5. Exploration Drilling	101
3.5.	Modelling	103

4. Application of Water Geochemistry to Geothermal Exploration and Reservoir Engineering

R. O. Fournier 109

4.1.	Introduction	109
4.2.	Hydrothermal Reactions	109
4.3.	Estimation of Reservoir Temperature	113
	4.3.1. The Silica Geothermometer	113
	4.3.2. The Na/K Geothermometer	118
	4.3.3. The Na–K–Ca Geothermometer	119
	4.3.4. The $\Delta O^{18}(SO_4^= - H_2O)$ Geothermometer	121
4.4.	Underground Mixing of Hot and Cold Waters	122
	4.4.1. Recognition of Mixed Waters	122
	4.4.2. The Silica Mixing Model	123
4.5.	Effects of Underground Boiling	124
	4.5.1. Calculation of Change in Concentration Resulting from Boiling	124
	4.5.2. Use of Enthalpy-chloride Diagrams for Estimating Reservoir Temperatures	125

	4.5.3. Effect of Boiling upon Geochemical Thermometers.	127
4.6.	Vapor-dominated Compared to Hot-Water Systems.	127
4.7.	Comparison of Thermal Waters from Springs and Wells	129
4.8.	Applications for Reservoir Engineering	136
	4.8.1. Early Indication of Aquifer Temperatures in Wells	136
	4.8.2. Monitoring Temperature Changes in Production Wells	136
	4.8.3. Confirmation of Production from Multiple Aquifers at Different Temperatures.	137
	4.8.4. Flashing in the Reservoir.	137
	4.8.5. Evidence of Higher Temperatures Elsewhere within a System	138
	4.8.6. Geochemical Evidence of Drawdown	139
4.9.	Summary.	140

5. Heat Extraction from Geothermal Reservoirs

I. G. Donaldson and M. A. Grant 145

5.1.	Introduction	145
	5.1.1. Some Background Considerations	146
5.2.	Idealized Systems.	148
	5.2.1. A Simple Basic Model	148
	5.2.2. The Single-phase (Hot Water) Reservoir	150
	5.2.3. The Two-phase System	152
5.3.	Heat Extraction from Some Typical Existing Fields	157
	5.3.1. The Warm-Water Reservoir	157
	5.3.2. Hot-Water Reservoirs	160
	5.3.3. Two-phase Systems	167
	5.3.4. Vapor-dominated Systems	171
5.4.	Other Systems.	173
	5.4.1. The Geopressured System	173
	5.4.2. The Hot Dry Rock System.	173
	5.4.3. Magma Chambers.	174
5.5.	Summary.	174

6. Geothermal Resource Assessment

L. J. P. Muffler. 181

6.1.	Introduction	181
6.2.	Resource Terminology.	181
6.3.	Geothermal Resource Terminology	184
6.4.	Methodology for Geothermal Resource Assessment.	185
6.5.	Selected Recent Geothermal Resource Assessments.	186
	6.5.1. Central and Southern Tuscany/Italy	187
	6.5.2. New Zealand.	187
	6.5.3. United States—1975.	188
	6.5.4. United States—1978.	190

6.6. Uncertainties and Areas that Need Investigation.	193
6.6.1. Undiscovered Hydrothermal Convection Systems.	193
6.6.2. Recovery of Geothermal Energy from Hydrothermal Convection Systems	193
6.6.3 Evaluation of Igneous-related Geothermal Energy.	194
6.6.4. Recovery of Energy from Geopressured-geothermal Fluids	194
6.6.5. Low-temperature Inventory	195

7. Environmental Aspects of Geothermal Energy Development

M. D. Crittenden, Jr.	199
7.1. Introduction	199
7.2. Evolution of Environmental Concern.	199
7.3. Air Quality.	201
7.4. Water Quality.	204
7.5. Landslides	206
7.6. Seismicity	208
7.6.1. Regional Seismicity	208
7.6.2. Surface Rupture.	209
7.6.3. Induced Seismicity.	210
7.7. Subsidence.	211
7.8. Conclusions.	213

CASE HISTORIES

8. The Low Enthalpy Geothermal Resource of the Pannonian Basin, Hungary

P. Ottlik, J. Gálfi, F. Horváth, K. Korim and L. Stegena	221
8.1. Introduction	221
8.2. Geological Framework.	223
8.2.1. Geological Overview.	224
8.2.2. Tectonism and Volcanism	228
8.3. Geothermics	228
8.3.1. Rock Temperature Measurements	230
8.3.2. Heat Flow Determinations.	230
8.4. Hydrogeology.	233
8.4.1. Water Migration Systems	236
8.4.2. Hydrochemistry.	238
8.5. Recovery and Utilization of the Geothermal Energy	239
8.5.1. The Main Branches of Thermal Water Utilization	241
8.5.2. Problems Related to the Operating of Geothermal Wells	243
8.5.3. High Enthalpy Reservoir Possibilities in the Pannonian Basin	244

9. Exploration and Development at Takinoue, Japan

H. Nakamura and K. Sumi 247

9.1. Introduction 247
9.2. Geological Framework 250
    9.2.1. General Geology of the Hachimantai Volcanic Region 250
    9.2.2 Geology in and around the Takinoue Geothermal Area 252
9.3. History of Exploration 252
    9.3.1. Cooperative Investigations with the Geological Survey of Japan . 252
    9.3.2. Exploration by Japan Metals & Chemicals Co. 256
9.4. Production Characteristics 263
    9.4.1. Production of Geothermal Steam 263
    9.4.2. Reinjection of Hot Water 263
    9.4.3. Well and Reservoir Testing 264
    9.4.4. Reinjection Testing 265
    9.4.5. Steam Supply System 266
9.5. Effectiveness and Economics 270
9.6. Future Development 272

10. The Krafla Geothermal Field, Northeast Iceland

V. Stefánsson 273

10.1. Introduction 273
10.2. Geological Framework 275
10.3. Exploration History and the Model of the Field 279
    10.3.1. Surface Exploration 279
    10.3.2. Subsurface Exploration 282
    10.3.3. The Model of the Field 282
10.4. Production Characteristics of the Field 285
    10.4.1. The Upper Zone 286
    10.4.2. The Lower Zone 290
    10.4.3. Interaction between the Zones 291
    10.4.4. Influence of Magmatic Activity on the Reservoir 291
10.5. Experience with Utilization 292
10.6. Current and Future Developments 293

11. The Geothermal System of the Jemez Mountains, New Mexico (U.S.A.) and Its Exploration

A. W. Laughlin 295

11.1. Introduction 295
11.2. Regional Geological and Geophysical Setting of the Jemez Mountains . . 296
11.3. Geology, Geophysics and Hydrology of the Jemez Mountains 300

11.4. Geothermal Exploration in the Jemez Mountains	304
11.4.1. High-temperature Hydrothermal Exploration	304
11.4.2. The Hot Dry Rock Geothermal Energy Extraction Method	305
11.4.3. Hot Dry Rock Exploration.	306
11.4.4. Exploration for Low to Moderate Temperature Hydrothermal Systems.	314
11.5. The Geothermal System of the Jemez Mountains	316

12. Extraction-reinjection at Ahuachapán Geothermal Field, El Salvador

G. Cuéllar, M. Choussy, and D. Escobar 321

12.1. Introduction	321
12.2. Geological Setting	321
12.3. Hydrogeology	325
12.3.1. Shallow Aquifer.	325
12.3.2. Saturated Aquifer	325
12.3.3. Saline Aquifer	325
12.4. Characteristics of the Ahuachapán Wells.	326
12.5. Reinjection Program.	326
12.5.1. Completion of the Reinjection Wells	326
12.5.2. Capacity of Reinjection Wells.	328
12.5.3. Reinjected and Extracted Mass	328
12.5.4. Reinjection Control	329
12.6. Effects of the Reinjection–Extraction Rate	329
12.6.1. Pressures	332
12.6.2. Temperatures	335
12.7. Conclusions.	335

Index   337

Preface

Geothermal energy is one of the so-called alternative energy sources whose use does not yet rival that of the main energy sources (oil, natural gas, coal, nuclear, or even hydropower). But there is abundant geothermal energy in the Earth's crust, and geothermal energy has the potential to impact the energy economies of many countries throughout the world, particularly as fossil fuels grow more expensive and scarce. But geothermal energy is not an exotic twenty-first century hope, demanding expensive and uncertain breakthroughs in physics. Instead, use of geothermal energy already has been demonstrated in many different settings throughout the world.

There is no argument that the geothermal energy in the Earth is immense, and there is likewise little argument that most of this energy is far too deeply buried ever to be of practical concern in terms of extraction and use by man. But even restricting our attention to the geothermal energy in only the Earth's crust (approximately the outer 10–50 km), the amount of energy is still huge. Of course, not all of this thermal energy can be considered to be a resource, any more than all the aluminum in the crust is considered to be a resource. But realistic assumptions as to the distribution and concentration of geothermal energy and as to the fraction that might be recovered indicate that geothermal energy at the very least can be a important supplementary energy source, and may become a major energy source (e.g., Armstead, 1978). Under any reasonable assumptions, it is clear that the magnitude of the geothermal resource of the world is far greater than present use.

Until recently, the growth of geothermal energy use has been slow but steady. For electrical generation alone, this growth has been about 7 per cent per year up to the mid-1970's. Recent years, however, have seen a striking increase in the growth rate, to approximately 19 per cent per year (see Figure 1 of Muffler, this book). Increase in direct use of geothermal energy (for space heating, process heat, agricultural use, etc.) shows a similar pattern. Clearly, the major factor affecting the geothermal growth rate is the petroleum crisis of 1973 and the resulting need for many countries to reduce or eliminate petroleum imports. This need is particularly acute for small, developing countries with no domestic reserves of petroleum. For such countries, geothermal energy can have a major impact on the energy economies. For example, 32 per cent of the electrical energy of El Salvador in 1977 was supplied by the 60 MWe installed in 1975 and 1976 at the Ahuachapán geothermal plant (Cuéllar, this book).

The growth in geothermal utilization has been paralleled by a growth in geothermal understanding. The 1950's and 1960's were decades of slow increase in understanding, with relatively few pioneer investigators who developed the fundamental principles of geology, geochemistry, and hydrology of hydrothermal convection systems. Case histories of geothermal fields were very sparce, notably the classical areas of Larderello, Monte Amiata, Wairakei, and The Geysers. The 1970's, however, have seen an explosion of data and investigations, in great part again in response to the petroleum crisis. As exploration and development increased, additional case histories were developed, for example at Broadlands, Cerro Prieto, and in the Imperial Valley of California. Furthermore, attention was no longer restricted to hydrothermal convection systems but was broadened to include geopressured reservoirs, hot dry rock, and even magma.

It is thus appropriate to summarize our geothermal understanding at the end of this decade of accelerated interest and growth. This book attempts to provide such a summary by means of topical chapters covering in detail selected major aspects of earth science investigations in geothermal energy, supplemented by case histories of systems chosen to represent a spectrum of current geothermal development around the world. The topical chapters are written by scientists closely involved with the development of their particular subject, and the case histories are written by persons intimately and personally knowledgable about the specific geothermal development. Together, the topical chapters and the case histories provide a overview of geothermal exploration and development from a variety of national and topical perspectives.

This book is restricted to the earth science aspects of geothermal energy, including geology, geophysics, geochemistry, hydrology, mathematical modelling, and environmental impact. For a discussion of the engineering, utilization, and economics of geothermal energy the reader is referred to the texts of Milora and Tester (1976), Wahl (1977), and Armstead (1978).

References

Armstead, H. C. H., 1978. *Geothermal Energy*, E. & F. N. Spon, London, 357 pp.
Milora, S. L., and Tester, J. W., 1976. *Geothermal Energy as a Source of Electric Power*, The MIT Press, Cambridge, Massachusetts, 186 pp.
Wahl, E. F., 1977. *Geothermal Energy Utilization*: John Wiley, New York, 302 pp.

LADISLAUS RYBACH
L. J. PATRICK MUFFLER

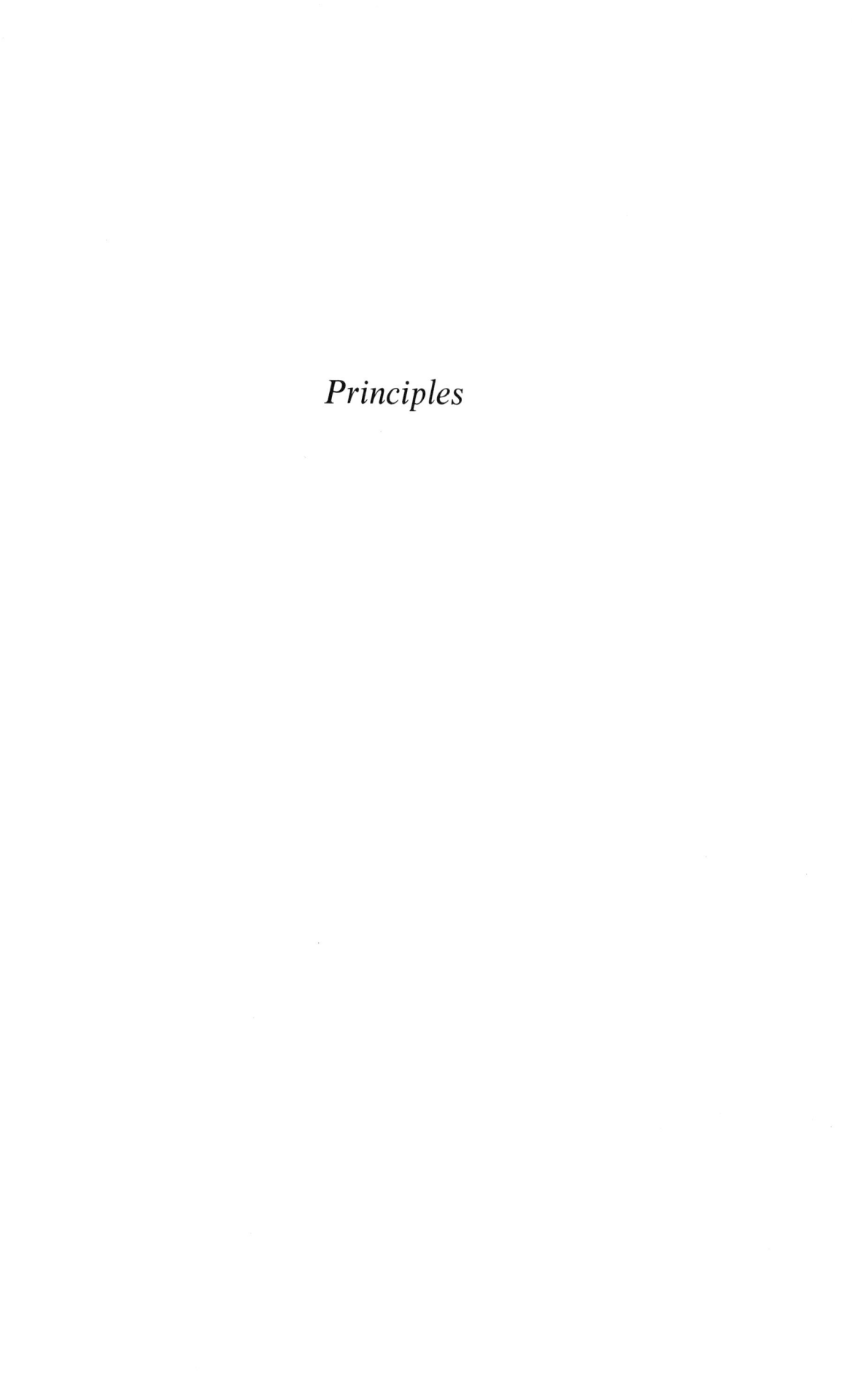

Principles

Geothermal Systems: Principles and Case Histories
Edited by L. Rybach and L. J. P. Muffler
© 1981 John Wiley & Sons Ltd.

1. Geothermal Systems, Conductive Heat Flow, Geothermal Anomalies

Ladislaus Rybach

Institute of Geophysics, ETH Zürich, CH-8093 Zürich, Switzerland

1.1. Geothermal Systems: Definition and Classification

Geothermal systems can be defined and classified on the basis of their geological, hydrological, and heat transfer characteristics. The term '*geothermal*' refers to the internal thermal energy of the Earth; in general it is employed to denote *systems* in which the Earth's heat is sufficiently concentrated to form an energy resource.

Since the thermal energy of the Earth's crust is stored predominantly in large rock masses, a *working fluid* (water, steam) is necessary to collect the diffuse heat and to transfer and concentrate it by forming, under favourable circumstances, a geothermal reservoir (for a detailed discussion of resources and reservoirs see Muffler, this volume). The driving force to move the working fluid is related to the density differnece between cold, downward-moving recharge water and hot geothermal water, the latter rising towards the Earth's surface by buoyancy.

The degree of concentration of thermal energy in a geothermal reservoir can be illustrated by comparing the average heat content (above surface temperature) of crustal rocks in the upper 10 kilometers, 85 kJ/kg, to the enthalpy of saturated steam (at 236°C and 3·2 MPa), 2790 kJ/kg: the factor of enrichment is thus in the order of 30 in the extractable fluid in a 'high-grade' geothermal resource. The concentration of geothermal energy requires, besides high porosity ('storage coefficient'), also high permeability ('hydraulic conductivity') of the reservoir rocks.

The concentration of thermal energy in geothermal systems, relative to adjacent terrain, is characterized by the existence of positive *geothermal anomalies*. Geologic and hydrologic factors as well as heat transfer phenomena can be of importance in forming geothermal anomalies: young magmatic heat sources at high crustal levels, hydrothermal circulation of meteoric water in fault/fracture systems, and irregularities in conductive heat transfer (differences in regional heat flow from deep crustal levels, contrasts in thermal conductivity and/or radiogenic heat production of the rocks in question; see e.g., Muffler et al., 1980). The most significant geothermal anomalies are related to upward moving magma.

The combination of locally variable factors like enthalpy/temperature, permeability distribution, reservoir depth, etc., leads to a great variety of different geothermal systems. At places, natural hydrothermal circulation might exist: this consists of deep-reaching recharge by meteoric water extracting heat from hot rocks at depth and subsequently rising to—and frequently recharging at—the surface; the working fluid is thus supplied by the system itself. The other end of the spectrum is represented by artificial, confined circulation loops like the man-made heat extraction system for Hot Dry Rocks where the working fluid has to be supplied (for a more detailed description of the Hot Dry Rock system see Laughlin, this volume). An intermediate position is taken by thick aquifers of great areal extent in sedimentary sequences, often in regions of normal heat flow. These aquifers can be tapped (in many cases pumping of the water might be necessary), with or without reinjection after direct utilization of geothermal heat.

Geothermal systems can be subdivided, according to their geologic environments and heat transfer regimes, into the following broad categories (see also Muffler, 1976a).

(i) Convective geothermal systems
    (a) Hydrothermal systems in high porosity/permeability environments related to shallow, young silicic intrusions.
    (b) Circulation systems in low porosity-fracture permeability environments, in areas of high to normal regional heat flow.
(ii) Conductive geothermal systems
    (a) Low temperature/low enthalpy aquifers in high porosity/permeability sedimentary sequences (including 'geopressured' zones) in regions of normal or slightly elevated heat flow.
    (b) Hot Dry Rock in a high temperature/low permeability environment.

Convective geothermal systems are characterized by natural circulation of the working fluid; thus most of the heat is transfered by circulating fluids rather than by conduction. Convection tends to increase the temperature in the upper part of the circulation system while temperatures in the lower part decrease. (For a detailed discussion of convective heat transfer see Garg and Kassoy, this volume).

The category of *hydrothermal systems* in high porosity/permeability environments, related to shallow, young silicic *intrusions* includes practically all geothermal systems which to date have actually been developed for commercial electric power production. The high porosity/permeability reservoir must be at depths shallow enough (<3 km) to be tapped by drill holes; the high permeability of the reservoir rocks enables inflow and production of geothermal fluids in substantial quantities.

The heat source is represented by shallow, young magmatic intrusions. Such intrusions occur primarily in specific geologic environments (see Section 1.4): spreading ridges, convergent plate margins, continental rifts, and intraplate melting anomalies. Long-lived thermal anomalies in the upper crust are likely to be supported by magmas that are silicic in composition rather than basaltic (Smith and Shaw, 1975; see also Section 1.4) since basaltic magma rises more quickly to the surface and has its heat dissipated in cooling of volcanic products, whereas silicic magma, due to its higher viscosity, is likely to lodge in the upper crust and thus act as a heat source of substantial duration (a few million years; old intrusive bodies will cool to ambient temperatures and no longer sustain a hydrothermal system; see Section 1.3.2 and Figure 1.5). In some cases, however, shallow basaltic magma chambers can have considerable geothermal potential in developing a hydrothermal system (see, e.g., Stefánsson, this volume).

Two subdivisions within this category are generally recognized, depending upon the heat:discharge balance of the system in question: vapor-dominated systems (e.g. The Geysers, California/U.S.A.; Larderello/Italy, see Figure 1.13) and the more common liquid-dominated systems (e.g. Wairakei and Broadlands/New Zealand; Imperial Valley, California/U.S.A.). Generally, both liquid water and steam exist to some degree in most of such reservoirs; the distinction is really only important with respect to utilization. The production characteristics of hot-water and vapor-dominated systems are discussed in detail by Donaldson and Grant (this volume). The chemical composition of the geothermal fluids (water/steam) is characteristic of the system in question (Ellis and Mahon, 1977).

Circulation systems can develop as well in regions devoid of young igneous intrusions and may result from deep circulation of meteoric water in the thermal regime of conductive regional heat flow. The prerequisite for such systems is the presence of fault/fracture zones with sufficient permeability for water circulation. The temperature attained by the water is dependent primarily upon the magnitude of the regional heat flow and the depth to which the water circulates: all other things being equal, higher temperatures will occur at shallower depths in regions of higher conductive heat flow. Regional heat flow provinces are discussed in detail in Section 1.3.3.

Examples of circulation systems of this kind are given by the numerous thermal springs spread all over the world. They occur in widely differing geologic settings and various rock types, the latter generally characterized by low primary porosity and permeability. While discharge areas are mostly of limited extent (usually at or near intersections of fault/fracture systems) recharge areas are larger by orders of magnitude. Residence times of thermal water in the circulation system can reach $10^2 - 10^3$ years.

Conductive geothermal systems are characterized by a thermal regime due to conduction alone, often in the steady state. The working fluid is either present (e.g. in deep aquifers in sedimentary basins) or must be supplied (Hot Dry Rock system, see Section 1.2.4). Convective circulation is prohibited by low temperature contrasts or, in the case of Hot Dry Rock systems, by low permeability.

Characteristic of deep sedimentary basins is the presence of extended *aquifers* at different depths (in layers with high porosity-permeability). Aquifer temperatures are usually below 150°C (= 'low enthalpy'). The aquifers may contain old, connate pore water (often with high salinity) or may be recharged in adjacent highland areas. If artesian pressure conditions prevail in a given aquifer, the geothermal water discharges freely to the surface upon drilling into the reservoir (see Ottlik et al., this volume). The productivity of aquifers can be defined by the quantity called transmissivity (the product of permeability and aquifer thickness).

Geopressured reservoirs are a special case of sedimentary aquifers in which the pore fluids are under pressure exceeding that of water column (= hydrostatic pressure). In fact, the pore fluids bear a large fraction of the total overburden (= lithostatic pressure); pore pressures of the order of 100 MPa are typical. Geopressured reservoirs are isolated by impervious and low-conductivity shales above and below them. Unlike 'normal' sedimentary regions where pore water (and heat) is expelled in the course of diagenetic compaction, geopressured zones represent 'heat traps'.

A typical example of geopressured zones is the Northern Gulf of Mexico basin U.S.A., both offshore and onshore. Geopressured reservoirs exist in the depth range of 3–7 kilometers and contain substantial amounts of natural gas as well (Wallace et al., 1979).

In the following section we focus on the geophysical and geochemical signatures of the main types of geothermal systems mentioned above. Geopressured resources are not further treated in this volume, except for the short section 5.4.1. on p. 173.

1.2. Geophysical and Geochemical Signatures of Geothermal Systems

1.2.1. *Convective Hydrothermal Systems Related to Shallow Intrusions*

These systems often manifest themselves at the surface by fumaroles, boiling springs, etc. In exploring such systems the primary aim is to locate and delineate the geothermal reservoir. A further goal is the determination of the temperatures which prevail in the reservoir.

Temperature logs in drillholes can exhibit *geothermal gradients* up to 200°C/km, indicating predominantly conductive heat transfer in impervious cap rock (cf. Section 1.3.3.2). Areas with low (or even negative) temperature gradients may mark zones for infiltration in which cold meteoric water seeps to depth to feed the hydrothermal system. The horizontal scale on which such changes of geothermal gradient occur in indicative of the lateral extension of the hydrothermal system.

Another indication of the location of convective hydrothermal systems related to shallow, young silicic intrusions can be given by *gravity mapping*: a negative gravity anomaly can exist, relative to adjacent terrain, due to the steam fraction in high-porosity reservoir rocks as well as to the lowered density caused by thermal expansion. Although there is a significant negative gravity anomaly above the vapor-dominated geothermal field of The Geysers, several factors like silicification by self-sealing may mask the negative anomalies and can even lead to positive gravity anomalies above convective hydrothermal systems (for a detailed discussion see Lumb, this volume).

Electrical methods have a great potential to locate convective geothermal systems despite the limitation that the measured rock resistivity depends on factors like salinity and porosity in a more pronounced manner than on temperature. Formation salinities are, however, rather low in volcanic rocks and thus resistivity mapping can directly be used to locate and delineate geothermal reservoirs in volcanic environments (see Nakamura and Sumi, Stefánsson, and a detailed discussion in Lumb; all this volume).

The shallow intrusion itself can in many cases by located by passive *seismic methods* like observing teleseismic P-wave delays or the attenuation of seismic waves (pronounced P-wave delays have been observed at the Yellowstone and Long Valley calderas as well as at The Geysers–Clear Lake area; Iyer 1975; Iyer et al., 1978). In favorable cases also the vertical extension of the anomalous body can be determined.

Chemical geothermometers, based on the composition of discharging hot (or cold) springs as well as of samples from drill holes, allow the estimation of reservoir temperatures, provided that a number of assumptions are valid in a given case (for detailed discussion see Fournier, this volume). Anomalously high $^3He/^4He$ ratios are indicative of fluid supply from mantle and/or deep crustal sources.

1.2.2. Circulation Systems with Meteoric Water in Areas of High to Normal Regional Heat Flow

This kind of geothermal system is characterized by hot spring discharges in fault/fracture zones, especially where these encounter topographic lows. Recharge of such systems occurs preferentially along fault zones as well. The hydrothermal circulation system is sustained by high to normal regional heat flow; the lifetime of such systems is in the order of $10^4 - 10^5$ years (Lacherbruch and Sass, 1977).

The general crustal setting of circulation systems in areas of high regional heat flow displays some peculiarities. The high heat flow often arises from anomalous thermal processes at the base of the crust. For example intrusion of hot mafic magmas from below occured in much of the crust in the western U.S., torn apart by extensional tectonism (cf. Section 1.4) over the past 20 million years.

Such regions are characterized by a number of geological and geophysical features like low average elevation and consequently high Bouguer gravity anomalies, high heat flow, thin crust and low P_n velocity (compressional wave velocity immediately below the crustmantle boundary), high electrical conductivity at depth and shallow Curie isotherm (Diment, 1980).

In some cases 'blind' high-temperature geothermal systems with no surface manifestations like hot springs can be delineated by conductive heat flow mapping (Marysville, Montana, U.S.A., see, e.g., Blackwell and Morgan 1976; Desert Peak, Nevada U.S.A., see Benoit 1978). The general feature is, however, significant heat transfer by *convection*. The conductive heat flow component can be estimated from the discharge of hot springs: the gain in heat content per unit volume (Q_v, in J/m^3) of the water at depth is

$$Q_v = \rho_f(T_r)\,[h(T_r) - h(T_s)] \tag{1.1}$$

where $\rho_f(T_r)$ is the fluid density at reservoir tempeature T_r, $h(T_r)$ and $h(T_s)$ are the fluid enthalpies at reservoir and surface temperatures. The convective heat flow component (in mW/m^2) is

$$q_{\mathrm{conv}} = \frac{Q_v \dot{d}}{A} \tag{1.2}$$

where $\dot{d}$ is the total spring discharge (in L/s) and A the discharge area. Example are given in Sorey and Lewis (1976) and Iriyama and Ōki (1978); see also Laughlin, this volume).

As mentioned above, *fault/fracture zones* (frequently kept open by seismic activity, see Jaffé et al., 1976) provide paths of circulation for these geothermal systems. Thus basement fracture zones represent attractive targets for exploration. In this respect, detailed structural geologic mapping along with *remote sensing techniques* (in favorable cases by thermal infrared imagery) can be used. For locating fault zones, active seismic techniques (reflection, refraction) have great potential. Local thermal anomalies can be detected by a dense network of shallow holes (5–15 m deep; see, e.g., Kappelmeyer, 1957); based on these results deep exploratory drillholes can be sited.

Spring discharge temperatures are often strongly influenced by the admixture of cold surface waters to the ascending thermal water. Drilling deeper into the system might recover hot water in large amounts. The degree of admixture can be estimated from *mixing models* based on the chemical composition of discharging springs and/or of water samples from drillholes (detailed discussion by Fournier, this volume). Isotope geochemistry like the D/H and $^{18}O/^{16}O$ ratios can help to answer questions concerning the origin and age of the thermal water as well as its residence time at depth (Edmunds et al., 1977). Oxygen isotopic composition of the waters reflect the altitude at which the system is fed by meteoric water (Panichi, 1977; Siegenthaler, 1979).

1.2.3. *Low Temperature Aquifers ($<150°C$) in High-porosity Environments*

Extended aquifers in sedimentary basins have considerable geothermal potential for direct applications (space heating, agricultural utilization, etc.; see Armstead, 1978). Classical examples for low enthalpy geothermal energy utilization are known from the aquifers of the Paris basin, France (La géothermie en France, 1978) and in the Pannonian basin, Hungary (see Ottlik et al., this volume). Such aquifers occur in regions of normal to slightly elevated heat flow. Aquifer temperature is a critical factor in the economy of utilization (La géotermie: Chauffage de logements, 1977). Regional geothermal mapping can supply the required information on the temperature distribution in aquifers (Rybach et al., 1980). Quite recently *chemical geothermometers* (SiO_2, Na/K), applied to water well samples, have provided valuable results in delineating heat flow distribution patterns

(Swanberg and Morgan, 1978; Wohlenberg and Haenel, 1978). For example, the SiO_2 content of groundwater is related to the heat flow q (in mW/m^2) according to

$$q = a[T(SiO_2) - b] \qquad (1.3)$$

where $a = 1.49$ mW/m^2,°C, $b = 13.2$°C and $T(SiO_2)$ the temperature calculated by the silica geothermometer,

$$T(SiO_2) = \frac{1315}{5 \cdot 205 - \log SiO_2} - 273.15 \qquad (1.4)$$

(SiO_2 is the dissolved silica, expressed in parts per million). Aquifers of this type (also those of the 'geopressured' kind), contain old, connate water (= water deposited at time of sedimentation). Geochemical techniques (rare gas analysis) give information on the age of formation water whereas *coalification studies* (measurement of the optical reflectivity R_m of microscopic coal particles) shed light on the thermal history of sedimentary formations in which the sedimentation rate as well as the paleo-heatflow are decisive parameters (the vertical gradient of reflectivity, $\Delta R_m / \Delta z$ is related to the paleogradient; Buntebarth, 1978).

The *salinity* of formation waters is another critical factor in direct geothermal utilization of aquifers (values as high as 100,000 mg/L can be encountered; sea water has 35,000 mg/L). Under favorable circumstances electrical resistivity measurements (vertical sounding) will supply information on the salinity-depth distribution. The measured quantity, the electrical resistivity of a given sedimentary layer ρ_s, depends on a number of parameters:

$$\rho_s = n \rho_w \phi^{-m} \qquad (1.5)$$

where ϕ is porosity, ρ_w the resistivity of the formation water and n and m are numerical constants (in first approximation 1·0 and 2·0, respectively). ρ_w in turn depends on temperature *and* salinity. Thus careful analysis of the resistivity data with respect to the above parameters is necessary to separate the salinity effect.

In some cases where high-salinity aquifers are charged from below by thermal waters with low salinity ascending along fault zones (a typical example is the Landau area, Upper Rhine graben, W. Germany; Werner et al., 1979) a resistivity high might indicate zones of ascending thermal water.

1.2.4. *Hot Dry Rock (HDR)*

In most 'normal' continental areas geothermal heat is present in great quantities at depths which are accessible with today's drilling technology. Since working fluids are absent due to the low natural permeability of the deep strata, heat can be extracted from such HDR resources only by establishing artifical fluid circulation. One possible man-made geothermal system is now under development at the field test stage in New Mexico, U.S.A., and is known as the 'Los Alamos HDR Concept' (see Laughlin, this volume). For electrical power production the main conditions for the HDR resource are: (a) temperature ≥ 200°C and (b) a very low permeability to fluid flow ($\leq 10^{-6}$ darcy). Thus a HDR reservoir can mainly be characterized by its natural permeability and heat content. The natural permeability at depth (= degree of fracturing) is rather difficult to be determined from the surface. A possibility to obtain information on the variation of fracture permeability with depth is offered by the inversion of seismic refraction data (Rybach et al., 1978). The potentially useful heat contained in a HDR reservoir can be evaluated from the temperature-depth curve, $T(z)$. For a given application (e.g. process heat or domestic heating by means of municipal heating networks) there is a lower temperature limit T_l, hence a minimum drilling depth z_l to reach this temperature in the subsurface. The maximum reservoir temperature is defined for a given $T(z)$ distribution by an assumed maximum economic drilling depth z_m; the heat content Q per unit surface area above a minimum usable temperature T_{min} is

$$Q = c_v \int_{z_l}^{z_m} [T(z) - T_l] dz - c_v T_{min}(z_m - z_l) \tag{1.6}$$

where c_v is the volumetric heat capacity of the rock.

In selecting HDR heat extraction sites a number of prerequisites must be considered. One possibility is to envisage border regions of hydrothermal systems where conduction is the predominant heat transfer mechanisms (e.g. the periphery of the Valles Caldera, New Mexico/U.S.A.; cf. Laughlin, this volume). Another target is represented by old plutons buried below a thick blanket of insulating sediments (see also Section 1.3.1.2). If radioactive heat generation in the (granitic) plutons is in the order of 4–8 $\mu W/m^3$ a temperature of 200°C can be attained at 5–6 km depth even in regions of normal heat flow (Costain, 1979; Rybach et al., 1978).

Since drill holes play a decisive role in most HDR systems and since drilling costs increase nearly exponentially with depth in crystalline rocks (Garnish, 1976), the economy of HDR heat extraction at a given site will strongly depend on the local $T(z)$ curve. The state of stress within the Earth's crust at the depths in question is of particular interest in all HDR operations. In the case of the Los Alamos technique, which applies hydraulic fracturing as the method for creating heat exchange surfaces, the knowledge of the stress-field orientation is essential. The orientation of a hydraulic fracture will be determined by the existing stress field in the rock (the fracture will develop perpendicular to the direction of the least principal stress); hence data on the *in situ* stress field are of basic interest in developing large, effective heat transfer areas. Furthermore the *in situ* stress level must be known in order to avoid man-made earthquakes, caused by fluid injection which may raise the pore fluid pressure to a critical level at which pre-existing tectonic stresses could be released. The effectiveness of the heat exchange circulation system will also depend on the *in situ* stress level: the injection pressure must be sufficiently high to keep the fracture(s) open. Special attention must be paid to the permeability of the HDR reservoir. The low-permeability environment is required to minimize fluid loss in the circulation system at depth. Fracture permeability is predominant in crystalline rocks. Thus the degree of fracturing and its variation with depth are other critical factors. Siting criteria with respect to the performance of HDR systems are discussed in Rybach et al. (1978) and Rybach (1979).

1.3. Conductive Heat Transfer and Regional Heat Flow

1.3.1. Fundamentals

In the upper part of the Earth's crust, the region of primary interest with respect to the utilization of geothermal resources, the transfer of thermal energy by conduction is usually the dominating process. Even in regions of strong geothermal anomalies the flow of heat through impermeable zones (e.g. the cap rock above vapor-dominated reservoirs) is by conduction alone. The distribution of heat flux is affected, besides convective effects, by boundary conditions (effects of topographic relief), by spatial variations of thermal properties (e.g. thermal conductivity) and by the variation of heat sources in space and time (transient sources: igneous intrusions; stationary source: heat generation by radioactive decay). In this section we focus on the conductive aspects of heat transfer. Convective aspects of heat transfer are discussed in detail in Chapter 2.

1.3.1.1. Heat flow. Heat can be transfered in a solid by conduction, convection and radiation. At temperatures relevant for geothermal systems the radiative component can be neglected, and, in the absence of mass movements, the conductive heat transfer can be described by the heat flow vector

$$\mathbf{q} = -K\nabla\mathbf{T} \qquad (1.7)$$

where K is the thermal conductivity tensor and the minus sign denotes the fact that heat flows *down* the temperature gradient $\nabla\mathbf{T}$. In isotropic solids the thermal conductivity is a scalar property; greater volumes of geologic materials can be considered as isotropic. K is in general a function of temperature, $K(T)$. In transient heat flow problems a further thermal property, the diffusivity κ is involved:

$$\kappa = K/\rho c \qquad (1.8)$$

where ρ is density and c specific heat.

Table 1.1 lists some typical values of thermal conductivity for common rock types at room temperature (for the temperature dependence of K see Section 1.3.3.2).

The heat flow is the surface expression of the geothermal conditions at depth. Two different steps are necessary to its determination: (a) measurement of the temperature gradient (e.g. in a borehole) and (b) experimental determination of thermal conductivity (on drill cores, cuttings or *in situ*). Several standard texts describe the methods of heat flow determination on land, including the corrections to be applied (e.g. Sass et al., 1971, Sass and Munroe, 1974).

The unit of heat flow is mW/m^2 (conversion to the 'old' heat flow unit, HFU: 1HFU = 41·8 mW/m^2). The average heat flow in thermally 'normal' continental regions is around 60 mW/m^2 (Jessop et al., 1976); values in excess of about 80–100 mW/m^2 indicate anomalous geothermal conditions in the subsurface. Thus heat flow mapping is one of the main exploration techniques for geothermal resources (see, e.g., an excellent example in Sass et al., 1977 and also Lumb, this volume). Special attention must be given, however, to

Table 1.1. Thermal conductivity of various rocks at room temperature

Rock type	Thermal conductivity (W/m, K)
Granite	2·5–3·8
Gabbro/basalt	1·7–2·5
Peridotite/pyroxenite	4·2–5·8
Limestone	1·7–3·3
Dolomite, salt	~5·0
Sandstone	1·2–4·2
Shale*	0·8–2·1
Volcanic tuffs†	1·2–2·1
Deep-sea sediments*	0·6–0·8
Water	0.6

* Depending on water content.
†Depending on porosity.

1. Geothermal Systems, Conductive Heat Flow, Geothermal Anomalies

possible effects of water migration. In fractured rocks with sufficient vertical permeability, convection of hot water may be the dominant heat transfer mechanism (see Chapter 2). If the temperature gradient exceeds 65°C/km, the specific volume of water increases with depth at a rate greater than the rate of volume reduction of water due to hydrostatic pressure. This results in an net buoyancy of deeper water and thus upward migration (Meidav and Tonani, 1976).

1.3.1.2. The geothermal gradient. In discussing the geothermal conditions at depth, the vertical geothermal dT/dz is frequently used; for practical purposes it is customary to consider the upward component of heat flow q and to reverse the sign convection (cf. equation 4.7); q can be considered as a scalar. In simple cases (e.g. a sedimentary sequence with horizontal strata where K varies with depth only and heat sources are neglected) the temperature at a given depth d will be

$$T(d) = T_0 + q \int_0^d \frac{dz}{K(z)} \tag{1.9}$$

or with n layers (denoting the thickness and constant thermal conductivity of the ith layer by h_i and K_i, respectively)

$$T(z) = T_0 + q \left(\sum_{i=1}^{n-1} \frac{h_i}{K_i} + \frac{z - \sum_{i=1}^{n-1} h_i}{K_n} \right) \tag{1.10}$$

where T_0 is the surface temperature. In each layer the product of the temperature gradient and the conductivity is constant:

$$\left(\frac{dT}{dz} \right)_i \cdot K_i = q \tag{1.11}$$

Thus a layer with relatively low thermal conductivity (e.g. a dry shale) is characterized by a correspondingly high temperature gradient.

Relatively high geothermal gradients can be encountered if a low-conductivity sediment (K_s) blanket is present, covering high-conductivity basement (K_b). Depending on the local heat flow q and the sediment thickness D, the temperature anomaly (as compared with uncovered basement) at the base of the blanket will be (Figure 1.1):

$$\Delta T = Dq \frac{K_b - K_s}{K_s + K_b} \tag{1.12}$$

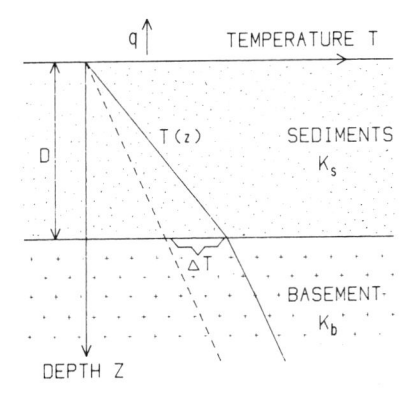

Figure 1.1. Blanketing effect of low-conductivity sediments. ΔT is the temperature anomaly at depth D

High geothermal gradients can also be encountered in impermeable rocks above zones with convective heat transfer. In the zone of convection itself the gradient is low due to the convective temperature adjustment. By assuming hydrostatic pressure in the subsurface, the conductive geothermal gradient above a vapor-dominated deposit will depend on the depth (cf. Figure 1.2) to the top of the reservoir.

1.3.1.3. Governing equations. The governing equations describe the generation and transport of heat. The differential equation describing heat transfer by conduction and convection can be written (by neglecting heat sources) as

$$\mathbf{v}(\mathbf{r}, t)\nabla \mathbf{T} + \frac{\partial T}{\partial t} = -\frac{1}{c(T)\rho} \nabla[K(T)\nabla \mathbf{T}] = 0 \tag{1.13}$$

which takes into account the dependence of the velocity $\mathbf{v}$ of mass movement on position and time as well as the temperature dependence of the thermal properties c and K. In the case of convection, $\mathbf{v}$ depends on the temperature distribution as well. The presence of the term $\partial T/\partial t$ indicates that the temperature distribution changes with time.

For conductive heat transfer alone, the first term of equation (1.13) will be zero and, neglecting the temperature dependence of c and K we have

$$\frac{\partial T}{\partial t} - \kappa \nabla^2 T = 0 \tag{1.14}$$

which equation is commonly used to calculate the cooling of igneous intrusions by conduction (see Section 1.3.3.2c and detailed review articles like Jaeger, 1964; Simmons, 1967; also Lachenbruch et al., 1976).

As a further step the presence of heat sources (A) can be considered. These may depend on position as well as on time, $A(\mathbf{r}, t)$; usually one is concerned with long-lived radiogenic heat sources (see Section 1.3.2.1), thus $\partial A/\partial t = 0$. In the stationary case ($\partial T/\partial t = 0$) the

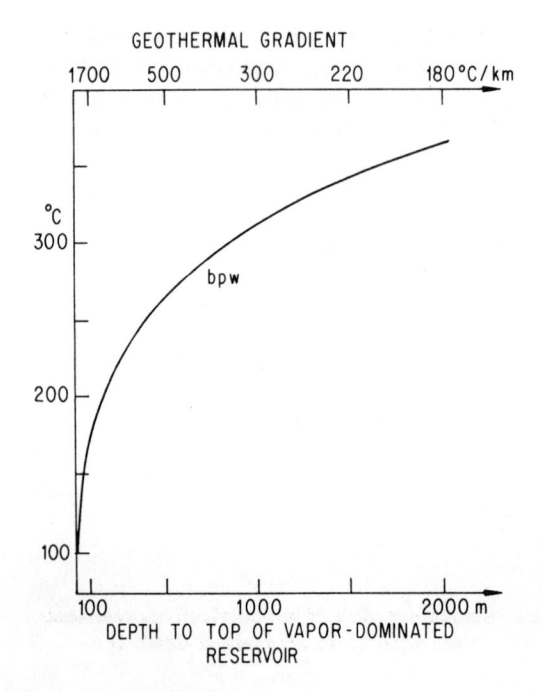

Figure 1.2. The boiling point of water (curve bpw) at different pressures (=depths), assuming hydrostatic pressure in the subsurface. The geothermal gradient in the zone above a vapor-dominated reservoir depends on the depth to reservoir top (after Kappelmeyer and Haenel, 1974 p. 76)

differential equation of heat conduction is

$$\nabla[K(\mathbf{r}, T)\nabla T + A(\mathbf{r}) = 0 \quad (1.15)$$

In the quantitative approach to the description of geothermal problems these equations are used to calculate the temperature field and to compare the results with observations on the Earth's surface, the heat flow. For the solutions, the initial and boundary conditions must be tailored to the specific problems in question. Customary representations of the temperature field are: in one dimension the temperature-depth curve ('geotherm') and in two dimensions the isotherms (examples are given in Figures 1.6 and 1.8, respectively). Analytical as well as numerical solutions of these governing equations are presented and discussed in Section 1.3.3.2.

1.3.2. Heat Sources

The supply of heat in the Earth's crust, both on the local and regional scale, is maintained by several processes. In the time domain, relatively short-lived (transient) sources like magmatic intrusions at high crustal levels are contrasted with heat sources of long duration (natural radioelements in crustal rocks with half lives of 10^9–10^{10} years, which can be considered as steady-state sources). Other heat sources like phase changes, chemical reactions are of limited importance in considering the thermal regime of the crust. Frictional heating is discussed in Section 1.2.4.3. A substantial amount of heat enters the crust at its base: the heat flow from the Earth's mantle (cf. Sections 1.3.2.1a and 1.3.3.1).

1.3.2.1. Steady-state source: radiogenic heat. 20 to 80 per cent of the surface heat flow arises from the heat generated by radioactive nuclei in the Earth's crust. Since heat production varies over orders of magnitude with rock type (see Tables 1.2–1.4), the distribution of radioactive heat sources plays a decisive role in shaping the temperature field of the crust.

(a) *Heat production by radioactivity*

Radioactive decay converts mass into radiation energy which in turn is converted to heat in the immediate vicinity of the decaying nucleus (Hurley and Fairbairn, 1953). All naturally radioactive isotopes generate heat to a certain extent. It can be shown, however, that the only significant contributions arise from the decay series of U^{238}, U^{235} and Th^{232} and from the isotope K^{40}. The revised heat production constants of the natural radioelements *uranium* (including the decay series of U^{238} and U^{235} in proportion of $U^{235}/U^{238}=1/139.6$ and in radioactive equilibrium), *thorium* (the Th^{232} decay series in equilibrium) and *potassium* (with 0.0118 per cent K^{40}) are $9.525 \cdot 10^{-5}$, $2.561 \cdot 10^{-5}$ and $3.477 \cdot 10^{-9}$ W/kg, respectively (Rybach, 1976a). Thus for a given rock with density ρ (kg/m^3) and radioelement contents c_U (ppm), c_{Th} (ppm) and c_K(%) the heat production A is given by

$$A(\mu W/m^3) = 10^{-5}\rho(9.52c_U + 2.56c_K + 3.48c_{Th}) \quad (1.16)$$

(b) *Variation of heat generation with rock type.*

It is evident from equation (1.16) that heat production of rocks is governed by the amounts of uranium, thorium, and potassium present, which vary greatly with rock type but exhibit certain regularities due to the similar geochemical behaviour of U, Th, and K during the processes which determine the distribution of the natural radioelements (magmatic differentiation, sedimentation, metamorphism). Characteristic average heat production values are listed in Tables 1.2–1.4 for the main rock types (after Rybach, 1976b). In igneous rocks heat production A depends on the bulk chemistry: A decreases from silicic (e.g. granites) through basic to ultrabasic rock types (e.g. peridotites). In metamorphic rocks, A depends, even for rocks with similar bulk chemistry, on the metamorphic grade (Table 1.3), due to selective depletion of rocks in radioactive elements by the upward-moving fluid phase during metamorphism. Sedimentary rocks, which make up only small portions of the Earth's crust, have in general—especially limestones and dolomites—very low A values. For a more detailed discussion of the geochemical aspects of heat production see Rybach (1976b).

(c) *Variation of heat generation with depth*

The thermal regime of the crust is governed on the regional scale by the heat inflow from below, that is, by the mantle heat flow. The latter can be deduced from measured surface heat flow by taking into account the integrated contribution of crustal radioactivity. The vertical distribution of heat sources clearly influences the temperature field in the continental crust (see e.g. Rybach, 1976c). There is direct (Lachenbruch and Bunker, 1971) and indirect (Swanberg, 1972; Hawkesworth, 1974; Ormaasen and Raade, 1978; Kissling et al., 1978) evidence for the decrease of heat production with depth according to the exponential law (Figure 1.3).

$$A(z) = A(0)\exp(-z/H) \qquad (1.17)$$

where $A(0)$ is the heat generation at the surface and H the logarithmic decrement (at $z=H$, $A(z)=A(0)/e$).

The geochemical processes which lead to this exponential distribution are not yet clear in detail, but the upward fractionation of the natural radioelements U, Th and K is

Table 1.2. Heat production of igneous rocks (after Rybach, 1976b)

Rock (intrusive/extrusive)	A ($\mu W/m^3$)	ρ (kg/m^3)	Δq^* (mW/m^2)
Granite/rhyolite	2·45	2670	2·5
Granodiorite/dacite	1·49	2720	1·5
Diorite, quartz diorite/ andesite	1·08	2820	1·1
Gabbro/basalt	0·31	2980	0·3
Ultramafic—peridotite	0·012	3230	0·01
—dunite	0·0019	3280	0·002

*Contribution to surface heat flow per 1 km thickness.

Table 1.3. Heat production of metamorphic rocks with granodioritic composition (after Heier and Adams 1965; Rybach, 1976b)

Metamorphic grade	A^* ($\mu W/m^3$)	ρ (kg/m^3)	Δq (mW/m^2)
Greenschist/low amphibolite facies	3·15	2700	3·2
High amphibolite facies	1·17	↓	1·2
Low granulite facies	0·73		0·7
High granulite facies	0·45	2900	0·4

*Calculated for $\bar{\rho} = 2750$ kg/m³.

1. Geothermal Systems, Conductive Heat Flow, Geothermal Anomalies

Table 1.4. Heat production of sedimentary rocks (after Rybach, 1976b)

Rock	A (μW/m^3)	ρ* (kg/m^3)	Δq (mW/m^2)
Carbonates		2600	
Limestone	0·62		0·6
Dolomite	0·36		0·4
Sandstones		2400	
Quartzite	0·33		0·3
Arkose	0·85		0·8
Graywacke	1·0		1·0
Shales	1·8	2400	1·8
Deep sea sediments	0·78	1300	0·8

*Broad average since ρ depends strongly on porosity.

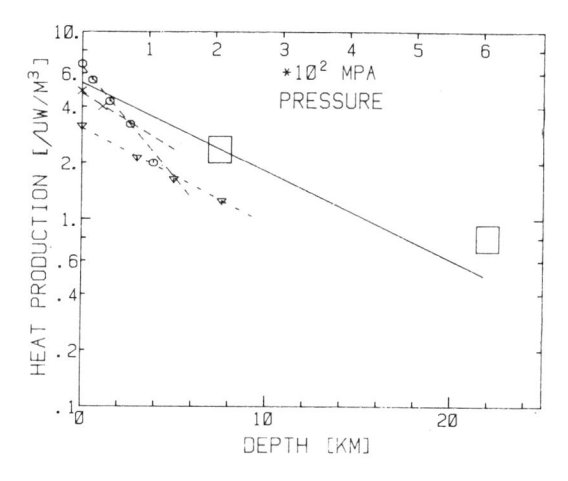

Figure 1.3. Depth dependence of heat production (straight lines indicate exponential dependence) according to: Swanberg (1972), Idaho Batholith/U.S.A. (solid line); Hawkesworth (1974). Tauern Crystalline/Austria (cricles); Rybach (1976a), basement, Northern Alpine Foreland (triangles) Ormaasen and Raade (1978), Precambrian rocks/Norway (boxes); Kissling et al. (1978, Rotondo granite/Switzerland (crosses)

certainly caused by some kind of melting and/or the migration of metamorphic fluids (Rybach, 1976b).

1.3.2.2. Transient heat sources. High regional heat flow can sustain hydrothermal convection systems, provided that the upper few kilometers of the Earth's crust is sufficiently fractured to allow fluid circulation. Such systems have relatively short lifetimes (10^3–10^5 years, Lachenbruch and Sass, 1977). Much longer lasting hydrothermal systems can be supported by local crustal heat sources like *shallow intrusions*. Such systems may persist for several million years (cooling by conduction alone). The age and solidification history of the intrusion are thus dominating factors.

According to Smith and Shaw (1975), basic magmas usually rise to the Earth's surface without forming magma chambers at high levels in the crust, whereas more silicic magmas stop at levels with the upper 10 km of the crust. We consider here the supply of heat to roof rocks by an intrusion of 'batholithic' type, i.e. with great thickness and lateral extension (for sake of simplicity we set thickness and extension infinite, see Figure 1.4). The heat flow q at the surface due to the intrusion varies with time t (Carslaw and Jaeger, 1959; p. 60) and is

$$q(t) = KT_i(\pi\kappa t)^{-1/2} \exp(-d^2/4\pi t) \qquad (1.18)$$

where d is the depth to the batholith top, T_i the excess temperature of the intrusion, K thermal conductivity and κ diffusivity. q has a maximum value of $q_{max} = 0.484 T_i K/d$ at the time $t_m = d^2/2\kappa$ (for simple estimations $t_{yr} \approx 16{,}000 d_{km}^2$). The heat production is $A(t) = q(t)/d$; the heat supplied per unit surface in the time interval from $t = 0$ (time of intrusion) to $t = t_m$ (maximum heat flow at the surface) is approx. $0.12(K/\kappa) T_i d$.

Young intrusions have not yet lost a significant portion of their heat and thus represent an interesting geothermal potential (Figure 1.5).

1.3.3. Conductive Heat Flow

The outflow of heat from the Earth's interior is a fundamental quantity in its energy balance. The surface heat flow thus signalizes the processes of heat generation, transport, and storage occurring at deeper levels in the Earth's crust and lithosphere. It is customary to evaluate suitable geothermal models for the subsurface by comparing the calculated heat flow distribution with the heat flow pattern as measured at the surface. Broad

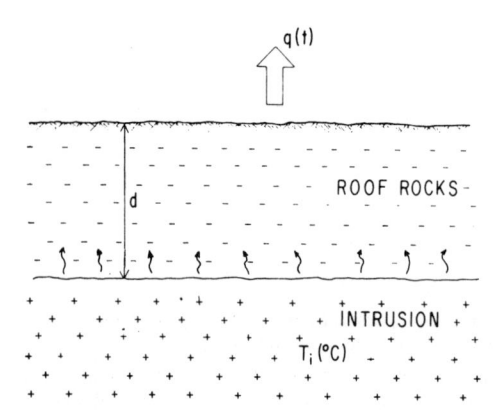

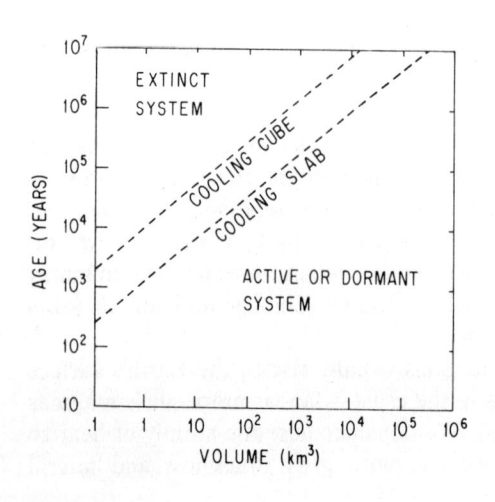

Figure 1.4. Batholith intrusion (one-dimensional model) with top at depth d

Figure 1.5. Cooling intrusions of different age and their geothermal potential (after Smith and Shaw, 1975). 'Age': total cooling time to cool centre of magma chamber from 800°C to 300°C

regional features of the heat flow field can be described by the concept of heat flow provinces.

1.3.3.1. Heat flow provinces. The systematic study of the relation between heat flow and radiogenic heat production lead to the recognition of heat flow provinces. The linear relationship between surface heat flow $q(0)$ and heat generation in crystalline rocks, A (Birch et al. 1968; Roy et al., 1968),

$$q(0) = q^* + Ah \qquad (1.19)$$

yields a reduced heat flow q^* (originating from lower crust and upper mantle) and a characteristic depth h, which are uniform within certain areas termed as 'heat flow provinces'. Such provinces are geologic units with uniform tectonic age and evolution. Table 1.5 lists some of these provinces and their characteristic q^* and h values.

The constant h (with the dimension of length) is less than normal crustal thickness and varies relatively little among the provinces. The uniform heat flow coming from below an upper layer of varied heat generation throughout the heat flow province is given by q^*; the local variations of surface heat flow can thus be attributed to variable heat generation in this upper layer (the Ah term in equation (1.19)).

Several vertical distributions of heat generation can satisfy the linear relationship equation (1.19). The exponential depth dependence (1.17) with $H = h$ is for various reasons the most satisfactory distribution model. Points which plot above the linear relationship (1.19) indicate additional heat sources in a given province (e.g. upward movement of magma); points which lie below the line indicate heat sinks (e.g. downward percolating surface waters).

In heat flow provinces where q^* reaches a certain value (~ 60 mW/m^2), the elevated regional heat flow may supply the heat of *hydrothermal convection systems* (Lachenburch and Sass, 1977). In a given case (e.g. the Jemez Mountains/New Mexico, U.S.A., see Chapter 11, p. 314) the convective component can be substantial.

1.3.3.2. Modeling of temperature field and heat flow. No direct methods are available so

Table 1.5. Typical continental heat flow provinces (after Jessop and Lewis, 1978)

Province	Geologic/geothermal characteristics	Mean surface heat flow, $q(0)$ (mW/m^2)	Reduced (mantle) heat flow, q^* (mW/m^2)	h (km)
Eastern U.S.A.	Tectonically stable continental area, conductive heat transfer	57	33	7·5
Basin and Range, U.S.A.	Area with active spreading tectonism, strong convective heat flow components	92	59	9·4
Sierra Nevada, U.S.A.	Heat flow transient due to former subduction tectonism	39	17	10·1
Precambrian Shields (average)	Stable continental shield	60	21	14·4

far to determine the sub-surface temperature distribution by measurements at the surface. Therefore model calculations are used to estimate the temperature field and its evolution in time for a given geologic situation. In mathematical terms, the heat equation (1.14) must be solved for initial and boundary conditions tailored to the problem in consideration. Since the first time derivative appears in (1.14), solutions can only be obtained for $t > 0$, i.e. the evolution of the initial temperature distribution at $t = 0$ can be calculated for later times, but it is not possible to 'work backwards in time', and to determine, starting with today's temperature distribution in a given area (e.g. in a sedimentary basin), the temperature field at an earlier time (say 2 million years ago).

The initial conditions like the initial temperature distribution $T_0(x, y, z)$ at $t = 0$ must be assumed on the basis of the available information. The temperature distribution at any later time t is (Carslaw and Jaeger, 1959; p. 56) at any given point (x, y, z)

$$T(x, y, z, t) = (1/8)(\pi\kappa t)^{-1/2} \int_{-\infty}^{\infty}\int_{-\infty}^{\infty}\int_{-\infty}^{\infty} T_0(x'y'z')$$
$$\exp\{-[(x-x')^2 + (y-y')^2 + (z-z')^2]/4\kappa t\}\, dx'\, dy'\, dz' \tag{1.20}$$

where κ denotes thermal diffusivity and x', y' and z' are the running coordinates. An example of this kind of calculation is given in Section 1.3.3.2c (Figure 1.10).

Another decisive parameter (boundary condition) is the surface heat flow $q(0)$. A primary objective of model calculations is the *downward continuation* of the geothermal information obtainable at the surface. Several simplifications are convenient in performing such model calculations: (i) assuming steady-state conditions, (ii) considering vertical heat transfer only, (iii) neglecting heat transfer by convection.

The model calculations described below proceed from simple to more complex cases (one–two dimensions, stationary–transient). For modeling with consideration of convective aspects of heat transfer see Chapter 2.

(a) One-dimensional steady-state models

For the one-dimensional steady-state case the heat equation (1.15) with conduction alone and considering the temperature dependence of K is

$$A(z) + \frac{d}{dz}\left[K(T)\frac{dT}{dz}\right] = 0 \tag{1.21}$$

Analytical solutions can be given for the following cases with the boundary conditions $T_0 = T_{\text{surface}}$ and $q_0 = K(dT/dz)_{z=0}$ (surface heat flow):

Case (1) No vertical variation of heat production A and thermal conductivity K: $A(z) = A_0$, $K(T) = K_0$

$$T(z) = T_0 + (q_0/K_0)z - (A_0/2K_0)z^2 \tag{1.22}$$

Case (2) No vertical variation of A; K depends on temperature

$$K(T) = K_0/(1 + \gamma T) \tag{1.23}$$

where γ is of the order of $10^{-3}\,°\text{C}^{-1}$ (see e.g. Balling, 1976)

$$T(z) = (1/\gamma)\{(1+\gamma T_0)\exp[(\gamma/K_0)(q_0 z - A_0 z^2/2)] - 1\} \tag{1.24}$$

Case (3) exponential decrease of heat production with depth, $A(z) = A_0 \exp(-z/H)$, no vertical variation of K

$$T(z) = T_0 + (q_0 - A_0 H)z/K_0 + A_0 H^2[1 - \exp(-z/H)]/K_0 \tag{1.25}$$

1. Geothermal Systems, Conductive Heat Flow, Geothermal Anomalies

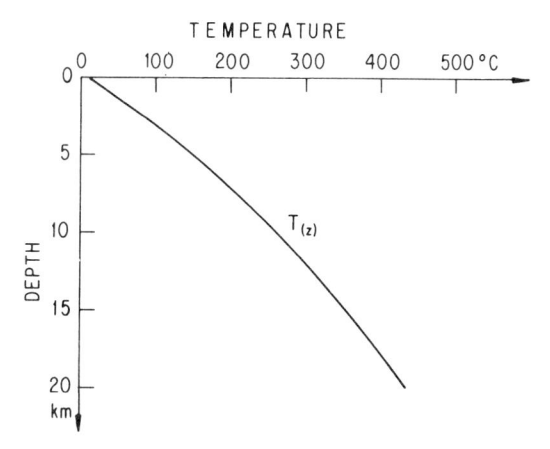

Figure 1.6. One-dimensional representation of temperature field (=geotherm) for the area of the Northern Foreland of the Alps. Calculated according to case 3 (see text) with $T_0 = 10°C$, $q_0 = 73.2$ mW/m², $H = 10$ km, $A_0 = 4.18$ μW/m³, $K_0 = 2.1$ W/m, K (from Rybach, 1976a)

Case (4) exponential decrease of A, temperature dependence of K according to (1.23)

$$T(z) = (1/\gamma)[(1 + \gamma T_0) \exp\{(\gamma/K_0)[(A_0 H^2)(1 - \exp(-z/H)) - A_0 Hz + q_0 z]\} - 1] \quad (1.26)$$

For n layers the recursion formulas of Pollack (1965 can be used. The one-dimensional representation of the temperature field, $T(z)$ is called geotherm. An example is given in Figure 1.6.

(b) Two-dimensional steady-state models

In two dimensions of the temperature distribution is considered on the (x, z) plane. The heat equation (1.15) can be written as

$$2A + \nabla^2(KT) + K\nabla^2 T - T\nabla^2 K = 0 \quad (1.27)$$

The use of numerical methods is customary to solve this partual differential equation. The region of interest is divided into a mesh with the distance b between nodal points $(i, k+1)$ and (i, k) of the mesh. In the numerical treatment the partial derivatives are approximated by

$$\frac{\partial^2 T}{\partial z^2} \approx \frac{T_{i-1,k} - 2T_{i,k} + T_{i+1,k}}{b^2} \quad (1.28)$$

and

$$\nabla^2 T \approx \frac{T_{i-1,k} + T_{i+1,k}}{n^2} + T_{i,k-1} + T_{i,k+1} - 2T_{i,k}\left(1 + \frac{1}{n^2}\right) = 0 \quad (1.29)$$

(cf. Figure 1.7)

By applying equations (1.29) to (1.27) one obtains the central difference approximation using the five-point formula

$$\frac{2A_{i,k}d^2}{K_{i,k}} + T_{i-1,k}\frac{K_{i-1,k} + K_{i,k}}{n^2 K_{i,k}} + T_{i+1,k}\frac{K_{i+1,k} + K_{i,k}}{n^2 K_{i,k}}$$

$$+ T_{i,k-1}\frac{K_{i,k-1} + K_{i,k}}{K_{i,k}} + T_{i,k+1}\frac{K_{i,k+1} + K_{i,k}}{K_{i,k}} -$$

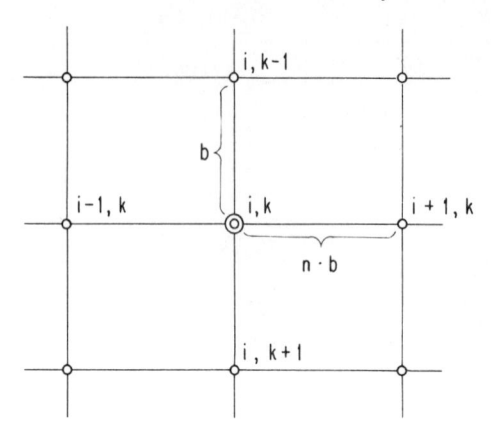

Figure 1.7. Rectangular mesh for the numerical (finite difference) approximation to the differential equation of heat conduction in two dimensions. n is an integer

$$-T_{i,k}\left(2+\frac{2}{n^2}+\frac{K_{i-1,k}+K_{i+1,k}}{n^2 K_{1,k}}+\frac{K_{i,k+1}+K_{i,k-1}}{K_{i,k}}\right)=0 \qquad (1.30)$$

where $A_{i,k}$ and $K_{i,k}$ are the heat production and thermal conductivity parameters to be assigned to the mesh under consideration.

The boundary values which must be fixed are the following:

1. Surface temperature $T_{i,0}$ ($i=0, 1, \ldots r$).
2. Horizontal heat flow $K(T)\partial T/\partial x$ which is zero in the mesh points $(0, k)$ and (r, k) ($k=0, 1, \ldots s$).
3. Heat flow at the base $q_b = K(T)\partial T/\partial z$ at the depth $z=bs$.

The calculation starts with an approximate temperature distribution $T'(i, k)$ which can be obtained by one-dimensional models (see previous section). The result will be a residual non-zero value on the right-hand side of equation (1.30),

$$R'_{i,k} = \frac{2A_{i,k}d^2}{K_{i,k}} + T'_{i-1,k}a_{i-1,k} + T'_{i+1,k}a_{i+1,k} + T'_{i,k-1}a_{i,k-1}$$
$$+ T'_{i,k+1}a_{i,k+1} - T'_{i,k}a_{i,k} \qquad (1.31)$$

where the a's are the coefficients of equation (1.30). From this the next estimate is

$$T''_{i,k} = T'_{i,k} + \frac{R_{i,k}}{a_{i,k}} \qquad (1.32)$$

for all the internal points of the mesh. Further approximations are calculated from equation (1.31) by iteration until a preselected accuracy is reached (for a more detailed treatment see, e.g., Buntebarth, 1975 or Balling, 1976). Figures 1.8 and 1.9 give examples of heat flow modeling on local and regional scale.

(c) *Non-stationary models, cooling intrusions*

In transient problems, the temperature distribution changes with time, i.e. $\partial T/\partial t \neq 0$. The most simple case is the one-dimensional situation for which the heat conduction equation (1.14) is (for heat transfer in the vertical direction only)

$$\frac{\partial T}{\partial t} = \kappa \frac{\partial^2 T}{\partial z^2} \qquad (1.33)$$

1. Geothermal Systems, Conductive Heat Flow, Geothermal Anomalies 21

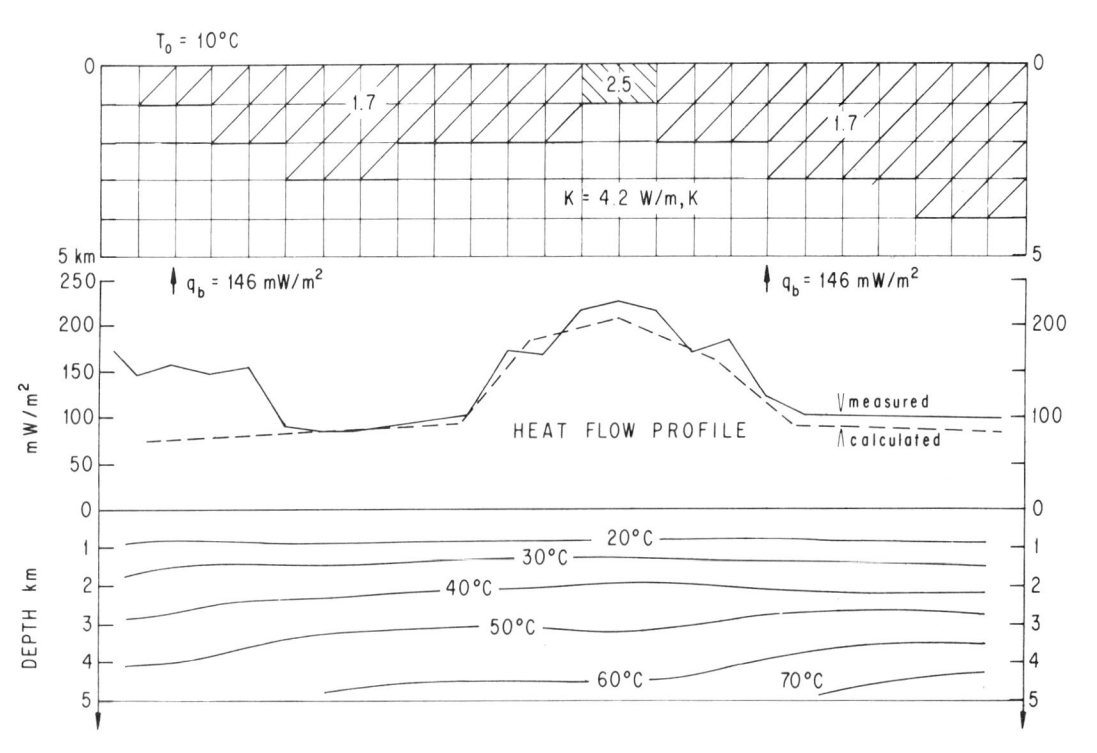

Figure 1.8. Modeling of heat flow anomaly, on the local scale. Upper part: thermal conductivity structure (1 × 1 km mesh). Lower part: heat flow profile and isotherms down to a depth of 5 km (Grass Valley, California; from Majer, 1978)

We consider a solid bounded by two parallel planes, $0 < z < l$ and illustrate the calculation procedure with a specific example: heat flow and temperature distribution above The Geysers Field, California (Urban et al., 1976). The evolution of the temperature profile above the steam deposit (its top is at $l = 1\cdot 5$ km depth; the reservoir temperature is 240°C) will be modeled. The initial and boundary conditions are

$$T = T_1 = 15°C \quad \text{for } z = 0 \text{ (surface temperature)}$$

$$T = T_2 = 240°C \quad \text{for } z = l = 1\cdot 5 \text{ km}$$

$$T_0(z) = 15°C + \frac{32°C}{\text{km}} z \quad \text{for } t = 0 \text{ (initial temperature distribution)}$$

The solution can be found in Carslaw and Jaeger (1959, p. 100)

$$T(z, t) = T_1 + (T_2 - T_1)\frac{z}{l} + \frac{2}{\pi}\sum_{n=1}^{\infty}\frac{T_2 \cos n\pi - T_1}{n} \sin \frac{n\pi z}{l}$$
$$\exp(-\kappa n^2 \pi t/l^2) + \frac{2}{l}\sum_{n=1}^{\infty} \sin \frac{n\pi z}{l} \exp(-\kappa n^2 \pi t/l^2)$$
$$\cdot \int_0^l T_0(z') \sin \frac{n\pi z'}{l} dz' \qquad (1.34)$$

The result is given in Figure 1.10.

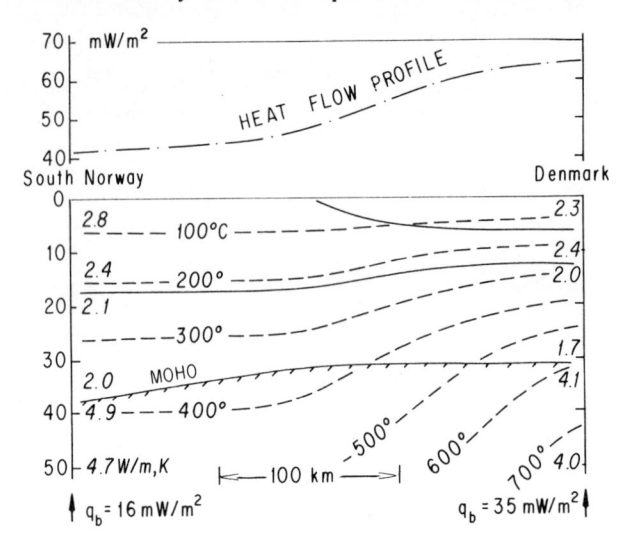

Figure 1.9. Modeling of regional heat flow distribution with temperature-dependent thermal conductivity (figures on left and right in diagram) and exponentail heat production-depth dependence (after Buntebarth, 1975, and Balling, 1976)

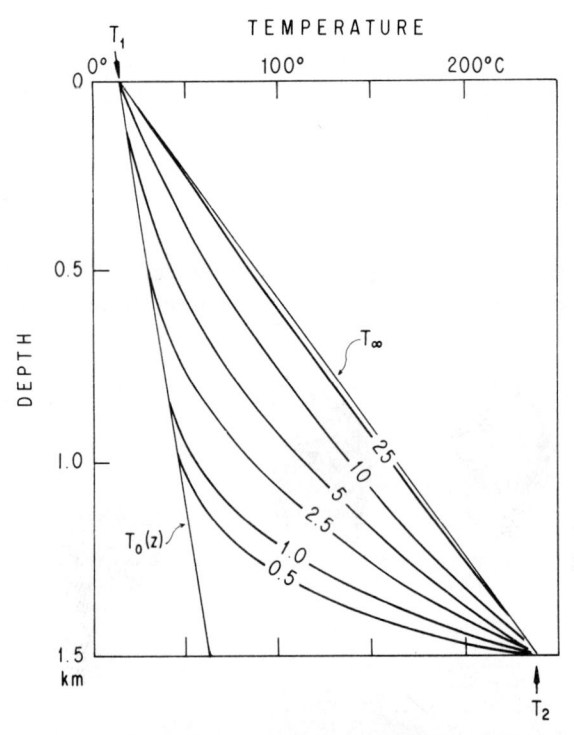

Figure 1.10. Evolution of the temperature profile above a steam reservoir. Curve parameter: time (in 10^3 years) after formation of steam zone (240°C at depth of 1·5 km; from Urban et al., 1976)

Cooling intrusions in the upper crust very often supply the heat for hydrothermal and/or vapor-dominated systems. The cooling of such hot igneous bodies involves both conduction and convection (see, e.g., Cathles, 1977; Norton and Knight, 1977), but valuable estimates of temperatures and times can be obtained by assuming that conduction is the only process: the temperatures and times so obtained give upper limits since additional convection will enhance the cooling process (and reduce especially post-solidification cooling times). (See detailed discussion in Smith and Shaw, 1979 and also in Chapter 2). A convenient model for cooling intrusive bodies is an instantaneous source in the form of a parallelepiped in a semi-infinite medium (Figure 1.11).

The upper surface ($x-y$ plane, Figure 1.11) is at $T=0$ for all times. The temperature is (Simmons, 1967)

$$T(x, y, z, t) = (T_0/8)E(x, x_1, x_2)E(y, y_1, y_2)[E(z, z_1, z_2) + E(z, -z_1, -z_2)] \quad (1.35)$$

where

$$E(\alpha, \alpha_1, \alpha_2) = \mathrm{erf}[(\alpha, -\alpha_1)/(4\kappa t)^{1/2}] + \mathrm{erf}[(\alpha - \alpha_2)/(4\kappa t)^{1/2}] \quad (1.36)$$

and

$$\mathrm{erf}\,\beta = \frac{2}{\sqrt{\pi}} \int_0^\beta e^{-x^2}\,dx \quad (1.37)$$

T_0 is the excess temperature (intrusion temperature minus temperature of the country rock); the effect of latent heat may be included by replacing T_0 in (1.35) by $(T_0 + L/c)$, where L (J/kg) is the latent heat of crystallization and c the heat capacity (J/kg, °C) of the intruding magma.

The heat flow at the surface is

$$q = \frac{KT_0}{4\sqrt{\pi\kappa t}} E(x, x_1, x_2)E(y, y_1, y_2)[\exp(-z_1^2/4\kappa t) - \exp(-z_2^2/4\kappa t)] \quad (1.38)$$

An example is given below for the Urach anomaly, South Germany. This feature manifests itself with pronounced anomalies of the geothermal gradient (Figure 1.12, maximum gradient ~ 80°C/km). Numerous volcanic necks are known from the same area.

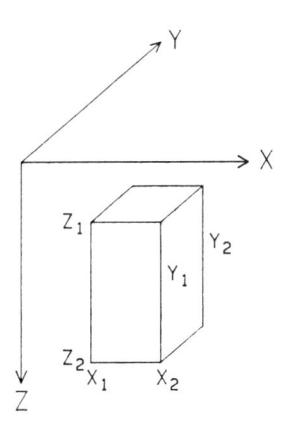

Figure 1.11. Rectangular parallelepiped (model intrusion) extends from x_1 to x_2, y_1 to y_2, z_1 to z_2

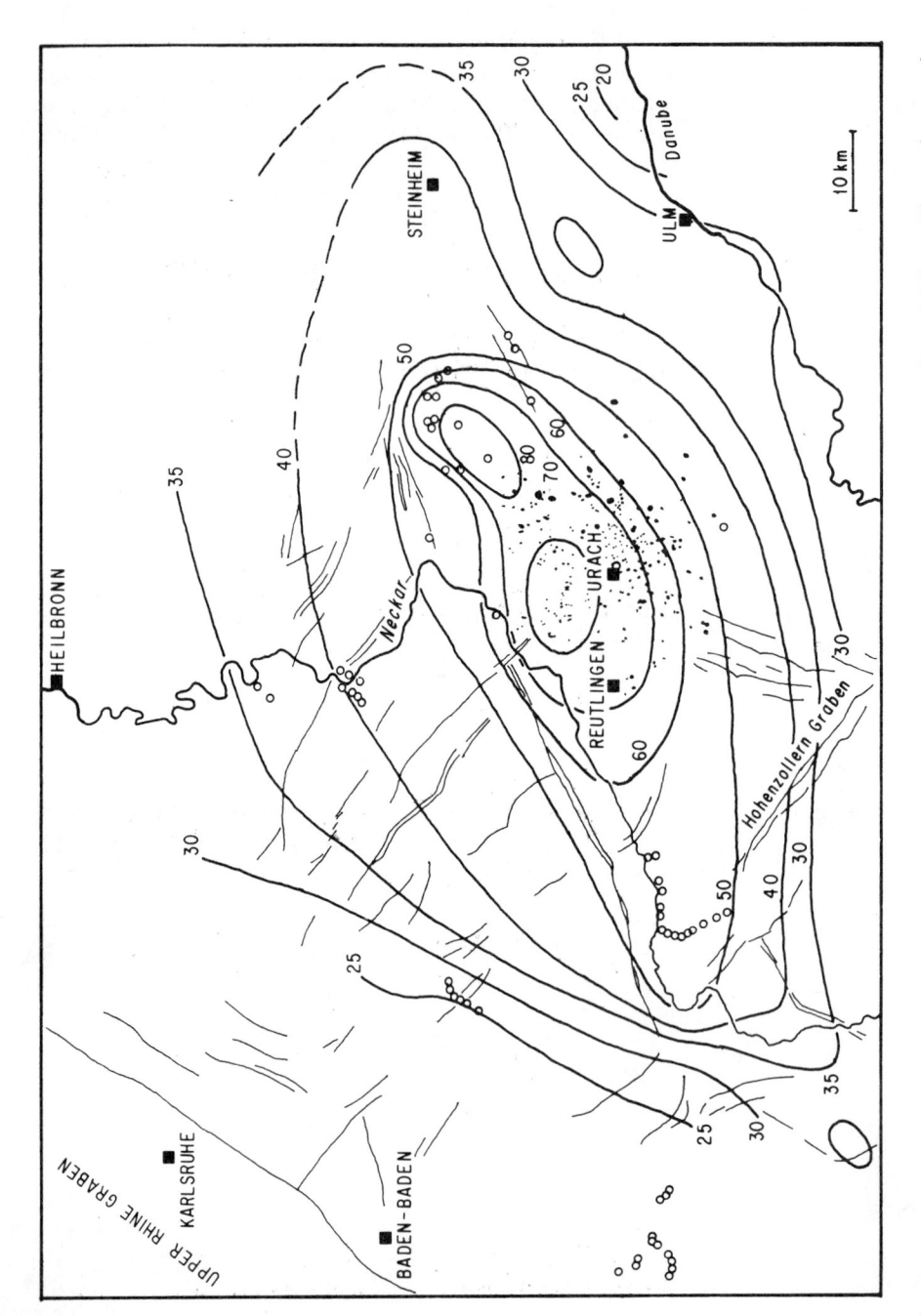

Figure 1.12. Urach geothermal anomaly (South Germany) with isolines of geothermal gradient (°C/km). Thermal anomaly coincides with area of numerous volcanic necks (black). Fine lines: faults, open circles: thermal/mineral springs (redrawn from Carlé, 1974)

The question is whether a deep-seated (now extinct) magma chamber has created the geothermal anomaly.

The volcanism has been dated to be 12 million years old (Carlé, 1974). Not much information is available on the depth of the magma chamber. By varying the depth to the top of the intrusion within the relatively wide limits of 5 to 15 km, the maximum heat flow appears 0·4 to 3·6 million years after emplacement of the pluton. Thus the volcanism is much too old to explain the recent thermal anomalies. Convective movement of thermal water along fracture zones (and possibly along the volcanic necks as well) is a likely explanation of the Urach anomaly (Carlé, 1974).

Detailed discussion of possible 'conductive' effects of magmatic bodies on the surrounding temperature field can be found in Lachenbruch et al., (1976).

1.3.4. Heat Contents of Rocks, Geothermal Resource Base

In attempts to estimate geothermal resources in broad regions a fundamental quantity is needed: the geothermal resource base. The resource base can further be subdivided into accessible and inaccessible, useful and residual, economic and subeconomic resources and also into resources and reserves. For a detailed discussion see Chapter 6. The resource base is defined as all the heat in the Earth's crust beneath a specific area, measured from local mean annual surface temperature. The depth range D must be specified; whereas the upper few kilometers of the crust are considered by White and Williams (1975) and Nathenson and Muffler (1975), the heat content of the whole crust is termed as resource base by Muffler and Cataldi (1978).

The heat content per unti surface area is

$$Q = c_v \int_0^D [T(z) - T_0] \, dz = c_v \left[\int_0^D T(z) \, dz - D T_0 \right] \quad (1.39)$$

where c_v is the volumetric heat capacity and $T(z)$ the geotherm characteristic of the area under consideration (e.g. heat flow provinces, Diment et al., 1975). The total heat content is then obtained by multiplying Q by the appropriate area.

An example is given below for the calculation of the resource base: for the area of the Northern Foreland of the Alps the $T(z)$ curve has been shown in Figure 1.6. For (1.39) the integral I is needed:

$$I = \int_0^D T(z) \, dz = \int_0^D \left\{ \frac{q_0 - A_0 H}{K} z + \frac{A_0 H^2}{K} [1 - \exp(-z/H)] \right\} dz$$

$$= \frac{D^2}{2K} (q_0 - A_0 H) + \frac{D}{K} \{ A_0 H^2 [1 + \exp(-D/H)] \} \quad (1.40)$$

thus

$$Q = c_v (I - D T_0) \quad (1.41)$$

With the data of Figure 1.6 and with $c_v = 2.5 \cdot 10^6$ J/m³, °C one obtains $7.5 \cdot 10^{18}$ J/km² for $D = 20$ km. Since c_v in general increases with temperature the above resource base figure is certainly on the conservative side.

1.4. Geothermal Anomalies and Their Plate Tectonic Framework

To understand the nature and occurrence of geothermal resources, one must consider the processes that lead to the formation of geothermal anomalies. In general, some sort of *upward transfer of heat by moving masses* (e.g. magma) leads to the development of geothermal anomalies. Geothermal anomalies related to magmatic activity orginate from transient storage of heat (transient' on the geological time scale) at high crustal levels (<10 km; Muffler, 1976a); in considering uprising magma in the crust, distinction must be made between basaltic and silicic magmas.

Basaltic magmas originate from mantle material by partial melting (Wyllie 1971, p. 186f). If they rise directly to the surface forming dykes and thin sheets they dissipate their heat content rapidly and thus no large shallow intrusive bodies will be formed. On the other hand, if they solidify near the base of the crust in extensional regimes like the Basin and Range Province U.S.A. (see, e.g., Lachenbruch, 1978), they greatly enhance the regional heat flow, which in turn can trigger hydrothermal convection along steep fault systems created by the extensional regime (Lachenbruch and Sass, 1977).

Silicic magmas can be generated by partial melting of mantle material as well as by differentiation of basalt (Robinson et al., 1976), but usually some kind of remelting of crustal material is involved. Due to the higher viscosity of silicic (relative to basaltic) melts, these magmas usually get trapped at several kilometers depth in the crust. On continents, geothermal resources are more likely to be associated with silicic volcanism than with basaltic volcanism (Smith and Shaw, 1973).

Since igneous intrusions have typical temperatures in the range of 700°–1200°C they can drastically heat their neighborhood upon emplacement. Corresponding surface heat flow can exceed several W/m^2. Due to conductive and convective cooling of the intrusion (cf. pp. 23 and 55) a considerable thermal influence can prevail after emplacement for a limited time only. As a general rule, only Quaternary intrusions (with ages in the range of 10^4 to 10^6 years) in the upper crust (<10 km) are still active thermally today (Healy, 1976).

Magmatic activity on the global scale can be located and delineated within the framework of *plate tectonics*, which allows one to delimit broad geographical areas of geothermal potential. The concept of plate tectonics tells us that magma upwelling is most likely along plate boundaries. Thus, geothermal resources related to magma intrusions are expected to occur mainly along spreading ridges, convergent margins (subduction zones) and intraplate melting anomalies. Figure 1.13 illustrates the general framework of the Earth's plates and the major known geothermal systems.

According to the principles of plate tectonics the Earth's rigid outer shell (=lithosphere, thickness about 80 km) is broken into several plates: there are seven very large plates (cf. Figure 1.13) and about a dozen small plates. All plates slide upon their more viscous base (=asthenosphere) relative to all others at velocities of several centimeters per year (for an excellent review see Hamilton, 1976). Interference occurs at plate boundaries: at divergent plate boundaries (spreading ridges) where upwelling molten material forms new crust, at convergent boundaries where one plate slides under the other subduction zone) and along transform faults where plates pass each other horizontally. Magmatic and thus geothermal activity is associated with the first two types only. In the following we address the thermal effects of magma generation within the framework of plate tectonics relevant to geothermal resources.

1.4.1. Spreading Ridges

It has been shown (Lubimova et al., 1976) that more than 30 per cent of the Earth's total heat loss occurs along the elongate ($\sim 5 \cdot 10^4$ km) system of submarine ridges which

Figure 1.13. Lithospheric plate boundaries provide the framework for the global distribution of major geothermal systems. Plate boundary types: Spreading ridge (1), subduction/trench (2), transform fault (3). Shaded areas (4): Plate interior undergoing active extensional, compressional or strike-slip faulting. Base map and plate boundaries after Hamilton (1976) and Panza and Mueller (1979), geothermal systems (dots) after Muffler (1976a, 1976b). Systems discussed in detail in this volume: A (Ahuachapán/El Salvador), K (Krafla/Iceland), P (Pannonian basin/Hungary), T (Takinoue/Japan), V (Valles Caldera, Jemez Mts/U.S.A.)

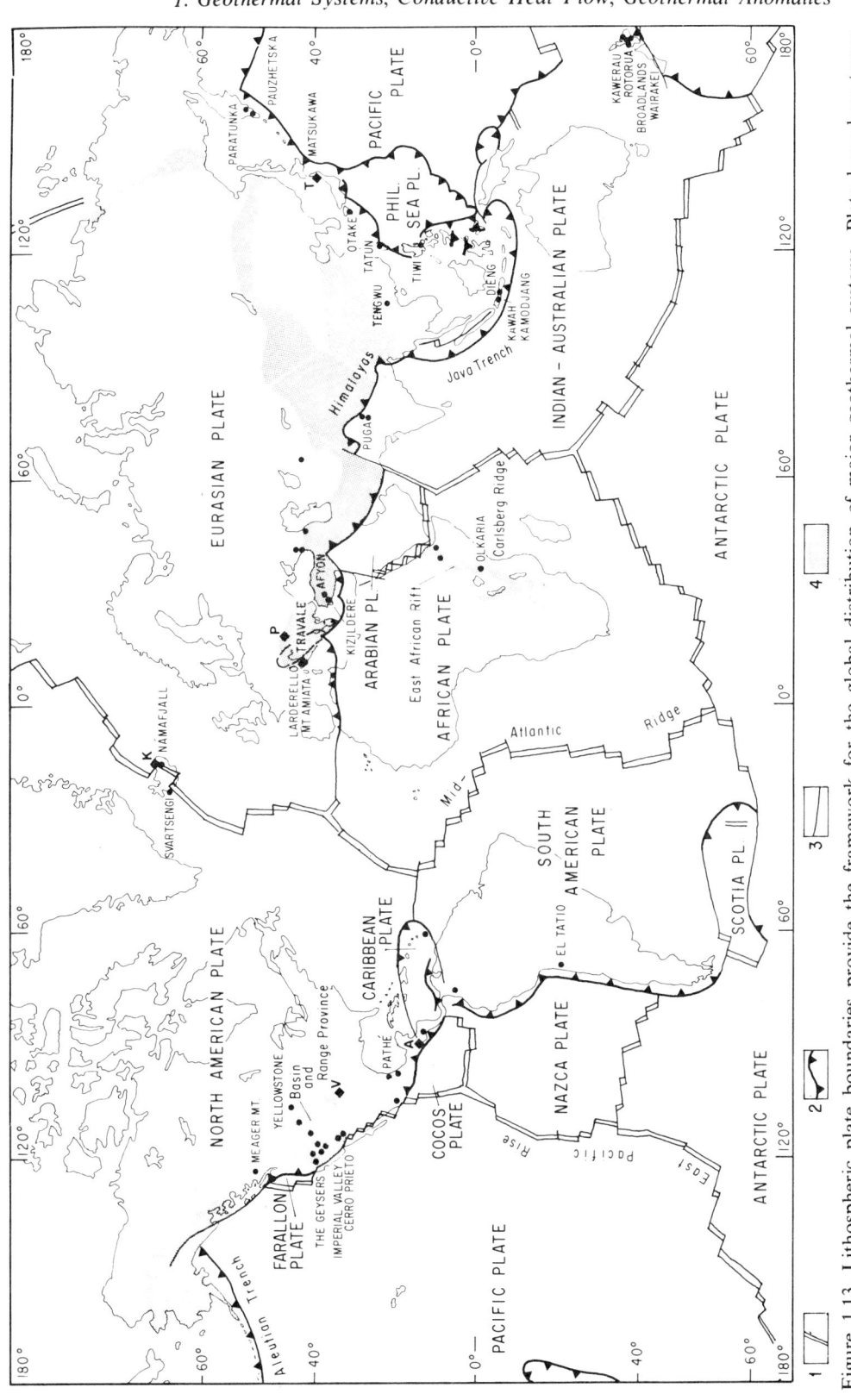

comprise less than 1 per cent of the Earth's surface. At ridges the lithospheric plates are pulling apart while upwelling basaltic melts from the mantle fill the gap to form new ('oceanic') lithosphere. The volcanism at ocean ridges can be attributed to pressure-release melting in the underlying mantle (Oxburgh and Turcotte, 1968). The topographic profile of the sea floor perpendicular to the ridge axis shows a general decline of the surface on both sides of the ridge: the hot (and less dense) lithosphere gradually cools on its way from the ridge. A similar trend is evident in heat flow: here again the heat flow decreases with increasing distance (and age) from the ridge (Figure 1.14). The isotherms are upwelling below the ridge axis; the shape of the isotherms depends, among other factors on the spreading velocity (Sleep, 1975).

Besides conductive cooling, hydrothermal convection occurs in oceanic crust, especially at and near the ridge axis (Crane, 1979; Fehn and Cathles, 1979). Depending on the permeability of oceanic crust, hydrothermal convection can affect (and thus cool) the uppermost 5–7 km (Lister, 1977). The hydrothermal mass flow through sea floor between 0 and 2 million years old was found to be in the order of 10^{11} g/yr per km of ridge (Fehn and Cathles, 1979).

In general terms, Iceland can be considered as a portion of a mid-oceanic spreading ridge which emerges above sea level (Pálmason and Saemundson, 1974), although there are distinct differences in the crustal structure of Iceland relative to that of an oceanic ridge (Mueller, 1978). In Iceland the heat transfer processes can be studied in detail: here heat is transported to the surface by volcanism (q_v), conduction (q_c) and hydrothermal activity (q_h). For the neovolcanic zone crossing Iceland a total output of 56 MWt per km ridge length has been estimated by Pálmason (1973), supplied by the following contributions of the three components: $q_v = 21$ MW/km, $q_c = 21$ MW/km and $q_h = 14$ MW/km. Iceland's geothermal potential is well known; see also Stefánsson (this volume).

Further areas where an oceanic spreading ridge extends on to a continental have

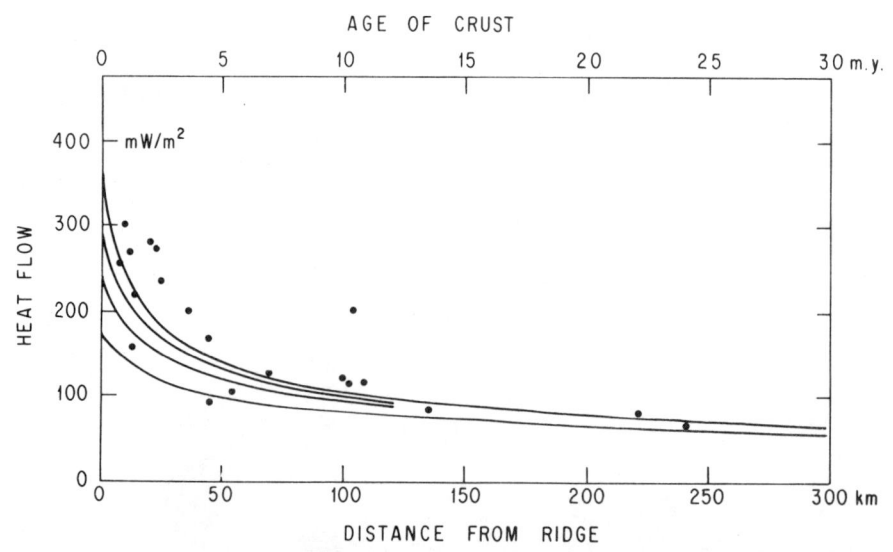

Figure 1.14. Decrease of heat flow with distance from the ridge axis (redrawn after Pálmason, 1973). Dots: heat flow determinations in Iceland, solid curves calculated for different initial lithospheric gradients in the range of 100°–250°C/km and for a spreading rate of 1 cm/y

equally favorable geothermal potential, e.g. the continuation of the East Pacific Rise through the Gulf of California towards the north (geothermal fields of Cerro Prieto/Mexico and Salton Sea/U.S.A.) and the extension of the Carlsberg Ridge into the Red Sea as well as into the Afar Depression, East Africa.

1.4.2. Continental Rifts

Continental rifts are formed by the break-up of continental lithosphere, upon upbulging of the mantle in the region of the fracture. The process which leads to graben formation in the fractured zone and to pressure release below it enables the ascent of molten material. Spreading rates are considerably lower than at oceanic spreading ridges. Geothermal activity is correspondingly more moderate; Lister (1976) estimates that the probability to find a major hydrothermal system is 0·025 per km rift length and cm/yr spreading rate.

The prerequisite of continental rifting, prior to extensional tectonics, is the presence of a zone of weakness and instability in the upper crust (layer of velocity and density inversion with lower viscosity; Mueller, 1978). Upon increased thermal input from below, this layer first undergoes plastic deformation until the first faults develop in the layer above. As rifting proceeds, a central faulted block is formed while at the base of the crust an anomalous mantle 'cushion' develops which will be 'impregnated' by basaltic melts generated below in the uppermost mantle due to the release of lithostatic pressure (Mueller, 1978). An extensive review on the origin of continental rifts was given by Mohr (1978).

The rifting process and especially the magmatic activity is episodic. Depending upon the depth of emplacement of the intrusions the geothermal activity will prevail for a certain time. The Rio Grande Rift/New Mexico, U.S.A. represents of a well-documented example for the magmatic-thermal history of such a rift system (Cook et al., 1978). The considerations involve reduced heat flow (see p. 17) as well as tectonic, seismic, and Bouguer gravity data. The analysis of the southern part of the rift reveals three episodic intrusions near the base of the crust at 35 million years, at 10 million years, and at 5 million years with excess temperature of the intrusions of 600°–800°C. The elevated temperatures in the crust subsequently gave rise to hydrothermal circulation in the fracture system of the rift. Due to such systems, surface heat flow values up to 0·6 W/m^2 can be found today in the Rio Grande Rift (Reiter et al., 1978). For further details see Laughlin (this volume).

Another example of a continental rift structure with pronounced geothermal anomalies is the Upper Rhine Graben, Central Europe. Here the rifting history is recorded by the sediment fill of the graben which also shows episodic development (Doebl et al., 1974); independent inference from paleogeothermal data (Buntebarth, 1978) confirms the episodic character of the rifting process. Geothermal anomalies are attributed to hydrothermal effects, driven by a deep-seated heat supply system (upwelling mantle material) underneath the graben (Werner, 1975).

Further examples for continental rift structures with geothermal potential are the East African Rift—with geothermal areas in Kenya (Noble and Ojiambo, 1976) and Uganda (Maasha, 1976)—which forms the continuation of the Afar rift zone (Stieltjes, 1976), the Baikal rift (see Lysak, 1976), and also the Basin and Range Province (western U.S.A.), which can be considered in general terms as an unusually broad zone of parallel-trending continental rifts (Hamilton, 1976).

1.4.3. Convergent Margins (Subduction Zones)

The oceanic lithosphere produced at spreading ridges moves laterally and is consumed at subduction zones where it slides under another plate, usually of 'continental' structure (in some cases the overriding plate is also of oceanic type, like the east side of the Philippine Sea). The trace of the downgoing plate is marked by earthquakes ('Benioff zone', exhibiting dips between 20° and 50°). The subducted material is in a state of thermal and gravitational instability since it is cooler and denser than the adjacent asthenosphere (for details see, e.g., Toksöz, 1975).

The spatial distribution of the earthquakes is indicative of the thermal state in subduction zones (Figure 1.15) since earthquakes occur only in brittle materials below a certain temperature threshold. Seismological studies reveal (Molnar et al., 1979) that the earthquakes occur in the coldest (interior) part of the downgoing plate rather than at its upper (or lower) surface undergoing friction. The extent of depression of the isotherms inside of the downgoing plate (cf. Figure 1.15) depends on the subduction velocity v and the lithosphere thickness; in subduction zones with high subduction velocity, earthquake foci can be located at depths as great as 650 km (Kurile Islands, for example) whereas in zones with low subduction rates (e.g. New Zealand) they disappear below 300 km.

Another factor to be considered in the thermal regime of subduction is the frictional heating along the surfaces of the downgoing plate. The heat Q generated by friction is $Q = \tau v/w$ where τ is the shear stress and w the width of the shear zone (the latter is in the order of 1 km in subduction zones but rather difficult to estimate; see Bird, 1978). Substantial heating can be expected only if the motion of the downgoing plate is resisted by very large shear stresses in the order of several hundred MPa. Such high stresses are unlikely in view

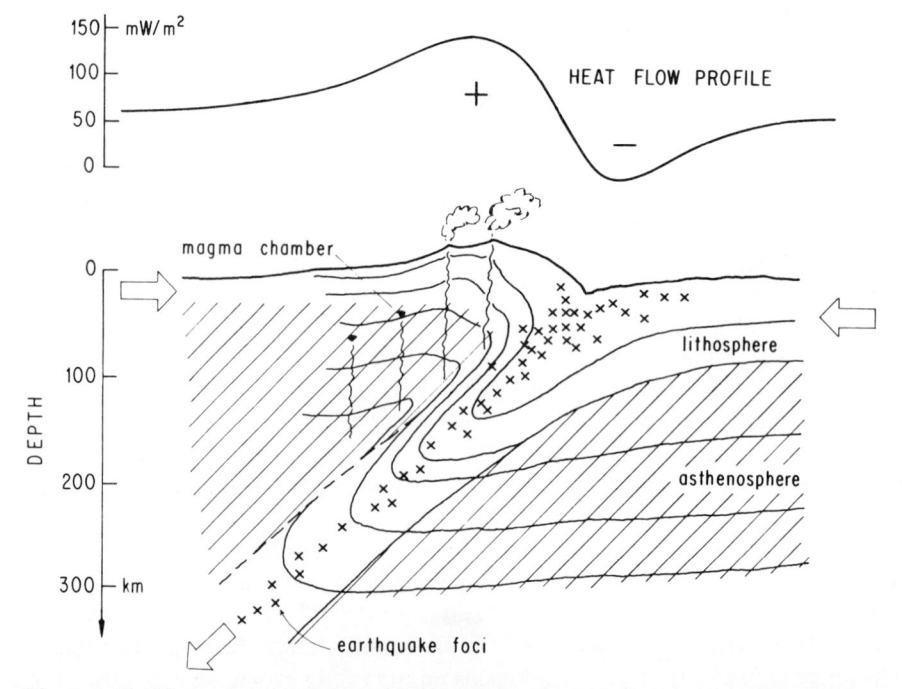

Figure 1.15. Thermal regime of a subduction zone. The downgoing lithospheric plate 'pulls down' the isotherms; the main volcanic axis develops approximately 125 km above the inclined zone of earthquakes which are located in the coldest part of the downgoing plate (partly after Uyeda, 1978)

of the rheological properties of the asthenosphere; in fact, stress values less than 20 MPa have been reported from the top 100 km of subduction zones (Bird, 1978). Direct melting caused by friction can thus be ruled out.

The predominant process of a magma formation is partial melting of the asthenosphere *above* the subducted slab due to water release below it. The water can be supplied by the pore fluids of subducted sediments and/or by metamorphic (dehydration/breakdown) reactions (Anderson et al., 1976). The magmas will appear in a volcanic belt paralleling the zone of subduction at a certain distance which depends on the dip of the downgoing slab. The extent of geothermal anomalies associated with this volcanism will depend on the chemical composition of magmas. The latter will vary according to whether the subduction zone is an oceanic-continental plate contact (this results in intermediate composition), oceanic-oceanic (basic magmas) or continental-continental (predominantly silicic magmas).

Geothermal areas related to subduction zones exist in

(a) Kawah Kamodjang, Indonesia; related to thrusting of the India plate under the China plate (Kartokusumo et al., 1976).

(b) Puga, Chumathang, and Parbati Valley of the Himalayas, north-west India; related to the same subduction zone as above, but in a complex zone of convergence between continental crust of each plate (Subramanian, 1976).

(c) El Tatio, Chile; related to subduction of oceanic lithosphere below the continental plate at the west coast of South America (Lahsen and Trujillo, 1976).

1.4.4. *Intraplate Thermal Anomalies*

Although most of the Earth's magmatic and geothermal activity occurs at or next to lithosphere plate margins, volcanism has repeatedly accurred within both oceanic and continental plates. Intraplate melting anomalies in oceanic lithosphere are known from the Hawaiian Islands, the Azores and from several volcanic chains and isolated islands ('seamounts') in the Pacific (Easter Island, Cook–Austral Islands). Within continental lithosphere, Africa exhibits a large number of Cenozoic volcanic centers which cannot be associated with plate margin processes (e.g. Tibesti, Hoggar, Bayuda; see Thrope and Smith, 1974).

The source of magma for these volcanoes is believed to be located deep below the lithosphere; while the lithosphere can move past such a magma-producing spot the source continues its activity from a fixed position relative to the mantle ('mantle plume'; Morgan, 1972). It seems that individual mantle plumes move very little with respect to each other and thus can be considered as a 'reference frame' for plate motions (Uyeda, 1978, p. 195).

Several processes can create thermal perturbations in the mantle: 'chemical' plumes resulting from radioelement inhomogeneities, convective upwelling, etc. Shear melting at the lithosphere/asthenosphere boundary, caused by mechanical plate-mantle interaction, has also been suggested (Shaw and Jackson, 1973). The lateral extent of such perturbations is in the range of 200–700 km, their heating effect is 200°–400°C and they cause a heat flow anomaly in the order of 30–70 mW/m^2 (Gass et al., 1978). The lithosphere above reacts with thermal expansion, doming/uplift, thinning, fracturing and, ultimately, with intraplate volcanism. African intraplate volcanism is alkaline in composition with varying degrees of silica-saturation and under-saturation (Gass et al., 1978).

Intraplate volcanism can be explained in terms of plate thickness and velocity since penetrative magmatism proceeds on a time scale one or two orders of magnitude greater than heat conduction. In cratonic parts of continental lithosphere (general thickness ~200 km, heat flow <45 mW/m^2; Chapman and Pollack, 1977), relatively slow plate movement (<2 cm/y) will already suppress the upward propagation of sublithospheric thermal anomalies and thus preclude intraplate volcanism. Thinner lithosphere and even slower plate movement are required to develop volcanism and thermal anomalies at the surface; the time scale will be in the order of 10 million years (Gass et al., 1978).

An alternative hypothesis has been put forward to explain continental intraplate volcanism by Turcotte and Oxburgh (1978): the ascent of magmas occur along propagating lithospheric fractures ('corrosion fracturing'). In fact, thermal plumes may mark the beginning of *rifting*: a genetic progression can be envisaged in Africa from unrifted domes (e.g. Tibesti, J. Marra) through rifted domes (Kavirondo, East African rift) leading to the spreading ridge-type (Afar).

The intraplate volcanism of the Western U.S. has a complex history, due to the interaction of *three* lithospheric plates (Pacific, North American, and Farallon/Juan de Fuca plates). Thermal events and periods of volcanic activity are related to changes in tectonic extension regional stress field (Christiansen and McKee, 1978).

An example of a geothermal system related to an intraplate melting anomaly is the Pahoa geothermal field now under development in Hawaii/U.S.A.

Thermal (gneiss) domes are characteristic features in metamorphic terranes: they consist of a granitoid core, enveloped by rocks of decreasing metamorphic grade with typical horizontal dimensions of several tens of km (Den Tex, 1975). Thus a well-focused heat source must have been responsible for their formation. Local release of CO_2-rich fluids from the mantle can be envisaged as a possible cause (Schuiling and Kreulen, 1979) since substantial amounts of CO_2 may be present deep in the mantle (Newton and Sharp, 1975). Upward migrating fluids with 'juvenile' CO_2 (the latter can be traced in numerous spring discharges) may be responsible for the thermal anomaly of Urach/W. Germany which exhibits large variations of the geothermal gradient over small horizontal distances (*cf.* Figure 1.12).

References

Anderson, R., Uyeda, S., and Miyashiro, A., 1976. 'Geophysical and geochemical constraints at converging plate boundaries—1. Dehydration in the downgoing slab', *Geophys. J. R. astr. Soc.*, **44**, 33–357.

Armstead, H. C. H., 1978. *Geothermal Energy*, Spon, London, 357 pp.

Balling, N. P., 1976. 'Geothermal models of the crust and uppermost mantle of the Fennoscandian, Shield in south Norway and the Danish Embayment', *J. Geophys.*, **42**, 237–256.

Benoit, W. R., 1978. 'The use of shallow and deep temperature gradients in geothermal exploration in northwestern Nevada using the Desert Peak thermal anomaly as a model', *Geothermal Resources Council Trans.*, **2**, 45–46.

Birch, F., Roy, R. F., and Decker, F. R., 1968. 'Heat flow and thermal history in New England and New York'. In *Studies of Appalachian Geology: Northern and Maritime*, E-an Zen, W. S. White, J. B. Hadley and J. B. Thompson, Jr. (Eds.), Interscience, New York. 437–451.

Bird, P., 1978. 'Stress and temperature in subduction shear zones: Tonga and Mariane, *Geophys. J.R. astr. Soc.*, **55**, 411–434.

Blackwell, D. D., and Morgan, P. 1976. 'Geological and geophysical exploration of the Marysville geothermal area, Montana, U.S.A.' In *Proc. Second U.N. Symposium on the Development and Use of Geothermal Resources*, San Francisco, 895–902.

Buntebarth, G., 1975. 'Geophysikalische Untersuchungen über die Verteilung von Uran, Thorium und Kalium in der Erdkruste sowie deren Ansendung auf Temperaturberechungen für verschiedene Krustentypen', *Ph.D. thesis*, Technical University Clausthal.

Buntebarth, G., 1978. 'The degree of metamorphism of organic matter in sedimentary rocks as a, paleo-geothermometer, applied to the Upper Rhine Graben', *Pure and Appl. Geophys.*, **117**, 83–91.

Carlé, W., 1974. 'Die Wärme-Anomalie der mittleren Schwäbischen Alb (Baden–Württemberg)'. In *Approaches to Taphrogenesis*, J. H. Illies and K. Fuchs (Ed.), Schweizerbart, Stuttgart, 207–212.

Carslaw, H. S., and Jaeger, J. C. 1959. *Conduction of Heat in Solids*, 2nd ed., Oxford University Press, New York, 510 pp.

Cathles, L. M., 1977. 'An analysis of the cooling of intrusives by ground-water convection which includes boiling', *Econ. Geol.*, **72**, 804–826.

Chapman, D. S., and Pollack, H. N., 1977. 'Regional geotherms and lithospheric thickness', *Geology*, **5**, 265–268.

Christiansen, R. L., and McKee, E. H. 1978. 'Late Cenozoic volcanic and tectonic evolution of the Great Basin and Columbia Intermontane regions', *Geological Society of America Memoir 152*, 283–311.

Cook, F. A., Decker, E. R., and Smithson, S. B., 1978. 'Preliminary transient heat flow, model of the Rio Grande rift in southern New Mexico', *Earth Planet, Sci Lett.*, **40**, 316–326.

Costain, J. K., 1979. 'Geothermal exploration methods and results—Atlantic Coastal Plain.' In *Proc. Symp. on Geothermal Energy and its Direct Uses in the Eastern United States*, Geothermal Resources Council Spec. Rept. **5**, 13–22.

Crane, K., 1979. 'Hydrothermal stress drops and convective patterns at three mid-ocean spreading centers', *Tectonophys.*, **55**, 215–238.

Den Tex, E., 1975. 'Thermally mantled gneiss domes: the case for convective heat flow in more or less solid orogenic basement.' In *Progress in Geodynamics*, C. J. Borradaile and A. R. Ritsema, (Eds.), R. Netherlands Acad. Arts and Sci., Amsterdam, 63–79.

Diment, W. H., Urban, T. C., Sass, J. H., Marshall, B. V., Munroe, R. J., and Lachenbruch, A. H., 1975. 'Temperatures and heat contents based on conductive transport of heat.' In *Assessment of Geothermal Resources of the United States*, D. E. White and D. L. Williams, (Eds.), U.S. Geol. Survey Circ. 726, 84–103.

Diment, W. H., 1980. 'Geology and Geophysics of geothermal areas. In *A Sourcebook on the Production of Electricity from Geothermal Energy*, J. Kestin, R. D. Pippo, H. E. Khalifa, and D. W. Ryley, (Eds.), U.S. Govt. Printing Office.

Doebl, F., Heling, D., Homann, W., Karweil, J., Teichmüller, M., and Welte, O., 1974. 'Diagenesis of clayey sediments and included dispersed organic matter in relationship to geothermics in the Upper Rhine Graden.' In *Approaches to Taphrogenesis*, J. H. Illies and K. Fuchs), (Eds.), Schweizerbart, Stuttgart, 192–207.

Ellis, A. J., and Mahon, W. A. J., 1977. *Chemistry and Geothermal Systems*, Academic Press, New York, 392 pp.

Edmunds, W. M., Burgess, W. G., and Andrews, W. G. 1977. 'Interim geochemical and hydrogeological results relating to U.K. geothermal resources.' In *Seminar on Geothermal Energy*, Vol. I, Commission of the European Communities, Bruxelles, 505–517.

Fehn, U., and Cathles, L. M., 1979. 'Hydrothermal convection at slow-spreading mid-ocean ridges', *Tectonophys*, **55**, 239–260.

Gass, I. G., Chapman, D. S., Pollack, H. N., and Thorpe, K. S., 1978. 'Geological and geophysical parameters of mid-plate volcanism', *Phil. Trans. R. Soc. London.*, **A288**, 581–597.

Garnish, J. D., 1976. 'Geothermal energy: the case for research in the United Kingdom', Energy Paper No. 9, Her Majesty's Stationery Office, London, 66 pp.

Hamilton, W., 1976. 'Plate tectonics and man', *U.S. Geol. Survey Annual Rept. Fiscal Year 1976*, 39–53.

Hawkesworth, C. J. 1974. 'Vertical distribution of heat production in the basement of the Eastern Alps', *Nature*, **249**, 435–436.

Healy, J., 1976. 'Geothermal fields in zones of recent volcanism.' In *Proc. Second U.N. Symposium on the Development and Use of Geothermal Resources*, San Francisco, 415–422.

Heier, K. S., and Adams, J. A. S., 1965. 'Concentration of radioactive elements in deep crustal material', *Geochim. Cosmochim. Acta*, **29**, 53–61.

Hurley, P. M., and Fairbairn, H., 1953. 'Radiation damage in zircon: a possible age method' *Bull. Geol. Soc. Amer.*, **64**, 659–673.

Iriyama, J., and Ōki, Y., 1978. 'Thermal structure and energy of the Hakone volcano, Japan.' *Pure and Appl. Geophys.*, **117**, 331–338.

Iyer, H. M., 1975. 'Anomalous delays of teleseismic P-waves in Yellowstone National Park', *Nature*, **253**, 425–427.

Iyer, H. M., Oppenheimer, D. H., and Hitchocock, T., 1978. 'Teleseismic P-delays at The Geysers-Clear Lake, California, geothermal region, *Geothermal Resources Council Trans.*, **2**, 317–319.
Jaeger, J. C., 1964. 'Application of the theory of heat conduction to geothermal measurements.' In *Terrestrial Heat Flow*, W. H. Lee (Ed.) American Geophysical Union Monograph No. 8, Washington, 7–23.
Jaffé, F. C., Rybach, L., and Vuataz, F., 1976. 'Thermal springs in Switzerland and their relation to seismotectonic features.' In *Proc. Int. Congr. on Thermal Waters, Geothermal Energy and Volcanology of the Mediterranean Area*, Vol. I, Athens, 275–285.
Jessop, A. M., Hobart, M. A., and Sclater, J. G., 1976. 'The world heat flow data collection 1975', *Geothermal Service of Canada, Geotherm. Ser.*, **5**, 175.
Jessop, A. M., and Lewis, T., 1978. 'Heat flow and heat generation in the Superior Province of the Canadian Shield', *Tectonophys.*, **50**, 55–77.
Kappelmeyer, O., 1957. 'The use of near-surface temperature measurements for discovering anomalies due to causes at depth', *Geophys. Prospect.*, **5**, 239–258.
Kappelmeyer, O., and Haenel, R., 1974. *Geothermics—with Special Reference to Application*, Borntraeger, Berlin, 238 pp.
Kartokusumo, W., Mahon, W. A., and Seal, K. E., 1976. 'Geochemistry of the Kawah Kamojang Geothermal System, Indonesia.' In *Proc. Second U.N. Symposium on the Development and Use of Geothermal Resources*, San Francisco, 757–759.
Kissling, E., Labhart, T. P., and Rybach, L., 1978. 'Radiometrische Untersuchungen am Rotondogranit', *Schweiz. Min. Petr. Mittg.*, **58**, 357–388.
Lachenbruch, A. H., and Bunker, C. M., 1971. 'Vertical gradients of heat production in the continental crust—2. Some estimates from borehole data', *J. Geophys. Res.*, **76**, 3852–3860.
Lachenbruch, A. H., Sass, J. H., Munroe, R. J., and Moses, T. H., Jr., 1976. 'Geothermal setting and simple heat conduction models for the Long Valley Caldera', *J. Geophys. Res.*, **81**, 769–784.
Lachenbruch, A. H., and Sass, J. H., 1977. 'Heat flow in the United States and the thermal regime of the crust.' In *The Earth's Crust*, J. G. Heacock (Ed.) American Geophysical Union Monograph No. 20, Washington, 626–675.
Lachenbruch, A. H., 1978. 'Heat flow in the Basin and Range Province and thermal effects of tectonic extension', *Pure and Appl. Geophys.*, **117**, 34–50.
'La géothermie: Chauffage de logement 1977. Délégation Génerale a la Recherche Scientifique et Technique publication, Paris, 24 pp.
'La géothermie en France', 1978. Publ. Délégation aux energies nouvelles, Paris, 72 pp.
Lahsen, A., and Trujillo, P., 1976. 'El campo geotermico de El Tatio, Chile,' In *Proc. Second U.N. Symposium on the Development and Use of Geothermal Resources*, San Francisco, 157–170.
Lister, C. R. B., 1976. 'Qualitative theory on the deep and of geothermal systems.' In *Proc. Second U.N. Symposium on the Development and Use of Geothermal Resources*, San Francisco, 459–463.
Lister, C. R. B., 1977. 'Qualitative models of spreading-center processes including hydrothermal penetration', *Tectonophys.*, **37**, 203–218.
Lubimova, E. A., Ljuboshits, V. M., and Nikitina, V. N., 1976. 'Effect of contrasts in the physical properties on the heat flow and electromagnetic profiles.' In *Geoelectric and Geothermal Studies*, A. Ádám (Ed.), Akadémiai Kiadó, Budapest, 72–99.
Lysak, S. V., 1976. 'Heat flow, geology and geophysics in the Baikal Rift zone and adjacent regions.' In *Geoelectric and Geothermal Studies* (Ed.), Adam, A. Akadémiai Kiado, Budapest, 455–462.
Maasha, N., 1976. 'Electrical resistivity and microearthquake survey of the Sempaya, Lake Kitagata, and Kitagata geothermal anomalies, Western Uganda.' In *Proc. Second U.N. Symposium on the Development and Use of Geothermal Resources*, San Francisco, 1103–1112.
Majer, E. L., 1978. 'Seismological investigations in geothermal regions', *Lawrence Berkeley Laboratory Report LBL-7054*, Berkeley, 225 pp.
Meidav, T., and Tonani, F., 1976. 'A critique of geothermal exploration techniques', In *Proc. Second U.N. Symposium on the Development and Use of Geothermal Resources*, San Francisco, 1143–1154.
Mohr, P., 1978. 'Origin and significance of continental rifts.' In *Int. Symp. on the Rio Grande Rift*, Los Alamos Sci. Lab. Symp. Proc. LA-7487c, p. 59.
Molnar, P., Freedman, D., and Shih, J. S., 1979. 'Length of intermediate and deep seismic zones and temperatures in downgoing slabs of lithosphere', *Geophys. J. R. Astr. Soc.*, **54**, 41–54.
Morgan, J. W., 1972. 'Plate motions and deep mantle convection.' In *Studies in Earth and Space Sciences*, R. Shagam (Ed.), Geol. Soc. Amer. Memoir 132, Boulder, 7–22.
Mueller, St., 1978. 'Evolution of the Earth's crust.' In *Tectonics and Geophysics of Continental Rifts*, I. B. Rambergand E.-R. Neumann (Eds.) Reidel Dordrecht 11–28.
Muffler, L. J. P., 1976a. 'Tectonic and hydrologic control of the nature and distribution of

geothermal resources.' In *Proc. Second U.N. Symposium on the Development and Use of Geothermal Resources*, San Francisco, 499–507.

Muffler, L. J. P. 1976b. 'Summary of Section I: Present status of Resources Development.' In *Proc. Second U.N. Symposium on the Development and Use of Geothermal Resources*. San Francisco, xxxiii–xliv.

Muffler, L. J. P., and Cataldi, R., 1978. 'Methods for regional assessment of geothermal resources.' *Geothermics* 7 (2–4), 53–90.

Muffler, L. J. P., Costain, J. K., Foley, D., Sammel, E. A., and Youngquist, W., 1980. 'Nature and distribution of geothermal energy', In Anderson, D. N., and Lund, J. W. (Eds.), *Direct Utilization of Geothermal Energy, a Technical Handbook*, Geothermal Resources Council Spec. Rept. no. 7, 1-1 to 1–15.

Natheson, M., and Muffler, L. J. P., 1975. 'Geothermal resources in hydrothermal convection systems and conduction-dominated areas.' In *Assessment of Geothermal Resources of the United States—1975*, D. E. White and D. L. Williams (Eds.), U.S. Geol. Survey Circ. **726**, 104–121.

Newton, R. C., and Sharp, W. E., 1975. 'Stability of forsterite $+CO_2$ and its bearing on the role of CO_2 in the mantle', *Earth Planet. Sci. Lett.*, **26**, 239–244.

Noble, J. W., and Ojiambo, S. B., 1976. 'Geothermal Exploration in Kenya.' In *Proc. Second U.N. Symposium on the Development and Use of Geothermal Resources*, San Francisco, 189–204.

Norton, D., and Knight, J., 1977. 'Transport phenomena in hydrothermal systems: Cooling plutons,' *Amer. J. of Sci.*, **277**, 937–981.

Ormaasen, D. E., and Raade, G., 1978. 'Heat generation versus depth of crystallization for Norwegian monzonitic rocks', *Earth Planet. Sci. Lett.*, **39**, 145–150.

Oxburgh, E. R., and Turcotte, D. L., 1968. 'Mid-oceanic ridges and geotherm distribution during mantle convection', *J. Geophys. Res.*, **73**, 2643–2661.

Pálmason, G., 1973. 'Kinematics and heat flow in a volcanic rift zone, with application to Iceland', *Geophys. J. R. astr. Soc.*, **33**, 451–481.

Pálmason, G. and Saemundsson, K., 1974. 'Iceland in relation to the Mid-Atlantic Ridge', *Ann. Rev. Earth Planet. Sci.* **2**, 25–50.

Panichi, C., D'Amore, F., Fancelli, R., Noto, P., and Nuti, S., 1977. 'Geochemical survey of the Siena Province—Interpretation.' In *Seminar on Geothermal Energy*, Vol. II, Commission of the European Communities, Bruxelles, 481–504.

Panza, G. F., and Mueller, St., 1979. 'The plate boundary between Eurasia and Africa in the Alpine area', *Mem. Sci. Geol. Padova*, **33**, 43–50.

Pollack, H. N., 1965. 'Steady heat conduction in layered mediums: the half-space and sphere', *J. Geophys. Res.*, **70**, 5645–5648.

Reiter, M., Shearer, C., and Edwards, C. L., 1978. 'Geothermal, anomalies along the Rio Grande rift in New Mexico', *Geology*, **6**, 85–88.

Robinson, P. T., Elders, W. A., and Muffler, L. J. P., 1976. 'Holocene volcanism in the Salton Sea geothermal field, Imperial Valley, California,' *Geol. Soc. America Bull.* **87**, 347–360.

Roy, R. F., Blackwell, D. D., and Birch, F., 1968. 'Heat generation, of plutonic rocks and continental heat flow provinces', *Earth Planet. Sci. Lett.*, **5**, 1–12.

Rybach, L., 1976a. 'Radioactive heat production in rocks and its relation to other petrophysical parameters', *Pure and Appl. Geophys.*, **114**, 309–318.

Rybach, L., 1976b. 'Radioactive heat production: A physical property determined by the chemistry of rocks.' In *The Physics and Chemistry of Minerals and Rocks*, R. G. J. Strens (Ed.), Wiley, London, 309–318.

Rybach, L., 1976c. 'Die Gesteinsradioaktivität und ihr Einfluss auf das Temperaturfeld in der kontinentalen Kruste' *J. Geophys.*, **42**, 93–101.

Rybach, L., Bodmer, Ph., Pavoni, N., and Mueller, St., 1978. 'Siting criteria for heat extraction from Hot Dry Rock: Application to Switzerland', *Pure Appl. Geophysics*, **116**, 1211–1224.

Rybach, L., 1979. 'Geothermal resources: an introduction with emphasis on low temperature reservoirs.' In *Proc. Symp. on Geothermal Energy and Its Direct Uses in the Eastern United States*, Geothermal Resources Council Spec. Rept. **5**, 1–7.

Rybach, L., Büchi, U. P., Bodmer, Ph., und Krüsi, H. R., 1980. 'Die Tiefengrundwässer des Schweizerischen Mittellandes aus geothermischer Sicht', *Ecologae geol. Helv.* **73**, 293–310.

Sass, J. H., Lachenbruch, A. H. Munroe, R. J., Greene, G. W., and Moses, T. H., Jr., 1971. 'Heat flow in the Western United States', *J. Geophys. Res.*, **76**, 6367–6413.

Sass, J. H., and Munroe, R. J., 1974. 'Basic heat-flow data from the United States', *U.S. Geol. Survey Open-file Rept.*, **74–79**, 450 pp.

Sass, J. H., Ziagos, J. P., Wollenberg, H. A., Munroe, R. J., Di Somma, D. E., and Lachenbruch, A.

H., 1977. 'Application of heat-flow techniques to geothermal exploration, Leach Hot Springs area, Grass Valley, Nevada', *U.S. Geol. Survey Open file rept.* **77-762**, 125 pp.

Schuiling, R. D., and Kreulen, R., 1979. 'Are thermal domes heated by CO_2-rich fluids from the mantle?', *Earth Planet. Sci. Lett.* **43**, 298–302.

Shaw, H. R., and Jackson, E. D., 1973. 'Linear island chains in the Pacific: result of thermal plumes or gravitational anchors?', *J. Geophys. Res.*, **78**, 8634–8652.

Siegenthaler, U., 1979. 'Stable hydrogen and oxygen isotopes in the water cycle.' In *Lectures in Isotope Geology*, E. Jäger and J. C. Hunziker (Eds.), Springer-Verlag, Berlin, 264–273.

Simmons, G., 1967. 'Interpretation of heat flow anomalies—2. Flux due to initial temperature of intrusives', *Rev. Geophys. Space Phys.*, **5**, 109–120.

Sleep, N. H., 1975. 'Formation of oceanic crust: Some thermal constraints', *J. Geophys. Res.*, **80**, 4037–4042.

Smith, R. L., and Shaw, H. R., 1973. 'Volcanic rocks as geologic guides to geothermal exploration and evaluation (abs.)', *EOS (Am. Geophys. Union Trans.)*, **54**, 1213.

Smith, R. L., and Shaw, H. R., 1975. 'Igneous-related geothermal systems.' In *Assessment of Geothermal Resources of the United States—1975*, D. E. White and D. L. Williams, (Eds.), U.S. Geol. Survey Circ. **726**, 58–83.

Smith, R. L., and Shaw, H. R., 1979. 'Igneous-related geothermal systems.' In *Assessment of Geothermal Resources of the United States—1978*, L. J. P. Muffler, (Ed.), U.S. Geol. Survey Circ. **790**, 12–17.

Sorey, M. L., and Lewis, R. E., 1976. 'Convective heat flow from hot springs in the Long Valley Caldera, Mono County, California', *J. Geophys. Res.*, **81**, 785–791.

Steeples, D. W., and Iyer, H. M., 1976. 'Low-velocity zone under Long Valley as determined from teleseismic events', *J. Geophys. Res.*, **81**, 849–860.

Stieltjes, L., 1976. 'Research for a geothermal field in a zone of oceanic spreading Example of the Asal Rift (French territory of the Afars and the Issas, Afar depression, East Africa).' In *Proc. Second U.N. Symposium on the Development and Use of Geothermal Resources*, San Francisco, 613–623.

Subramanian, S. A., 1976. 'Present status of geothermal resources development in India.' In *Proc. Second U.N. Symposium on the Development and Uuse of Geothermal Resources*, San Francisco, 269–271.

Swanberg, C. A., 1972. 'Vertical distribution of heat generation in the Idaho batholith', *J. Geophys. Res.*, **77**, 2508–2513.

Swanberg, C. A., and Morgan, P., 1978. 'The linear relation between temperatures based on the silica content of groundwater and regional heat flow: a new heat flow map of the United States', *Pure Appl. Geophys.*, **117**, 227–241.

Thorpe, R. S., and Smith, K., 1974. 'Distribution of Cenozoic Volcanism in Africa,' *Earth Planet. Sci Lett.*, **22**, 91–95.

Toksöz, N. M., 1975. 'The subduction of the lithosphere', *Scientific American*, February 1975, 98–98.

Turcotte, D. L., and Oxburgh, E. R., 1978. 'Intra-plate volcanism', *Phil Trans. R. Soc. Lond.*, **A288**, 561–579.

Urban, T. C., Diment, W. H., Sass., J. H., and Jamieson, I. M., 1976. 'Heat flow at The Geysers, California U.S.A.' In *Proc. Second U.N. Symposium on the Development and Use of Geothermal Resources*, San Francisco, 1241–1245.

Uyeda, S., 1978. *A New View of the Earth*, Freeman, San Francisco, 217 pp.

Wallace, R. H., Kraemer, T. F., Taylor, R. E., and Wesselman, J. B., 1979. 'Assessment of geopressured-geothermal resources in the northern Gulf of Mexico basin' In *Assessment of Geothermal Resources of the United States—1978*, L. J. P. Muffler (Ed.), U.S. Geol. Survey Circ, **790**, 132–155.

Werner, D., 1975. 'Probleme der Geothermik am Beispiel des Rheingrabens', *Ph.D. Thesis*, University of Karslruhe.

Werner, D., Scriba, H., and Sieber, Ch., 1979. 'Electrical resistivity in the sediments of the Rhinegraben in the region of the Landau thermal anomaly, (abs.)', *EOS, Trans. Am. Geophys. Un.*, **60**, 595.

White, D. E., and Williams, D. L., (Eds). (1975). Assessment of Geothermal Resources of the United States—1975. *U.S. Geol. Survey Circ.* **726**, 155 pp.

Wohlenberg, J., and Haenel, R., 1978. Kompilation von Temperature-Daten für den Tempertur-Atlas der Bundesrepublik Deutschland', *Statusreport 1978—Geotechnik und Lagerstätten, Projektleitung Energieforschung, KFA Jülich*, 1–12.

Wyllie, P. J., 1971. *The Dynamic Earth*, Wiley, New York, 416 pp.

Geothermal Systems: Principles and Case Histories
Edited by L. Rybach and L. J. P. Muffler
© 1981 John Wiley & Sons Ltd.

2. Convective Heat and Mass Transfer in Hydrothermal Systems

SABODH K. GARG

Systems, Science and Software, La Jolla, California 92038, U.S.A.

DAVID R. KASSOY

Department of Mechanical Engineering, University of Colorado, Boulder, Colorado 80309, U.S.A.

2.1. Mathematical Modelling of Hydrothermal Systems

Practically all geothermal energy which to date has actually been tapped for commercial power production is of the hydrothermal variety. In hydrothermal systems, nearsurface water seeps through porous permeable channels to great depth where it encounters hot rock. The heated fluid (water and/or steam) is then driven upward by bouyancy. Thus, convection cells are created in which leakage to the surface or out of the formation is replenished by inflow of meteoric groundwater. Occasionally, these systems have spectacular surface manifestations, such as geysers or hot springs.

There exist two types of hydrothermal systems: vapor-dominated systems (e.g., The Geysers, California, U.S.A.; Larderello, Italy) and the much-commoner liquid-dominated systems (e.g. Wairakei and Broadlands, New Zealand; Heber, East Mesa, and Salton Sea in Imperial Valley, California, U.S.A.). Generally, of course, both liquid water and steam exist to some degree in most such reservoirs, so the distinction is really only important as concerns utilization of the reservoir (see Donaldson and Grant, this volume).

In hydrothermal systems, most of the heat (as well as mass) is transported by convection of liquid water and/or steam through porous permeable rocks. These systems are thus essentially different from dry rock (prior to exploitation) and geopressured systems in which conduction is the dominant mechanism for heat transfer.

The major know hydrothermal systems are associated with seismically active regions with extensive faulting. Although some hydrothermal systems are associated with porous and permeable sedimentary/volcanic rocks, a larger number are in fractured rocks which are otherwise more or less impervious. The fractures act as conduits for geothermal fluids.

The Wairakei geothermal field is traversed by several large northeast-striking normal faults and some smaller northwest cross-faults; the heated water rises from depth along the Waiora and Wairakei faults to charge the main aquifer (Grindley, 1965; Pritchett et al., 1976). Furthermore, the most successful wells drilled at Wairakei intersected one or more fault zones (Grindley, 1965). The East Mesa geothermal field is supplied by heated fluid rising from depth along faults which are known from remote sensing and microearthquake monitoring (Combs and Hadley, 1973). The presence of faults has also been documented at many other geothermal fields (e.g., the Salton Sea geothermal field; the Cerro Prieto field, Mexico; The Geysers and Larderello steam fields; Long Valley, California; Broadlands, New Zealand).

There exists now a large body of circumstantial evidence that faulted regions are frequently associated with geothermal systems. The exact role that the faults play in determining the structure of a geothermal system is, however, a matter of speculation since insufficient information is available on the detailed structure of fault regions at depth. Grant (personal communication, 1978) suggests that the connection between faults and geothermal anomalies may be only indirect inasmuch as faulting may lead to fracturing and hence enhanced permeability. In view of the fact that the water in geothermal systems appears to be mainly of meteoric origin (see e.g., White, 1961), it has been suggested that surface water gradually seeps down in permeable sediments and/or volcanic rocks where it comes close to a deep hot source. The region of downflow is generally thought to be considerably larger than the area of the geothermal anomaly. Since the heated water is buoyant relative to the cool recharge water, the heated water will rise through the faulted region which is characterized by a relatively high permeability (Kassoy and Zebib, 1978). If the fault intersects a horizontal aquifer of high permeability, then the hot water rising through the fault will charge the aquifer. The association of geothermal systems with active seismic regions assures that continuous mechanical fracturing (e.g., that resulting from earthquake activity) will counteract any tendency of the fault zone to be blocked by solids precipitated from the rising and cooling column of saline water.

In the preceding, we briefly reviewed the geological setting and hydrology of geothermal anomalies. We shall now turn our attention to the mathematical modelling of the transport of mass and energy in convective hydrothermal systems. We start out with a discussion of the governing equations for fluid flow in deformable porous/fractured media. This is followed by a description of classical analyses for convection of heat and mass in porous media (Section 2.2). In Section 2.3, we present some plausible models of geothermal systems; special attention will be devoted to the recharge of aquifers. Pre-production models for some selected geothermal anomalies are finally discussed in Section 2.4. The response of geothermal systems under production is discussed elsewhere in this volume by Donaldson and Grant.

2.1.1. Mass and Energy Transport in Fractured Media

Transport of mass and energy in fractured media is of great interest to groundwater hydrologists and to civil and petroleum engineers. Laboratory and field studies have been directed towards defining (1) the geometry of fracture systems, and (2) their response under a hydrodynamic gradient (see e.g., Wittke, 1973). Fluid flow in fissured rocks differs in several important respects from the percolation of fluid through competent porous rocks. Firstly, fracture-induced permeability is usually much greater than the permeability due to pore space. For example, Louis (1970) suggests that the matrix permeability of rock becomes important only in the absence of continuous joints or for joint apertures less than 10 μm. Secondly, fracture permeability is usually anisotropic. Thirdly, fracture porosity (and hence permeability) is much more sensitive to fluid pressure and rock stresses than the matrix porosity (Wittke, 1973; Duncan et al., 1972).

Spacing of discontinuities (fractures or joints) is one of the more important variables in any realistic mathematical description of flow through fissured rocks. If the characteristic distance between discontinuities is comparable to the dimensions of the investigated rock mass, it is necessary to consider the details of the joint (fracture) geometry (see, e.g., Noorishad et al., 1971). In geothermal systems, fracture geometry is usually an unknown, and details of the flow through discrete fractures are of little practical interest (One exception to this remark is the Kawerau geothermal system in New Zealand discussed by Donaldson and Grant elsewhere in this volume). Here we will limit our attention to the opposite case wherein the spacing of the discontinuities is small in comparison to the reservoir being considered and, consequently, the fissured rock can be treated as a continuous medium with anisotropic permeability.

A continuum model for fluid flow through fractured rock masses is discussed by Gray et al. (1976). These authors assume that the fracturing is extensive and well-distributed such that it is reasonable to consider a superficial discharge through the fractures as well as the pores. Although it is relatively straightforward to write down the balance laws for mass and energy transport, the practical utility of models of the type advanced by Gray et al., is limited by one's inability to prescribe *a priori* the exchange of heat and mass between the fluid in the pores and that in the fractures. A more practical approach for treating fluid flow through fractured systems is to represent the fissured rock mass by an 'equivalent' porous media. The equivalent porous media will have, in general, anisotropic and spatially-varying permeability. Such representations of porous fractured media have been successfully employed by Mercer et al. (1975) and by Pritchett et al. (1976) in modeling the behavoir of the Wairakei geothermal reservoir. At Wairakei, most of the aquifer permeability ($\sim 10^{-13}$ m^2) is due to the presence of fractures; the matrix (or pore) permeability is less than 5×10^{-17} m^2 (Pritchett et al. 1976).

2.1.2. Balance Laws for Fluid Flow Through Porous Media

The pore fluid in hydrothermal systems is predominantly saline water (salinities range from less than one-tenth of sea water at Wairakei to ten times that of sea water at the Salton Sea field) with small amounts of noncondensable gases. For our present purposes, it will suffice to regard the pore fluid as pure water. Governing equations expressing the balance of mass, momentum and energy in a deformable porous matrix have been developed among others by Donaldson (1968a), Mercer et al. (1974), Brownell, et al. (1977), and Garg and Pritchett (1977). Brownell, et al. (1977) consider a fully interacting rock-fluid system; particular examples of these interactions include the relationship between pore collapse and fluid extraction rates, the reinjection of collapsed pores by condensate injection, and the effects of variations in natural mechanical forces (e.g., overburden stress) on the state of the geothermal system. In these cases there would appear to be significant interactions between the porous matrix and the fluid, and it is therefore necessary to employ general interaction theories in order to assess the effect of the motion of the matrix on the motion of the fluid and vice versa.

In the following, we will restrict our attention to systems wherein the matrix remains more or less stationary and does not significantly (except through variations in porosity) influence the fluid motion. For most practical purposes, the fluid motion can be adequately described by Darcy's law and the various phases (i.e., rock matrix, liquid water, and water vapor) can be regarded as in local thermal equilibrium. The latter statement implies that we need consider only the mixture (rock/liquid/vapor) energy balance. We will also assume that (1) the rock porosity ϕ at any point $\mathbf{x}$ depends only upon the local fluid pressure, and (2) the liquid and the vapor are in local pressure equilibrium ($p_l = p_g = p$) such that the capillary pressure is negligible. With these assumptions, the balance equations for mass, momentum and energy can be written as follows (Garg and Pritchett, 1977; Brownell et al. 1977):

Mass
Fluid (liquid–vapor mixture):

$$\frac{\partial}{\partial t}(\phi\rho)+\nabla\cdot[\rho_l\mathbf{q}_l+\rho_g\mathbf{q}_g]=0 \tag{2.1}$$

Momentum
Liquid:

$$\mathbf{q}_l=-\frac{R_l\mathbf{k}}{\mu_l}\cdot(\nabla p-\rho_l\mathbf{g}) \tag{2.2}$$

Vapor:

$$\mathbf{q}_g=-\frac{R_g\mathbf{k}}{\mu_g}\cdot(\nabla p-\rho_g\mathbf{g}) \tag{2.3}$$

Energy
(Rock–liquid–vapor mixture):

$$\frac{\partial}{\partial t}[(1-\phi)\rho_r E_r+\phi\rho E]+\nabla\cdot[\rho_l E_l\mathbf{q}_l+\rho_g E_g\mathbf{q}_g+p\mathbf{q}_l+p\mathbf{q}_g]$$
$$=\nabla\cdot(K_m\nabla T)+[\rho_l\mathbf{q}_l+\rho_g\mathbf{q}_g]\cdot\mathbf{g} \tag{2.4}$$

where

E = Specific internal energy for liquid–vapor mixture ($=[(1-S)\rho_l E_l+S\rho_g E_g]/\rho$)
E_i = Specific internal energy for ith phase ($i=r$, rock matrix; $i=l$, liquid, $i=g$, vapor)
$\mathbf{g}$ = Acceleration due to gravity
$\mathbf{k}$ = Permeability tensor
K_m = Mixture (solid–liquid–vapor) thermal conductivity
p = Pressure
$\mathbf{q}_i$ = Volume flux vector for ith phase ($i=l, g$)
R_i = Relative permeability for ith phase ($i=l, g$)
S = Relative vapor volume ($=$ vapor volume/pore volume)
t = Time
T = Temperature
$\mathbf{x}$ = Space vector
μ_i = Viscosity of ith phase ($i=l, g$)
ρ = Density for liquid–vapor mixture ($=(1-S)\rho_l+S\rho_g$)
ρ_i = Density of ith phase ($i=r, l, g$).
ϕ = Porosity

It is necessary to adjoin the system of balance equations (2.1)–(2.4) with suitable constitutive relations and with initial and boundary conditions. In the remainder of this section, we will consider the prescription of constitutive data.

The balance equations (2.1)–(2.4) contain liquid and vapor viscosities, densities, and internal energies in addition to relative vapor volume, fluid density ρ, internal energy E, temperature T, and pressure p. Furthermore, specification of mixture thermal conductivity

K_m requires the knowledge of liquid K_l and vapor K_g conductivities. At present, there exist several tabular or semi-analytic equations of state for water; one such equation of state is described by Brownell et al. (1977). Given fluid density ρ and internal energy E, the equation of state subroutines yield pressure p, steam volume fraction S, fluid temperature T, and liquid and vapor densities, internal energies, viscosities, and thermal conductivities.

We need to prescribe the following functions for the rock matrix:

$T_r(E_r) \sim$ dependence of rock temperature on rock internal energy;
$K_m(K_i, \phi, S) \sim$ dependence of the mixture (solid–liquid–vapor) conductivity on K_i, ϕ, and S.
$K_r(T) \sim$ dependence of rock grain thermal conductivity on temperature;
$\mathbf{k}(\phi, p, T) \sim$ dependence of permeability on porosity ϕ and fluid state
$R_{l,g}(S, T) \sim$ dependence of relative liquid and vapor permeabilities on steam saturation and temperature;
$\phi(p, T) \sim$ dependence of rock porosity upon the fluid pressure and temperature.

For geothermal applications it is sufficient to assume that

$$E_r = c_r T_r \tag{2.5}$$

where c_r is the constant volume heat capacity of the solid. Ramey et al. (1974) present a review of the measurements and empirical formulas for the thermal conductivities of dry and fluid-saturated rocks. The thermal conductivity of most rocks decreases with an increase in temperature. Thermal conductivities of fluid-saturated rocks are 2–5 times greater than those of dry rocks.

The permeability $\mathbf{k}$ depends in a complex manner on porosity ϕ and the fluid pressure and temperature. In geothermal applications, it is usual to regard rock permeability $\mathbf{k}$ as a function of porosity ϕ only.

$$\mathbf{k} = \mathbf{k}(\phi) \tag{2.6}$$

Sufficient data for steam-water systems are presently unavailable to determine the dependence of relative permeabilities $(R_{l,g})$ on steam saturation and temperature. In the absence of detailed experimental data, relative permeabilities for steam-water systems are often approximated by expressions for oil-gas reservoirs (see e.g., Corey et al., 1956).

Porosity ϕ will in general depend on the current state of stress (rock and fluid stresses), stress history, temperature, and the rock type (Brownell et al., 1977). In geothermal applications it is often assumed that the reservoir undergoes only uniaxial compaction and that the overburden remains essentially constant. Under these assumptions, porosity ϕ is given by (Garg et al., 1977)

$$\frac{\partial \phi}{\partial t} = (1-\phi) C_m \frac{\partial p}{\partial t} + 3(C_m \eta K - \eta_s) \frac{\partial T}{\partial t} \tag{2.7}$$

where

$C_m =$ Uniaxial compaction coefficient
$\eta(\eta_s) =$ Coefficient of linear thermal expansion for porous rock (rock grain)
$K =$ Bulk modulus of porous rock

The uniaxial compaction coefficient C_m, and bulk modulus K depend upon the pore pressure, the loading direction (increase or decrease in pore pressure) and the past stress history. In field applications, small changes in pore pressure can often result in extremely large changes in C_m (and K). Thus for example at Wairakei, an average reservoir pressure drop of less than 3 MPa has apparently resulted in a fifteen-fold change in C_{μ} (Pritchett et al., 1976).

2.1.3. Boussinesq Approximation

The governing equations for two-phase fluid flow through porous media as outlined in the preceding section are quite complicated. Fortunately, for many special applications (e.g., single-phase flow) it is possible to considerably simplify the basic system of equations. We shall now consider thermally driven flow in single-phase (liquid) geothermal systems. In this case, the only mass-moving forces present are those due to thermal expansion effects (Wooding, 1957). Following Wooding, we assume that

1. Rock matrix is homogeneous and isotropic in its physical properties including porosity, permeability and thermal conductivity. Furthermore, we will require the rock physical properties to be independent of temperature.
2. The fluid (water) is an incompressible liquid whose density ρ and kinematic viscosity v depend upon temperature according to the laws:

$$\rho = \rho_0 \left[1 - \alpha(T - T_0) - \beta(T - T_0)^2\right] \quad (2.8)$$

$$v = v_0 \sigma(T) \quad (2.9)$$

where ρ_0, α, T_0 and β are constants, and $\sigma(T)$ is a prescribed function of temperature.
3. Pressure work and viscous dissipation are negligible, and fluid (solid) internal energy $E(E_r)$ is given by

$$E = c_v T \quad (2.10)$$

$$E_r = c_r T$$

where c_v (c_r) denotes the constant volume heat capacity of the fluid (rock matrix).

With the above assumptions, the balance laws (equations (2.1), (2.2) and (2.4)) become:

Mass
Fluid (liquid)

$$\nabla \cdot \mathbf{q} = 0 \quad (2.11)$$

Momentum
Fluid (liquid)

$$\left(\frac{\nabla p}{\rho_0} - \mathbf{g}\right) + \alpha(T - T_0)\left[1 + \frac{\beta}{\alpha}(T - T_0)\right]\mathbf{g} + \frac{v_0 \sigma}{k}\mathbf{q} = 0 \quad (2.12)$$

Energy
(Rock–Fluid Mixture)

$$[(1-\phi)\rho_r c_r + \phi \rho_0 c_v]\frac{\partial T}{\partial t} + \rho_0 \mathbf{q} c_v \cdot \nabla T = \nabla \cdot (K_m \nabla T) \quad (2.13)$$

We note that pressure p in this formulation is no longer a thermodynamic variable; it is rather determined by the imposed boundary conditions. The balance laws (equations (2.11)–(2.13)) form the basis of numerous studies of thermally driven convection in porous media (See Section 2.2).

2.2. Idealized Convective Heat and Mass Transport

The study of heat and mass transfer in a geothermal system should be initiated by analysis and interpretation of field data. The information obtained can then be used to develop a physically viable conceptual model which consists of a well defined geometrical system with boundaries on which appropriate thermal and hydrological conditions can be applied, an internal structure consistent with the local geology, and the material and chemical properties of rock and water. Subsequently a mathematical model is constructed, based on porous media flow, which can provide quantitative estimates of the physical processes encompassed by the conceptual model. Assessment of the modeling procedure is carried out by comparing theoretical predictions of near-surface heat flux as well as temperature and formation pressure distributions with available field data.

A fully descriptive model of a geothermal system should describe the source of water, mechanisms for water transport to depth, the process of heating in the deeper sections of the system, the subsequent rise of buoyant hot liquid, its dispersion into chargeable aquifers and the cooling of aquifer liquids by near-surface effects (surface emanations, mixing with colder groundwater, conduction). Detailed modeling of this type requires input data from a wide spectrum of geophysical, geological and geochemical measurements as well as a profound understanding of the hydrodynamics of liquids convecting in rock formations. Vigorous convection will occur only in geological formations having adequate permeability derived from intergranular spaces and fracture distributions. The latter are certainly present in known systems located in tectonically active areas where volcanism and seismicity provide physical mechanisms for rock deformation and failure. For most geothermal areas, data from peripheral areas or the deep segments of the system are not available. Little is known about transport of near-surface water to depth, the characteristics of the heat source, or the geochemical processes in the hottest sections of the system.

To date no single model, based on flow of a liquid in a saturated porous medium, is capable of describing all facets of heat and mass transfer occurring in a given geothermal system. Rather, specific elements of the problem are considered in terms of either hypothetical idealized or plausible models. In the former, elementary geometrical configurations, simplified porous medium properties, and mathematically simple thermal boundary conditions are considered in order to emphasize the hydrodynamics of convection. The knowledge gained can be employed in more physically plausible models which do include realistic configuration, appropriate geological structure and material properties.

2.2.1. Nondimensional Equations and Parameters

Convective processes in an enclosed liquid-saturated porous medium can be described in terms of appropriate nondimensionalized equations. These may be derived from equations (2.11)–(2.13) by using the following variables

$$\mathbf{x} = \mathbf{x}'/L, \quad t = t'v'_R/L, \quad \mathbf{v} = \mathbf{q}'v'_R \qquad (2.14)\text{–}(2.16)$$

$$T = T'/T'_0, \quad p = (p' - p'_H(\mathbf{x}))/p'_R, \quad \beta = \frac{\beta' T'_0}{\alpha'} \qquad (2.17)\text{–}(2.19)$$

$$\mu = \mu'/\mu'_0, \quad k = k'/k'_0, \quad \hat{\mathbf{g}} = \mathbf{g}'/g' \qquad (2.20)\text{–}(2.22)$$

$$v'_R = g'k'_0\alpha' \Delta T'/v'_0, \quad p'_R = \rho'_0 g'\alpha' \Delta T' L' \quad (2.23, 2.24)$$

$$\tau = \Delta T'/T'_0; \quad (2.25)$$

where L', v'_R, p'_R and $\Delta T'$ are the characteristic values of system depth, convection-induced velocity, convection induced over-pressure and maximum impressed temperature difference respectively. (Note that in Sections 2.2 and 2.3 primed quantities denote dimensional values. Dimensional equations (2.1)–(2.13) do not have primes.) The subscript zero implies that variables are evaluated at a specific reference location, usually the upper boundary of the system. In equations (2.22)–(2.24) the magnitude of the gravity vector is represented by g'. The reference hydrostatic pressure distribution $p'_H(\mathbf{x}')$ is described by the equation

$$\nabla' p'_H = \rho'_0 \mathbf{g}' \left\{ 1 - \alpha'(T'_H - T'_0) \left[1 + \frac{\beta'}{\alpha'}(T'_H - T'_0) \right] \right\} \quad (2.26)$$

in which the reference hydrostatic steady temperature field $T'_H(\mathbf{x}')$ is given by the solution to

$$\nabla' \cdot K'_m \nabla' T'_H = 0 \quad (2.27)$$

Both equations are subject to boundary conditions appropriate to a specific problem.

The complete nondimensionalized system is given by

$$\nabla \cdot \mathbf{v} = 0 \quad (2.28)$$

$$\frac{\mu}{k} \mathbf{v} = -\nabla p - \frac{(T - T_H)}{\tau}[1 + \beta(T + T_H - 2)]\hat{\mathbf{g}} \quad (2.29)$$

$$R\left[c_m \frac{\partial T}{\partial t} + c_v \mathbf{v} \cdot \nabla T \right] = \nabla \cdot K_m \nabla T \quad (2.30)$$

where

$$c_m = \left[\phi c'_v + (1-\phi) \frac{\rho'_r}{\rho} c'_r \right]/c'_{p0}, \quad T_H = T'_H/T'_0$$

$$K_m = K'_m/K'_{m0}, \quad c_v = c'_v/c'_{p0}$$

The Rayleigh number,

$$R = \left(\frac{g'k'_0 \alpha' \Delta T' L'}{v'^2_0} \right) \frac{c'_{p0}\mu'_0}{K'_{m0}} = \left(\rho'_0 v'_R c'_{p0} \Delta T' \right) / \left(K'_{m0} \frac{\Delta T'}{L'} \right) \quad (2.31)$$

is the ratio of the convective heat transport to that conducted across the system. Estimates of the characteristic convection velocity, the overpressure, the overheat ratio τ and the Rayleigh number can be obtained from equations (2.23)–(2.25) and equation (2.31) if values of the dependent variables, typical of liquid dominated geothermal systems, are used. If $k'_0 \approx 10^{-13}$ m^2 (100 md.) and the thermodynamic variables are evaluated at $T'_0 = 298$ K then

$$10^{-2} \text{ m/day} \lesssim v'_R \lesssim 10^{-1} \text{ m/day}$$

$$0.3 \text{ MPa} \lesssim p'_R \lesssim 1.5 \text{ MPa}$$

$$0.2 \lesssim \tau \lesssim 1$$

$$200 \lesssim R \lesssim 10{,}000$$

represent typical values of the quantities. Although the vertical convection velocity is relatively small, the associated mass flux over 10^6 m^2 is 10^7 kg/day. The overpressure should be compared with the typical hydrostatic pressure of several tens of MPa. Deviations in the actual pressure of the magnitude of p'_R are measurable with contemporary logging equipment.

For two-dimensional planar calculations in the $x-z$ plane, the velocity field is represented by $\mathbf{v} = (u, w) = (\Psi_z, -\Psi_x)$ where Ψ is a stream function. The temperature can be written as $T = T_H + \tau\theta$. When the flow is steady and $k = c_v = K_m = 1$, $\beta = 0$, then $T_H = 1 - \tau z$ and equations (2.28)–(2.30) reduce to

$$\mu \nabla^2 \Psi + \mu_z \Psi_z + \mu_x \Psi_x = -\theta_x \tag{2.32}$$

$$R[\Psi_z \theta_x + \Psi_x(1 - \theta_z)] = \nabla^2 \theta \tag{2.33}$$

where $\mu = \mu(T)$ must be specified. Subscripts x and z denote derivatives.

Heat transfer results can be expressed conveniently in terms of the local Nusselt number

$$\mathrm{Nu}_l(x) = \frac{K'_m(\partial T'/\partial z')}{K'_{m0}(\Delta T'/L)} = \frac{K_m}{\tau}(\partial T/\partial z) = K_m(-1 + \partial\theta/\partial z) \tag{2.34}$$

or the spatially averaged Nusselt number

$$\mathrm{Nu} = \frac{1}{x^*}\int_0^{x^*} K_m\left(-1 + \frac{\partial\theta}{\partial z}\right)dx \tag{2.35}$$

where x^* is the horizontal extent of the region of interest. In each case the vertical temperature gradient is evaluated on an appropriate horizontal plane.

Analogous forms can be written for axisymmetric flows.

2.2.2. Idealized Modeling

Traditional studies of convection in saturated porous media emphasize the flow of a constant property fluid in an isotropic, homogeneous matrix. A complete review of this literature has been given by Combarnous and Bories (1975) and by Cheng (1978). Here, we will consider idealized convection problems in which some element of the physics can be related to flow in the geothermal environment. Emphasis will be placed on the effects of variable liquid properties, particularly viscosity, associated with large temperature variations, simple anisotropy of the matrix properties, impressed background flows (forced convection), and on the appearance of nonsteady and three-dimensional flows.

Linear stability theory. In the most elementary models, one considers the conditions required for the onset of convection in a water saturated homogeneous, isotropic medium confined between two horizontal impermeable infinite plates maintained at uniform, but different temperatures. When $\Delta T' = O(10^2 \text{ K})$, typical of geothermal systems, the variation of viscosity with temperature is significant (Keenan et al., 1969). Morrison et al. (1949) were the first to suspect that the critical value of the Rayleigh number, R_c, in a variable viscosity system, was reduced from the classical constant property value $4\pi^2$ (Horton and Rogers, 1945; Lapwood, 1948). In a series of papers, Rogers and Morrison (1950), Rogers et al. (1951), Morrison and Rogers (1952), and Rogers (1953) developed linear stability solutions for a liquid with a viscosity decreasing with temperature. Estimates of eigenvalues, based on techniques that are crude by contemporary standards, showed reduced critical Rayleigh numbers.

Precise values of R_c were found by Kassoy and Zebib (1975) from the linearized form of equations (2.32) and (2.33). The viscosity variation $\mu(T)$ is described by Wooding's (1957) empirical formula for liquid water where, for linear theory, $T \approx 1 - \tau z$. When $T_0' = 298$ K and $\Delta T' < 225$ K, the viscosity decreases from top to bottom of the system by as much as a factor of 8. The results presented in Figure 2.1 show that the critical Rayleigh number, R_c, decreases by almost a factor of 4·5. Subsequently, Morland et al. (1977) extended the calculation to include a variation of thermal expansion coefficient with temperature represented by the β-term in (2.29). The associated density decrease caused a further reduction in R_c shown in Figure 2.1. The definitive stability calculation of this type, carried out by Straus and Schubert (1977), includes extremely accurate variable viscosity and expansion coefficient formulas and the effect of a pressure-work term in the energy equation (2.30). The variation of R_c with τ found by Straus and Schubert, shown in Figure 2.1, differs from that of Morland et al. because the $\mu(T)$ relations are slightly different. Depth effects on the critical Rayleigh number were found for systems where $L' \gtrsim 4 \times 10^3$ m when $\Delta T' \lesssim 250$ K. Since most geothermal reservoirs are less than 4×10^3 m deep and have temperature variations within the stated limits, it appears unnecessary to include the additional pressure-work term in equation (2.30) when considering reservoir heat and mass transfer calculations.

For a model geothermal system where T_0' and $\Delta T'$ are specified and the material properties of the liquid and porous matrix are known, it is clear from equation (2.31) that a reduction in the critical value implies a concomitant decrease in k_0' or L' or a combination of the product. When τ is relatively large, convection processes will occur in thinner, less permeable formations.

In order to account for the finite extent of any real geothermal system Zebib and Kassoy (1976a) examined the stability problem in a rectangular parallelepiped with insulated vertical boundaries when $\Delta T' \lesssim 200$ K. In a thin vertical slab which might model the shear zone associated with a fault, the preferred cellular motion is two-dimensional. Onset of motion for $\tau = 0.34$ and 0.68 occurs at $R_c = 1.6\pi^2$ and π^2 respectively compared to

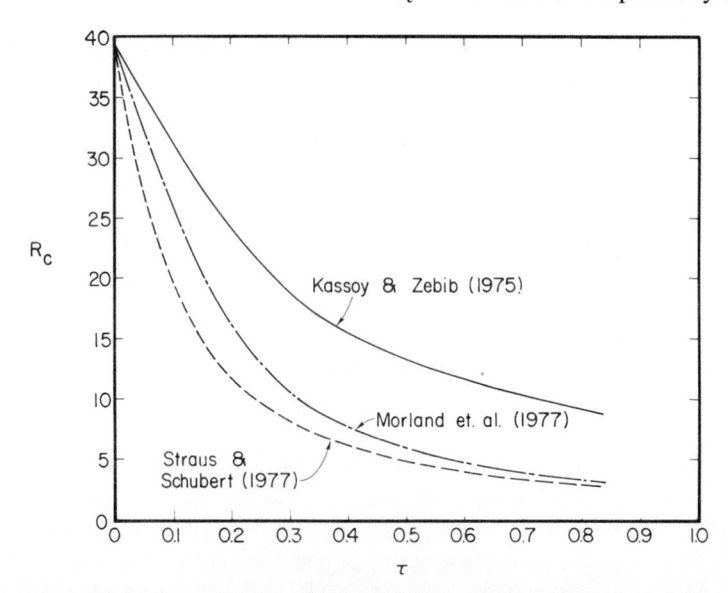

Figure 2.1. The variation of the critical Rayleigh number R_c with over-heat ratio τ from Kassoy and Zebib (1975), Morland et al (1977) and Straus and Schubert (1977)

classical result $R_c = 4\pi^2$ valid for $\tau \ll 1$. Three-dimensional modes can be expected when the box dimensions are nearly the same and when the height is less than the length and width, conditions which probably apply to most geothermal reservoirs.

The applicability of these results to real systems is limited because geological conditions do not conform to the assumed properties of the model: most notably, constant temperature horizontal boundaries, homogeneous isotropic material properties and a steady state conduction temperature profile T_H. However, in a qualitative sense the results show that the heat and mass transport will be sensitive to the viscous properties of the water.

Rogers and Morrison (1950) noted that the conduction temperature distribution in the rest state ($\mathbf{v} = 0$) was probably unsteady in the geophysical context. For instance, if the temperature beneath a cold, water-saturated, horizontal slab of porous material is raised suddenly by the intrusion of a sill, the resulting conduction temperature profile $T_H(t, \mathbf{x})$ found from equation (2.30) with $\mathbf{v} = 0$ will exhibit large local gradients near the hot bottom boundary. Rogers and Morrison (1950) showed that local instability would occur in these regions. It follows that the overall critical Rayleigh number is reduced from the value based on the steady state linear profile for T_H. Improved eigenvalue solutions were given by Rogers et al. (1951). Apparently, this interesting phenomena has not been considered further.

The complex lithology found in many geothermal systems can be modelled by assuming that the horizontal and vertical permeabilities are distinct. In mathematical terms, equation (2.29) must be generalized to include a different k-value in each component of the vector-equation. Wooding (1976) has examined the effect of anisotropic permeability on linear stability in a horizontal slab configuration when the ratios μ/k_h and μ/k_v are constant. This implies that the decrease in viscosity with depth, caused by increasing temperature, is counteracted by a permeability decrease with depth caused by compaction. As k_v/k_h decreases from unity there is a substantial decrease in both R_c and the wave number α. Compared to the classical case $k_v/k_h = 1$ where $R_c = 4\pi^2$ and $\alpha = \pi$, Wooding (1976) finds for instance that when $k_v/k_h = 0.1$ then $R_c = 17$ and $\alpha = 1.75$. In the former the convection cell spacing is $2L'$ where L' is the slab (geothermal reservoir) depth whereas in the latter the larger spacing $3.58L'$ is found. For $L' = 3 \times 10^3$ m, representative of depth in the Taupo volcanic zone of New Zealand, the anisotropic case gives a spacing of about 1.1×10^4 m, which is representative of spacing of well defined zones of anomalously high heat flow at that location. This suggests that the pattern of heated liquid upwelling at that location could be attributed solely to hydrodynamical processes occurring in an anisotropic material rather than to internal geological structure like fault zones.

Large scale hydrological flow, driven by artesian pressure gradients, occurs frequently in permeable rock formations. The sudden intrusion of a magma body beneath such a region would introduce thermal gradients into the system that could induce buoyant convection. One may examine the conditions required for the onset of motion in this context by considering perturbations from simple prescribed background velocity fields. Prats (1966) showed that the value of R_c for the infinite horizontal aquifer configuration was unaffected by addition of a uniform horizontal flow. The flow pattern and the specific temperature distribution are dependent on the magnitude of the sweep rate.

A vertical background flow has been found to have a stabilizing effect on the convection instability. Homsy and Sherwood (1976) found that R_c increases with the magnitude of the vertical velocity.

Weber (1974) has considered the stability of a constant viscosity fluid in an infinite horizontal slab when the surface temperatures differ by a constant value but increase linearly with distance. At the top and bottom boundaries, $T = 1 \mp (\tau/2) - \tau \delta x$ where δ represents the non-dimensional horizontal temperature gradient. Consistent with this boundary temperature distribution is a horizontal background flow $u = \delta z$ and a background temperature field that varies with x as well as z. Infinitesimal disturbances grow when

$$R = R_c = 4\pi^2 (1 + \delta^2 + O(\delta^4)), \qquad \delta < 1 \tag{2.36}$$

implying that the horizontal boundary temperature variation stabilizes the flow relative to the case $\delta = 0$. Conceivably, this configuration might model onset of convection in an aquifer embedded in a rock formation with a spatially varying temperature associated with a nearby intrusive.

Nonlinear theory. Relatively high near-surface heat fluxes observed in the vicinity of known geothermal systems are cuased by enhanced heat transfer associated with liquid convection in the subsurface structure. The presence of motion implies that the Rayleigh number of the system is larger than some critical value. If the system can be represented by an idealized model then the critical value, $R_c > 0$, is that discussed in the previous section. When $R > R_c$ one must contend with the complete nonlinear equations (2.28)–(2.30). The solutions are used to find the magnitudes of heat and mass transfer in the system and the spatial distribution of velocity, temperature and pressure. For geothermal reservoir applications, the near-surface heat flux distribution and the variation of temperature and pressure with depth at all horizontal locations are quantities that can be compared with field data.

Holst and Aziz (1972a) developed a numerical solution for convection rolls in a square with insulated vertical boundaries and constant temperature horizontal surfaces. For values of $\tau \leq 0.885$, the density, specific heat, thermal conductivity and viscosity were assumed to vary with temperature according to empirical laws for the hydrocarbon n-heptane. Asymmetric steady convection patterns were found in which relatively large velocities occured along the hot bottom boundary where the viscosity is minimized. The extent of isotherm distortion was far larger in the region of upflow than where the colder, more viscous fluid decends.

More recently Zebib and Kassoy (1976b) considered the heat and mass transfer in a water saturated porous medium for a configuration like that of Holst and Aziz (1972a). Weak nonlinear theory and finite difference calculations based on equations (2.32) and (2.33) were used to describe the convection process and heat transfer. The nondimensional water viscosity was assumed to vary according to Wooding's (1957) empirical law. Results from the numerical computations show that the spatially averaged Nusselt number, defined in equation (3.35) for $x^* = 1$, can be displayed as single curve when plotted against R/R_c for $0 < \tau \leq 0.67$ and $R/R_c \leq 3.25$ as shown in Figure 2.2. In the latter ratio, R_c is the critical value of the Rayleigh number for a specific $\Delta T'$. Typical streamlines (solid lines) and isotherms (dashed lines) are shown in Figure 2.3 for $\Delta T' = 200$ K and $R/R_c = 2.75$ where $R_c = 10.76$ *from* Figure 2.1. Although the absolute value of R is not large, one may observe the strong thermal and velocity boundary layers in the flow field.

In the geothermal context, where $R/R_c = O(10^2)$ would be a conservative estimate, the idealized model implies that one would observe narrow plumes of rapidly rising hot water in the convection cells. When the vertically moving fluid approaches an impermeable horizontal boundary, it turns, convecting energy with it. It is clear from the mushroom shaped isotherms in Figure 2.3 that convection persists over conduction losses for some

2. Convective Heat and Mass Transfer in Hydrothermal Systems

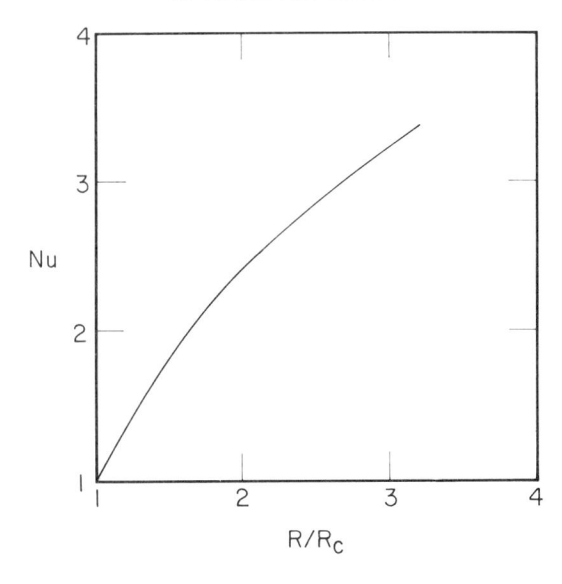

Figure 2.2. The Nusselt number Nu as a function of the Rayleigh number ratio R/R_c obtained from finite difference calculations of the nonlinear convection equations (Zebib and Kassoy, 1976b)

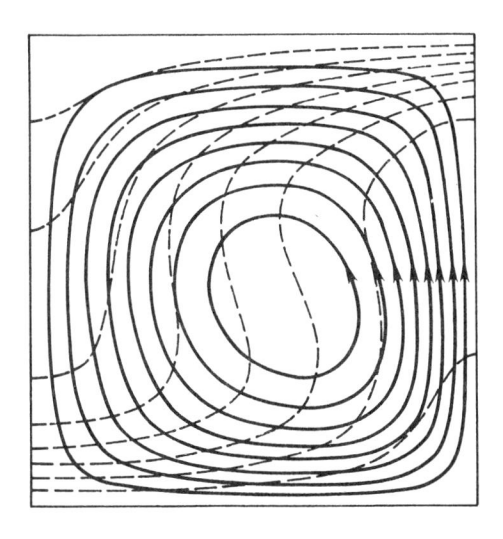

Figure 2.3. The streamline (solid lines) and isotherm (dashed lines) pattern for $\Delta T' = 200$ K and $R/R_c = 2.75$. The difference between streamlines is $\psi_{MAX}/9$ where $\psi_{MAX} = 0.205$ while that between isotherms is $\tau/9$ where $\tau = 0.67$

distance in the horizontal direction. The effect will be more pronounced at higher values of R.

The results in Figure 2.2 indicate that convection enhances conduction heat transfer by a factor of nearly 3·5 in terms of averaged surface heat flux when $R/R_c = 3.25$. The corresponding local value of the Nusselt number defined in equation (2.34) at the surface just above the center of the rising plume is about 5.5. This enhancement factor is within the range observed at some geothermal sites where one finds near-surface fluxes at 200–600 mW/m^2 compared to conduction-controlled background values of between 60–80 mW/m^2.

Heat and mass transfer modeling, based on a homogeneous isotropic porous matrix, cannot simulate the effects of observed geological structure. In sedimentary and alluvial geothermal systems, stratigraphic properties may determine major features of the flow process. Donaldson (1962) considered constant-property convection in a system composed of an upper permeable layer beneath which lies an impermeable zone of the same thickness. Fluid motion in the former induced a small vertical distortion of the isotherms in the lower layer. Much larger variations occur in the convection zone.

A three layer model, consisting of a permeable zone sandwiched between the impermeable layers, was considered by Sorey (1975). At a given Rayleigh number the Nusselt number was lower for the three-layer model than for Donaldson's two-layer model. Overall heat transfer is reduced as the fraction of a system composed of low permeability and pure conduction zones is increased.

Rana et al. (1979) examined convection in a rectangular reservoir consisting of three horizontal layers with recharge on the upper surface. Although most of the water circulates through the system once, entering and leaving at the upper surface, the barrier effect of the relatively impermeable middle layer causes small regions of closed streamline patterns in the lowest layer. The appearance of this unexpected phenomenon suggests that lithology effects can be substantial in certain models of convective flows.

The influence of compaction increase with depth on the flow pattern in a rectangle was considered by Ribando and Torrance (1976). The permeability decreases exponentially with depth by a factor of 280. At $R=200$ and $Nu=1\cdot76$, the most vigorous convection occurs in the upper half of the system. Although small amounts of liquid are able to penetrate into the deep impermeable sections, conduction is the dominant heat transfer mechanism there because the rest state isotherm pattern is hardly disturbed. This implies that conduction in the deepest formations in a geothermal system, peripherally involved with water convection, will control the heat supply to the convection system above.

Field studies indicate that geothermal systems are frequently transitory in nature. An estimate of the hydrodynamic start-up time for a model system can be obtained from Elder's (1967a) study of transient convection in a rectangle with a centrally heated lower bottom boundary. When $R=400$ the flow evolves to a steady state for $t \approx 40$. Equation (2.15) can be used to show that for a configuration of geophysical interest where $L'=3 \times 10^3$ m and $v'_R = 10^{-2}$ m/day, the relaxation time is typically 4×10^4 years. A comparison with the shorter time scales associated with other physical processes that alter hydrodynamic transients (earth tides, hydrothermal matrix alteration and intermittant tectonic events) suggests that it will be difficult to determine whether a geothermal system is in a steady state.

Caltagirone's (1975) calculation for convection in a rectangle with constant temperature horizontal boundaries showed a time to steady state of $t=60$ when $R=200$. It was also shown that the convection process became irregularly oscillatory for sufficiently high values of R. The characteristic time between cycles in oscillatory convection when $R=800$ is given by $t \approx 10$, corresponding to 10^4 years for the geophysical parameters mentioned above. Horne and O'Sullivan (1974), considering convection in a square, with a partially heated lower bottom boundary, found oscillation periods decreasing from $t=11\cdot8$ to $7\cdot6$ as R increased from 375 to 1000. These values are similar to Caltagirone's, although the configuration is different. The numerical calculations also showed that the long-time solution to the convection problem depends on the initial conditions imposed. Elder (1967b) suspected this evolutionary phenomenon at an earlier date, and Sorey (1975) has found similar effects. These results suggest that transient processes are characterized by time scales of 10^4 years. It follows that steady-state modelling will be adequate for 1–50 year time scale if shorter time scale chemical deposition effects do not alter the material properties of the system.

There is no reason to suspect that convection processes in geothermal systems are confined to two-dimensions. The geometrical configuration of a typical system, as

2. Convective Heat and Mass Transfer in Hydrothermal Systems

determined by geological structure and thermal conditions imposed by heat-transfer characteristics in the crust, are probably compatible with three-dimensional flows. Hydrodynamical considerations by Beck (1972), Straus (1975), and Straus and Schubert (1978) indicate that such flows are likely in both infinite parallel-plate and box-like configurations.

Three-dimensional fully nonlinear flow in a rectangular parallelepiped was considered by Holst and Aziz (1972b). Both two- and three-dimensional motion could be obtained by appropriate choice of the initial conditions. For three-dimensional motion in a cube when $R=120$, the time to steady state was $t=66$, similar in character to two-dimensional values.

Weakly nonlinear theory was used by Zebib and Kassoy (1978) to show that both two- and three-dimensional steady flow patterns were possible in the box configuration when each of the horizontal dimension is an integer multiple of the height. The existance of both steady modes implies that initial conditions on the transient problem would determine which configuration appears.

Straus and Schubert (1979) used a three-dimensional Galerkin procedure to consider the transient development of convection from randomly specified initial conditions in a cube for $45 \leq R \leq 150$. Both two- and three-dimensional steady solutions were obtained for all values of R, further emphasizing that initial conditions determine the steady flow configuration. The Nusselt number results in Figure 2.4 indicate that at $R=97$ the heat flux of two- and three-dimensional modes is equal.

Horne (1979) developed an efficient finite difference scheme for computing the transient development of a constant property flow in a cube. In present notation $T = 1 + \tau\theta$ and the initial condition is given by

$$\theta = 1 - z - f \cos m\pi x \cos l\pi y \cos n\pi z \tag{2.37}$$

where z is the vertical variable and the amplitude f is set arbitrarily between ± 1.

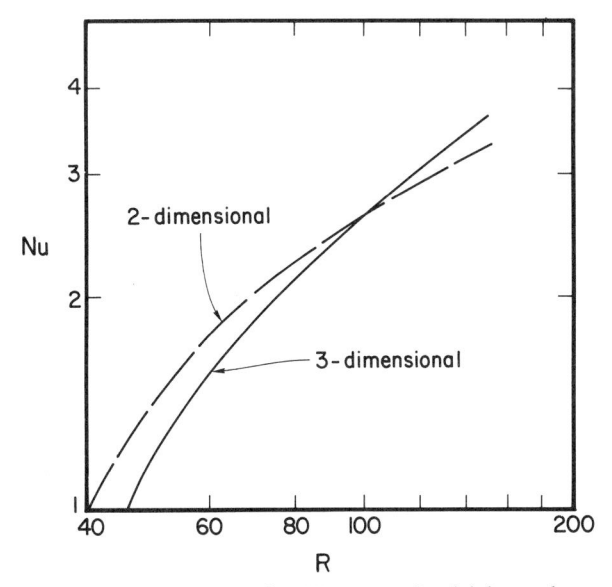

Figure 2.4. Nusselt number Nu versus Rayleigh number R for two (dashed) and three (solid) dimensional convection in a cubic box of fluid saturated porous media (adopted from Straus and Schubert, 1979)

Calculations were carried out for the initial modes and Rayleigh numbers shown in Table 2.1. The nature of the long-time flow pattern and the associated Nusselt number is given as well. Once again, it is clear that the long-time configuration is dependent upon initial conditions for a given R-value. The Nusselt number values at $R=75$ and 100 are in resonable agreement with those of Straus and Schubert (1979).

These three-dimensional considerations for highly idealized models of geothermal systems imply that the flow configuration of a given system may depend upon the geological and geophysical history of the area.

2.3. Plausible Models of Geothermal Reservoirs

Further insight into the heat and mass transfer processes occurring within a typical liquid-dominated geothermal system can be obtained from a study of hypothetical, but physically plausible models. The mathematical description should emulate the physical properties of a real system. Geometrical configurations, structural properties, and conditions specified at boundaries (on velocity, temperature, heat flux or pressure) should resemble field situations.

2.3.1. *Pipe Models*

The earliest of the models used to describe the gross features of an entire geothermal area was based on the pipe system formulated by Einarsson (1942). Qualitative properties of this model have been discussed by Bodvarsson (1961), White (1961), and Elder (1966). The first fully quantitative treatment, presented by Donaldson (1968b), is based on the configuration in Figure 2.5. Cold water at temperature T'_0 charges a vertical, porous channel of width A'. Heating occurs at depth in a connecting horizontal channel where the bottom boundary temperature increases linearly to T'_1. The central vertical porous channel of width $2a'$ and height h' represents the upwelling region of an entire geothermal area. Convection is driven by the horizontal pressure gradient resulting from the difference between the cold and hot hydrostatic head in the two channels. The vertical convection causes upward distortion of high temperature isotherms and large increases in total surface heat transfer.

Table 2.1. Flow configurations and Nusselt numbers predicted by Horne's (1979) numerical calculations for several Rayleigh numbers and different initial conditions represented by the wave numbers m, n and l

R	m	n	l	Obs. Flow	Nu
75	1	1	0·1	Steady 2-D	2·25
75	1	1	0·8	Steady 3-D	2·14
75	1	1	1	Steady 3-D	2·1
100	1	1	0	Steady 2-D	2·8
100	1	1	1	Steady 3-D	2·9
300	1	1	0	Fluct. 2-D	4·97 (avg.)
300	1	1	0·1	Fluct. 3-D	6–7
300	1	1	0·5	Steady 3-D	6·45
400	1	1	0·1	Fluct. 3-D	7–8

2. Convective Heat and Mass Transfer in Hydrothermal Systems

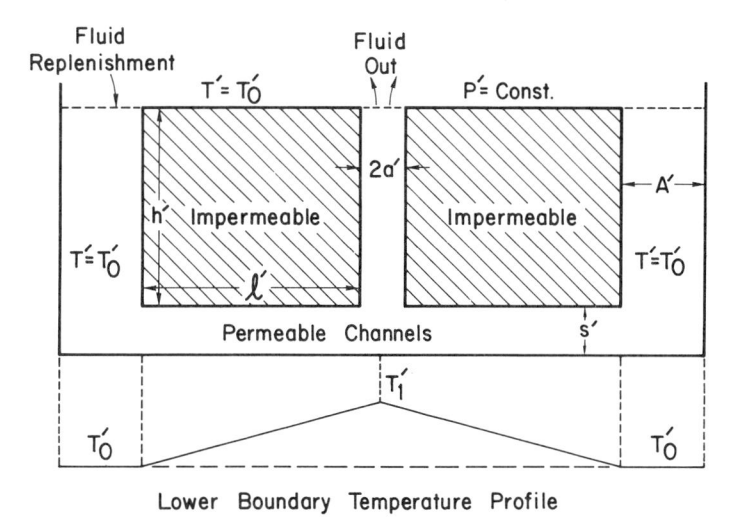

Figure 2.5. The pipe model used by Donaldson (1968b)

The quantitative analysis of the coupled heat and mass transfer problem described generally by equations (2.28)–(2.30) is simplified by assuming spatially invariant mean values for the velocity field, constant viscosity and uniform porous medium properties. The resulting linear energy equation is solved for the thermal boundary conditions shown in Figure 2.5. A nonlinear variation of the Nusselt number on the upper surface with the aspect ratio a'/h' was found. Apparently an increase in a' leads to increased vertical convective transport through the upper surface of the central channel, and upward isotherm distortion in the impermeable blocks, which causes an enhancement in surface conductive heat flux as well.

Donaldson (1968b) also noted that the upward isotherm distortion due to convection caused the temperature gradient in the lower 85 per cent of the central channel to be reduced considerably from the pure conduction value. A substantial portion of the temperature rise occurred in the upper 15 per cent. These steep near-surface gradients are typical of many geothermal field observations. The predicted gradients are probably overestimated because the temperature of the upwelling fluid at the surface is required to be the ambient value T'_0. In reality the liquid usually exits (typically from a hot spring) at a considerably higher temperature. In fact, for systems in which the Rayleigh number is significantly larger than the range considered by Donaldson ($R < 89$), the temperature of the surface environment will have little effect on the hot spring efflux temperature.

In the case of a central channel with a significantly lower permeability than that in the horizontal channel, Donaldson (1970) demonstrated that the flow is like that described above. When the permeability in the entire system is sufficiently large, a flow instability develops which leads to enhanced heat transfer by cellular convection superimposed on the through flow. Increasing the mass flow number M, which is the ratio of mass flow rate to the characteristic value q'_R, decreases the percentage of enhancement while increasing the overall level of heat transfer.

Should the flow resistance of the connective system become much larger than the central channel, due, say, to some tectonic event or hydrothermal alteration, then the net mass flux is diminished. If the Rayleigh number is large enough, the cellular convection process will remain. In this case of small M, the heat transfer enhancement effect is maximized. These results were applied to the Broadlands geothermal field to show that roll convection could be responsible for about half the surface heat output.

The pipe model can be used to develop good estimates of mass transfer through a system and overall surface heat flux. The former depends directly on the properties of the rock system connecting the cold charge zone with the upflow region. These properties are rarely observable because bore-hole data is not usually obtained outside the geothermally active area. Furthermore, Donaldson's results show that a variety of factors (aspect ratio, Rayleigh number) in addition to the mass flow rate, affect the overall heat flux. Hence the mass flow rate is not related directly to the overall heat flux. The model cannot be used to investigate the spatial distribution of mass transfer and temperature within an active geothermal area because geological details (e.g., definable aquifers, fault zones) are absent. In addition, the velocity field was decoupled *a priori* from the energetics of the problem, precluding the nonlinear interaction between fluid flow and energy transfer normally occurring in natural convection.

2.3.2. Large Scale Convection Cell Models

Convection cell studies of the type discussed in Section 2.2 have been used by Cheng and co-workers (1974, 1975, 1977) to model heat and mass transfer in an entire island geothermal reservoir. A survey of this work has been given by Cheng (1978). The initiation and evolution of such a reservoir has been treated by Cheng and Teckchandani (1977). An initial value problem is formulated for the convection process in a confined island aquifer, with a constant temperature impermeable upper boundary, due to the sudden appearance of a distributed temperature (hot spot) at the lower boundary which

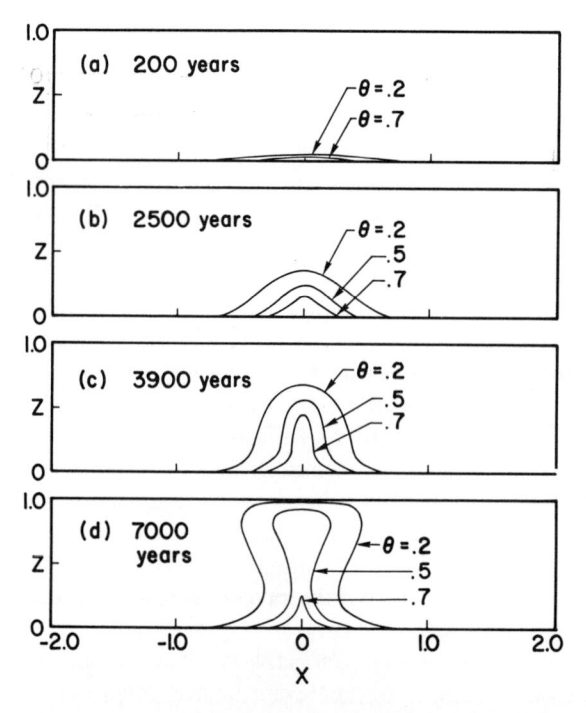

Figure 2.6. Development of isotherm convection patterns in a geothermal reservoir at $R=200$. (Cheng and Teckchandani, 1977)

2. Convective Heat and Mass Transfer in Hydrothermal Systems

might model the effects of an intrusive. In Figure 2.6 we see the isotherm development for a system with $R=200$. A steady state is reached at 7×10^3 years. In this case $\theta=(T'-T'_a)/(T'_M-T'_a)$, where T'_a and T'_M are the upper boundary and the lower boundary maximum temperatures respectively. When the Rayleigh number is increased to 350, oscillatory effects are observed. The first maximum is isotherm distortion occurs at 4×10^3 years. A second maximum occurs when 12×10^3 years have passed. The second episode does not replicate the first in detail. This result implies that the steady state solutions calculated by Cheng et al. (1975) will not be observed beyond some sufficiently large value of R.

Norton (1977) and Norton and Knight (1977) have examined the convection system initiated when an intrusive is emplaced rapidly in a deep, water-saturated fractured rock system. The flow is thought to occur primarily in cracks generated by regional tectonic stresses and by the intrusive activity itself. Cracks and fault zones caused by rock failure mechanisms are estimated to exist to depths of 1.5×10^4 m. In a typical situation, the intrusive is modeled as a block of hot solid material 4×10^3 m high and 6×10^3 m wide embedded in a rectangular reservoir 10^4 m deep and 2.4×10^4 m wide. Numerical solutions of the porous media equations are obtained by finite difference methods. Realistic models of the thermodynamic and material properties of the saturating liquid, water, are used. Typical Rayleigh number values of the configurations studied can be obtained from equation (3.31) with $L'\approx10^4$ m, $v\approx2\times10^{-7}$ m^2/s, $\Delta T'\approx10^3$ K, $c_p\approx4.186\times10^3$ J/kg K, $\alpha\approx3\times10^{-3}$/K and $K_m\approx2.51$ W/m K. It follows that $R\approx10^{18}$ k'/m^2 which suggests that convective heat transfer can be significant, compared to that due to conduction only when the crystal fracture permeability is greater than 10^{-18} m^2. Norton and Knight's (1977) detailed calculations showed that when a 1143 K pluton is emplaced in a reservoir for which $k=10^{-21}$ m^2, conduction controlled energetics ($R\approx10^{-3}$) produces a characteristic cooling time of 8×10^5 years. This value can be compared favorably with the characteristic conduction time of 3×10^6 years calculated from $t'_{\text{COND}}=\rho'c'_pL'^2/K'_m$. In the case of a reservoir permeability of 10^{-15} m^2, for which $R\approx10^3$, convective heat transfer will predominate. When the emplaced pluton has a temperature of 1193 K, a vigorous confined convection plume appears directly above the pluton, with weak downflow extending over a vastly larger region. After about 10^5 years convective heat flux just above the pluton maximizes at about 1670 mW/m^2 compared to 210 mW/m^2 due to conduction. Superimposed oscillatory convection processes appear to be occurring on a scale of about 5×10^4 years, a characteristic value for large Rayleigh number systems regardless of detailed boundary conditions. Vertical volumetric flow rates above the pluton range between 3×10^{-5} m/day and 4×10^{-4} m/day, which is typical of the characteristic convection velocity v'_R defined in equation (2.23). These flow rates will cause mass circulation rates of 10^7–10^8 kg/m^2 over the characteristic cooling period of 10^5 years. One may expect that thermal metamorphism and hydrothermal alteration will be significant in rock-water systems with such large circulation rates. The chemical effects will be confined primarily to fracture systems offering the primary permeability rather than appearing in the intergranular spaces. The lack of extensive surface area in the fractures, when compared to that in an analogously permeable intergranular system, implies that the liquid will be out of equilibrium with the bulk of the rock.

Norton (1978) found that pluton cooling was accelerated if water was able to circulate through internal cracks generated by thermal stressing as temperatures were reduced. In a typical example, an impermeable pluton 4×10^3 m in height and 6×10^3 m in width, initially at temperature of 1573 K, is emplaced in a reservoir of dimensions 10^4 m by 2.4×10^4 m, in which $k=10^{-14}$ m^2 in the upper 3×10^3 m, $k=10^{-15}$ m^2 in the middle 3.5×10^3 m, and $k=10^{-16}$ m^2 in the lower 3.5×10^3 m. The initial temperature distribution corresponds to the normal geothermal gradient determined by a heat input of 50 mW/m² through the bottom boundary. When the pluton cools to 973 K, the permeability of the pluton is set at 10^{-17} m². The convection process causes a 10-fold increase in surface heat flux above the regional value over a period of 8×10^4 years. At great depth, where critical state phenomena occur, convection is enhanced significantly. Since the thermodynamic anomalies in density, heat capacity and viscosity are significant over an extensive region of the P-T plane, it is possible that the convection process is affected in a large geometrical volume. Further elucidation of fluid motions in these deep circulating systems has been given in Norton (1978).

In a significant related study, Cathles (1977) has calculated the heat and mass transfer process induced in a fractured, rectangular, saturated reservoir, 10^4 m wide and 5×10^3 m deep, when an intrusive, at 973 K, 1.5×10^3 m in width and 5×10^3 m high, enters the system. Initally the reservoir has a normal geothermal gradient based on a prescribed bottom surface heat flux of 63 mW/m² and a top temperature of 293 K. The vertical boundaries are insulated. Transient numerical calculations are carried out for the porous media equations with uniform permeability. Realistic fluid properties of pure water are included. Phase changes are permitted, although no regions of co-existing phases are possible. True boiling or condensation occurs across a surface in the geometrical space.

If the fracture permeability of the reservoir is 2.5×10^{-16} m² so that $R \approx 250$ and the upper surface is impermeable, then the maximum conductive surface heat flux peaks at nearly 1970 mW/m² about 10^4 years after pluton emplacement. A rapid decline to about 410 mW/m² occurs during the next 2×10^4 years. Cooling is basically completed in 10^5 years. If surface mass flux is permitted, the net (conductive+convective) surface flux peaks at about 7120 mW/m² after 7.75×10^3 years. An examination of Figures 2c and 3a of Cathles (1977) suggests that this is due almost entirely to convection. These heat flow values may be compared to Elder's (1965) Wairakei estimate of an average heat flow of 2100 mW/m² existing over about 5×10^7 m² with much larger maxima associated with localized hot springs.

The hydrodynamic patterns obtained by Cathles (1977) indicate that fluid percolates downward over an extensive region. The path followed in the P–T diagram takes the fluid around the critical point, so that a discrete boiling surface cannot exist. When salinity effects are included phase transition does occur at depth. Early in the cooling process (5×10^3 to 5×10^4 years) a super-critical region is formed above the pluton. Fluid rising through this region cools so that eventually a discrete condensation surface is encountered. This zone, which might make an excellent geothermal resource, lasts for only 5×10^3 to 10^4 years, disappearing before there is much surface expression of high heat flow. The region would be difficult to find, unless a geophysical anomaly could be associated with its properties.

The pressure field in the reservoir is found to be nearly hydrostatic. Variations associated with the convective flow process are of the order of 1–2 MPa as predicted from an evaluation of the reference pressure quantity p'_R in (2.24) for the physical conditions considered. Cathles (1977) calculated anomalously low pressure at great depth ($\lesssim 10$ MPa below the hydrostatic value) when the pressure is specified on the impermeable upper boundary. This difficulty is associated with variability of average system density in a constant volume configuration. It appears more useful physically to either specify the pressure at depth, or to allow for uplift of a confining surface-water table, which acts as a barrier to prevent geothermal waters from exiting at the surface while permitting thermal expansion. Cathles' extensive investigation also includes a sensitivity study of the effects of permeability, upper surface boundary conditions, and pluton location and size, on the heat and mass transfer in the system.

2.3.3. Hot Springs and Fault Zones

In many geothermal areas, near-surface thermal properties are affected profoundly by energy transfer from hot spring systems. Water moving upward from a deep hot reservoir through a narrowly delineated zone of fractured rock will lose heat to adjacent formations. If a hydraulic connection exists between the upflow zone and the intersecting rocks, then convective charging may occur. When an impermeable barrier exists between the two regions, due for instance to hydrothermal alteration on the perimeter of the vertical conduit, heat transfer will be strictly by conduction. Sorey (1975) has developed an analytical model of the latter problem. He considers the cooling of a rising column of hot water which originates at the top of a deep reservoir. The mass flow rate and reservoir temperature are specified. Heat is conducted out of the conduit, through an impermeable barrier, into an adjacent near-surface aquifer. The water exits at the surface (the hot spring) at a temperature higher than the neighboring surface value but lower than the original reservoir value.

The near-surface aquifer has a bottom temperature equal to that in the deep reservoir, and a distant vertical boundary which is insulated. At the upper surface either a specified temperature value or a convective-radiative heat loss condition is specified. In this steady state model, all the heat conducted into the aquifer along the hot vertical boundary must ultimately be lost through the upper surface.

In a typical example, Sorey considers a mass flow rate of 1.1×10^5 kg/day from a reservoir at 453 K occuring in a planar vertical fault zone, 10 m wide, 10^3 m long and 10^3 m deep. When the adjacent aquifer is impermeable and the surface temperature is specified at 283 K, the heat loss from the upflow zone is 5.44×10^5 W, and the water temperature at the surface is 349 K. In this effectively high Rayleigh number modeling of convection in the fault zone, the temperature drops 55 K in the lower 88 per cent of the 10^3 m deep fault compared to about 50 K in the upper 12 per cent. Surface heat flux adjacent to the spring is about 3430 mW/m^2.

The specified surface temperature tends to over-constrain the energy transfer process in the upper portion of the aquifer, which leads to excessive surface heat flux. When the surface heat loss boundary condition is used, with a heat transfer coefficient of 4.186×10^{-2} W/m^2 K, then the surface spring temperature rises to 364.8 K. At the surface of the aquifer the temperature and heat flow varies from 326 K and 1790 mW/m^2 at 5 m from the fault plane to 286.5 K and 150 mW/m^2 beyond 10^3 m. The vertical temperature distribution is even more distorted, with a gradient of 0.06 K/m in the lower 92 per cent of the fault compared with 0.41 K/m in the upper 8 per cent.

If the aquifer permeability is set at $k = 10^{-14}$ m^2, then the effective Rayleigh number is 200. Natural convection processes occurring in the aquifer are affected by heat addition from the fault zone for about 5×10^2 m inward from the vertical boundary. The spring temperature is now 369 K because the heat loss from the fault plane is reduced to about 4.186×10^5 W from the previous value of 5.46×10^5 W. In the fault the vertical temperature distribution is further skewed. The 55 K drop occurs over about 95 per cent of the fault dimension. A thermal boundary layer, encompassing only about 5 per cent of fault dimension, contains the remaining drop. The corresponding gradients are 0.058 K/m and 0.58 K/m. These results demonstrate the extraordinary influence of convective mass and heat transfer in a geothermal system.

In a rather different hot spring model, Turcotte et al. (1977) considered two-dimensional convection rolls occurring in a thin vertical slab of saturated porous media representing the fracture field associated with a fault zone. The vertical boundaries of the slab are impermeable and insulated. At the lower impermeable boundary, the heat flux is prescribed. Surface water entering the convection cell through the upper surface arrives with a prescribed ambient temperature while that exiting (the hot spring emanation) must satisfy a condition that the vertical gradient of the conductive heat flux vanishes. This boundary condition permits an exit temperature higher than the surface rock value which is specified as ambient. The problem is formulated in term of distinct rock and water temperatures. Solutions for the appropriate porous media equations are developed numerically using input data from the Steamboat Springs, Nevada, hydrothermal system. Nearly equal rock and water temperatures are found in the bottom 90% of the sytem because the local characteristic conduction time is short compared to the analogous convection time. A temperature differential does appear in the upper portion driven by the disparate cooling conditions on the rock and water at the upper surface. Field measurements imply that the temperature difference is far less than that predicted.

In a third model of heat and mass transfer in a fault zone, Kassoy and Zebib (1978) have considered the cooling of a rising column of liquid water in a vertical, slender channel of saturated porous material with impermeable walls on which the temperature increases linearly with depth. This is used to model the fracture permeability system associated with the fault zone in a region of continual seismic activity like that depicted in Figure 2.7. It is hypothesized that the fault zone is charged at depth with hot liquid flowing out of an extensively fractured basement rock system. If the fracture permeability is large compared to the intergranular value for the adjacent unfractured sediment above the basement (which may contain extensive clay deposits and/or be hydrothermally

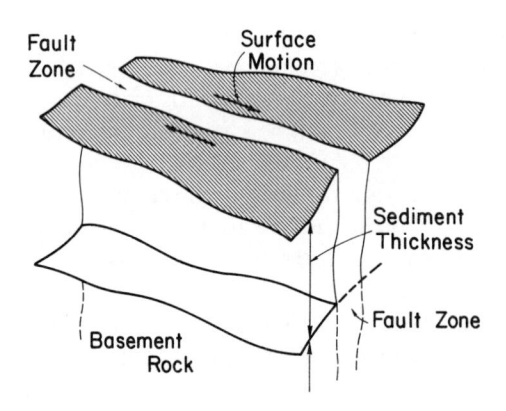

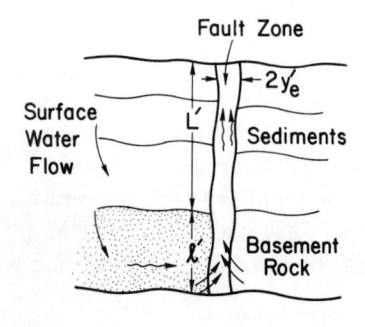

Figure 2.7. A conceptual view of a vertical fault zone extending through sediments of thickness L' and basement rock of magmatic origin. (Kassoy and Zebib, 1978)

altered), then the rising liquid will pass through the fault zone without mass loss, exiting at the surface (a hot spring). The specified temperature boundary condition, implying that the thermal resistance of the adjacent rock is considerably less than that of the saturated material in the fault system, represents an idealization of actual conditions. Solutions to equations (2.28)–(2.30) are developed for large Rayleigh numbers. The results show that two-dimensional upflow in the fault zone can exist only if the quantity $\gamma = R^{1/2} y_e < (\pi/2)$, where the Rayleigh number is based on the fault depth L' and the temperature difference in the sedimentary sequence, and y_e is the ratio of the fault half width y'_e to the fault depth L'. As the critical value of γ is approached, natural convection processes in the plane of the fault begin to enhance the forced two-dimensional convection occurring across the fault zone width. When $R = 10^3$ and the fault depth $L' = 3 \times 10^3$ m, two-dimensional flow will prevail if the fault zone width is less than 300 m.

The surface temperature value in the center of the fault is found to be substantially larger than the surface value of the adjacent impermeable material. In a case where $R = 2 \cdot 28 \times 10^2$, based on $L' = 3 \times 10^3$ m, $k'_0 = 10^{-13}$ m^2, $\Delta T' = 300$ K and typical thermodynamic data, where the impermeable surface temperature is 300 K, $y'_e = 100$ m and where a mass flux of $1 \cdot 22 \times 10^6$ kg/day flows through the horizontal cross-section of 10^3 m of the fault, then $\gamma = 0 \cdot 5$ and the maximum surface water temperature is 339 K. Values greater than 373 K are found for sufficiently wide faults suggesting that the upper portions of such systems really include 2-phase flow. The typical vertical convection volumetric flow rate is about 6×10^{-3} m/day. While conduction accounts for approximately 272 mW/m^2 of vertical heat transport, convection of hot liquid gives, typically 126,000 mW/m^2. Conductive losses through the walls of the fault amount to 29,300 mW/m^2. These values examplify the effectiveness of energy transport by convection, even at flow rates of only 10^{-2} m/day. The conclusions reached and estimates obtained are valid only for high Rayleigh number systems in which a particle of fluid loses little heat by conduction as it moves through the system.

2.3.4. Fault Zone Controlled Charging of a Geothermal Reservoir

Goyal and Kassoy (1977) and Goyal (1978) have developed a conceptual model for the geothermal reservoir at East Mesa, Imperial Valley, California, following an extensive analysis and interpretation of the complete spectrum of available field data. The conceptual model of the system is based on a fracture zone (the fault) of finite width which extends downward nearly vertically through a clay-rich region (the cap), through the interbedded sediments of the reservoir, and finally into the basement rock. It is postulated that the fault is charged at depth by liquid which has been heated in an extensive basement fracture system. The rate of charge cannot be specified *a priori* without a global analysis of the convection process. Liquid rises in the reservoir section of the fault. The presence of clays in the cap suppresses vertical transport there. Water pushed out of the fault by artesian overpressure is assumed to flow horizontally in the reservoir aquifer. Vertical transport should be less important because of the presence of interbedded shales.

For mathematical purposes, the fracture zone is idealized as a vertical slab of porous media. The adjacent reservoir aquifer is represented as a porous medium with horizontal permeability only. Finally the overlying clay cap is assumed to be impermeable. The configuration is similar to that in the lower portion of Figure 2.7. The cap occupies a thin section below the upper surface.

Spatially uniform temperature boundary conditions are imposed on the cold cap surface and at the hot bottom boundary of the reservoir. On the lateral boundary far from the fault, the temperature distribution is controlled purely by vertical conduction, the pressure distribution is hydrostatic, and mass flux is permitted to conserve matter.

In the high Rayleigh number approximation, liquid rises adiabatically in the fault with the exception of a very thin zone just below the clay cap. Charging of the adjacent aquifer occurs along the entire length of the fault. On a horizontal length scale proportional to the fault depth, the cooling effect of the cold surface, through the conductive clay cap, is confined to a narrow but growing thermal boundary layer in the aquifer adjacent to the interface. Below that layer, the fluid moves horizontally at the high temperature of the source fluid in the fault. Sufficiently far from the fault, typically 5–10 fault depths, the entire aquifer is affected by the cold upper surface.

Due to the adiabatic flow in the fault, nearly all the temperature drop occurs in the overlying clay cap. The conductive surface heat flux is maximized there. The flux declines slowly as one moves away from the fault, because an increasingly large fraction of the temperature change can occur in the growing thermal boundary layer in the top part of the aquifer. As a result, the temperature variation with depth shows a large gradient in the cap followed by a quickly decreasing gradient in the boundary layer. Eventually the surface flux drops to the background value near the far boundary, where purely conductive heat flux prevails. In this far-field region, the temperature variation with depth is distributed over the entire cap and aquifer.

Goyal (1978) developed the nondimensional temperature distribution shown in Figure 2.8 based on parameter values representative of the East Mesa. The temperature profile is within a few percent of the purely conductive value at $1·9 \times 10^4$ m from the fault. It should be noted that fairly low gradient profiles in the reservoir prevail close to the fault although the flow is purely horizontal in the aquifer. As one moves from right to left in Figure 2.8 the temperature profiles make a transition from those observed in geothermally active zones to those observed on the margins.

The maximum surface heat flux is found to be about four times the background value, which is similar to the field observation at East Mesa. In addition the distribution of surface heat flux with distance from the fault is like that seen in the field. The results of this model suggest that fault zones can provide a mechanism for charging adjacent geothermal aquifers.

2.4. Pre-production Models of Hydrothermal Systems

For geothermal reservoirs it is necessary to predict both the quantity of fluid that can be produced and its temperature, in order to estimate the total usable energy. A hydrothermal system consists of a flowing convective fluid heated at depth and rising towards the surface as a result of the reduced density. The system is not only non-isothermal but also dynamic, as a consequence of buoyant flow. (As stated in Section 2.2, transient processes associated with the initiation of convection in a geothermal system occur over time scales of more than 10^4 years. Therefore, it follows that steady-state treatment will be adequate for describing the physical processes occurring in an unexploited geothermal system over a 10–50 year time scale.) For realistic prediction of hydrothermal reservoir performance, it is necessary first to establish the pre-production temperature and flow fields. During the exploration and initial development stage, the natural pre-production flow of the fluid within the system will be dominant, except in the immediate vicinity of any exploratory wells. As the development of the system takes place,

2. Convective Heat and Mass Transfer in Hydrothermal Systems

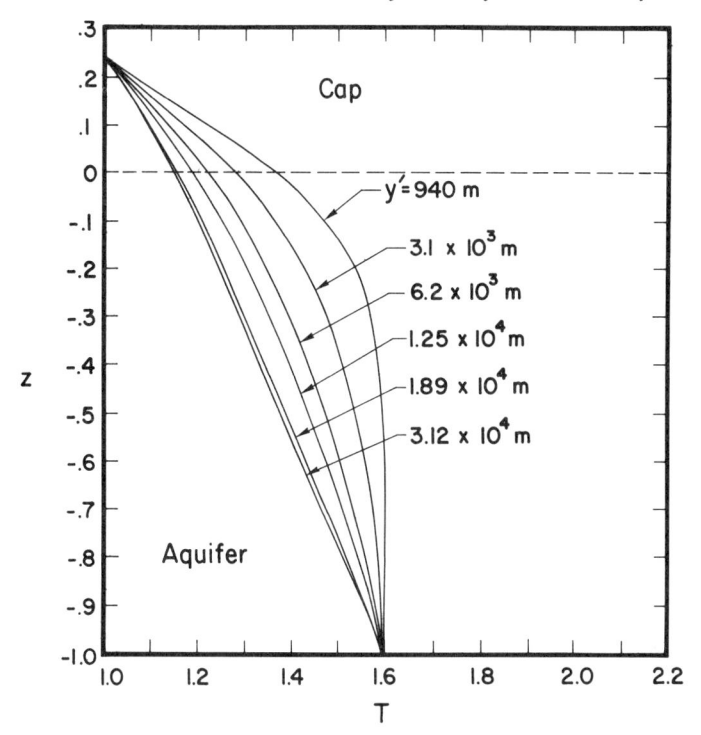

Figure 2.8. Nondimensional temperature T ($= T'/300$ K) in the Clay Cap (thickness $= 800$ m) and the Aquifer (thickness $= 3 \cdot 35 \times 10^3$ m) as a function of nondimensional depth z ($= z'/3 \cdot 35 \times 10^3$ m) for several horizontal distances, y', from a fault of width 200 m

the effect of the natural flow will likely be swamped by the perturbations induced by the production and injection wells.

The development of a mathematical model for describing the pre-production flow in a geothermal system requires the solution of the governing equations outlined in Section 2.1. This sytem of governing equations must, in general, be solved numerically. During the past few years, there has been excellent progress (see Kruger and Ramey, 1975; 1976; 1977) in developing computer programs which solve the equations of heat and mass transport in geothermal reservoir systems in one, two or three dimensions. Several of the computer programs possess considerable flexibility as far as fluid and rock properties, problem geometry and boundary conditions are concerned. Thus, for example, the MUSHRM reservoir simulator (Garg et al., 1977) can treat multi-phase multi-species (water/steam, water with dissolved methane/free methane, water with dissolved salt/ steam/precipitated salt) fluid flow in one, two or three dimensions. In MUSHRM, each computational zone in the finite difference grid may contain a different rock type. Provision is also made for all practical boundary conditions.

During exploration and initial development stage, only scattered temperature, material properties (fluid and rock thermomechanical properties, rock porosity and permeability, etc.), surface heat flow, and fluid flow measurements are available. It may also be possible, at least tentatively, to specify the stratigraphy from various geophysical measurements (e.g., seismic surveys). A computer based simulator of the pre-production fluid and heat flow in the geothermal reservoir system offers the tool for synthesizing these evolving data sets (presumably as a result of drilling production wells) into an integrated geohydrologic model. Because the data available will always be incomplete (and possibly very sparse), however, it will necessarily involve a variety of hypotheses concerning geology, temperature distribution, groundwater flow, convective flow, heat sources, etc. It is possible that some of the hypotheses will prove to be untenable with the availability of additional data (We shall return to this point when discussing pre-production models of the Wairakei geothermal system.) The simulation developed will not be unique, and the pre-production model may need to be updated as additional data become available. The pre-production model can be used to locate initial exploration and/or production wells. Assessment of the model could be carried out by comparing theoretical predictions of mass/heat production and pressure drop with field data from the wells. Planning of well tests for reservoir verification could be based on resolving major uncertainties in the evolving model.

There have been several attempts in the literature (see, e.g., Sorey (1976), Riney et al. (1977), Goyal and Kassoy (1977), Mercer et al. (1975), and Mercer and Faust (1979)) to develop a quantitative model of the pre-production fluid and heat transport in a real geothermal system. Although the scope, objectives and details of these studies are different, each attempts to develop a pre-production model of the reservoir system consistent with the available geologic, geophysical, geochemical and hydrologic data base. In the following, we review several of these studies. Our main goal in doing this will be to illustrate the use of the available data base to construct a particular pre-production model. The additional data required to resolve the major uncertainties in the various models will also be discussed.

2.4.1. *Long Valley Caldera*

Long Valley Caldera, an elliptical depression covering approximately 4.5×10^8 m^2, lies on the eastern front of the Sierra Nevada in east-central California. The caldera contains a hot-water convection system with several hot springs. Spring measurements and temperature profiles in wells indicate a total heat discharge of 2.89×10^8 W. Geochemical mixing models have been used to estimate that 190–300 kg/s of water at 483–555 K discharges upward from the reservoir towards the hot springs, with the highest reservoir temperature corresponding to the lowest mass flux (Sorey, 1976).

Sorey (1976) employed a three-dimensional conceptual model consisting of five horizontal layers, covering 4.5×10^8 m^2 and extending to a depth of 6×10^3 m (Figure 2.9). The five layers correspond to the major rock units identified by the seismic refraction and geologic studies. The upper layer of 10^3 m thickness, consisting of the post-caldera sedimentary and volcanic rocks, contains the shallow groundwater system. Layer 1 is considered to be an impermeable cap except along parts of the caldera rim, where recharge takes place, and in the Hot Creek Gorge area, where hot water flows upward along faults to discharge in the gorge springs. The deep hot water system (hydrothermal reservoir) is assumed to occur in layers 2 and 3; these layers contain welded and fractured tuff. Layers 4 and 5 correspond to impermeable but thermally conductive basement rocks. The presence of magma below 6×10^3 m in the western part of the caldera—indicated by seismic and heat-flow studies—is simulated by a temperature distribution at the base of the model held constant in time.

The evolution of temperature in the system over geologic time was computed under assumed fluid mass flux boundary conditions (i.e., fluid recharge and discharge) by solving

2. Convective Heat and Mass Transfer in Hydrothermal Systems 63

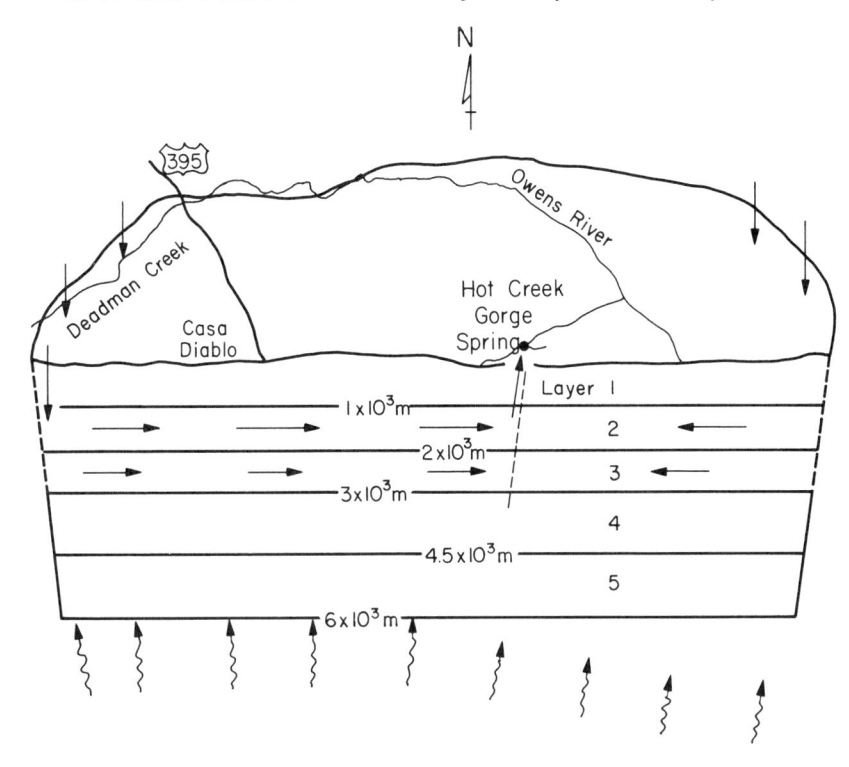

Figure 2.9. Block diagram showing conceptual model of Long Valley hydrothermal system. Straight arrows indicate ground-water flow; wavy arrows indicate heat flow. (Sorey, 1976)

numerically the fluid and heat transport equations using a Boussinesq approximation. In the numerical model each horizontal layer is subdivided into 82 grid blocks or nodes (Figure 2.10). Hot water is assumed to discharge only over the surface of the grid-block that includes the springs in Hot Creek gorge, and through the southeast rim; recharge takes place along the western and the northeastern rim of the caldera. Figure 2.10 also shows the locations of the principal faults within the caldera. Fractures in the welded tuff associated with faulting are assumed to provide the channels for flow in the hydrothermal system. As discussed by Sorey, the apparent lack of faulting in the eastern part of the caldera is not necessarily associated with low permeability.

Since only one deep well (depth $\sim 2 \times 10^3$ m) was completed in the caldera, there existed little or no information on the thermomechanical properties (porosity, permeability, specific heat, etc.) of reservoir rocks. Groundwater flow is from the highest altitudes along the west and northeast rim to discharge areas at lower altitudes in Hot Creek gorge and at depth through the southeast rim of the caldera. An additional driving force causing flow is provided by buoyancy (i.e., density differences between hot and cold parts of the flow system). Sorey evaluated the effective reservoir permeability ($3 \times 10^{-14} - 5 \times 10^{-14}$ m^2) by specifying pressures based on water table altitudes in recharge and discharge areas and adjusting the permeability distribution to yield the desired mass flux of water (~ 250 kg/s). Other reservoir properties employed in the simulation are given in Sorey (1976).

Although the model is unavoidably speculative, it serves to provide a better under-

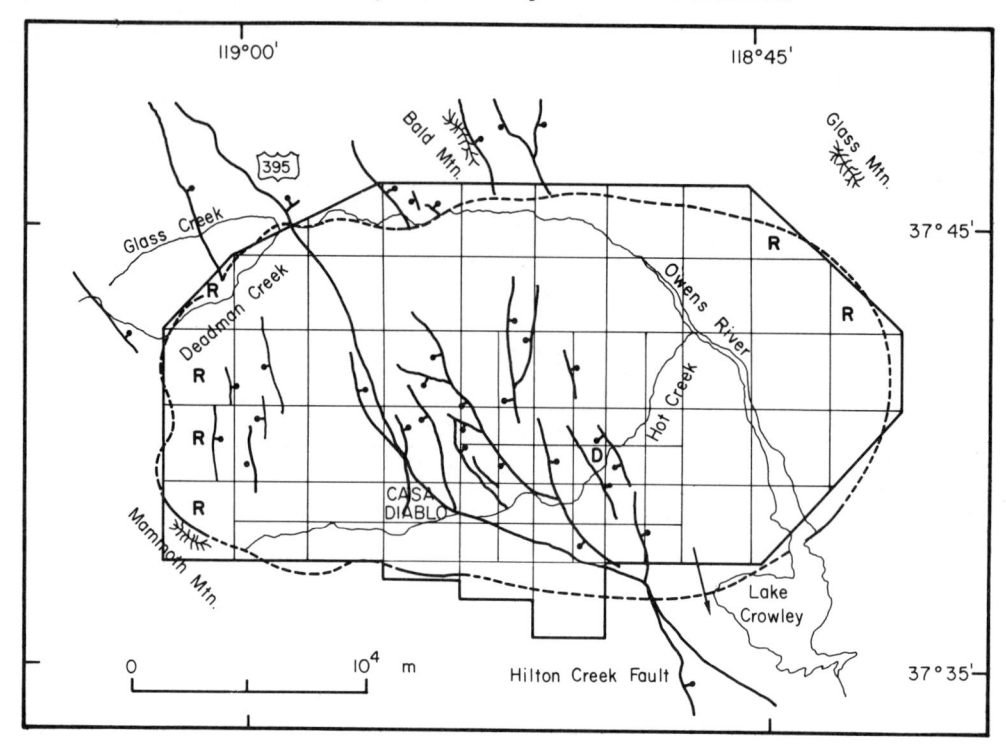

Figure 2.10. Sketch map of Long Valley caldera showing nodal configuration. R denotes recharge node; D denotes discharge node covering Hot Creek gorge. Principal faults are shown as solid heavy lines with ball on downthrown side. Arrow denotes discharge at depth through southeastern caldera rim. (Sorey, 1976)

standing of heat and mass transport in the system. The model has been used to estimate the depths of fluid circulation for which the underlying magma chamber could supply the required heat flow (~ 630 mW/m^2) over the area of the caldera for various periods of time. The depth of fluid circulation was found to increase with period of time for which the fluid circulation is assumed to occur. Thus, for example, model simulations for a period of $3\cdot 5 \times 10^4$ years show that present-day heat discharge could have been sustained for this period by fluid circulation to $1\cdot 5 \times 10^3$–$2\cdot 5 \times 10^3$ m depth. Figure 2.11 shows that the simulated rock temperatures at a depth of $1\cdot 5 \times 10^3$ m under the Hot Creek gorge area are near 473 K. Cooler temperatures east of Hot Creek, resulting from recharge along the northeast rim, are consistent with those reported from the only deep test hole drilled in the caldera. Hotter temperatures indicated by the simulation in the western part of the caldera as well as the adequacy of the numberical model can only be confirmed by deep drilling in this part of the caldera.

2.4.2. Salton Sea Geothermal Reservoir

The Salton Sea geothermal field (SSGF) is a high-salinity, high temperature geothermal system. The San Diego Gas & Electric Company has constructed a nominal 10 MWe Geothermal Loop Experimental Facility (GLEF). Intermittant brine production/injection have been performed since May 1976, but no associated flow data have been published. The Lawrence Livermore Laboratory, however, has correlated the data available from

2. Convective Heat and Mass Transfer in Hydrothermal Systems

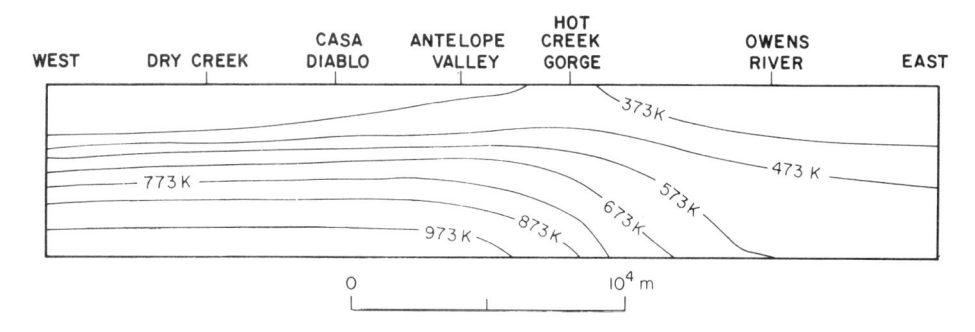

Figure 2.11. Diagrammatic east-west cross-section of Long Valley caldera showing isotherms in model after 3.5×10^4 years with hot-spring discharge of 250 kg/s, south-east-rim outflow of 110 kg/s, and reservoir depth of $10^3 - 2 \times 10^3$ m (Sorey, 1976). No vertical exaggeration

surface measurements and logs from various wells in the SSGF. Riney et al. (1977) used this limited data base and a reservoir simulator (MUSHRM) to synthesize a preproduction reservoir model for a portion of the SSGF which contains the GLEF site. Nine geothermal wells have been drilled within this area of approximately 3.4×10^7 m^2.

The main sequence reservoir rock in the SSGF is sandstone with shale lenses and layers, overlain with a relatively impermeable shale (caprock), and is believed to be separated into 'upper' and 'lower' reservoirs by a relatively thick and continuous shale layer (Towse, 1975; Schroeder, 1976). From studies of cores, cuttings and logs from wells drilled in the SSGF, Towse (1975) determined the approximate depths to the top of the upper reservoir and to the major shale break separating the upper and lower reservoirs. The geologic layers dip in a northwesterly direction essentially parallel to the Brawley fault zone. The region investigated by Riney et al. (1977) is shown in Figure 2.12. A cross-section was constructed by projecting the data onto a vertical plane parallel to the surface trace of the Brawley fault zone (Figure 2.13). The interfaces between the geologic layers were taken to be planes dipping to the northwest. The temperature-depth profiles measured in the geothermal wells (Palmer, 1975) were projected to construct the approximate temperature contours shown in Figure 2.13. The GLEF production wells (W1, M1) are perforated almost entirely within the upper reservoir whereas the injection wells (M2, M3) are performed mostly within the lower reservoir.

Schroeder (1976) analyzed the sparse data available from drillstem test records from M1 and W1 and concluded that the horizontal permeability of the reservoir sands in the upper reservoir shale/sand sequence exceeds 5×10^{-13} m^2. The sands comprise over 50 per cent of the sequence and their porosity exceeds 0.3. For the upper reservoir sequence, Riney et al. (1977) assumed the rock porosity and horizontal permeability to be 0.20 and 5×10^{-13} m^2 respectively.

Riney et al. (1977) employed a two-dimensional (2D) areal model covering the area shown in Figure 2.12 (the dipping and thickening upper reservoir is treated by including the component of gravity along the direction of the dip and by varying the rock properties to offset variations in thickness). The Brawley and Red Hill faults were assumed to prevent any fluid flow across the side boundaries (Figure 2.12). The fluids produced by wells on opposite sides of the Brawley fault appear to have a different origin, but there is no definite evidence that the Red Hill fault is a sealing fault.

Figure 2.13 shows that the temperature at the mid-plane of the upper reservoir is

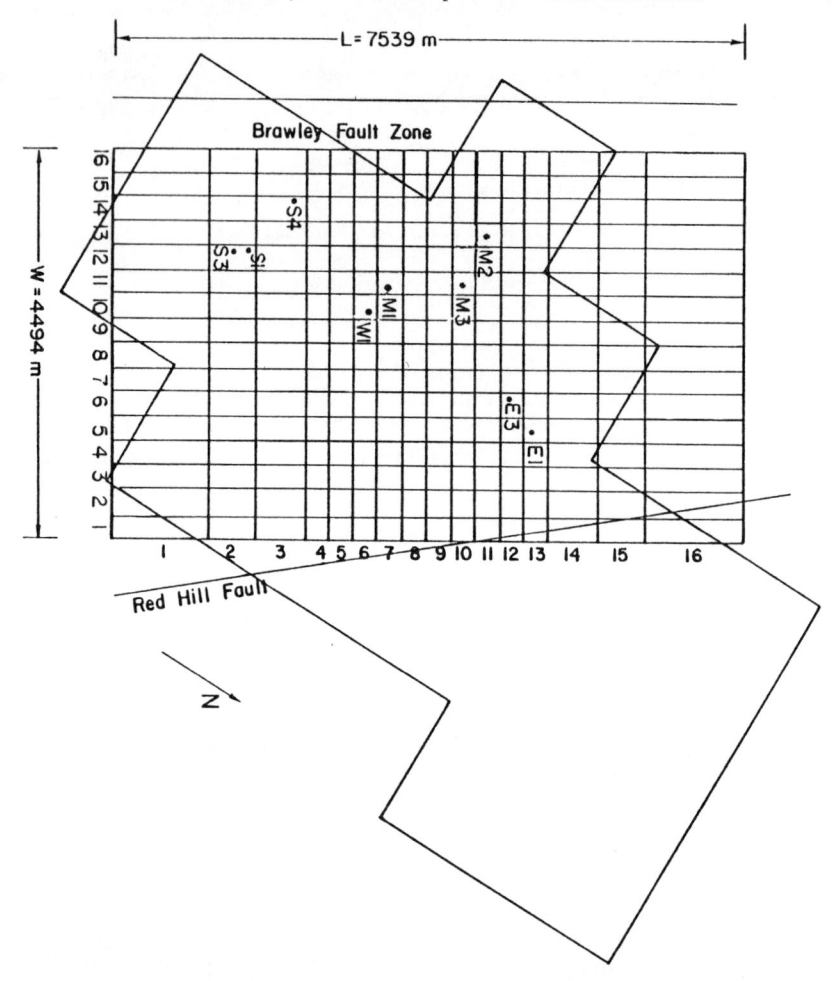

Figure 2.12. Portion of the Salton Sea geothermal field chosen for simulation. Development wells within the region are also shown

much less at the southeastern end (left, $y=0$) than at the northwestern end (right, $y=L$). Using a brine equation of state and the temperature-depth profiles at the two ends, the corresponding mid-plane hydrostatic pressures were computed to be $P(0) = 3\cdot802 \times 10^6$ Pa and $P(L) = 8\cdot507 \times 10^6$ Pa. By considering the temperature variation and dip angle along the length of the reservoir, it was found that if there were no lengthwise flow, the value of $P(L)$ would need to be $8\cdot824 \times 10^6$ Pa. The lengthwise pressure drive, $\Delta P \sim 3\cdot17 \times 10^5$ Pa, apparently causes an influx of ~ 323 K groundwater from the southeast end ($y=0$) which would cool the upper reservoir if hot brine infusion from the lower reservoir were completely precluded by the shale barrier.

These boundary conditions and reservoir properties were incorporated into MUSHRM and a series of calculations performed until a satisfactory match with the mid-plane pre-production temperatures in the upper reservoir was obtained. A 1D version was first applied to the dipping and thickening upper reservoir with the provision that for each zone there is infusion of hot brine at the rate required to obtain the corresponding projected mid-plane pre-production temperature. The total influx rates and the length-

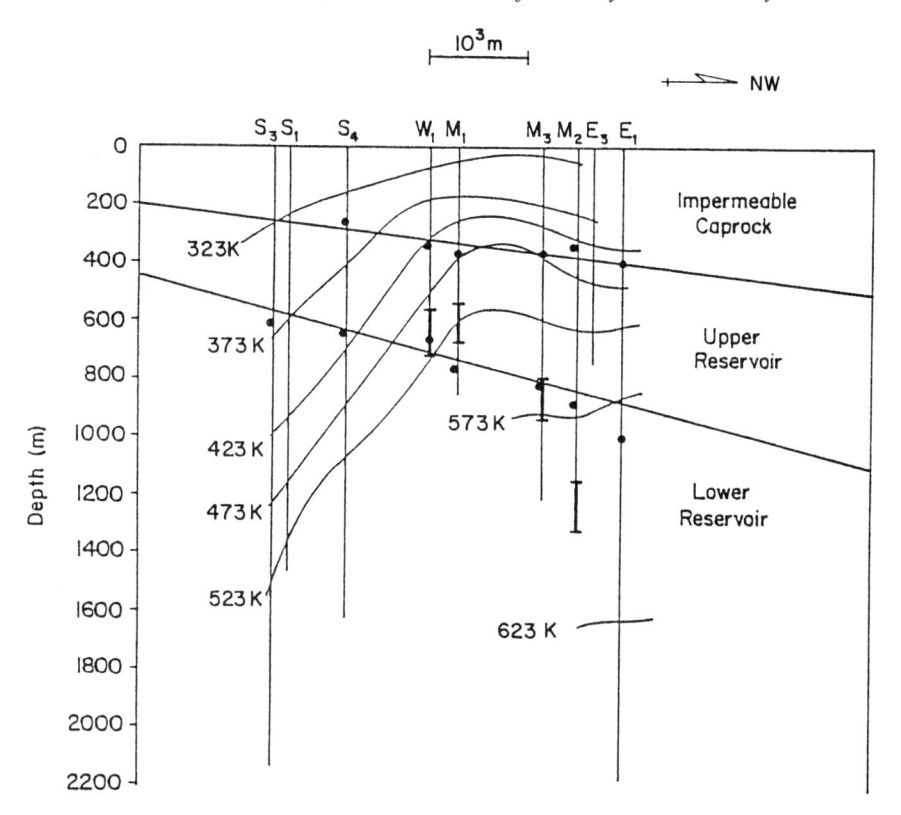

2. Convective Heat and Mass Transfer in Hydrothermal Systems

Figure 2.13. Vertical section and projected data from development wells. Points due to Towse (1975) show the location of the top and the bottom of the upper reservoir. Bars indicate perforated intervals

wise variation of the influx (computed from 1D calculations) were maintained, but the lateral distribution of the influx rate was allowed to vary in a subsequent series of 2D areal calculations. A symmetric distribution with maximum at the center was found to best fit the lateral variation of the mid-plane temperatures measured in the wells. The desired mid-plane temperatures for the well locations are satisfactorily matched by the steady-state temperature contours calculated with the pre-production model (Figure 2.14). The velocity plot, Figure 2.15, shows that the infusion of hot brine from the lower reservoir pushes a large part of the incoming cold groundwater to the edges of the upper reservoir, producing the lower temperatures there.

Because of the limited data base, Riney et al. (1977) had to invoke a variety of hypotheses regarding geology (Red Hill Fault as a sealing boundary, break in shales to allow influx of hot water into upper reservoir), temperature distribution, rock properties, groundwater flow (from the southeast to the northwest), convective flow, etc. Tracer tests may be employed to determine the direction of groundwater flow, and perhaps the amount and location of the hot brine influx. As the SSGF moves from the exploration and assessment stage of development to the exploitation and utilization stage, more accurate information on the reservoir boundaries and the thermomechanical properties of reservoir rocks and fluids should become available. It may be that the preproduction model of Riney et al. (1977) will require revision to include this new information.

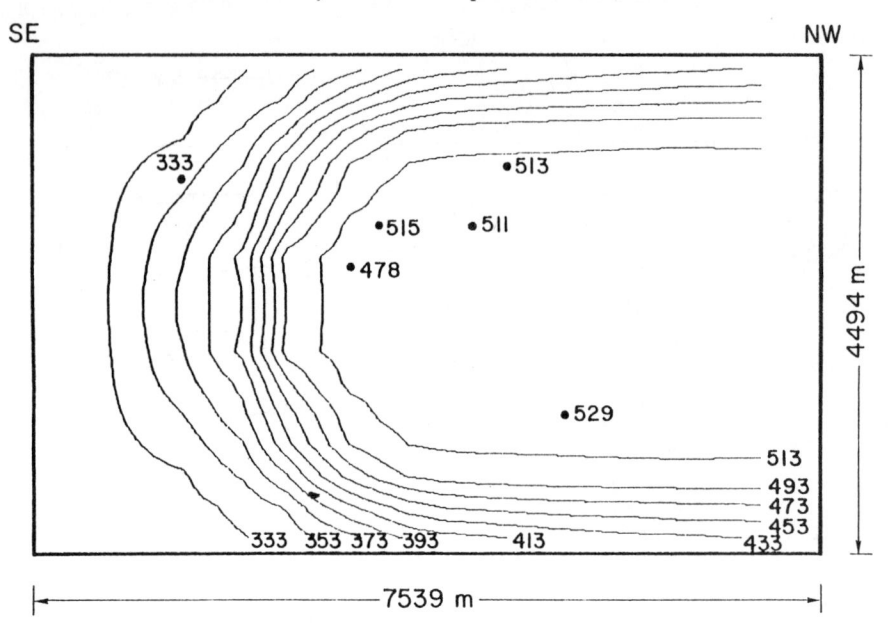

Figure 2.14. Pre-production model temperature contours (K) compared with mid-plane temperatures (at half thickness of the upper reservoir) measured at well locations

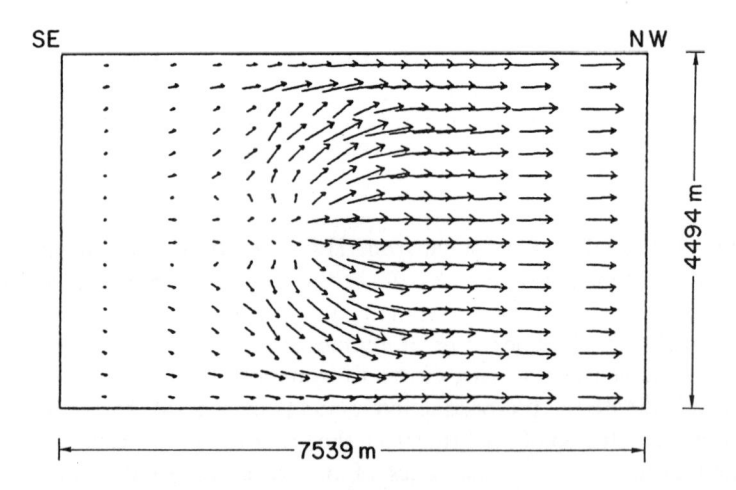

Figure 2.15. Pre-production model velocity field in upper reservoir (longest vector is 0.91×10^{-6} m/s)

2.4.3. Wairakei Geothermal Field

The Wairakei geothermal system is of particular interest since it was the first (and until 1973 the only) liquid-dominated geothermal system to be exploited for electrical power. Although the Larderello field in Italy is older and larger, and The Geysers field in California is larger, both are vapor-dominated reservoirs and therefore are atypical of prospective geothermal resources generally. Drilling began at Wairakei in 1950 and ceased in 1968; mass production rates reached a peak in the mid-1960's and have been slowly declining since. Pritchett et al. (1978) present a summary of the data pertaining to the character, performance and response to production of the Wairakei geothermal field.

Centered in the North Island of New Zealand, the Wairakei geothermal system is located north of Lake Taupo and west of the Waikato river; it occupies a surface area of approximately 1.5×10^7 m^2 (Grindley, 1965) and extends westward of the river approximately 5×10^3 m. Although the actual field only covers 1.5×10^7 m^2, there exists considerable evidence (resistivity surveys, pressure response of outlying wells) which indicates that the impermeable boundaries of the entire system enclose a much larger area (Pritchett et al., 1978). Figure 2.16 shows the spatial relationships among the various peripheral wells, the main bore field and the 'resistivity boundary' of the Wairakei/Tauhara field. For purposes of this figure, the resistivity boundary is taken as the region between the 10 and 20 ohmmeter contours (Pritchett et al., 1978). As discussed by Pritchett et al., pressure response in the 200 series wells indicates that any impermeable boundary of the field extends beyond the resistivity boundary in the west. The bores in the Tauhara field were drilled late in the history of the Wairakei development program—the first was completed in June 1964. At present, there exists little doubt that the two reservoirs (Wairakei and Tauhara) are hydrologically connected (Pritchett et al., 1978). The latter observation is important to note in that existing studies of Wairakei exclude the Tauhara system.

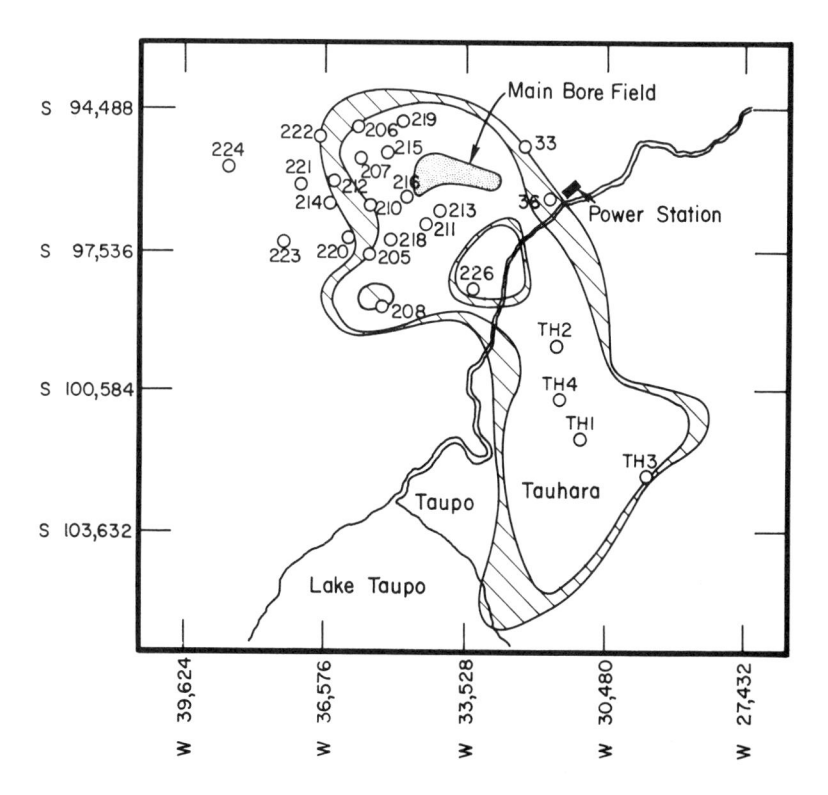

Figure 2.16. Resistivity boundary (10 to 20 ohm-meters) of the Wairakei/Tauhara field, and the peripheral wells. Co-ordinates are meters (converted from original feet) with respect to 1949 Maketu datum. (Pritchett, et al., 1978)

The geology of the Wairakei/Tauhara geothermal fields has been described in detail by Grindley and his co-workers (1965, 1966). The Wairakei hydrothermal field is underlain by a nearly horizontal Quaternary acidic volcanic sequence. The generalized stratigraphic sequence consists of the following formations (in order of increasing depth): Holocene pumice cover, Wairakei breccia, Huka Falls formation, Haparangi rhyolite, Waiora formation, Waiora Valley andesite, Wairakei ignimbrites and the Ohakuri group. There exist at least two aquifers in the above sequence: the Wairakei breccia and the Waiora formation. A deeper third aquifer may also exist in the Ohakuri group. Figure 2.17 shows a geologic section passing through the main production and subsidence areas. The bulk of mass production is generally assumed to come from the Waiora aquifer and from the Waiora/Ignimbrites interface region. Although most of the production wells at Wairakei are located in regions of high permeability associated with major faults, the reservoir as a whole behaves as a porous medium (Mercer et al., 1975). The Waiora aquifer is overlain by Huka Falls formation of relatively low permeability. Both recharge to and discharge from the Waiora aquifer takes place through the Huka Falls formation. The boundary between the Waiora aquifer and the Wairakei ignimbrites is not well defined; fracture zones and the irregular surface of the unconformity between the ignimbrites and the aquifer provides a region of locally high permeability. The Wairakei ignimbrite, a rock of low primary permeability, is complexly faulted. It is generally believed that these fault zones in the ignimbrites serve as conduits for hot water recharge to the Waiora aquifer.

Numerous natural heat flow assessments of Wairakei have been carried out since 1951. Estimates for the total heat flux lie between 3.431×10^8 to 6.820×10^8 W (Fisher, 1964). Fisher concludes that the best value for heat flux (with 285 K as reference temperature) is approximately 4.184×10^8 W. Although no measurements for mass discharge are available, it can be estimated from the heat flux data since the bulk of natural heat flow is believed to be associated with natural mass discharge. Assuming a mean enthalpy of 1.025×10^6 J/kg, Fisher obtained a mass discharge of 440 kg/s for the 1951–1952 period. There are indications that the total natural heat flux has increased and the mass flow has

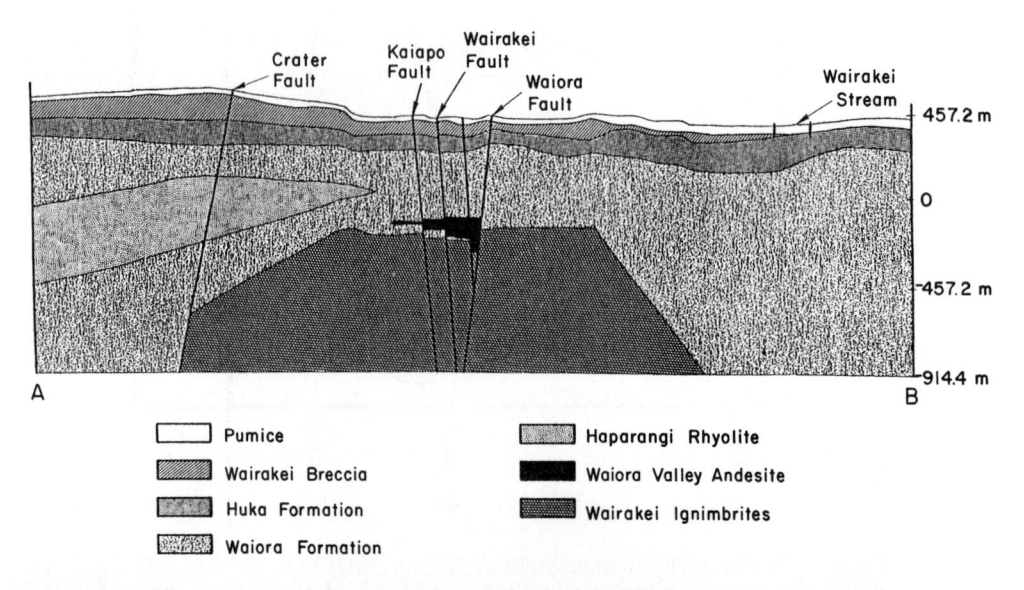

Figure 2.17. A geologic cross-section of Wairakei field passing through the main production and subsidence areas (Pritchett et al., 1978)

2. Convective Heat and Mass Transfer in Hydrothermal Systems

decreased over the life of the field. Temperatures at Wairakei have been measured periodically. Although the temperature measurements are subject to considerable uncertainty (Pritchett et al., 1978), it is possible to deduce general trends. Temperatures increase rapidly with depth down to the top of the Waiora aquifer where the temperature in the hotter regions is around 473 K. In the aquifer, the temperature gradient is reduced; the maximum recorded temperature is about 523 K. Since in the 1950's mass discharge was very small, the temperatures and pressures measured in the original wells can be taken as representative of the pre-exploitation conditions. Grant (personal communication, 1979) points out that in early days, the well head pressures on wells standing shut and full of water were employed to calculate pressures at depth since no downhole pressures were measured. The well head pressures declined rapidly in some cases with production. This makes the determination of undisturbed pressures very difficult. As a result of production, over the years, both the temperatures and the pressures in the Waiora formation have declined.

Mercer et al. (1975) employed a two-dimensional areal reservoir simulator to model the pre-production mass and heat transfer at Wairakei. Figure 2.18 shows the portion of the Wairakei geothermal system considered by Mercer et al.; note that the Tauhara region is excluded. Mercer et al. assumed that the Huka Falls formation transmits both mass and heat vertically from the Waiora aquifer. The Wairakei ignimbrites were assumed to be, however, impermeable and to only transmit heat through conduction. The boundaries of the field (Figure 2.18) were taken to be impermeable with temperature specified as a known function of the space variables. Details of the numerical solution may be found in Mercer et al.; in the following, we shall merely outline the basic model.

Since the problem is treated as two-dimensional in the areal plane, it is necessary to use vertically averaged temperature values. Figure 2.19 shows the measured steady state temperature distribution in the Waiora aquifer. Mercer et al. (1975) assumed the reservoir to be single-phase prior to production. In order to match single-phase flow conditions and

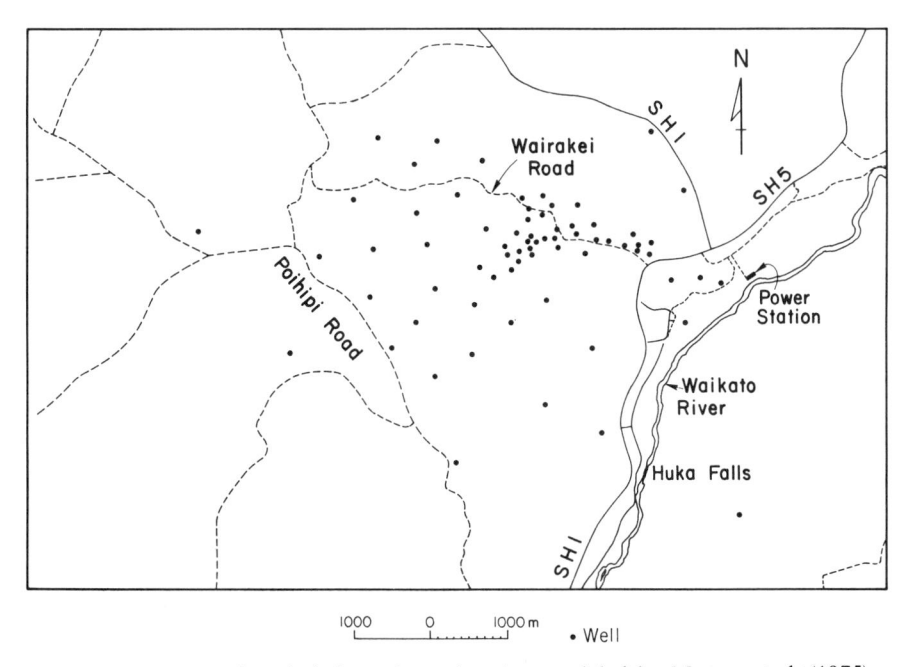

Figure 2.18. Map of Wairakei geothermal region modeled by Mercer et al. (1975)

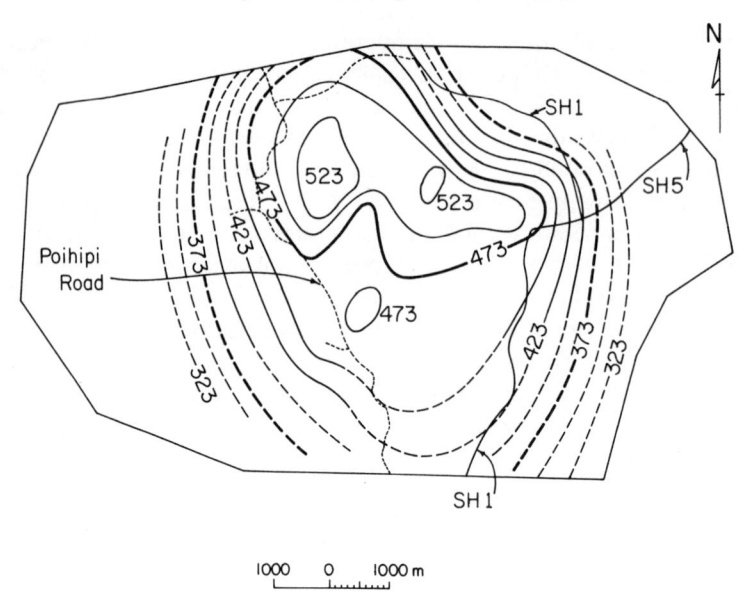

Figure 2.19. Measured steady state (vertically averaged) temperature distribution in the Waiora aquifer of Wairakei (in Kelvin). Contour interval is 25 K. (Mercer et al., 1975)

the observed temperature distribution, Mercer et al. varied the heat source at the base of the Waiora aquifer (i.e., top of the ignimbrites). The calculated temperature distribution was in fair agreement with that shown in Figure 2.19.

The direction and magnitude of fluid transport through the Huka Falls formation is governed by the pressure distribution on the two sides of this formation. The pressure distribution at the top of the Huka Falls formation was determined by treating the Wairakei breccia and the Holocene pumice as a single aquifer of variable thickness and assuming that the water table extends to the surface. The total calculated heat and mass discharges are 9.75×10^8 W (using a reference temperature of 273 K) and 1133 kg/s, respectively. The estimated values are larger than most observed values. Mercer et al. suggest that the larger area used in the model may have been responsible for the larger calculated flux.

Mercer et al. used the single-phase model to reproduce history data up to 1962. The results of this simulation indicate little or no change in reservoir temperatures. The calculated pressure drops, on the other hand, displayed only a qualitative agreement with the observed pressure drops.

Pritchett et al. (1978) present conclusive evidence that the single-phase model will most likely lead to too high an estimate for the observed pressure drop. It is now generally believed that the Wairakei geothermal field was either two-phase or close to flashing prior to production. The original model of Mercer et al. has been criticized on at least two other grounds (Mercer and Faust, 1979). Mercer et al. allowed no recharge from the ignimbrites, whereas both the gravity data and the pressure drop history provide a strong justification for the belief that ignimbrites serve as a major source of hot water. Finally, Mercer et al. assumed a rather large value for the vertical permeability of the Huka Falls formation ($\sim 10^{-14}$ m^2); this fact is probably most responsible for the discrepancy between the calculated and measured mass discharge values.

Recently, Mercer and Faust (1979) have developed a new areal model of the Wairakei geothermal system. This model is similar to that of Mercer et al. except in so far as (1) mass leakage is allowed through the ignimbrites, (2) the vertical permeability of Huka Falls formation is reduced ($\sim 10^{-15}$ m^2) and (3) two phase flow is allowed in the reservoir. This improved model predicts a surface mass discharge of 424 kg/s—in fair agreement with the generally accepted value of 440 kg/s. Mercer and Faust also reproduced the historical production data up to 1974. The leakage from the ignimbrites was calculated so as to match the pre-production temperatures, and the observed pressure drops during production. Because of the uncertainty associated with the treatment of the leakage terms, there is also some uncertainty in predictions of reservoir behavior after 1974.

The Wairakei modeling experience clearly illustrates the difficulties in developing viable models of unexploited geothermal systems. Many of the early assumptions regarding reservoir behavior (e.g., single-phase flow, no recharge through the ignimbrites) had to be modified in light of later production data. It is likely that similar experience will recur in the case of other geothermal systems. The pre-production model needs to be regarded as an evolving framework for synthesizing the available information, for indicating where more data are necessary to resolve the uncertainties regarding reservoir behavior, and for making future predictions of reservoir response under production.

Acknowledgements

We would like to thank I. G. Donaldson and M. A. Grant for their comments on an earlier draft of the manuscript.

References

Beck, J. L., 1972. 'Convection in a box of porous material saturated with fluid', *Physics of Fluids*, **15**, 1377–1383.

Bodvarsson, G. 1961. 'Physical characteristics of natural heat resources in Iceland', *Proceedings U.N. Conference on New Sources of Energy, Rome, Italy*, **2**, 82–89.

Brownell, D. H., Jr., Garg, S. K., and Pritchett, J. W., 1977. 'Governing equations for geothermal reservoirs', *Water Resources Research*, **13**, 929–934.

Caltagirone, J. P., 1975. 'Thermoconvective instabilities in a horizontal porous layer', *Journal of Fluid Mechanics*, **72**, 269–287.

Cathles, L. M., 1977. 'An analysis of the cooling of intrusives by ground-water convection which includes boiling', *Economic Geology*, **72**, 804–826.

Cheng, P., 1978. 'Heat transfer in geothermal systems.' In *Advances in Heat Transfer*, **14**, T. F. Irvine, Jr. and J. P. Hartnett (Eds.), Academic Press, New York, 1–105.

Cheng, P., and Lau, K. H., 1974. 'Steady state free convection in an unconfined geothermal reservoir', *Journal of Geophysical Research*, **79**, 4425–4431.

Cheng, P., and Teckchandani, L., 1977. 'Numerical solutions for transient heating and fluid withdrawal in a liquid dominated geothermal reservoir.' In *The Earth's Crust*, J. G. Heacock, (Ed.), Geophysical Monograph #20, American Geophysical Union, Washington, D.C., 705–721.

Cheng, P., Yeung, K. C., and Lau, K. H., 1975. 'Numerical solutions for steady free convection in island geothermal reservoirs', *Technical Report #8*, Hawaii Geothermal Project, University of Hawaii, Honolulu, Hawaii.

Combarnous, M. A., and Bories, S. A., 1975. 'Hydrothermal convection in saturated porous media', *Advances in Hydroscience*, **10**, 231–307.

Combs, J., and Hadley, D. M., 1973. 'Microearthquake investigations of the mesa geothermal anomaly, Imperial Valley, California', *EOS-Transactions American Geophysical Union*, **54**, 1213–1214.

Corey, A. T., Rathjens, C. H., Henderson, J. H., and Wyllie, M. R. J., 1956. 'Three-phase relative permeability', *Transactions AIME*, **207**, 349–351.

Donaldson, I. G., 1962. 'Temperature gradients in the upper layers of the Earth's crust due to convective water flows', *Journal of Geophysical Research*, **67**, 3449–3459.

Donaldson, I. G., 1968a. 'The flow of steam water mixtures through permeable beds: a simple simulation of a natural undisturbed hydrothermal region', *New Zealand Journal of Science*, **11**, 3–23.

Donaldson, I. G., 1968b. 'A possible model for hydrothermal systems and methods of studying such a model', *Proceedings Third Australasian Conference on Hydraulics and Fluid Mechanics*, 200–204.

Donaldson, I. G., 1970. 'The simulation of geothermal systems with a simple convective model', *Geothermics*, Special Issue **2**, 649–654.

Duncan, J. M., Witherspoon, P. A., Mitchell, J. K., Watkins, D. J., Hardcastle, J. H., and Chen, J. C., 1972. Seepage and groundwater effects associated with explosive cratering', *University of California, Berkeley, California Report TE-72-2*.

Einarsson, T., 1942. 'The nature of the springs of Iceland', *Rit. Visind. Isl.*, **26**, 1–92. (In German.)

Elder, J. W., 1965. 'Physical processes in geothermal areas.' In *Terrestrial Heat Flow*, W. H. K. Lee (Ed.), *Geophysical Monograph # 8*, Americal Geophysical Union, Washington, D. C., 211–239.

Elder, J. W., 1966. 'Heat and mass transfer in the Earth: hydrothermal systems', *New Zealand Department of Scientific and Industrial Research Bulletin # 169*.

Elder, J. W., 1967a. 'Transient convection in a porous medium', *Journal of Fluid Mechanics*, **27**, 609–623.

Elder, J. W., 1967b. 'Steady free convection in a porous medium heated from below', *Journal of Fluid Mechanics*, **27**, 29–48.

Fisher, R. G., 1964. 'Geothermal heat flow at Wairakei during 1958', *New Zealand Journal of Geology and Geophysics*, **7**, 172–184.

Garg, S. K., and Pritchett, J. W., 1977. 'On pressure-work, viscous dissipation and the energy balance relation for geothermal reservoirs', *Advances in Water Resources*, **1**, 41–47.

Garg, S. K., Pritchett, J. W., Rice, M. H., and Riney, T. D., 1977. 'U.S. Gulf Coast geopressured geothermal reservoir simulation', *Systems, Science, and Software, La Jolla, California, Report SSS-R-77-3147*.

Goyal, K. P., 1978. 'Heat and mass transfer in a saturated porous medium with application to geothermal reservoirs', *Ph.D. Thesis*, Mechanical Engineering Department, University of Colorado, Boulder.

Goyal, K. P., and Kassoy, D. R., 1977. 'A fault-zone controlled model of the Mesa anomaly', *Proceedings Third Workshop Geothermal Reservoir Engineering, Stanford University, Stanford, California*, 209–213. See also 'Fault-zone controlled charging of a liquid dominated geothermal reservoir', *Journal of Geophysical Research 85*, 1867–1875 (1980).

Gray, W. G., O'Neill, K., and Pinder, G. F., 1976. 'Simulation of heat transport in fractured, single-phase geothermal reservoirs', *Summaries Second Workshop Geothermal Reservoir Engineering, Stanford University, Stanford, California*, 222–228.

Grindley, G. W., 1965. 'The geology, structure and exploitation of the Wairakei geothermal field, Taupo, New Zealand' *New Zealand Geological Survey, Bulletin No. 75*.

Grindley, G. W., Rishworth, D. E., and Watters, W. A., 1966. 'Geology of the Tauhara geothermal field, Lake Taupo', *New Zealand Geological Survey Geothermal Report No. 4*.

Holst, P. H., and Aziz, K., 1972a. 'A theoretical and experimental study of natural convection in a confined porous medium', *Canadian Journal of Chemical Engineering*, **50**, 232–241.

Holst, P. H., and Aziz, K., 1972b. 'Transient three-dimensional natural convection in confined porous media', *International Journal of Heat and Mass Transfer*, **15**, 73–90.

Homsy, G. M., and Sherwood, A. E., 1976. 'Convective instabilities in porous media with through flow', *AIChE Journal*, **22**, 168–174.

Horne, R. N., 1979. 'Three-dimensional convection in a confined porous medium heated from below', *Journal of Fluid Mechanics*, **92**, 751–766.

Horne, R. N., and O'Sullivan, M. J., 1974. 'Oscillatory convection in a porous medium heated from below', *Journal of Fluid Mechanics*, **66**, 339–352.

Horton, C. W., and Rogers, F. T., Jr., 1945. 'Convection currents in a porous media', *Journal of Applied Physics*, **16**, 367–370.

Kassoy, D. R., and Zebib, A., 1975. 'Variable viscosity effects on the onset of convection in porous media', *Physics of Fluids*, **18**, 1649–1651.

Kassoy, D. R., and Zebib, A., 1978. 'Convection fluid dynamics in a model of a fault zone in the Earth's crust', *Journal of Fluid Mechanics*, **88**, 769–792.

Keenan, J. H., Keyes, F. G., Hill, P. G., and Moore, J. G., 1969. *Steam Tables* (International Edition—Metric Units), Wiley, New York.

Kruger, P., and Ramey, H. J., Jr. (Eds.), 1975. 'Geothermal reservoir engineering', *Proceedings First Workshop Geothermal Reservoir Engineering*, Stanford University, Stanford, California.

Kruger, P., and Ramey, H. J., Jr. (Eds.), 1976. *Summaries second workshop geothermal reservoir engineering*, Stanford University, Stanford, California.

Kruger, P., and Ramey, H. J., Jr. (Eds.), 1977. *Proceedings Third Workshop Geothermal Reservoir Engineering*, Stanford University, Stanford, California.

Lapwood, E. R., 1948. 'Convection of a Fluid in a Porous Medium', *Proceedings Cambridge Philosophical Society*, **44**, 508–521.

Louis, C., 1970. 'Water flows in fissured rock and their effects on the stability of rock massifs', Lawrence Livermore Laboratory, Livermore, California, Report UCRL–Trans–10,469 (English translation of a dissertation, Universität (TH), Karlsruhe, West Germany, 1967).

Mercer, J. W., and Faust, C. R., 1979. 'Geothermal reservoir simulation III: Application of liquid- and vapor-dominated hydrothermal techniques to Wairakei, New Zealand', *Water Resources Research*, **15**, 653–671.

Mercer, J. W., Jr., Faust, C., and Pinder, G. F., 1974. 'Geothermal reservoir simulation', *Proceedings NSF/RANN Conference on Research for the Development of Geothermal Energy Resources*, Jet Propulsion Laboratory/California Institute of Technology, Pasadena, California, 256–267.

Mercer, J. W., Pinder, G. F., and Donaldson, I. G., 1975. 'A Galerkin-finite element analysis of the hydrothermal system at Wairakei, New Zealand', *Journal of Geophysical Research*, **80**, 2608–2621.

Morland, L. W., Zebib, A., and Kassoy, D. R., 1977. 'Variable property effects on the onset of convection in an elastic porous matrix', *Physics of Fluids*, **20**, 1255–1259.

Morrison, H. L., and Rogers, F. T., Jr., 1952. 'Significance of flow patterns for initial convection in porous media', *Journal of Applied Physics*, **23**, 1058–1059.

Morrison, H. L., Rogers, F. T., Jr., and Horton, C. W., 1949. 'Convection currents in porous media–II. Observation of conditions at onset of convection', *Journal of Applied Physics*, **20**, 1027–1029.

Noorishad, J., Witherspoon, P. A., and Brekke, T. L., 1971. 'A method for coupled stress and flow analysis of fractured rock masses', University of California, Berkeley, California, Report No. 71-6.

Norton, D., 1977. 'Fluid circulation in the Earth's crust.' In *The Earth's Crust*, J. G. Heacock (Ed.), Geophysical Monograph #20, American Geophysical Union, Washington, D.C., 693–704.

Norton, D., 1978. 'Sourcelines, sourceregions and pathlines for fluids in hydrothermal systems related to cooling plutons', *Economic Geology*, **73**, 21–28.

Norton, D., and Knight, J., 1977. 'Transport phenomena in hydrothermal systems: cooling plutons', *American Journal of Science*, **277**, 937–981.

Palmer, T. D., 1975. 'Characteristics of geothermal wells located in the Salton Sea geothermal field, Imperial County, California', Lawrence Livermore Laboratory, Livermore, California, Report UCRL-51976.

Prats, M., 1966. 'The effect of horizontal fluid flow on thermally induced convection currents in porous mediums', *Journal of Geophysical Research*, **71**, 4835–4838.

Pritchett, J. W., Garg, S. K., Brownell, D. H., Jr., Rice, L. F., Rice, M. H., Riney, T. D., and Hendrickson, R. R., 1976. 'Geohydrological environmental effects of geothermal power production—Phase IIA', Systems, Science and Software, La Jolla, California, Report SSS-R-77-2998.

Pritchett, J. W., Rice, L. F., and Garg, S. K., 1978. 'Reservoir engineering data: Wairakei geothermal field, New Zealand', Systems, Science and Software, La Jolla, California, Report SS-R-78-3597-1.

Ramey, H. J., Jr., Brigham, W. E., Chen, H. K., Atkinson, P. G., and Arihara, N., 1974. 'Thermodynamic and hydrodynamic properties of hydrothermal systems', Stanford Geothermal Program Report SGP-TR-6, Stanford University, Stanford, California.

Rana, R., Horne, R. N., and Cheng, P., 1979, 'Natural convection in a multi-layered geothermal reservoir', *Journal of Heat Transfer* **101**, 411–416.

Ribando, R. J., and Torrance, K. E., 1976. 'Natural convection in a porous medium: Effects of confinement, variable permeability, and thermal boundary conditions', *Journal of Heat Transfer*, **98**, 42–48.

Riney, T. D., Pritchett, J. W., and Garg, S. K., 1977. 'Salton sea geothermal reservoir simulations',

Proceedings Third Workshop Geothermal Reservoir Engineering, Stanford University, Stanford, California, 178–184.

Rogers, F. T., Jr., 1953. 'Convection currents in porous media—V. Variational form of the theory', *Journal of Applied Physics*, **24**, 877–880.

Rogers, F. T., Jr., and Morrison, H. L., 1950. 'Convection currents in porous media—III. Extended theory of critical gradients', *Journal of Applied Physics*, **21**, 1177–1180.

Rogers, F. T., Jr., Schilberg, L. E., and Morrison, H. L., 1951 'Convection currents in porous media––IV. Remarks on the theory', *Journal of Applied Physics*, **22**, 1476–1479.

Schroeder, R. C., 1976. 'Reservoir engineering report for the Magma–SDG&E geothermal experimental site near the Salton Sea, California', *Lawrence Livermore Laboratory, Livermore, California, Report UCRL-52094*.

Sorey, M. L., 1975. 'Numerical modelling of liquid geothermal systems', *U.S. Geological Survey, Menlo Park, California, Open-File Report, 75-613*.

Sorey, M. L., 1976. 'A model of the hydrothermal system of Long Valley, Caldera, California', *Summaries Second Workshop Geothermal Reservoir Engineering*, Stanford University, Stanford, California, 324–338.

Straus, J. M., 1975. 'Large amplitude convection in porous media', *Journal of Fluid Mechanics*, **64**, 51–63.

Straus, J. M., and Schubert, G., 1977. 'Thermal convection of water in a porous medium: Effects of temperature- and pressure-dependent thermodynamic and transport properties', *Journal of Geophysical Research*, **82**, 325–333.

Straus, J. M., and Schubert, G., 1978. 'On the existence of three-dimensional convection in a rectangular box of fluid-saturated porous material', *Journal of Fluid Mechanics*, **87**, 385–394.

Straus, J. M., and Schubert, G., 1979. 'Three-dimensional convection in a cubic box of fluid-saturated porous material', *Journal of Fluid Mechanics*, **91**, 155–166.

Towse, D. F., 1975. 'An estimate of the geothermal energy resource in the Salton Trough, California and Mexico', *Lawrence Livermore Laboratory, Livermore, California, Report UCRL-51851*.

Turcotte, D. L., Ribando, R. J., and Torrance, K. E., 1977. 'Numerical calculation of two-temperature thermal convection in a permeable layer with application to the Steamboat Springs thermal system, Nevada.' In *The Earth's Crust*, J. G. Heacock, (Ed.), Geophysical Monograph # 20, American Geophysical Union, Washington, D.C., 722–736.

Weber, J. E., 1974. 'Convection in a porous medium with horizontal and vertical temperature gradients', *International Journal of Heat and Mass Transfer*, **17**, 241–248.

White, D. E., 1961. 'Preliminary evaluation of geothermal areas by geochemistry, geology and shallow drilling', *Proceedings U.N. Conference on New Sources of Energy, Rome, Italy*, **2**, 402–408.

Wittke, W., 1973. General Report on the Symposium, 'Percolation Through Fissured Rock', *Bulletin of the International Association of Engineering Geology, Krefeld, No. 7*, 3–28.

Wooding, R. A., 1957. 'Steady state free thermal convection of liquid in a saturated permeable medium', *Journal of Fluid Mechanics*, **2**, 273–285.

Wooding, R. A., 1976. 'Influence of anisotropy and variable viscosity upon convection in a heated saturated porous layer', *New Zealand Department of Scientific and Industrial Research, Technical Report 55*.

Zebib, A., and Kassoy, D. R., 1976a. 'Onset of natural convection in a box of water saturated porous media with large temperature variation', *Physics of Fluids*, **20**, 4–9.

Zebib, A., and Kassoy, D. R., 1976b. 'Steady nonlinear convection in saturated porous media with large temperature variation', *CUMER 76-5*, Mechanical Engineering Department, University of Colorado, Boulder.

Zebib, A., and Kassoy, D. R., 1978. 'Three-dimensional natural convection motion in a confined porous medium', *Physics of Fluids*, **21**, 1–3.

Geothermal Systems: Principles and Case Histories
Edited by L. Rybach and L. J. P. Muffler
© 1981 John Wiley & Sons Ltd.

3. Prospecting for Geothermal Resources

J. T. LUMB

*Department of Scientific and Industrial
Research, Wellington, New Zealand*

3.1. Introduction

In recent years, a great deal of study in many parts of the world has been devoted to the detection, measurement and understanding of the various characteristics of geothermal fields and has been reported upon in numerous technical papers and conferences. Of particular note are the two United Nations symposia on the development and use of geothermal resources held in 1970 and 1975. The proceedings of these meetings form an essential part of the library of any geothermal prospector. Many of the references for this paper are to be found in these proceedings and have been quoted because of their accessibility and the generally excellent guides to further reading that they give.

Earlier reviews of this topic (e.g. Banwell, 1970; Combs and Muffler, 1973) have dwelt upon exploration as the means of detecting and delineating economic reservoirs of geothermal energy. Exploration must go further than this and provide much of the information which forms the background to our deeper general understanding of geothermal processes. Exploration does not stop once a field has been delineated. The same methods continue to be used in a monitoring role as the exploitation of a field proceeds, and indeed, it is frequently only after deep drilling has taken place that some of the results of exploration come to be understood. Information gained during the exploration of a field is also of importance in assessing some of the environmental effects and the technical difficulties that may be experienced during the development and use of the resource.

Although the Earth is an immense source of heat, most of this heat either is buried too deeply or is too diffuse to be exploited economically. However, in some places the heat is concentrated as a result of certain geological and hydrological processes, and it is in these geothermal regions that conditions sometimes occur which allow the heat to be extracted through drill-holes sunk to modest depths in the Earth's crust. The purpose of this chapter is to summarize and critically appraise the various techniques that are available for finding concentrations of geothermal energy at locations where it is (a) accessible and (b) in a form that can be used with our present technology and that which is being developed now. Such geothermal 'fields' are characterized by various properties that are described elsewhere (e.g. Ellis and Mahon, 1977; Kruger and Otte, 1973) as well as in other chapters of this book. It is these properties that give rise to the features which are detectable at or near the ground surface and which form the basis for prospecting.

The objects of geothermal prospecting can thus be summarized as:

- the location of geothermal phenomena;

- the indication that a useful geothermal production field exists by:
  (a) roughly estimating the amount of surface discharge (heat and areal extent), and
  (b) estimating the temperature of the source;
- the estimation of the size of the resource;
- the determination of the nature of the field (e.g. dry steam, steam-water mixture);
- the location of productive zones;
- the determination of the enthalpy of the fluids that will be discharged by the wells in the field;
- the formation of a body of basic data against which the results of future monitoring can be viewed;
- the determination of pre-exploitation values of environmentally sensitive parameters; and
- the gaining of knowledge of any characteristics that might cause difficulties in the development of a field (e.g. low pH, high gas content, likelihood of subsidence).

3.2. Prospecting Strategies

The approach to be taken to prospecting for geothermal energy in any particular instance will depend on a number of factors. There will be constraints such as finance and manpower, as well as physical limitations, such as the terrain or climate under which the work is to be performed. The approach must be designed to suit the type of geothermal field being explored. Then, the objectives of the study must be considered, for example whether they are basically academic or concerned with proving a resource for exploitation. Finally, the experience and expertise of those planning and actually doing the prospecting is a particularly important factor.

Except in the case of a purely academic study, the objectives will normally be those set out in the introduction, but the relative importance of each objective will depend upon a number of things which will affect the exploration programme. For example, the preliminary location of geothermal phenomena assumes a much greater importance in a remote, unexplored jungle area than in, say, a well-known thermal region frequented by tourists. Estimating the size of the resource may be less important if the use to which it is to be put is a small-scale application that obviously requires much less heat than is already discharging naturally. If the energy is to be used for district heating or some other application where low-grade heat is required, the requirement of high-enthalpy fluid is less important.

One of the objectives of the exploration programme will normally be to determine the nature of the field, i.e. to establish whether it is vapour- or water-dominated. As the nature of the field is a determining factor in the design of the exploration programme, it is clear that if this information is not available before prospecting begins, the programme must be flexible enough to take it into account at a later stage. In most cases, preliminary geochemical reconnaissance will provide the data necessary to resolve this point.

The constraining effects of finance and manpower will normally force a decision as to whether it is better to try to cover a large area with inexpensive techniques or to concentrate many studies on a very small area. Particular circumstances will normally dictate the answer to this dilemma, but in general the first option will be most profitable. If by choosing to survey a small area the boundaries of all sectors of the geothermal field are not even roughly located, then no matter how intense the study, the influence of the rest of the field on the area studied will be unknown.

The *terrain* in which the survey is to take place will limit not only the locations at which observations, sample collection and measurements can be made, but it will also affect the choice of geophysical methods to be applied. In considering the logistics of the fieldwork, the availability of manpower must be taken into account at the same time as the terrain. With sufficient labour it is possible to carry cumbersome equipment into places where access is difficult, and the cutting of lines through dense vegetation is less of a handicap. On the other hand, with a shortage of labour, very steep or densely forested country could prevent the use of some geophysical techniques such as Schlumberger or Wenner resistivity surveys which require long cables to be laid out.

The effect of terrain on the interpretation of data must also be considered, although in most geophysical methods appropriate corrections can be made if the topography is defined in sufficient detail. However, when the terrain is very rough, the necessary corrections for resistivity techniques are difficult to compute and ambiguous data may result. The presence of tall trees in an area that is particularly susceptible to strong winds may restrict the use of seismic methods.

One of the main constraining effects of *climate* will be to restrict the time available for fieldwork through days lost because of heavy rain or to determine the time of year in which the fieldwork can be undertaken. If because of other considerations, the fieldwork must be done at a certain time of year, some exploration methods may be very difficult to interpret, for example, geochemical analyses on samples collected during or immediately after the rainy season in tropical countries. Also, climatic conditions may affect the operation of some instruments.

The influence on a prospecting strategy of the experience and expertise of those who will do the work is highly significant and in the worst situation could lead to the complete failure of the survey to achieve its end. The greatest potential danger in this regard occurs when the exploration programme is designed under contract by a company that specializes in a particular technique. If the employing company or utility does not have the necessary skills to evaluate proposals for the different approaches that are submitted by the various tendering companies, some very expensive mistakes can be made.

The person responsible for the exploration programme must therefore consider what his objectives are and balance these and the various other factors to get the best out of the situation. Above all it must be appreciated that the chosen strategy is to be designed to find heat in an accessible and useable form, and, as was stressed by Combs and Muffler (1973), the most useful methods are not necessarily those which have proved most successful in petroleum or mineral prospecting.

3.3. Hydrological Aspects

The geological environments in which geothermal activitity is to be found are generally restricted to areas that are on or close to the active margins of tectonic plates where crustal material is being produced or consumed (Muffler, 1976; Rybach, this volume). There are exceptions, such as in Hawaii, but in all cases there is a heat source which can be explained in terms of the regional geology and which creates the conditions needed to sustain the geothermal field.

Although the location of a geothermal field is controlled by the regional geology, its characteristics, as recognized at the surface or in drill-holes and which are the subject of geothermal exploration, are essentially hydrological phenomena. They depend not only on a suitable heat source but also on the presence of a medium (normally water) in which the heat can be transported to the surface and on a permeable path along which the medium can travel.

Exploration methods should therefore be able to detect, either directly or indirectly, the presence of anomalously high temperatures, underground fluids and variations in permeability. Methods which can be used to determine the first two of these parameters will establish the existence of a geothermal field and will indicate its size. However, hot fluids can accumulate over a long period in rocks that are porous but are not very permeable, and any attempt to tap such fluids will be relatively unsuccessful because there will be no path along which they can flow readily into the drill-hole. Geothermal exploration methods which can detect permeability therefore assume considerable importance in delineating those parts of the field from which good production can be expected. At present, the only certain method of finding if the rocks are permeable is to drill.

With increasing depth, the vertical lithostatic pressure will become greater and reduce the likelihood of horizontal permeability. In such conditions near-vertical fissures in relatively competent rocks will be the principal paths along which fluids can migrate, suggesting that in deep wells, some degree of deviation from vertical might improve the chances of intersecting a fissure.

Thermal waters may boil during their passage to the surface, and because the resulting steam is less viscous than the water (except in the two-phase boiling zone itself), it will flow more readily through rocks of lower permeability. However, as the successful exploitation of a geothermal field is normally dependent on obtaining a substantial mass flow from the wells, it will still be as important to locate high-permeability zones in vapour-dominated systems as in hot-water systems. The margins of such a vapour-dominated field are often more difficult to delineate than those of a hot-water field because they may be more diffuse as a result of the steam's greater mobility and because the steam has a very high electrical resistivity.

In recent years interest has grown in 'hot dry rock' geothermal fields (e.g. Smith, 1978) where the normally essential elements of water and permeability are absent. Any prospecting for such fields will therefore have to concentrate only on the detection of thermal anomalies.

Another type of geothermal field that has attracted attention in the last five years or so is that which occurs in 'geopressured' zones (Wallace et al., 1979). These are most likely to occur in very deep sedimentary basins where fossil groundwater has become locked in the sediments and, as the rocks became depressed, has encountered higher temperatures; greatly increased pore-pressures have resulted. To maintain the pressure such fields require a deep cap-rock, the location of which may be the most appropriate approach to prospecting. Those geopressured zones that have been investigated to date have been associated with deep oil fields, for example in the American gulf states, and their discovery was a 'by-product' of petroleum exploration.

3.4. Exploration Techniques

The exploration techniques described below have all been used, some with success, others without, to meet the objectives listed in the introduction to this chapter. Each method is described independently of the others, but this is not to infer that they are to be used that way. Geothermal fields are complicated geological and hydrological phenomena, and very rarely will a single prospecting method be able to define a field sufficiently well to indicate the best targets for drilling. The process of exploring a geothermal field is an iterative one calling on the skills of the geologist, the hydrologist, the geochemist and the geophysicist as well as the driller and the reservoir engineer. Each is continually re-assessing his own findings in the light of those of the others.

The prospecting methods that will yield the most useful information depend on the characteristics of the field being investigated. The exploration programme therefore begins with an assessment of what is already known about the area. This will not only give a lead to the most suitable methods to use, but also will indicate existing studies that will not require duplication. At this stage a reconnaissance survey by a geologist, a geochemist and a geophysicist should be undertaken. In most cases the greatest value will result from this survey if the three scientists work together in the field, perhaps in the company of an engineer.

The *geologist* undertaking the reconnaissance study will use basic mapping techniques and will enumerate and locate all the thermal features and view the geothermal activity against the background of the regional geology. A generalized picture of the hydrology will be developed and an assessment made of detailed geological studies to be done later. Apart from making a rough estimate of the natural heat loss from the area, mainly from surface waters, the *geophysicist* will assess the usefulness of the various prospecting methods available to him and consider the field conditions which may limit what can be done in practice. Many geophysical methods require some clearance of groundcover along lines to facilitate fieldwork and an accurate topographic survey. Plans for these should be formulated during the reconnaissance. *Geochemical* information gathered at this early stage can be quite critical. The results of chemical geothermometry studies will give some idea of deep temperatures, and the salinity of the surface discharges will indicate to the geophysicist the likely success of resistivity prospecting.

Because geochemical and geological studies are so much less expensive than most types of geophysics, they should be taken to a fairly advanced stage and evaluated before any detailed geophysical surveys are performed. There will be occasions, for example in remote rural areas where modest quantities of heat are needed for, say, crop drying, when the reconnaissance survey alone is able to indicate both the value of the geothermal field for the intended use and suitable drilling sites.

Combs and Muffler (1973) recommended against using sophisticated techniques until one has worked through a progression of preliminary steps. This advice is enthusiastically reiterated here: many geophysical methods, especially some of the dipole and electromagnetic surveys, are capable of very elegant interpretation but frequently require some prior knowledge of the location of boundary zones to yield interpretable results. They are, therefore, less useful as reconnaissance methods than their simple logistical attributes would suggest.

3.4.1. *Literature Survey and Preliminary Investigations*

A thorough search of the relevant literature is an essential first step in any investigation. There usually exists some mapped information—topographical, geological and also perhaps geophysical—such as gravity or magnetic. Other information on these topics and others—such as geochemistry, hydrology, meteorology—is likely to have been reported upon and should be very carefully assessed. For the geothermal prospector who is about to take up an assignment overseas, this stage is extremely important because if not done thoroughly a good deal of valuable time on the site may otherwise be wasted. Any gaps in the existing topographical maps should be noted and, if possible, steps should be taken to have the necessary surveys done in order to fill them. It is useful at an early stage to find out information on land ownership and whether there will be problems gaining access to undertake surveys. It may be the exploration manager's job to ensure that statutory procedures are properly completed, and he therefore needs to know what those procedures are as early as possible because they could well be the major causes of delays later on.

In addition to studying the information on the area to be investigated, even the most experienced prospector will do well to make a reappraisal of the exploration methods that are available. It is at this stage that the investigation programme is formulated, and a programme that does not include the most appropriate exploration techniques can be a very expensive one.

In recent years work has progressed on the production of maps, such as those accompanying USGS Circular 790 (Muffler, 1979) which present a summary of basic geothermal energy resource information. McEuan et al. (1976) have produced a map of geothermal potential on which they contoured numerical values assigned to various geothermal features. The features to which the values were assigned were all identified by searching a computer file.

The preliminary investigations should include a study of any aerial photographs that may be available. If none exist, and local regulations permit, aerial photography should be instigated. In the hands of a good photo-geologist, high quality aerial photographs can yield much structural and other geological information and, in the absence of adequate maps, are useful for location during field surveys. If aerial photography has to be done, an infra-red imagery survey should be considered at the same time. If no good large-scale maps exist, aerial photography undertaken at this stage should be done to full photogrammetric standards with proper ground control.

The value of an aeromagnetic survey is questionable and, unless it can be combined with any photography that needs to be done, a detailed survey of this type is probably best omitted until ground-based observations suggest that it would serve some purpose.

3.4.2. Geological, Hydrological and Mineralogical Studies

Geological studies, in their broadest sense, play an important part in all phases of geothermal exploration and development from the initial identification of an area worth investigating to the evaluation of individual production wells and the observation of the geological effects of exploitation. The aim of geological evaluation at the earliest stage is to estimate the likelihood of obtaining steam or hot water in usable quantities (Healy, 1970). Sometimes, at this initial stage, a number of geothermal areas will be studied together and one will be chosen for detailed investigation. Unless there is some overriding logistical reason, the choice will be made on this early geological assessment. In addition to basic geological mapping, the first stage will include the mapping of surface geothermal activity such as fumaroles, hot springs, hot seepages, and hot ground. A very rough estimate of the heat discharging through these surface features will indicate an absolute minimum value for the sustainable output of the field and will provide a useful means of comparison between fields. A preliminary chemical analysis (see below) of some of the discharging fluids will give information indicative of physical conditions at depth in the field and of the rock types present there.

An appreciation of the regional geology and tectonic setting is important as this will help in determining whether future detailed investigations should, for example, be directed towards locating permeable aquifers that are essentially horizontal, or fault zones and other areas of fractured rocks. A study of the deposits and hydrothermal alteration around surface features may give an indication of the age of the thermal area, whether it has been more, or less, active in the past, and whether the main centre of activity has shifted.

The location of thermal springs should be compared with the geology (Healy, 1970) and the topography. Concentrations of the hottest features need to be recognized and correlations with other phenomena, such as faulting, sought. In addition to studies of thermal waters, the initial hydrological survey should cover cold springs, streams, local rainfall patterns, water levels in wells and any information that is indicative of underground water movements (Combs and Muffler, 1973). In the initial stages, before deep wells are drilled, it may be possible to obtain information only on the relatively shallow hydrology of the field. As deep exploratory wells are drilled, and water known to originate at a particular depth can be examined, a picture of the deepwater hydrology can be built up. At a later stage when draw-down effects around producing wells can be observed, it may be possible to draw conclusions about the inter-relationship between the hydrology of the shallow, generally cold, water and that of the deep, thermal water.

Geothermal resources can be divided between two broad categories (Healy, 1970; Muffler, 1976) depending upon whether or not they are related to recent (or active) volcanic or intrusive activity, and a different geological approach to each is necessary. In a volcanic or intrusive environment, high-enthalpy surface features such as fumaroles and boiling springs are found, but these are less common in other areas. In areas where there has been no volcanism or intrusion, hydrological investigation will normally centre upon very deep water circulation.

An aspect of the geological investigation of a geothermal field which assumes particular importance at the exploration drilling stage is the location of adequate permeability. Browne (1970) has shown that studies of hydrothermal alteration in core samples from exploration wells can aid the identification of the hotter, more permeable zones during drilling. A knowledge of the conditions in which particular minerals are in equilibrium with the ground water and the surrounding rocks can indicate changes which may have taken place since the mineral was formed. Grindley and Browne (1976) have suggested that some secondary permeability may be associated with natural hydraulic fracturing which could occur if pore fluid pressures exceed the tensile strength of the rocks, and they have described petrological studies of drill cores which can be used to identify this phenomenon.

Subsidence is a common feature in exploited geothermal areas (Hatton, 1970) and the recognition of weak rocks from a study of the geomechanical properties of cores may, with a good distribution of samples, allow the identification of areas most likely to be affected. The measurement of other physical parameters, such as density and magnetic properties of core sample from exploration bores, is also useful because it provides data for the reassessment of geophysical survey interpretation and the refinement of theoretical models designed to represent the geothermal field. Through their role in providing data for improving earlier interpretations, geological studies contribute to the continuing exploration of a field through all its stages of development.

3.4.3 Geochemistry (including Isotope Chemistry)

At a very early stage in the investigation of a geothermal field the analysis of samples of water and gas from hot and cold surface waters and springs, fumaroles and steaming ground can yield information on a number of the field's characteristics, at relatively low cost. As an aid to assessing the energy content of the field, the most important of these is the estimated sub-surface temperature, a parameter which cannot normally be obtained other than by very costly drilling. The same samples will also be a source of information as to the type of system (e.g. vapour-dominated, hot water), the origin of the fluids, geology and mineralogy at depth, possible scaling and corrosion problems in any future development and zones of high vertical permeability, as well as the range of composition and the economic value (if any) of the fluid constituents. The chemistry of geothermal systems is the subject of a detailed review by Ellis and Mahon (1977).

Chemical sampling. The value of any chemical analysis depends on the care and skill with which the various analytical techniques are applied. Equally, in the case of analyses of field samples, the final result is critically determined by the care taken in the original sampling and the subsequent transport and storage of the samples. Ellis and Mahon (1977) make the point that it is always desirable for the analyst to be associated with the original sampling in the field.

Diagnostic chemical information from a geothermal field will normally originate in samples from hot springs and fumaroles. The waters of freely-flowing, boiling springs are likely to have suffered least modification during their passage to the surface and should be sampled whenever possible. Similarly, the largest fumaroles are likely to give the best information because the steam they discharge is less likely to have suffered any change on its way to the surface. Samples collected from hot, but not boiling, springs, smaller fumaroles and steaming ground are much more likely to be contaminated, but with the aid of mixing models (see Fournier, p. 122, this volume) and samples of cold ground water, information indicative of conditions at depth can often be obtained.

The techniques of sampling that have proved to be most useful to New Zealand geochemists have been described by Ellis and Mahon (1977) who also gave details of the appropriate treatment of samples immediately after collection so that they arrive at the laboratory in good condition. The same authors also presented a very useful summary of analytical methods.

Sub-surface temperatures. The estimation of the temperature of deep geothermal waters using the methods of chemical geothermometry relies upon the temperature dependence of the concentration of certain constituents, chemical equilibria between minerals and water, gas solubilities, chemical reactions and isotope distribution between water phases and mineral phases.

The paragraphs which follow provide a summary of geothermometric techniques which are described in detail by Fournier (this volume). On page 114 Fournier gives a table (Table 4.1) summarizing the equations used in the more successful geothermometers.

Before any deep wells are drilled, these methods are applied to samples collected from surface discharge features, and it is important either that these samples have not been altered in their passage to the surface from the hot source or, if they have been altered, that some appropriate correction can be made. When a saturated solution is cooled (as when a sample of deep origin is brought to the surface), unless it displays retrograde solubility, the solute is normally precipitated reflecting a lower solubility at the lower temperature. However, with some materials, such as silica, even though its solubility shows the normal variation with temperature, this does not happen, or it happens very slowly. The solution at the lower temperature thus has a concentration which is the same or only slightly lower than it had at its hotter origin. Thus, assuming that the solution was saturated to begin with, it is possible, in an ideal situation, to estimate the original temperature. In practice, such ideal conditions do not exist. The presence of other chemicals may affect the silica solubility, the hot water may have been cooled by mixing with cold water, or the water may have travelled very slowly to the surface and some precipitation may have occurred. These effects generally reduce the silica concentration, and uncorrected silica temperatures are often low. However, when the pH is high, silica solubility is increased and care must be taken in interpreting the results. If it is possible to trace the history of the water as it rises from its hot source, then some corrections can be made (Truesdell and Fournier, 1976; Fournier, this volume) and a better estimate of temperature obtained.

If, instead of depending on the concentration of a single component, the ratios of the concentrations of a number of components can be used, then, ideally, the problems associated with dilution can be overcome. One such ratio that has yielded consistent temperatures during 15 years of monitoring at Wairakei is that of sodium to potassium (Ellis and Mahon, 1977). The Na/K method makes use of the very strong temperature dependence of the partitioning of these two elements between aluminosilicates and solutions, and the ratio is related to exchange between Na- and K-feldspars (Ellis and Mahon, 1977; Fournier and Truesdell, 1973). With waters of low temperature and high calcium content, anomalous results have been reported, and Fourier and Truesdell (1973) have proposed an empirical Na–K–Ca relationship which takes into account the role of calcium in aluminosilicate reactions (see also Fournier, this volume). This method has also produced some surprising results, and Goff and Donnelly (1978) have suggested that Na–K–Ca temperatures may be dependent on the salinity of the water and the nature of the bedrock with which the waters attained equilibrium. Fournier and Potter (1979) have discussed the adverse effects of magnesium on the Na–K–Ca geothermometer and have suggested corrections that can be made. Laboratory studies by Weissberg and Wilson (1977) have shown that the presence of montmorillonite can cause elevated Na/K ratios, and hence the indicated temperatures may be too low.

Other temperature-dependent chemical equilibria which have been suggested to estimate deep temperatures include the magnesium concentration, the ratio of molal concentration of bicarbonate to carbon dioxide, the calcium sulphate concentration, the ratio of sodium to rubidium and the equilibrium reaction between carbon dioxide and methane. These methods are all qualitative rather than quantitative and can generally be used only for indicating that a particular spring has a broadly hot or cold origin (Ellis and Mahon, 1977). The carbon dioxide–methane method is unique among those listed in that it involves gases and the equilibrium is highly pressure dependent (Hulston, 1964). It is most likely to be reliable only for samples from wells in steam-producing fields.

Another type of chemical geothermometer that has been suggested involves the solid solution of aluminium in quartz crystals (Dennen et al., 1970). A recent field test of this method at Broadlands, New Zealand, (Browne and Wodzicki, 1977) yielded temperatures ranging from 177 to 645°C indicating that, at least in that situation, it is not reliable.

The temperature dependence of the fractionation of the isotopes of a number of the lighter elements provides another potentially very useful group of geothermometers. The isotope ratios that have been used for this purpose are $^{13}C/^{12}C$, $^{18}O/^{16}O$, D/H and $^{34}S/^{32}S$, Summaries of their application and laboratory studies have been given by Ellis and Mahon (1977) and Hulston (1977), and the reader is again referred to Fournier's chapter, this volume.

As is the case with chemical geothermometers, success with the use of isotopes depends on equilibria that are established at depth being maintained during the passage of the fluid to the surface. Unfortunately, the time taken for the reestablishment of a new equilibrium is so very long in some cases (Hulston and McCabe, 1962; Hulston, 1977) that the equilibria observed are believed to be those relating to conditions much deeper in the crust than is of interest is geothermal exploration, and the calculated temperatures are therefore anomalously high. For example, Hulston (1973) estimated the half-time for sulphur isotope exchange between H_2S and $H^{34}SO_4^-$ to be more than six years. This could possibly explain the anomalously high temperature ($370\pm70°C$) calculated by Rafter et al. (1958) from the sulphur isotopic composition of hydrogen sulphide and sulphate at Wairakei. An additional complication with this isotope exchange reaction is that oxidation processes which may take place at the surface could limit its value when applied to hot-spring discharges (Ellis and Mahon, 1977). Further, Kusakabe (1974) in obtaining temperatures similar to those of Rafter et al. (370–400°C) suggested that a kinetic rather than an equilibrium isotope fractionation is involved. More recent studies by Giggenbach (1977) suppose that the equilibration does take place and conclude that, by taking account of the formation of ion pairs at elevated temperatures and the consequent reduction in activity of the sulphate ion, an equilibrium temperature of 260°C can be postulated. It is clear that the processes involved are complex and are not fully understood. Such methods should, therefore, be applied with considerable care, and attention should be paid to the many factors that could possibly affect the results.

The problem of the very long times involved in the reestablishment of equilibrium has been addressed by Hulston (1973, 1977) who has suggested two exchange reactions with acceptably short half-times. One of these is the oxygen isotope exchange between bisulphate and water which has a temperature- and pH-dependent half-time in Wairakei water of about four months. This exchange reaction gives a temperature range of 250–300°C for Wairakei. The other reaction involves hydrogen isotope exchange with water and a very short half-time of approximately one week is indicated (Hulston, 1973). Temperatures of Wairakei water using this method are about 260°C—similar to those measured. The use of oxygen isotopes is described by McKenzie and Truesdell (1977), who also indicate corrections for overcoming the adverse effects of mixing with other waters.

The $^{13}C/^{12}C$ ratio in carbon dioxide and methane has been used as a geothermometer, although as early as 1962 Hulston and McCabe reported a very slow exchange rate, and temperatures from this reaction are commonly too high. The same carbon isotope ratio in carbon dioxide and bicarbonate ions may yield more reasonable temperatures, but carbon dioxide, exsolved when bicarbonate present in deep waters reacts with other weak acids, may present difficulties (Ellis and Mahon, 1977).

Other observations from geochemistry. In addition to indicating temperatures deep in a geothermal system, chemical analysis will also provide information on the various characteristics listed at the beginning of this section. In general these are specified in qualitative terms in the early investigation stages and are used as indicators to the most appropriate course for further investigations. For example, if hot springs are found to be high in chloride, a hot-water system is indicated, and an electrical survey should reveal a low resistivity anomaly.

While broadly indicating the type of a geothermal field, chloride content as observed in surface features is not related in a simple way to subsurface temperatures (Mahon, 1970), and its use as a geothermometer is not common, although Truesdell and Fournier (1976) have suggested a method combining chloride content and enthalpies of the fluids from thermal, cold and mixed springs. Variations of chloride concentration within geothermal fields have been plotted (e.g. Lloyd, 1959), and isochloride contours can be useful in picking out the major areas of upflow of hydrothermal solutions and hence the zones of greatest vertical permeability in the field. When more than one such zone occurs the relative contents of other soluble constituents may indicate whether or not there is more than one hot water source. Multiple sources may indicate different potential production horizons, and an early indication of any possible differences in chemistry is very valuable.

In dealing with geothermometers I have already indicated that, during the passage to the surface, the chemistry of thermal waters may change through dilution or mixing. These are only two of a number of factors which can affect concentrations (see Figure 3.1). To some extent, these variations in concentration can be neglected by using the atomic ratios of chloride to other constituents such as boron, cesium, arsenic or bromine. These ratios are sometimes indicative of the rocks with which the water has had prolonged contact.

3.4.4. Geophysical Methods

Geophysical exploration examines the physical phenomena of the Earth, measures their associated parameters and is applied to the solution of geological problems. In the case of geothermal exploration, the problems are associated with the delineation of geothermal fields and the suitable location of drill-holes through which hot fluids at depth can be tapped. Almost all the methods of geophysics (see any geophysics textbook such as Dobrin, 1976) have been applied to geothermal prospecting, but their effectiveness was greatly increased when the emphasis was shifted from prospecting the geology and the structures that contain the geothermal fluids to prospecting the fluids themselves and concentrating on determining those parameters which are most sensitive to changes in the temperature. The advantages of this change in direction were pointed out by Hatherton et al. (1966) describing the successes achieved in New Zealand when electrical resistivity measurements were used to map variations in the deep groundwater. The assumption that

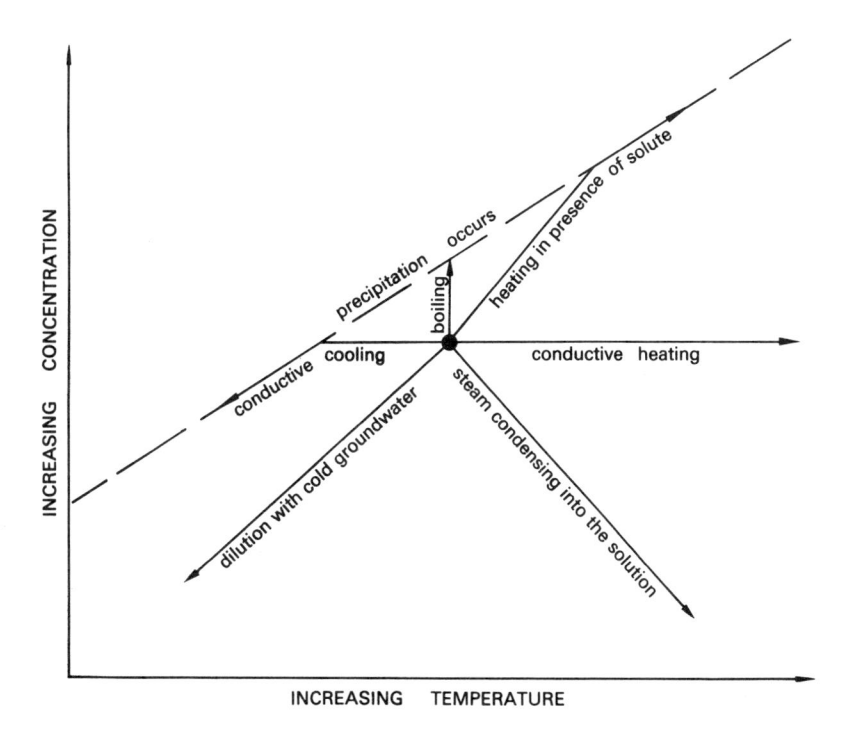

Figure 3.1. Generalized diagram showing the dependence of concentration on temperature under different conditions

hot, chloride waters would be characterized by anomalously low resistivity has proved to be valid (although the converse is not necessarily true), and it is now a standard practice in areas where hot water-dominated fields occur to undertake electrical resistivity surveys using one or more of the currently favoured methods.

Heat, in an accessible and usable form, is the ultimate object of geothermal prospecting, and the measurement of temperature, heat flow or geothermal gradient must form an essential part of any survey. Thermal processes taking place in the Earth's crust are sometimes accompanied by phenomena such as seismic activity and by changes in the magnetic properties and density of the rocks, all of which can be used, with greater or lesser degrees of success, as indicators of the presence of geothermal fields.

Geophysical techniques traditionally employed in the exploration for petroleum such as seismic, gravity and magnetic surveying have been used to detect structures in which hot fluids might accumulate, but generally they neither indicate the actual presence of those fluids nor assist in defining the reservoir (Combs and Muffler, 1973). In recent years interest has grown in geophysical methods which may be capable of detecting, within a known geothermal field, zones where permeability is high and from which good production might be expected.

The methods of the geophysicist also have a role beyond that of defining drilling targets. For example, micro-earthquake surveys can be used to detect fissuring when it is artificially induced by techniques such as hydraulic fracturing. Gravity methods can detect the net mass loss due to discharge from the wells drilled in a geothermal field. Although not strictly 'prospecting', these uses are mentioned here because they depend upon standard exploration techniques and because they need to be applied in the early stages of the investigation of a potential geothermal field to provide the base data against which later results can be assessed.

Combs (1976) makes the point that geophysical methods can often provide information just as effectively and certainly at a lower cost than can a borehole. However, it should always be recognized that any particular technique is not universally applicable, and the methods used should be chosen carefully to suit the situation.

Most geophysical methods display a progressive reduction of resolving capacity as they are extended to probe deeper into the Earth, and this should always be borne in mind when seeking to match the capability of the drill. Several authors have discussed the penetration depth of various electrical prospecting methods (e.g. Keller and Frischknecht, 1966; Roy and Apparao, 1971). Roy and Apparao concluded that the depth of investigation is much smaller than is generally assumed, but they had defined this depth as that at which a thin horizontal layer contributes the *maximum* amount to the total signal at the surface. For most practical purposes it is sufficient to define an effective probing depth below which a boundary cannot be detected (Risk et al., 1970). This depth is typically between three and five times that given by Roy and Apparao (1971) and depends on whether the low resistivity lies above or below the boundary (for a simple two-layer case). A single apparent resistivity observation is, of course, affected by actual resistivities at all levels down to the effective probing depth. It is thus incorrect to say, for example, that a constant separation traverse using the dipole–dipole method indicates resistivity variations *at* a depth equal to half the separation. Standard plotting of 'pseudosections' is no more than a way of presenting data and should not be interpreted to represent the distribution of resistivity at depth.

As for electrical methods, depth of penetration with the 'active' seismic techniques is limited by the geometry of the field set-up and also by the amount of energy put into the ground. Their resolving power, however, is less dependent on the depth of penetration, except that high frequency seismic signals are often preferentially attenuated in geothermal areas and the remaining, deep-penetrating, long wave-length signals will pass around smaller bodies and will not be reflected or refracted by them.

The depths of investigation of microearthquake and other passive seismic methods depend, of course, on the location of the seismic source, but the precision with which its depth is determined relies, as with active seismic methods, upon a detailed knowledge of the velocity structure of the overlying geological sequence.

The only prospecting tool whose depth of penetration is known accurately is the drill, and all deep geophysical results should be re-assessed in the light of drill-hole information. But drilling is also very expensive, and as much information as possible should be gleaned from scientific studies to ensure that all wells are sited most profitably.

Heat flow measurements. It is hardly surprising that geothermal areas are generally characterized by surface heat flows that are in excess of the world-wide average value of about 63 mW/m^2 and that the measurement of this phenomenon usually forms part of a geothermal exploration program (Burgassi et al., 1970; Dawson and Dickinson, 1970; Robertson and Dawson, 1964 and many others). Such surveys can provide information on the areal extent of the geothermal field and, from the total heat flow, on the field's potential thermal output.

As White (1969) has shown, in fields where exploitation on a large-scale is feasible, heat flows up to several thousands times the normal can be expected, but such high values are generally restricted to fairly small areas often around natural discharge features. At the Tauhara Field, New Zealand, Dickinson (1976) has estimated that the heat flow over 85 per cent of the anomalous area is about 2 W/m^2 (32 times normal), and only over 4 per cent of the area does it exceed 1000 times normal.

Conductive heat flow must be obtained from measurements of temperature gradient and thermal conductivity, the latter usually being measured on appropriate rock samples in the laboratory. In a geologically complex area it may be necessary to make many such laboratory measurements if truly representative values are to be obtained, and Goss and Combs (1976) have attempted to avoid such a time-consuming and expensive approach by predicting thermal conductivity from standard geophysical bore-logs. When heat flow rate is determined just below the ground surface, direct measurement of thermal conductivity on appropriate samples is difficult, but if convection is neglected it can be estimated from measurements of diurnal or seasonal temperature variations at different depths (Ingersoll et al., 1955).

In an area where the rocks are uniform and horizontally bedded, the thermal conductivity is normally equally uniform, and the temperature gradient itself therefore displays an areal distribution idential to that of the heat flow. Where such homgeneity does not exist, Sestini (1970) has demonstrated the value of measuring thermal conductivities so that heat flow can be obtained.

Most reported heat flow and temperature gradient studies (e.g. Sestini, 1970, Burgassi et al., 1970, Shanker et al., 1976, Baldi et al., 1976, Blackwell and Morgan, 1976) have been based on data from drill-holes whose depths ranged from 15 to over 100 metres, and it has been argued (Combs and Muffler, 1973) that such depths are necessary to avoid the disturbing near-surface affects of insolation, topography, precipitation and movement of groundwater. However, a temperature gradient survey using measurements made at depths of 15 cm and 1 m (e.g. Thompson et al., 1964) is much less expensive, can provide many more data points in considerably less time and, in an area where large-scale development is feasible, is likely to produce an equally useful definition of the extent of the field. Because of the acknowledged shortcomings of such shallow measurements, no one has ever considered the extrapolation of temperatures to great depths. On the other hand, extrapolation has been done from deep temperature gradient holes and, in the case of a prospect for hot dry rock at Marysville, Montana, with extremely disappointing results (Blackwell and Morgan, 1976).

All inferences as to temperature distributions at depth, whether quantitative or simply indicative, based only on temperature gradients or heat flows (determined in any depth range) require the observed high values to lie directly above the heat source. Geothermal investigations in Chile and elsewhere (Healy and Hochstein, 1974) have shown that the surface expression of geothermal activity is not necessarily directly above the heat source and that hydrothermal waters can flow horizontally for considerable distances. It should always be recognized that geothermal gradients are meaningful only for conductive heat transfer and that vertical, as well as horizontal, convection can upset the extrapolation of temperature information (see e.g. Eckstein, 1979). For example, there may be large heat transfer but zero thermal gradient.

The techniques used in the measurement of temperatures in gradient holes are normally straightforward, and special high temperature equipment is rarely needed. Where temperature gradients are not too high and water stands in the hole without convecting it is possible to obtain a continuous temperature profile. If a high gradient occurs within the hole, it is necessary to prevent convection by filling the hole with bentonite mud or other fluid of a sufficiently high viscosity (Sestini, 1970). The most reliable temperatures are obtained by a fixed array of thermocouples or thermistors left in the hole until it has stabilized before they are read.

If the purpose of the heat flow survey is to estimate the total heat lost by the geothermal system under natural conditions, the conductive contribution estimated from the measurements described above is likely to be of little significance. Dawson and Dickinson (1970) showed that at Wairakei only 3 per cent of the heat loss from the field was by conduction from warm ground while 87 per cent was by convection, evaporation and direct discharge from steaming ground and features such as hot pools, springs and seepages. The remaining 10 per cent was from fumaroles.

Thompson et al. (1964) described techniques for determining heat flow in steaming ground and fumaroles from measurements made with a venturi meter and pitot tube respectively. Evaporative losses from smooth water surfaces are calculated from water and air temperatures using relationships given in the International Critical Tables. Provided that wind speeds are low, Banwell et al. (1957) found this approach to be quite accurate. For boiling pools Dawson (1964) showed that extra heat loss is related directly to the height of boiling. Heat losses through springs and seepages into streams are calculated from temperatures and water flows. Underground flows remain unmeasured, and total heat flow, based on surface discharge only, can be seriously underestimated for this reason.

Remote-sensing methods using imagery in the near or intermediate infra-red region have been applied in a number of geothermal fields (Dickinson, 1976; Noble and Ojiambo, 1976; McNitt, 1976; Friedman et al., 1969; Gomez Valle et al., 1970). Most of these studies have employed infra-red scanners detecting radiated (rather than reflected) energy in the 3- to 5-micron or 8- to 14-micron bands and have been restricted to mapping areas where the temperature is detectably above ambient. Dickinson (1976) has, with the use of suitable ground control, related these surface temperatures to heat flow. The principal value of infra-red lies in its ability to detect previously unknown hot areas in unsurveyed territory as was the case in Kenya (Noble and Ojiambo, 1976), and in such cases it is unlikely that Dickinson's extrapolation could be made. Earlier studies by Hochstein and Dickinson (1970) suggested that aerial infra-red surveying may be a useful method for monitoring changes in geothermal activity in urban areas. Recent work in New Zealand has shown that this might be unreliable because natural variations are sometimes very rapid, and in these cases misleading trends might be inferred unless very frequent surveys have been made. This type of monitoring is probably best performed with a small number of carefully sited holes where temperatures and water levels can be measured much more frequently and cheaply.

Undisturbed underground temperatures are difficult to obtain in completed drill-holes because of convection and flow up or down the hole. Attempts to measure undisturbed temperatures have been made at Yellowstone by White et al. (1975) who recorded bottom-hole temperature at intervals during drilling. Such methods normally prolong the drilling process too much for commercial exploration, but a recent method of analyzing bottom-hole thermal recovery data developed by Albright (1976) may overcome this difficulty in some instances.

Electrical and electromagnetic methods. Of all the geophysical methods that have been used to prospect for geothermal energy, particularly in areas where hot-water fields dominate, those which measure the electrical resistivity at depth in the ground have been the most useful. One of the main reasons for this success is that the resistivity variations observed are related directly to properties of the actual object of the search (hot water) and not only to the host rocks. Thus a resistivity map can indicate the areal extent of a hot water reservoir over whatever depth range the survey technique is able to probe. The resistivity of a body of rock containing an electrolyte depends on the resistivity of the rock itself, the resistivity of the electrolyte and the temperature. The measured resistivity also depends on the porosity and the extent to which the voids are filled with electrolyte.

In a geothermal field the electrolyte involved is the geothermal fluid, and its resistivity is inversely related to the concentration of ions that it carries. The resistivity of the electrolyte is generally much lower than that of the host rock, and it varies inversely with temperature more rapidly than does that of the rock. The temperature dependence of the resistivity decreases with increasing temperature (Keller and Frischknecht, 1966) and according to Risk et al. (1970) can be neglected above 150°C, by comparison with the changes one would expect from variations in porosity. Steam usually displays a high resistivity. Unless a geothermal field is fed by two or more sources with very different chemistries, the differences of resistivity due to ion concentration variations will also be very very small, except near the margins and in those other parts of the field where mixing with less saline groundwater might occur.

Although the interstitial fluids usually have a much lower resistivity than the host rock, this is not necessarily so when clays are abundant. Risk et al. (1970) suggested that at Broadlands the resistivity of some clays might be as low as 0.5 ohm-metres and comparable with values expected for many geothermal fluids (Meidav, 1970). This is especially important in tropical countries where very deep weathering of the rocks may produce a thick clay mantle. The results of an electrical resistivity survey are, therefore, not always unambiguous, but in an area known to contain geothermal phenomena an interpretation can usually be achieved with a high degree of confidence.

In vapour-dominated systems, resistivity data can be more difficult to interpret, and because dry steam has a very high resistivity, sometimes no anomaly at all may be present. However, in such a field, the steam zone may be overlain by a condensate layer as is the case at Kawah Kamojang, Indonesia, where a low electrical resistivity anomaly has been observed by Hochstein (1976) and related to the presence of various sulphates in the waters. This phenomenon had been recognized earlier at Yellowstone National Park by Zohdy et al. (1973) who attributed the low resistivity to the abundance of hot liquid water in the condensate layer.

Most of the early resistivity surveys, especially those done in New Zealand, used direct current in either the *Wenner* or the *Schlumberger* electrode configuration (Banwell and Macdonald, 1965; Hatherton et al., 1966; Macdonald and Muffler, 1972). The usual procedure is to make traverses with a fixed electrode spacing to map resistivity variations down to some (nearly) constant depth below the ground surface. The effective probing depth is related not only to the spacing between the current electrodes but also to the distribution of different resistivities in the ground. It is therefore useful to select a number of critical sites at which to make a series of soundings (measurements with increasing electrode spacing) so that resistivity variations with depth can be determined. These observations will provide an indication of the effective probing depth as well as true resistivity values against which the apparent resistivities, obtained from the constant spacing traverse measurements, can be compared. In an area where the geology displays a marked 'grain' or where many parallel fissures occur, some anisotropy is to be expected in the observed resistivity values (Risk, 1976b). Field observations made with the electrode spreads at right angles and centred on the same spot may detect such anisotropic variation of resistivity, and even if it cannot be fully interpreted, a knowledge of its existence, if large, will draw attention to possible dependence of apparent resistivities upon the alignment of the electrodes.

Because the depth of penetration with Wenner and Schlumberger surveys depends mainly on the current electrode separation, laying out the necessary lengths of wire becomes difficult when very deep probing is required. Electrode separations of several kilometres are possible, but considerable problems of logistics have to be overcome, large power supplies are needed and progress is likely to be very slow. In good field conditions, with conveniently aligned roads or tracks and nearly level topography, a Schlumberger array with a distance between the current electrodes of 2 km ($AB/2 = 1$ km)* can be laid out with comparative ease. Such an array would penetrate to about 1000–1500 m, which is about twice the depth than can easily be achieved in more typical field conditions.

In order to map the resistivity at depths more comparable with those at which productive horizons are commonly encountered in drill-holes in geothermal fields, methods have been used which can probe deeper yet do not suffer the logistical disadvantages of the Wenner and Schlumberger arrays. One such family of methods makes use of the electric field distribution about a dipole source. Two dipole configurations are in common use: one, the *dipole–dipole* in which two similar pairs of closely spaced electrodes (one to carry the current and the other to measure the voltage) are traversed along profiles while all the electrodes are kept collinear; the other, the *bipole–dipole* or *roving dipole* in which one electrode pair (the bipole) usually with a spacing of hundreds of metres is kept fixed while the other pair (with a much smaller spacing of only tens of metres) is shifted around the prospect area.

Risk et al. (1970) used the *roving dipole* technique at Broadlands in an attempt to obtain deep penetration and to assess its usefulness as a reconnaissance prospecting tool. At the outset it appeared very suitable for reconnaissance because the larger current electrodes could remain in place while only the potential electrodes need be moved. They found, and

---

* Schlumberger array electrode spacings are usually given as half the distance between the current electrodes which are conventionally labelled A and B.

it was later described theoretically by Bibby and Risk (1973), that the results could be misleading because they were affected by the positions of the current bipole relative to the field boundaries (Figure 3.2). If a knowledge of the location of a geothermal field's boundaries is to be a prerequisite for the satisfactory siting of the current electrodes, then the method's use for reconnaissance is severely limited.

However, if the boundaries are known, even approximately, the logistic advantages of the method make it attractive for follow-up studies such as the more precise location of the boundaries. Risk (1976a) applied the method to locating as precisely as possible the boundary of the Broadlands geothermal field by setting up a large source bipole straddling the boundary and traversing a receiver array along profiles at right angles to the boundary. The receiver array consisted of two dipoles at right angles to one another so that, as well as an apparent resistivity, an electric field direction could be measured for each array location. It is thought that measurements of this type are sufficiently precise for the method to be used for monitoring (by later re-measurement) lateral movements of the boundary of a geothermal field during exploitation.

The Broadlands geothermal field is one of a very small number where sufficient subsurface temperature data exist to allow a proper comparison with precisely located boundary zones defined by resistivity measurements. Risk (in press) shows a good agreement between the boundary zone and temperatures measured at 500 m depth. There is less agreement with temperatures at 1000 m, the effective penetration depth which might normally be associated with Risk's electrode configuration. This suggests that the depth of penetration may not be as great as is usually assumed and may be closer to those values obtained by Roy and Apparao (1971).

An adaptation of the bipole–dipole method which has also been described by Risk (1976b) and which has the potential for indicating permeable zones at depth, makes use of a large multiple (four electrode) bipole source array. A small rectangular pair of receiving dipoles is shifted around inside the quadrilateral formed by the current electrodes. For each receiver position, the six different current bipole combinations are used in turn so that each measurement is related to a different current path through the ground. This procedure allows the estimation of electrical resistivity anisotropy, and in the survey described by Risk the results were interpreted as being related to fissuring formed during the extrusion of a now buried rhyolite body. A major problem with the interpretation of data derived from the use of this technique is that the six measurements at each receiver site are related to the resistivities at six different places in the area being surveyed. Risk has plotted them all at the receiver site, and although a coherent pattern of resistivity anisotropy can be seen, considerable care must be exercised in any attempt to relate this to the detail of the inferred fissuring. The development of methods capable of indicating permeability at depth is a very high priority for geothermal geophysicists, and this method must be regarded as the most promising electrical one to date.

The bipole–dipole method has been used by Jiracek and Gerety (1978) at Los Alturas Estates, New Mexico, but in this case one of the current electrodes was situated below the ground surface in a well and the other at a distance of over 3 km. Apparent resistivities obtained with a roving receiving dipole with this set-up were compared with those obtained when the down-hole electrode was replaced by one on the surface near the well-head and also with the results of dipole–dipole and Schlumberger surveys. The assumption upon which the technique is based is that if the electric current can be introduced directly into a subsurface low-resistivity layer, then it will be channelled in the layer which will then be easier to locate. The various methods all produced comparable results, but Jiracek and Gerety were unable to make a complete assessment because an adequate

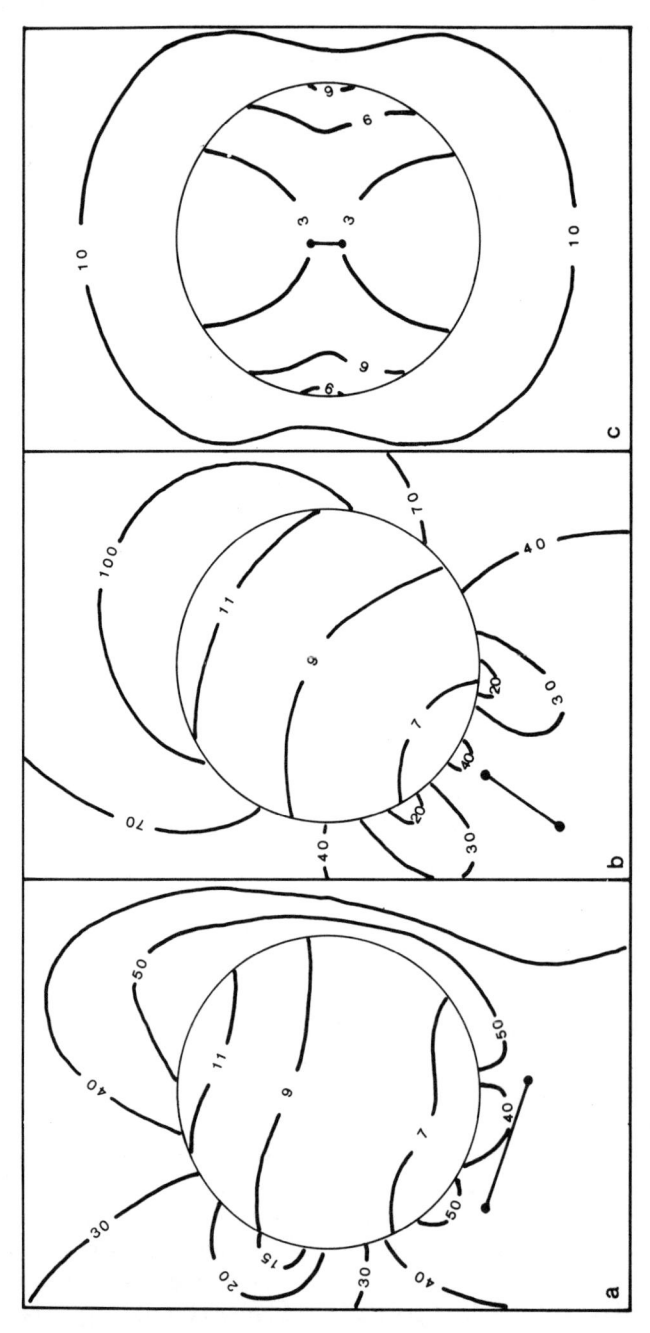

Figure 3.2. Theoretical bipole-dipole resistivity anomalies for different locations of the source bipole relative to field boundaries. The model is a hemispheroid with resistivity = 3 ohm-metres in a surrounding high resistivity. The bipole locations are shown. After Bibby and Risk (1973), reproduced by permission of the Society of Exploration Geophysicists

three-dimensional model was lacking, and no deep temperature measurements were available with which to compare the results.

More orthodox applications of the bipole–dipole method have been made at Raft River, Idaho, by Williams et al. (1976), at Grass Valley, Nevada, by Beyer et al. (1976) and on Oahu, Hawaii by Souto (1978). In most surveys of this type, the data are presented as apparent resistivities which are then interpreted qualitatively, often by comparison with other, more easily interpreted results. Williams et al. (1976), however, presented apparent resistivities normalized to calculated values for a simple model. They then interpreted the residual anomalies as representing differences in resistivities from those of the model.

Apparent resistivity maps drawn from bipole–dipole data frequently appear to bear little visual resemblance to the shape of the electrically anomalous body (Figure 3.2), and it can be dangerous to interpret them intuitively. The interpretation of such data requires very sophisticated methods, and a number of modelling techniques have been applied to the problem (e.g. Bibby and Risk, 1973; Bibby, 1978; Frangos et al., 1978). Of these, the finite element method proposed by Bibby is probably the most versatile although he has restricted his models to those with axial symmetry. Bibby (1977) has also described a way of collating data from several bipole–dipole surveys using anistotropy tensors to give an apparent resistivity map that is least influenced by the relationship of the source bipole to the boundaries of the field.

The collinear *dipole–dipole* traversing method is frequently used for geothermal reconnaissance because it is, logistically, relatively simple, having no great lengths of cable to be laid out, and because considerable theoretical depth penetration can be achieved if the two dipoles are far apart. The method has been applied in many areas including Chile and El Salvador to confirm field boundary locations (McNitt 1976), in Kenya, where its ability to discriminate between vertical and horizontal boundaries made it more suitable than the Schlumberger method (McNitt, 1976), and in Mexico where Garcia (1976) plotted apparent resistivities to a depth of 2000 m. In the Jemez Mountains, New Mexico, Jiracek et al. (1976) made dipole–dipole soundings. Data from dipole–dipole surveys are usually presented in the form of a 'pseudosection' with apparent resistivity values plotted at a point mid-way between the two dipoles and at a depth equal to half the distance between the centre points of the dipoles. Although this conventional approach provides a very simple method of presenting data which can be compared between profiles, it is not clear that the observed apparent resistivity is best plotted in this position. Because the values plotted are not *true* resistivities it is essential to realise that the 'pseudosections' do not represent the real resistivity distribution along the line of the profile. The proper procedure is to use the 'pseudosection' only as an intermediate step for comparing observed and computer-generated apparent resistivity data. The model that has generated the best fitting computer pseudosection is taken as the best interpretation.

In addition to the logistic simplicity of dipole methods, Combs and Muffler (1973) quoted Harthill (1971) as saying that they were insensitive to rugged topography. None of the studies referred to above appear to have made any attempt to make corrections for topographic effects, but Hohmann et al. (1978) have shown from computer studies that a symmetrical valley whose width is about seven times and whose depth is about three times the minimum dipole separation will generate a 'pseudosection' which shows a low resistivity zone in the valley. The low apparent resistivities are approximately one-third of those found outside the valley (400 ohm-metres, compared with over 1200 ohm-metres) and a similar, but reversed effect is noted over a ridge.

In an attempt to obtain deep penetration, Shore (1978) used the two electrode configuration described by Roy and Apparao (1971) as the deepest probing of the various d.c. resistivity arrays. One current and one potential electrode is set up in the prospect area, and the other current and potential electrodes are installed permanently at some very remote location. Shore plotted results at the mid-point between the current and potential electrodes in the prospect area, but it would appear that this procedure is very questionable and it is most unlikely that vertical boundaries can be located reliably.

The usefulness of controlled-source *electromagnetic methods* for geothermal prospecting has been demonstrated in a number of areas. One particular advantage, not possessed by d.c. methods, is that the depth of penetration can be varied by altering the signal frequency and without changing the geometry of the field set-up. Although obtaining only shallow penetration (about 30 metres), Lumb and Macdonald (1970) showed the two-loop technique to be a very satisfactory substitute for shallow, conventional d.c. resistivity and near-surface temperature surveys with the advantages of low cost and high speed. Much deeper penetration is possible with this technique, but because low frequencies are needed, the equipment is rather bulky.

Electromagnetic soundings can be made by using the transient, time-domain technique, TEM (Keller, 1970), in which a step-current is passed through the ground and the resulting transient magnetic field is observed and analyzed. The magnetic field's 'tail', produced by the decay of the transient, is related to resistivities at greater depths as time increases from the onset of the signal. With this technique a sounding can be made from a single observation, but signal-to noise ratios, particularly at the end of the tail, are so low that many repeat readings must be made. TEM methods are able to take advantage of recent advances in sensitive magnetometer design and can use programmable microprocessors for controlling the stacking and processing of the received transient signal (Morrison et al., 1978). These methods show great promise.

Magnetotelluric (MT) methods share many of the logistical advantages of controlled-source electromagnetic methods, but they use naturally occurring magnetic fields and Earth currents. These methods generally fall into one of two categories: low frequency methods which can investigate large-scale features such as deep heat sources (because hot rock is conductive) under geothermal fields (e.g. Hermance, 1973) and methods using signals in the audio-frequency range (AMT) which do not have such deep penetration but can be measured rapidly in reconnaissance surveys (e.g. Whiteford, 1976; Hoover and Long, 1976).

A very high degree of complexity can be applied to the interpretation of magnetotelluric signals if all five natural components (two horizontal telluric and three magnetic) are recorded. Theoretical curves simulating data from complex horizontally stratified rock sequences have been given by Cormy and Musé (1976) and show that relatively small variations in resistivity at depth can be detected. Cormy and Musé do not, however, address the inverse problem of interpreting field data to give an unambiguous solution.

Telluric current measurements alone have also been used for geothermal exploration, and Beyer et al. (1976) have described a convenient three-electrode technique which lends itself to reconnaissance studies. This method, and the magnetotelluric methods, depend on naturally occurring, time-variant magnetic fields, and their interpretation assumes that these fields are normal to the Earth's surface. This is not always true as the location of the source of the electromagnetic waves is not always the same and the repeatability of observations is not always very good in areas of complex structure (Whiteford, 1976a).

Another method which depends on naturally occurring electric fields is the self-potential (SP) method. Use of the method has been described in several recent studies (e.g: Corwin; 1976, Anderson and Johnson, 1978; Zablocki, 1976), but the actual source of the S–P anomalies is not fully understood. Most authors are agreed that they are related to the flow of underground water, but the effect observed appears to vary from field to field. For example, Anderson and Johnson (1978) have found high-amplitude >(1 volt) positive anomalies in Hawaii, but low-amplitude (<100 millivolts) negative anomalies in Nevada, Zablocki (1976) mapped a prominent, positive SP anomaly at East Puna, Hawaii, which formed the basis for the siting of a well that encountered hot fluids at depth, but in many areas 'noise' of amplitude up to 50 mV and of unknown source may confuse the field data.

Microearthquake and other 'passive' seismic observations. Abnormally high levels of groundnoise and the attenuation of seismic waves in geothermal areas have, since the early days of geophysical prospecting for geothermal energy, been noted for the difficulties they create in the interpretation of explosive-type seismic records (Studt, 1958; Modriniak and Studt, 1959). However, it was not until Clacy (1968) made observations of groundnoise which he recorded on a slow-speed tape recorder, designed by Dibble (1964), that any quantitative attempt was made to exploit either of these phenomena. Clacy observed that his records were dominated by a signal whose frequency was about 2 Hz which he ascribed to boiling or some other thermally-associated process in the geothermal field. Although many fields appear to have characteristic ground-noise 'signatures' (e.g. Whiteford, 1970; Douze and Sorrells, 1972; Goforth et al., 1972; Iyer and Hitchcock, 1974; del Pezzo et al., 1975) and mechanisms for their generation have been suggested, this type of microseismicity is not quantitatively understood (Ehara and Yuhara, 1978). Indeed, Butler and Brown (1978) have referred to formidable difficulties in testing the hypothesis that geothermal resources are, in fact, subsurface seismic emitters.

The attenuation and source depth of seismic ground-noise have been estimated by Whiteford (1976) from data for Wairakei and Waiotapu by comparing observed amplitudes of ground particle velocity with those predicted by a theoretical equation. The greatest depth that he obtained was 200 ± 100 m at Waiotapu, which suggests that the ground-noise might be associated with some relatively near-surface thermal process, perhaps boiling. An alternative mechanism suggested by Iyer and Hitchcock (1976), who found that the noise originated 'some depth beneath the surface', is the circulation of fluids at high temperature and pressure. At Long Valley, California, Iyer and Hitchcock (1976) concluded that the seismic noise anomaly is controlled by wave amplification in an alluvial basin. However, such amplification is well-known where seismic waves travel through thick, unconsolidated strata and cannot be considered diagnostic of a geothermal field.

At East Mesa, California, microearthquake studies, as well as ground-noise surveys (Combs and Hadley, 1977), have been interpreted to locate an active fault which has played an important part in the sitting of production and injection wells (Swanberg, 1976). Additional microearthquake studies made since 1976 have failed to show any activity on the East Mesa fault, and McEvilly et al. (1978) have suggested that the shocks used to define the fault might in fact be associated with the Brawley fault. Whether this, or simply a cessation of activity, is the true explanation of the discrepancy, it is clear that ambiguity may exist in those areas where microearthquake activity appears to be associated with thermal areas. At Ahuachapán, El Salvador, Ward and Jacob (1971) have identified a zone of active faulting in which successful production wells have been drilled. Some geothermal areas such as Broadlands, New Zealand, have been reported as being aseismic (Evison et al., 1976), but the East Mesa experience could indicate that such a lack of seismic activity is not permanent.

In contrast with the low level, almost continuous seismic ground noise which is throught to originate in some process associated with thermal fluids, the small, discrete, less frequent microearthquakes are generally identified with small movements or fracturing of the rocks at depth. In most studies microearthquakes are identified with movement along faults (e.g. Hamilton and Muffler, 1972; Ward and Björnsson, 1971; Ward and Jacob, 1971; Combs, 1974), although Quillin and Combs (1978) have suggested that some very small events (nanoearthquakes) at Radium Springs, New Mexico, may be caused by natural hydraulic fractures occurring when pore pressures increased by rising temperatures in areas of very high thermal gradient exceed the tensile strength of the rock.

The use of teleseismic P-wave delays in geothermal exploration has been discussed by Steeples and Iyer (1976) and Iyer et al. (1979) who were able, with this technique, to produce conceptual models of the large-scale sub-surface structure showing large zones extending through the Earth's crust into the mantle where seismic velocities may be reduced by 10 or 25 per cent. Steeples and Iyer (1976) have used P-wave delays to support a gravity-based model of crustal thinning at Imperial Valley, California, but detailed interpretations of this type, using small delay-time differences ($<0 \cdot 1$ s) require a much greater knowledge of the geology than is usually available at the exploration stage of geothermal development. Perhaps the greatest value of this method is that it may indicate large volumes of rock which may be interpreted as being partially melted and hence act as a substantial source of heat for any overlying geothermal field.

The attenuation of seismic waves provides another means of obtaining information on the large-scale structure of geothermal areas. Young and Ward (1978) and Combs and Jarzobek (1978) have observed this phenomenon but have not been able to produce completely unambiguous results. As with P-wave delays, a detailed interpretation of attenuation data may require too great a knowledge of the geology to make it useful during exploration.

Because the large-scale withdrawal or injection of geothermal fluids may be capable of increasing the seismicity of a geothermal area, seismic surveys are sometimes established during well testing (e.g. Cameli and Carabelli, 1976). Although a seismic (or microseismic) survey during the early investigations may not be able to indicate the likelihood of increased seismicity during production, it will provide a pre-development base with which later surveys can be compared. Similarly, observations made at this early stage will also form a useful background against which to view seismic records made during attempts to induce hydraulic fracturing.

Refraction and reflection seismic surveys. Seismic prospecting surveys using explosive sources have been described for a number of fields (e.g. Modriniak and Studt, 1959; Hayakawa and Mori, 1962; Hayakawa, 1970; Hochstein and Hunt, 1970), but have not been used to any great extent since the emphasis in exploration was shifted from 'structural' to 'geothermal' mapping (Hatherton et al., 1966). Ground noise and attenuation of the seismic signal made interpretation difficult and these were further contributing factors in the loss of favour experienced by these methods. Hochstein and Hunt (1970) at Broadlands, New Zealand, found that seismic reflections were difficult to interpret but that refraction records were more satisfactory, and they were able to map the upper surfaces of two buried rhyolite flows. However, Hayakawa (1970), after first experiencing difficulties with reflection data, applied digital stacking techniques to remove noise and multiple reflections and obtained a suitable interpretation for the Matsukawa geothermal area, Japan. Hayakawa's success with this approach suggests that today's more sophisticated techniques using coherent, controlled energy sources, such as Vibroseis, might be applied to advantage in those geothermal fields where a knowledge of the geological structure would be an aid to defining areas where the most productive wells might be sited. In particular, such techniques might allow faults to be accurately located.

Hochstein and Hunt (1970) and Hayakawa (1970) have investigated the attenuation of seismic waves in geothermal areas, and Combs and Jarzabek (1978) made spectral analyses of records from explosive shots which they used for calibration in observations of the attenuation of seismic waves from earthquakes. Although Hochstein and Hunt (1970) found that the attenuation constant within the Broadlands field was generally greater than that outside (2×10^{-3} m^{-1} compared with $<2 \times 10^{-3}$ m^{-1}), they concluded that little use could be made of this parameter because different values were obtained when shot point and geophone were interchanged. Hayakawa (1970) found that higher frequencies suffered the greater attenuation, a result confirmed by Combs and Jarzabek (1978).

Gravity and magnetic methods. The role of gravity surveys in geothermal prospecting appeared to become rather insignificant when exploration studies began to look at the geothermal fluids rather than the geological structure that contained them (see Hatherton et al., 1966). In one of the few gravity surveys described at the United Nations geothermal symposium held in Pisa in 1970, Hochstein and Hunt (1970) concluded that their results cast doubts on the utility of structural concepts as a guide to the finding of geothermal fluid. However, they did observe that part of the positive residual gravity anomaly at Broadlands was due to anomalously dense rocks lying between the ground surface and the greywacke basement and that the higher density was probably due to hydrothermal alteration. Meidav (1970) also noted positive residual gravity anomalies in the Imperial Valley, California, ascribing them to the metamorphism of loosely consolidated sediments by rising plumes of hot water.

At the second United Nations geothermal symposium no fewer than fourteen exploration studies included gravity surveys, and with one exception, positive gravity residuals were located in or near the geothermal areas. The exception was at The Geysers–Clear Lake region where Isherwood (1976) attributed negative gravity residuals to a magma chamber at depth and a steam-saturated reservoir structure within 1·5 km of the surface. In several geothermal regions, such as north-central Kyushu (Yamasaki and Hayashi, 1976), Cesano, Italy (Baldi et al., 1976) and Kizildere, Turkey (Tezcan, 1976) gravity anomalies were associated with the general background geology or with buried topography and were not given any special geothermal significance. At the East Mesa geothermal field, California, Swanberg (1976) attributed a residual gravity 'high' to possible silicification and low-grade metamorphism of sediments. In Japan, in the Kurikoma field, the surface manifestations of geothermal activity were also related to positive gravity *residual* anomalies but were found to lie on the flanks of negative *Bouguer* anomalies (Baba, 1976).

The range of associations between gravity and features associated with geothermal areas, combined with the knowledge that similar anomalies can occur in non-geothermal areas, also suggests that considerable care must be exercised in the interpretation of these surveys, and it is unlikely that a gravity survey alone will ever delineate a geothermal field.

Following the drilling of an extremely good well (capable of producing 250 MW of heat) which taps a fissure in the greywacke basement at Kawerau, New Zealand (Kear et al., in press) there has been some renewed interest in re-assessing old gravity surveys. In fields where such good production may be found by drilling through basement fissures, 'structural' methods such as gravity surveys assume a greater importance, but only if features associated with the fissures are sufficiently large or near to the surface that they can be detected. Hochstein and Hunt (1970) found at Broadlands that, although they could obtain a generalized picture of the basement relief, they were unable to identify any detail. They also found that density variations in the overlying volcanic rocks not only masked the detail in the basement but also made more difficult the determination of the depth to the basement surface.

One of the most detailed recent gravity studies in a geothermal area is of The Geysers–Clear Lake region described by Isherwood (1976) and was accompanied by an aeromagnetic survey. It was, therefore, possible to construct a pseudogravity' anomaly map from the magnetic data with which could be tested the hypothesis that both gravity and magnetic anomalies are caused by the same structures. Isherwood found this not to be so and concluded, as noted above, that part of the negative gravity anomaly is caused by a stream-filled zone.

The magnetic survey at The Geysers–Clear Lake area appears to have responded to the complex geology and shows the same dominant NW–SE trend. Hochstein and Hunt (1970) found a similar complexity in the Broadlands data. When the near-surface effects were removed (for both Broadlands and The Geysers) the resulting amomalies still could not be related exactly to the geothermal fields. In the Broadlands magnetic results (as at Kawerau; Studt, 1958) a negative magnetic anomaly occurs which might in part be explained by the presence of rocks whose magnetic properties have been modified as a result of the hydrothermal alteration of the magnetic minerals. However there is no clear correlation with the field as defined by resistivity, and this cannot simply be explained by the dipolar nature of the magnetic anomaly.

Negative magnetic anomalies have been identified with geothermal fields in Iceland at Krafla and Námafjall (Pálmason, 1976). Such anomalies are not present in all Icelandic fields, and Pálmason advised against undertaking an extensive aeromagnetic survey without first making a pilot survey on the ground. Pálmason also referred to the use of magnetic surveys for tracing dykes and faults which control the flow of low-temperature thermal water to the surface.

In general, one must conclude that gravity and magnetic surveys are still unlikely to provide definitive information. In many instances the results will be complex and impossible to interpret fully without a detailed knowledge of the geology. However, both surveys can be done fairly cheaply, and the more expensive topographic levelling that is a necessary adjunct to the gravity survey should be undertaken in any case as a basis for future engineering work.

Even though a gravity survey may supply little useful information for the exploration stage, this initial study will provide the necessary base with which future surveys might be compared during exploitation. Hunt (1970, 1977) has demonstrated the great value of gravity monitoring, combined with repeated precise levelling, for determining the net mass loss from the Wairakei geothermal field. This application of gravity surveying is associated with geothermal reservoir management rather than with exploration, but it appears to be such a useful technique that it must justify a gravity survey at the exploration stage. As Hunt (1977) found, the gravity changes between surveys may be very small (0·1 μN/kg at some sites), and it is essential that repeat readings are made at exactly the same place and that elevations are precisely measured on each occasion. In order to ensure this, it is preferable to make gravity observations over a network of benchmarks.

3.4.5. Exploration Drilling

The drilling of exploratory holes is the most important phase in the prospecting for geothermal energy, and, because the scientific studies described above were primarily directed towards the sitting of these holes, this phase marks the end of the prospecting. On the other hand, the drilling of the first well marks the beginning of the main exploratory stage when the geographical limits of the field are tested and the earlier predictions of the field's capacity to produce steam or water are confirmed (or rebutted). Drilling exploration wells is also the most expensive phase of the prospecting and it is important that every well is drilled for a purpose and that the maximum possible amount of information is gained from it.

It is often desirable, for what might broadly be called 'political reasons', that the first well to be drilled in an area should demonstrate that the field is capable of producing steam or water in suitable quantities. The first well is therefore drilled at one of the most promising locations, and if it is a failure, then clearly a re-evaluation of the scientific data is called for. Subsequent wells (following a successful first well) are sited to confirm the findings of the first well in other potentially productive parts of the field, to test for the boundaries of the field as defined by the scientific studies and finally to test hypotheses that certain areas within the field will provide the best wells. Wells in the first two groups are sited on the basis of the scientific information, but the siting of those in the last group should also take account of the information from the earlier holes.

The most useful guide to the siting of an initial exploration well is the coincidence of high temperatures indicated by resistivity or, in some cases, heat flow surveys, and the possibility of permeability indicated by the presence of faults or hot springs whose chemistry indicates that the water has travelled rapidly to the surface.

In New Zealand, the standard practice is to drill and complete exploration wells in the same way as production wells, the main reason being that if such a well is successful it can be used for production later. From their experience with slim exploration holes in Iceland, Arnórsson et al. (1976) found that many technical problems can arise with such holes and they are not, in fact, as inexpensive as would be expected by comparison with production-sized holes. In particular they found that drilling times were extended because of slow penetration rates at the top of the hole when the shorter, light drill stem does not allow the drill-bit to be fully loaded and because they experienced difficulty in stopping circulation losses of the drilling fluid. At Raft River, Idaho, where geothermal fluid temperatures only reached 150°C, Miller et al. (1978) ran and cemented only the surface casing of their exploratory wells unless hot water was found, thus reducing the costs of non-productive wells. If a resource was located, the well was cooled and the production casing run and cemented in place. Because prospecting wells are intended to provide valid information on conditions at depth in the geothermal field, it is important that narrow fissures in the production horizons are not permanently clogged with mud. Although such mud damage may not be a big problem where high permeabilities are found, in less permeable areas it could lead to a serious misinterpretation of the potential productive capacity of a field.

Exploration well assessment. During drilling three types of information are obtained from a well: geological, from cuttings and cores; temperature from returned mud temperatures and mud flow rates; and permeability from circulation losses. Coring at regular intervals and at lithological boundaries as indicated by cuttings or changes in drilling rate is a routine practice in New Zealand, and samples are immediately subjected to mineralogical analysis from which is gained additional information on likely temperatures and permeabilities as well as a picture of the geological sequence (Browne, 1970).

Immediately after drilling and when drilling mud has been flushed out of the well, the overall permeability of the well is tested by measuring the pressure either at the well-head or at some fixed depth while cold water is injected over a range of flow rates (Wainwright, 1970). If the pressure rises appreciably as the flow rate increases a poor permeability is indicated; if it remains nearly constant then the well is highly permeable. The location of the permeable zone is determined from down-hole temperatures recorded during the cold water injection. Temperatures and pressures measured in the well over a period of about 30 days while it is heating up after the injection test in a static (shut-in) condition also indicate the location of the more permeable zones. These measurements also show whether the well has penetrated a zone of boiling or single-phase water or if steam or gas is present in large quantities (Ellis and Mahon, 1977). As White et al. (1975) clearly demonstrated, down-hole pressures may not be representative of pre-drilling conditions because the well may have provided an extra channel of communication between aquifers and circulation may take place in a shut-in well. If this occurs, the effect may be seen in a constant temperature over the interval where flow takes place (Wainwright, 1970) or it may be seen in down-hole flow measurements which are now being made in geothermal wells in New Zealand (M. Syms, personal commun.). A flow of this nature within the well would also upset the measurements of temperature gradient described by Albright (1976).

Tests of the well's output, made after it has heated up, consist of mass flow and heat measurements over a range of discharge pressures. Output measurements in wells which discharge a single phase, either steam or water, are easily made, but with two-phase mixtures methods such as James's (1970) lip-pressure method or other techniques involving steam separation or calorimeter tests (Wainwright, 1970) must be applied. Measurements of pressure drawdown from observations while a well is discharging are indicative of permeability in the vicinity of the well and, in some cases, can differentiate between fissure- and porous-bed permeability (James, 1976).

As exploratory drilling proceeds, down-hole measurements from an increasing number of wells can be combined to form a picture of pressure and temperature distribution in the field, and the siting of the later prospecting wells will be guided by this information.

With the drilling of wells comes the opportunity of testing many of the earlier interpretations of geophysical and geochemical evidence and of refining them where necessary. In addition to the re-assessment of resistivity and various 'structural' surveys following the measurement of physical properties of cores, the down-hole temperatures make possible a re-appraisal of the surface heat flow and temperature gradient studies. The chemical analysis of waters uncontaminated by near surface groundwaters is possible once the well has discharged and the drilling fluids and cold water from the completion tests have been flushed out. Samples for chemical analysis can be collected either at the surface from a flowing well or at selected depths using a down-hole sampling bottle (e.g., Klyen, 1973). Chemical analyses of down-hole samples permit a fresh look at the underground temperature predictions and indicate whether the well has indeed penetrated the hottest part of the field.

Many of the well-logging methods, such as electrical, neutron-gamma, density and sonic logging, all of which could be used to advantage in geothermal wells, often cannot be used without first cooling down the wells. As soon as techniques and materials are developed to overcome these problems, well-logging will offer an alternative to measuring the various properties of core samples and, furthermore, should provide more representative data than those currently available from single cores at arbitrarily chosen depths.

3.5. Modelling

From the very earliest stage the geothermal prospector begins to consider a conceptual model of the area being explored, and it undergoes progressive refinement as more information becomes available. Generally, the early model will be very simple and based on analogies with geothermal fields that are better understood. By the time a resistivity or other survey is at the stage when the outline of a field can be defined approximately and geothermometers have indicated a possible reservoir temperature, a very simple model can yield an estimate of the potential power-generating capacity such as that presented by Bolton and Studt (1977). Little advance can be made on simple models until drilling has commenced and output tests performed on the wells. From this point, predictive modelling becomes possible and can be applied to the siting of further wells and to improving estimates of the field's capacity.

Measurement of temperature and pressure in both static and discharging well conditions along with draw-down, pressure transient and interference tests provide the data necessary for the formulation of models showing reservoir characteristics. From these models optimum discharge rates and pressures can be suggested and recommendations made for the optimum siting and spacing of production wells.

After a number of exploration wells have been discharged for some time (preferably months) their run-down characteristics become apparent, and sometimes the nature of the interaction between geothermal and cold groundwaters can be evaluated. This information, incorporated in a model of the field, can be used to suggest, in conjunction with the geological evidence, the most suitable sites, for example, for the large-scale injection of separated water when the field goes into full production. The field models formulated during the exploration drilling phase will continue to be refined as more information is gleaned from production experience and will become the main tool in continuing evolution of a field management programme (see also Garg and Kassoy, this volume). In this sense, geothermal exploration comes to an end when the final production well is shut in.

Acknowledgements

I would like to thank Tony Mahon, George Risk, Lynette Clelland, Frank Studt and Dick Dale for the critical comments and helpful suggestions they made at various stages during the preparation of the manuscript. I must also express gratitude to my family for their patience during the writing of some of the more difficult parts. The manuscript was typed by Mrs Pat Gibbons who gave it very high priority in the face of much competition.

References

Albright, J. N., 1976. 'A new and more accurate method for the direct measurement of earth temperature gradients in deep boreholes', *2nd U.N. Symposium, Proceedings*, 2, 847–851.

Anderson, L. A., and Johnson, G. R., 1978. 'Some observations of the self-potential effect in geothermal areas in Hawaii and Nevada', *Geothermal Resources Council, Transactions*, 2, 9–12.

Anórsson, S. Björnsson, A., Gíslason, G., and Gudmundsson, G., 1976. 'Systematic exploration of the Krisuvik high temperature area, Reykjanes Peninsula, Iceland', *2nd U.N. Symposium, Proceedings*, 2, 853–864.

Baba, K., 1976. 'Gravimetric survey of geothermal areas in Kurikoma and elsewhere in Japan', *2nd U.N. Symposium, Proceedings*, 2, 865–870.

Baldi, P., Cameli, G. M., Locardi, E., Moutón, J. and Scandellari, F., 1976. 'Geology and Geophysics of the Cesano geothermal field', *2nd U.N. Symposium, Proceedings*, 2, 871–881.

Banwell, C. J., Cooper, E. R., Thompson, G. E. K. and McCree, K. J., 1957. 'Physics of the New Zealand Thermal Area', *N.Z. Department of Scientific and Industrial Research, Bull.* 123, 109 pp.

Banwell, C. J. and Macdonald, W. J. P., 1965, 'Resistivity surveying in New Zealand thermal areas', *Eighth Commonwealth Mining and Metall. Congress*, New Zealand section, Paper No. 213, 1–7.

Beyer, H., Morrison, H. F. and Dey, A., 1976. 'Electrical exploration of geothermal systems in the Basin and Range valleys of Nevada', *2nd U.N. Symposium, Proceedings*, **2**, 889–894.

Bibby, H. M., 1977. 'The apparent resistivity tensor', *Geophysics*, **42**, 1258–1261.

Bibby, H. M., 1978. 'Direct current resistivity modelling for axially symmetric bodies using the finite element method', *Geophysics*, **43**, 550–562.

Bibby, H. M., and Risk, G. F., 1973. 'Interpretation of dipole-dipole resistivity surveys using a hemispheroidal model', *Geophysics*, **38**, 719–736.

Blackwell, D. D., and Morgan, P., 1976. 'Geological and geophysical exploration of the Marysville geothermal area, Montana, U.S.A.', *2nd U.N. Symposium. Proceedings*, **2**, 895–902.

Bolton, R. S., and Studt, F. E., 1977. 'Investigation and development of New Zealand's geothermal resources', Paper presented to the 10th World Energy Conference, Istanbul.

Browne, P. R. L., 1970. 'Hydrothermal alteration as an aid in investigating geothermal fields', *Geothermics*, special issue 2, **2**, 564–570.

Browne, P. R. L., and Wodzicki, A., 1977. 'The aluminium-in-quartz geothermometer: a field test.' In *Geochemistry 1977*, A. J. Ellis (Comp.), *Dept. Sci. and Industr. Res. Bull*, **218**, New Zealand, 35–36.

Burgassi, P. D., Ceron, P., Ferrara, G. C., Sestini, G., and Toro, B., 1970. 'Geothermal gradient and heat flow in the Radicofani region (east to Monte Amiata, Italy)', *Geothermics*, special issue 2, **2** 443–449.

Butler, D., and Brown, P. L., 1978. 'Ambient ground-noise measurements and exploration for geothermal resources', *Geothermal Resources Council, Transactions*, **2**, 55–58.

Cameli, G. M., and Carabelli, E., 1976. 'Seismic control during a reinjection experiment in the Viterbo region (Central Italy)', *2nd U.N. Symposium, Proceedings*, **2**, 1329–1334.

Clacy, G. R. T., 1968. 'Geothermal ground noise amplitude and frequency spectra in the New Zealand volcanic region', *Journ. Geophysical Research*, **73**, 5377–5383.

Combs, J., 1974. 'Microearthquake investigation of the Mesa geothermal anomaly, Imperial Valley, California—Final Report', USDI Bureau of Reclamation, 33 pp.

Combs, J., 1976. 'Summary of section IV: Geophysical techniques in exploration', *2nd U.N. Symposium, Proceedings*, **1**, lxxxi–lxxxvi.

Combs, J., and Hadley, D., 1977. 'Microearthquake investigation of the Mesa geothermal anomaly, Imperial Valley, California', *Geophysics*, **42**, 17–33.

Combs, J., and Jarzabek, D., 1978. 'Seismic wave attenuation anomalies in the East Mesa geothermal field, Imperial Valley, California: preliminary results', *Geothermal Resources Council, Transactions*, **2**, 109–112.

Combs, J., and Muffler, L. J. P., 1973. 'Exploration for geothermal resources.' In *Geothermal Energy*, Kruger and Otte, (Eds.), Stanford, California, 95–128.

Cormy, G., and Musé, L., 1976. 'Utilisation de la MT-5-EX en prospection géothermique', *2nd U.N. Symposium, Proceedings*, **2**, 929–933. (English translation of text only; 933–935).

Corwin, R. F., 1976. 'Self-potential exploration for geothermal reservoirs', *2nd U.N. Symposium, Proceedings*, **2**, 937–945.

Dawson, G. B., 1964. 'The nature and assessment of heat flow from hydrothermal areas', *N.Z. Journal of Geology and Geophysics*, **7**, 155–171.

Dawson, G. B., and Dickinson, D. J., 1970. 'Heat flow studies in thermal areas of the North Island of New Zealand', *Geothermics*, special issue 2, **2**, 466–473.

del Pezzo, E., Guerra, I., Luongo, G., and Scarpa, R., 1975. 'Seismic Noise measurements in the Mt Amiata geothermal area, Italy', *Geothermics*, **4**, 40–43.

Dennen, W. H., Blackburn, W. H., and Quesada, A., 1970. 'Aluminium in quartz as a geothermometer', *Contributions to Mineralogy and Petrology*, **27**, 332–342.

Dibble, R. R., 1964. 'A portable slow motion tape recorder for geophysical purposes', *N.Z. Journ. Geology and Geophysics*, **7**, 445–465.

Dickinson, D. J., 1976. 'An airborne infrared survey of the Tauhara geothermal field, New Zealand', *2nd U.N. Symposium, Proceedings*, **2**, 955–961.

Dobrin, M. G., 1976. *Introduction to Geophysical Prospecting*, 3rd Ed., McGraw-Hill, New York, 630 pp.

Douze, E. J., and Sorrells, G. G., 1972. 'Geothermal ground-noise surveys', *Geophysics*, **37**, 813–824

Eckstein, Y., 1979. 'Heat flow and the hydrologic cycle: Examples from Israel.' In *Terrestrial Heat Flow in Europe* V. Čermak and L. Rybach (Eds.), Springer-Verlag, Berlin, 88–97.

Ehara, S., and Yuhara, K., 1978. 'Seismic noise measurements at some geothermal areas in Japan', *Geothermal Resources Council, Transactions*, **2**, 171–172.

Ellis, A. J., and Mahon, W. A. J., 1977. *Chemistry and Geothermal Systems*, Academic Press, New York, 392 pp.

Evison, F. F., Robinson, R., and Arabasz, W. J., 1976. 'Microearthquakes, geothermal activity and structure, central North Island, New Zealand', *N.Z. Journ. Geology and Geophysics*, **19**, 625–637.

Fournier, R. O., and Potter, R. W. II, 1979. 'Magnesium correction to the Na–K–Ca chemical geothermometer', *Geochim. Cosmochim. Acta*, **43**, 1543–1550.

Fournier, R. O., and Truesdell, A. H., 1973. 'An empirical Na–K–Ca geothermometer for natural waters', *Geochim. Cosmochim. Acta*, **37**, 1255–1275.

Frangos, W., Hohmann, G. W., and Jiracek, G. E., 1978. 'Three-dimensional roving dipole interpretation', *Geothermal Resources Council, Transactions*, **2**, 195–198.

Friedman, J. D., Williams, R. J. Jr., Pálmason, G., and Miller, C. D., 1969. 'Infrared surveys in Iceland—preliminary report', *U.S. Geol. Survey Prof. Paper* 650-C, C89–C105.

Giggenbach, W. F., 1977. 'The isotopic composition of sulphur in sedimentary rocks bordering the Taupo Volcanic zone.' In *Geochemistry 1977*, A. J. Ellis (Compiler), Dept. Sci. and Industr. Res. Bull, **218**, New Zealand, 57–64.

Goff, F. E., and Donnelly, J. M., 1978. 'The influence of P_{CO_2}, salinity, and bedrock type on Na–K–Ca geothermometer as applied in the Clear Lake geothermal region, California', *Geothermal Resources Council, Transactions*, **2**, 211–214.

Goforth, T. T., Douze, E. J., and Sorrells, G. G., 1972 'Seismic noise measurements in a geothermal area', *Geophysical Prospecting*, **20**, 76–82.

Gómez Valle, R. G., Friedman, J. D., Gawarecki, S. J., and Banwell, C. J., 1970. 'Photogeologic and thermal infrared reconnaissance surveys of the Los Negrito—Ixtlan de los Hervores geothermal area, Michoacan, Mexico', *Geothermics*, special issue 2, **2**, 381–398.

Goss, R., and Combs, J., 1976. 'Thermal conductivity measurements and prediction from geophysical well-log parameters with borehole application', *2nd U.N. Symposium, Proceedings*, **2**, 1019–1027.

Grindley, G. W., and Browne, P. R. L., 1976. 'Structural and hydrological factors controlling, the permeabilities of some hot-water geothermal fields', *2nd U.N. Symposium, Proceedings*, **1**, 377–386.

Hamilton, R. M., and Muffler, L. J. P., 1972. 'Microearthquakes at The Geysers geothermal area, California', *Journ. Geophysical Research*, **77**, 2081–2086.

Harthill, N., 1971. 'Geophysical prospecting for geothermal energy', *The Mines Magazine*, June 1971, Colorado School of Mines. 13–18.

Hatherton, T., Macdonald, W. J. P., and Thompson, G. E. K., 1966. 'Geophysical methods in geothermal prospecting in New Zealand', *Bull. Volcanologique*, **29**, 485–498.

Hatton, J. W., 1970. 'Ground subsidence of a geothermal field during exploitation', *Geothermics*, special issue 2, **2**, 1294–1296.

Hayakawa, M., 1970. 'The study of underground structure and geophysical state in geothermal areas by sesmic exploration', *Geothermics*, special issue 2, **2**, 347–357.

Hayakawa, M., and Mori, K., 1962. 'Seismic prospecting at Matsukawa geothermal area, Iwate prefecture', *Bull. Geological Survey of Japan*, **13**, 643–648.

Healy, J., 1970. 'Pre-investigation geological appraisal of geothermal fields', *Geothermics*, special issue 2, **2**, 571–577.

Healy, J., 1976. 'Geothermal fields in zones of recent volcanism', *2nd U.N. Symposium, Proceedings*, **1**, 415–422.

Healy, J., and Hochstein, M. P., 1974. 'Horizontal flows in geothermal systems', *Journ. Hydrology (New Zealand)*, **12**, 71–81.

Hermance, J. F., 1973. 'An electrical model for the sub-Icelandic crust', *Geophysics*, **38**, 3–13.

Hochstein, M. P., 1976, 'Geophysical exploration of the Kawah Kamojang geothermal field, West Java', *2nd U.N. Symposium, Proceedings*, **2**, 1049–1058.

Hochstein, M. P., and Dickinson, D. J., 1970. 'Infrared remote sensing of thermal ground in the Taupo region, New Zealand', *Geothermics*, special issue 2, **2**, 420–423.

Hochstein, M. P., and Hunt, T. M., 1970. 'Seismic, gravity and magnetic studies, Broadlands geothermal field', *Geothermics*, special issue 2, **2**, 333–346.

Hohmann, G. W., Fox, R. C., and Rijo, L., 1978. 'Topographic effects in resistivity surveys', *Geothermal Resources Council, Transactions*, **2**, 287–290.

Hoover, D. B., and Long, C. L., 1976. 'Audio-magnetotelluric methods in reconnaissance geothermal exploration' *2nd U.N. Symposium, Proceedings*, **2**, 1059–1064.

Hulston, J. R., 1964. 'Isotope geology in the hydrothermal areas of New Zealand', *Proc. U.N. Conference on New Sources of Energy*, Rome, **2**, 259–264.

Hulston, J. R., 1973. 'Estimation of underground temperature distributions in active hydrothermal systems using stable isotope equilibria (Abstract)', *2nd Conf. Australian and New Zealand Soc. for Mass Spectrometry*, Melbourne, 14.2.

Hulston, J. R., 1977, 'Isotope work applied to geothermal systems at the Institute of Nuclear Sciences, New Zealand', *Geothermics*, **5**, 89–96.

Hulston, J. R., and McCabe, W. J., 1962, 'Mass spectrometer measurements in the thermal areas of New Zealand—2. Carbon isotopic reations', *Geochim. Cosmochim. Acta*, **26**, 399–410.

Hunt, T. M., 1970. 'Net mass loss from the Wairakei geothermal field, New Zealand', *Geothermics*, special issue 2, **2**, 487–491.

Hunt, T. M., 1977, 'Recharge of water in Wairakei geothermal field determined from repeat gravity measurements', *N.Z. Journ. Geology and Geophysics*, **20**, 303–317.

Ingersoll, L. R., Zobel, O. J., and Ingersoll, A. C., 1955. *Heat Conduction*, Thames and Hudson, London, 325 pp.

Isherwood, W. F., 1976, 'Gravity and magnetic studies of The Geysers-Clear Lake geothermal region, California, U.S.A.', *2nd U.N. Symposium, Proceedings*, **2**, 1065–1073.

Iyer, H. M., and Hitchock, T., 1974. 'Sesmic noise measurements in Yellowstone National Park', *Geophysics*, **39**, 389–400.

Iyer, H. M., and Hitchcock, T., 1976, 'Seismic noise as a geothermal exploration tool: techniques and results', *2nd U.N. Symposium, Proceedings*, **2**, 1075–1083.

Iyer, H. M., Oppenheimer, D. H., and Hitchcock, T., 1979. 'Abnormal P-wave delays in The Geysers-Clear Lake geothermal area, California', *Science*, **204**, 495–497.

James, R., 1970. 'Factors controlling borehole performance', *Geothermics*, special issue 2, **2**, 1502–1515.

James, R., 1976. 'Drawdown test results differentiate between crack-flow and porous bed permeability', *2nd U.N. Symposium, Proceedings*, **3**, 1693–1696.

Jiracek, G. R., and Gerety, M. T., 1978. 'Comparison of surface and downhole resistivity mapping of geothermal reservoirs in New Mexico', *Geothermal Resources Council, Transactions*, **2**, 335–336.

Jiracek, G. R., Smith, C., and Dorn, G. A., 1976. 'Deep geothermal exploration in New Mexico using electrical resistivity', *2nd U.N. Symposium, Proceedings*, **2**, 1095–1102.

Kear, D., Lumb, J. T., and Studt, F. E., 1978. 'New Zealand geothermal exploration and exploitation developments', *Proceedings Circum-Pacific Energy and Mineral Resources Conference, Honolulu, Hawaii*, July 30–Aug. 6, 1978, American Association of Petroleum Geologists, Tulsa.

Keller, G. V., 1970. 'Induction methods in prospecting for hot water', *Geothermics*, special issue 2, **2**, 318–332.

Keller, G. V., and Frischknecht, F. C., 1966. *Electrical Methods of Geophysical Prospecting*, Pergamon Press, Oxford, 519 pp.

Klyen, L. E., 1973, 'A vessel for collecting subsurface water samples from geothermal drillholes', *Geothermics*, **2**, 57–60.

Kusakabe, M., 1974. 'Oxygen and sulphur isotope study of Wairakei geothermal well discharges', *N.Z. Journal of Science*, **17**, 183–191.

Lloyd, E. F., 1959. 'The hot springs and hydrothermal eruptions of Waiotapu', *N.Z. Journ. Geol. Geophys.*, **2**, 141–176.

Lumb, J. T., and Macdonald, W. J. P., 1970. 'Near-surface resistivity surveys of geothermal areas using the electromagnetic method', *Geothermics*, special issue 2, **2**, 311–317.

Macdonald, W. J. P., and Muffler, L. J. P., 1972. 'Recent geophysical exploration of the Kawerau geothermal field, North Island, New Zealand', *N.Z. Journ. Geology and Geophysics*, **15**, 303–317.

McEuan, R. B., Birkhahn, P. C., and Pinckney, C. J., 1976. 'Predictive regionalization of geothermal resource potential', *2nd U.N. Symposium, Proceedings*, **2**, 1121–1125.

McEvilly, T. V., Schechter, B., and Majer, E. L. 1978. 'East Mesa seismic study.' In *Geothermal Exploration Technology: Annual Report 1978*, Lawrence Berkeley Laboratory, University of California, 23–25.

McKenzie, W. F., and Truesdell, A. H., 1977. 'Geothermal reservoir temperatures estimated from oxygen isotope compositions of dissolved sulfate and water from hot springs and shallow drill holes', *Geothermics*, **5**, 51–61.

McNitt, J. R. (1976), 'Summary of United Nations geothermal exploration experience, 1965 to 1975', *2nd U.N. Symposium, Proceedings*, **2**, 1127–1134.

Mahon, W. A. J. 1970. 'Chemistry in the exploration and exploitation of hydrothermal systems', *Geothermics*, special issue 2, **2**, 1310–1322.

Meidav, T., 1970. 'Application of electrical resistivity and gravimetry in deep geothermal exploration', *Geothermics*, special issue 2, **2**, 303–310.

Miller, L. G., Prestwick, S. M., and Gould, R. W., 1978. 'Drilling and directional drilling a moderate temperature geothermal resource', *Geothermal Resources Council, Transactions*, **2**, 455–456.

Modriniak, N., and Studt, F. E., 1959. 'Geological structure and volcanism of the Taupo–Tarawera district', *N.Z. Journ. Geology and Geophysics*, **2**, 654–684.

Morrison, H. F., Goldstein, N. E., Hoversten, M., Oppliger, G., and Riveros, C., 1978. 'Controlled-source electromagnetic system.' In *Geothermal exploration technology: Annual Report 1978*, Lawrence Berkeley Laboratory, University of California, 9–12.

Muffler, L. J. P., 1976. 'Tectonic and hydrologic control of the nature and distribution of geothermal resources', *2nd U.N. Symposium, Proceedings*, **1**, 499–507.

Muffler, L. J. P. (Ed.) 1979. 'Assessment of Geothermal Resources of the United States—1978', *U.S. Geol. Survey Circ. 790*, 163 pp.

Noble, J. W., and Ojiambo, S. B., (1976) 'Geothermal exploration in Kenya', *2nd U.N. Symposium, Proceedings*, **1**, 189–204.

Pálmason, G., 1976. 'Geophysical methods in geothermal exploration', *2nd U.N. Symposium, Proceedings*, **2**, 1175–1184.

Quillin, B., and Combs, J., 1978. 'Microearthquake survey of the Radium Springs KGRA, South-central New Mexico', *Geothermal Resources Council, Transactions*, **2**, 547–550.

Rafter, T. A., Wilson, S. H., and Shilton, B. W., 1958. 'Sulphur isotopic variations in nature—5. Sulphur isotopic variations in New Zealand geothermal bore waters', *N. Z. Journal of Science*, **12**, 54–59.

Risk, G. F., 1976a. 'Monitoring the boundary of the Broadlands geothermal field, New Zealand', *2nd U.N. Symposium, Proceedings*, **2**, 1185–1189.

Risk, G. F., 1976b. 'Detection of buried zones of fissured rock in geothermal fields using resistivity anisotropy measurements', *2nd U.N. Symposium, Proceedings*, **2**, 1191–1198.

Risk, G. F. (in press), 'Defining the boundaries and structure of the Broadlands geothermal field, New Zealand, by geophysical methods—a case history', *Proc. Circum Pacific Energy and Mineral Resources Conference*, Honolulu, Hawaii, July 30–Aug. 4, 1978, American Association of Petroleum Geologists, Tulsa.

Risk, G. F., Macdonald, W. J. P., and Dawson, G. B., 1970. 'D.C. resistivity surveys of the Broadlands geothermal region, New Zealand', *Geothermics*, special issue 2, **2**, 287–294.

Robertson, E. I., and Dawson, G. B., 1964. 'Geothermal heat flow through the soil at Wairakei', *N.Z. Journ. Geology and Geophysics*, **7**, 134–143.

Roy, A., and Apparao, A., 1971. 'Depth of investigation of direct current methods', *Geophysics*, **36**, 943–959.

Sestini, G., 1970. 'Heat-flow measurements in non-homogeneous terrains', *Geothermics*, special issue 2, **2**, 424–436.

Shanker, R., Padhi, R. N., Arora, C. L., Prakash, G., Thussu, J. L., and Dua, K. J. S., 1976. 'Geothermal explorations of the Puga and Chumathang geothermal fields, Ladakh, India', *2nd U.N. Symposium, Proceedings*, **1**, 245–258.

Shore, G. A., 1978. 'Meager Creek geothermal systems, British Columbia, part III: Resistivity methods and results', *Geothermal Resources Council, Transactions*, **2**, 593–596.

Smith, M. L., 1978. 'Heat extraction from hot, dry crustal rock', *Pure and Applied Geophysics*, **117** 290–296.

Souto, J. M., 1978. 'Oahu Geothermal exploration', *Geothermal Resources Council, Transactions*, **2**, 605–607.

Steeples, D. W., and Iyer, H. M., 1976. 'Teleseismic P-wave delays in geothermal exploration', *2nd U.N. Symposium, Proceedings*, **2**, 1199–1206.

Studt, F. E., 1958. 'Geophysical reconnaissance at Kawerau, New Zealand', *N.Z. Journ. Geology and Geophysics*, **1**, 219–246.

Swanberg, C. A., 1976. 'The Mesa geothermal anomaly, Imperial Valley, California; a comparison and evaluation of results obtained from surface geophysics and deep drilling', *2nd U.N. Symposium, Proceedings*, **2**, 1217–1229.

Tezcan, A. K., 1976. 'Geophysical studies in Sarayköy-Kizildere geothermal field, Turkey', *2nd U.N. Symposium, Proceedings*, **2**, 1231–1240.

Thompson, G. E. K., Banwell, C. J., Dawson, G. B., and Dickinson, D. J., 1964. 'Prospecting of hydrothermal areas by surface thermal surveys', *Proc. U.N. Conf. on New Sources of Energy*, Rome 1961, **2**, 386–401.

Truesdell, A. H., 1976. 'Summary of Section III: Geochemical techniques in exploration', *2nd U.N. Symposium, Proceedings*, **1**, liii–lxxix.

Truesdell, A. H., and Fournier, R. O., 1976. 'Calculations of deep temperatures in geothermal systems from the chemistry of boiling spring waters of mixed origin', *2nd U.N. Symposium, Proceedings*, **1**, 837–844.

Wainwright, D. K., 1970. 'Subsurface and output measurements on geothermal bores in New Zealand', *Geothermics*, special issue 2, **2**, 764–767.

Wallace, R. H., Jr., Kraemer, T. F., Taylor, R. E., and Wesselman, J. B., 1979. 'Assessment of geopressured-geothermal energy resources in the northern Gulf of Mexico basin.' In *Assessment of Geothermal Resources of the United States—1978*, L. J. P. Muffler (Ed.), *U.S. Geol. Survey Circ. 790*. 132–155.

Ward, P. L., and Björnsson, S., 1971. 'Microearthquakes, swarms, and the geothermal areas of Iceland', *Journ. Geophysical Research*, **76**, 3953–3982.

Ward, P. L., and Jacob, K. H., 1971. 'Microearthquakes in the Ahuachapán geothermal field, El Salvador, Central America', *Science*, **173**, 328–330.

Weissberg, B. G., and Wilson, P. T., 1977. 'Montmorillonites and the Na/K geothermometer.' In *Geochemistry 1977*, A. J. Ellis, (Comp.), Dept. Sci. and Industr. Res. Bull., **218**,, New Zealand, 31–34.

West, R. C., and Pritchard, J. I., 1978. 'Combined electromagnetic and galvanic (d.c.) electrical resistivity soundings', *Geothermal Resources Council, Transactions*, **2**, 713–716.

White, D. E., Fournier, R. O., Muffler, L. J. P., and Truesdell, A. H., 1975. 'Physical results of research drilling in the thermal areas of Yellowstone National Park, Wyoming', *U.S. Geol. Survey, Prof. Paper 892*, 70 pp.

White, D. E., Muffler, L. J. P., and Truesdell, A. H., 1971. 'Vapor-dominated hydrothermal systems compared with hot water systems', *Economic Geology*, **66**, 75–97.

Whiteford, P. C., 1970. 'Ground movement in the Waiotapu geothermal region, New Zealand', *Geothermics*, special issue 2, **2**, 478–486.

Whiteford, P. C., 1976a. 'Assessment of the audiomagnetotelluric method for geothermal resistivity surveying', *2nd U.N. Symposium, Proceedings*, **2**, 1255–1261.

Whiteford, P. C., 1976b. 'Studies of the propagation and source location of geothermal seismic noise', *2nd U.N. Symposium, Proceedings*, 2, 1263–1271.

Williams, P. L., Mabey, D. R., Zohdy, A. A. R., Ackermann, H., Hoover, D. B., Pierce, K. L., and Oriel, S. S., 1976. 'Geology and geophysics of the southern Raft River Valley geothermal area, Idaho, U.S.A.', *2nd U.N. Symposium, Proceedings*, **2**, 1273–1282.

Yamasaki, T., and Hayashi, M., 1976. 'Geologic background of Otake and other geothermal areas in north-central Kyushu, Southwestern Japan', *2nd U.N. Symposium, Proceedings*, **2**, 673–684.

Young, C. Y., and Ward, R. W., 1978, '2-D inversion of seismic attenuation observations in Coso hot springs, KGRA', *Geothermal Resources Council, Transactions*, **2**, 743–746.

Zablocki, C. J., 1976. 'Mapping thermal anomalies on an active volcano by the self-potential method, Kilauea, Hawaii', *2nd U.N. Symposium, Proceedings*, **2**, 1299–1309.

Zohdy, A. A. R., Anderson, L. A., and Muffler, L. J. P., 1973. 'Resistivity, self-potential, and induced-polarization surveys of a vapor-dominated geothermal system', *Geophysics*, **38**, 1130–1144.

Geothermal Systems: Principles and Case Histories
Edited by L. Rybach and L. J. P. Muffler
© 1981 John Wiley & Sons Ltd.

4. Application of Water Geochemistry to Geothermal Exploration and Reservoir Engineering

ROBERT O. FOURNIER

U.S. Geological Survey, Menlo Park, California 94025

4.1. Introduction

The application of geochemistry to the development of geothermal energy is a big subject encompassing many facets. In this chapter emphasis will be placed upon the use of water chemistry to determine underground temperatures and boiling and mixing relations in the exploration and production phases of geothermal energy utilization. Although knowledge of the composition and behavior of gas associated with hot water is of very great interest and practical importance, gas geochemistry will not be discussed here.

Thermal energy serves as an engine that sets water in the Earth's crust into convective motion, forming what we call hydrothermal systems. Water may become heated owing to deep circulation along favorable structures in regions of normal geothermal gradient or by interaction with magmas and cooling igneous rocks that have intruded to relatively shallow levels in the crust.

The explored parts of most presently active hydrothermal systems are dominated by meteoric or ocean water that has changed composition during underground movement in response to changing temperature, pressure, and rock type. Changing temperature has a major effect upon the ratios of cations in solution and the concentration of dissolved silica. Changing pressure is an important factor in regard to evaporative concentration and partitioning of volatile components between water and steam during boiling. Variations in rock type strongly influence the total salinity and particularly the chloride concentration that a hydrothermal solution is likely to attain.

4.2. Hydrothermal Reactions

The compositions of geothermal fluids are controlled by temperature-dependent reactions between minerals and fluids. In order to understand and model hydrothermal systems both the fluid and solids must be characterized.

Browne (1978) summarized the factors affecting the formation of hydrothermal minerals as follows: (a) temperature, (b) pressure, (c) rock type, (d) permeability, (e) fluid composition, and (f) duration of activity. He further stated that the effects of rock type were most pronounced at low temperatures and generally insignificant above 280°C. Above 280°C and at least as high as 350°C the typical stable mineral assemblage found in active geothermal systems is not dependent on original rock type and includes albite, K-feldspar, chlorite, Fe-epidote, calcite, quartz, illite, and pyrite. At lower temperatures many different zeolites and clay minerals also are found. Apparently epidote does not form below about 240°C, although in some places it may persist metastably at lower temperatures. Where permeability is low, equilibrium between rocks and reservoir fluids is seldom achieved, and unstable primary minerals or glass can persist at high temperatures. Metastable minerals also form and persist in some geothermal systems, particularly where glassy rocks are present and temperatures are below about 200°C.

In geothermal reservoirs where permeabilities are relatively high and water residence times are long (months to years), water and rock should reach chemical equilibrium, especially where temperatures exceed 200°C. At equilibrium, ratios of cations in solution are controlled by temperature-dependent exchange reactions such as

$$\underset{\text{Albite}}{NaAlSi_3O_8} + K^+ = \underset{\text{K-feldspar}}{KAlSi_3O_8} + Na^+, \tag{4.1}$$

$$K_{eq} = \frac{[Na^+]}{[K^+]}, \tag{4.2}$$

and

$$\underset{\text{Wairakite}}{CaAl_2Si_4O_{12} \cdot 2H_2O} + \underset{\text{Quartz}}{2SiO_2} + 2Na^+ = \underset{\text{Albite}}{2NaAlSi_3O_8} + Ca^{++} + 2H_2O, \tag{4.3}$$

$$K_{eq} = \frac{[Ca^{++}][H_2O]^2}{[Na^+]^2}. \tag{4.4}$$

Hydrogen ion activity (pH) is controlled by hydrolysis reactions, such as

$$\underset{\text{K-feldspar}}{3KAlSi_3O_8} + 2H^+ = \underset{\text{K-mica}}{KAl_3Si_3O_{10}(OH)_2} + \underset{\text{Quartz}}{6SiO_2} + 2K^+, \tag{4.5}$$

$$K_{eq} = \frac{[K^+]}{[H^+]} \tag{4.6}$$

$$\underset{\text{Albite}}{2.33NaAlSi_3O_8} + 2H^+ = \underset{\text{Na-Montmorillonite}}{Na_{.33}Al_{2.33}Si_{3.67}O_{10}(OH)_2} + \underset{\text{Quartz}}{3.33SiO_2} + 2Na^+, \tag{4.7}$$

$$K_{eq} = \frac{[K^+]}{[H^+]} \tag{4.8}$$

and

$$\underset{\text{Wairakite}}{1.17CaAl_2Si_4O_{12} \cdot 2H_2O} + 2H^+ = \underset{\text{Ca-Montmorillonite}}{Ca_{.17}Al_{2.33}Si_{3.7}O_{10}(OH)_2} + \underset{\text{Quartz}}{SiO_2} +$$
$$+ 2H_2O + Ca^{++} \tag{4.9}$$

$$K_{eq} = \frac{[Ca^{++}][H_2O]^2}{[H^+]^2} \tag{4.10}$$

In the above equations K_{eq} is the equilibrium constant for the given reaction, assuming unit activities of the solid phases, and square brackets indicate activities of the dissolved species. The extent to which natural water-rock systems approach chemical equilibrium can be tested by comparing actual compositions of fluids and minerals found in drilled

4. Application of Water Geochemistry to Geothermal Exploration

geothermal reservoirs with theoretical compositions calculated using thermodynamic data (Helgeson, 1969; Helgeson et al., 1969; Robie et al., 1978).

The first step in testing water-rock equilibration is to collect and analyze representative fluid and rock samples (Ellis and Mahon, 1977; Watson, 1978) from a reservoir at a known temperature. Cuttings and core are collected during drilling, although sidewall cores can be collected later. Core is preferred over cuttings for mineralogic work because paragenetic relations are clearer, and minerals that occur only in veins can be distinguished from those only in the rock matrix. Also, with cuttings, one cannot be sure that all the fragments come from the same interval of rock. Fragments from higher in the hole may become incorporated in the drilling mud along with cuttings from the bottom of the well.

Fluid samples are collected after drilling mud and makeup water have been flushed from the well. The best way to collect a water sample is with a downhole sampler, but wellhead samples can be used provided the water in the well comes from only one permeable zone. From the results of the chemical analyses combined with data on the conditions of sample collection, the original thermodynamic state of the fluid in the reservoir can be calculated using computer programs (Truesdell and Singers, 1974; Arnórsson, 1978) that correct for the presence of complex ions and the separation of steam. The natural and theoretical systems can be compared by noting the degree of saturation of the fluid in respect to each of the various minerals found in core and cuttings from the reservoir rocks, using a computer program (Kharaka and Barnes, 1973). A major limitation of this approach is that thermochemical data are not available for many of the zeolites, clays, and micas commonly found in geothermal systems. Also, many of the minerals are solid solutions which presently available computer programs treat by assuming pure end-member compositions. Results of this type of procedure usually show that the solutions are saturated with respect to the observed minerals (those for which thermochemical data are available) within the limits of error of the method.

Another method of comparison is to determine where the activity ratios of aqueous species (calculated by computer) plot on theoretical activity diagrams (Helgeson et al., 1969) that depict mineral stability fields in chemically restricted systems. Ellis (1969), Browne and Ellis (1970) and Ellis and Mahon (1977) plot water compositions from Wairakei and Broadlands, N.Z., on a variety of activity diagrams. The diagram for Broadlands in terms of $[Na^+]/[H^+]$ and $[K^+]/[H^+]$ activity ratios at 260°C is shown in Figure 4.1, and for $[K^+]/[H^+]$ and $[Ca^{++}]/[H^+]^2$ in figure 4.2. The Broadlands water composition shown in figure 4.1 plots close to the triple point of K-feldspar, K-mica, and albite, suggesting an approach to water-rock equilibrium involving those minerals at 260°C. In Figure 4.2, the Broadlands water plots close to a triple point of K-feldspar, K-mica, and wairakite. Calcite also is present with $0.15\ mCO_2$ in the system. The maximum temperatures in wells at Broadlands usually are in the range 270° to 290°C. If Figures 4.1 and 4.2 had been drawn for 290°C, the water composition would have plotted even closer to the triple points. Reasons, other than temperature, for slight discrepancies between actual solution compositions and theoretical equilibrium compositions involve uncertainties in the structural states of the reacting solids and limitations in our knowledge of all the possible complex ions that may be present at high temperatures in natural waters. The influence of structural state is illustrated in Figure 4.3 which shows expected $[Na^+]/[K^+]$ activity ratios in solutions equilibrated with different alkali feldspar pairs at various temperatures. At 100°C, the value of $\log([Na^+]/[K^+])$ for the assemblage low-albite plus microcline is about 0.2 units lower than for the assemblage low-albite plus adularia and

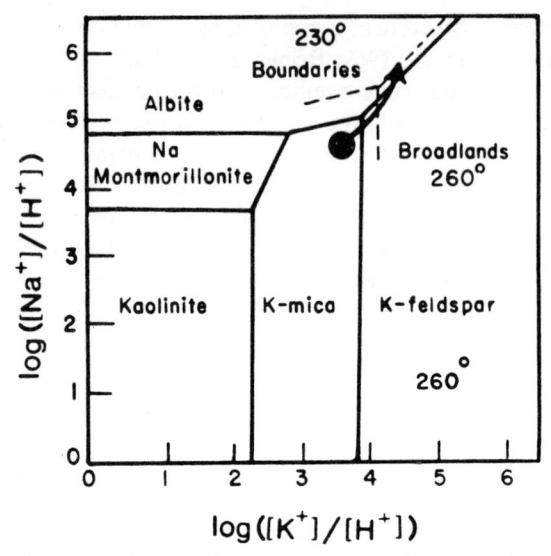

Figure 4.1. Phase diagram showing estimated stability fields for various minerals at different $[Na^+]/[H^+]$ and $[K^+]/[H^+]$ activity ratios in the presence of quartz at 260°C. Dashed lines show a portion of the diagram at 230°C (from Browne and Ellis, 1970)

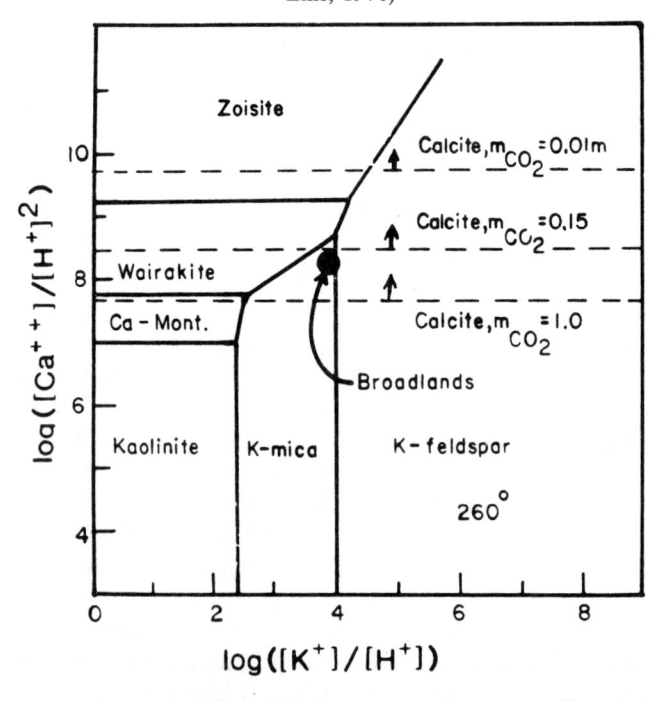

Figure 4.2. Phase diagram showing estimated stability fields for various minerals at different $[K^+]/[H^+]$ and $[Ca^{++}]/[H^+]^2$ activity ratios in the presence of quartz. Also shown are activity ratios of $[Ca^{++}]/[H^+]^2$ at which calcite will precipitate for m_{CO_2} values of 0·01, 0·15, and 1·0 (from Ellis, 1969)

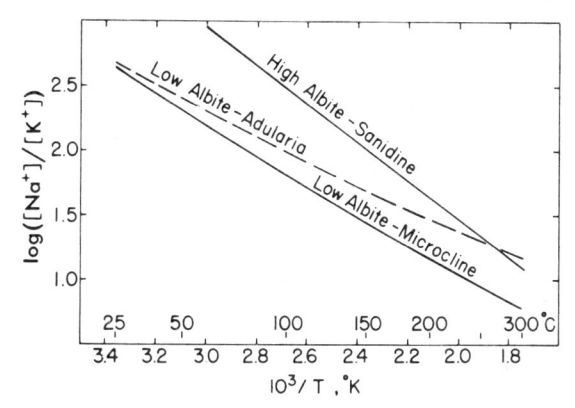

4. Application of Water Geochemistry to Geothermal Exploration

Figure 4.3. Theoretical variation of [Na$^+$]/[K$^+$] activity ratio in solutions equilibrated with alkali feldspar pairs with indicated structural states at 25°C to 300°C (data from Helgeson et al., 1969)

about 0.7 units lower than for the assemblage high-albite plus sanidine. Differences are still significant at 300°C; about 0.3 units.

The activity diagram method of presentation allows one to focus upon a few of the more important components in the system and to visualize the effects of changing one or more parameters. In Figure 4.1, lowering the temperature while maintaining water-rock equilibrium would cause the solution to follow the path shown by the arrow. In Figure 4.2 the effect of increasing CO_2 concentration is shown by the expanding field of calcite that completely replaces zoisite when $m_{CO_2} \cong 0.03$ and completely replaces wairakite when $m_{CO_2} \cong 1.0$.

4.3. Estimation of Reservoir Temperature

Many different chemical and isotopic reactions might be used as geochemical thermometers or geothermometers (Ellis and Mahon, 1977; Fournier 1977; Truesdell and Hulston, 1980) to estimate reservoir temperatures. At present the most widely used are silica (Fournier and Rowe, 1966; Mahon, 1966), Na/K (White, 1965; Ellis and Mahon, 1967; White, 1970; Truesdell, 1976a; Fournier, 1979a), Na–K–Ca (Fournier and Truesdell 1973), and sulfate oxygen isotope, $\Delta^{18}O(SO_4^= - H_2O)$, which uses the fractionation of oxygen isotopes between water and dissolved sulfate (Lloyd, 1968; Mizutani and Rafter, 1969; Mizutani, 1972; McKenzie and Truesdell, 1977). Equations expressing the temperature dependence of the above geothermometers are listed in Table 4.1. It is relatively easy to compute temperatures using the equations in Table 4.1. Interpreting the results of these computations, however, requires consideration of the geologic and hydrologic setting and the physical nature as well as complete chemical analysis of the fluid that was sampled. Some people find it dismaying that different geothermometers give different estimated temperatures. However, shrewd investigators will use that kind of information to their advantage when formulating a model of the hydrothermal system, because different geothermometers may indicate temperatures in different parts of the system.

4.3.1. The Silica Geothermometer

Bodvarsson (1960) suggested an empirical, qualitative geothermometer based on the silica concentrations in natural thermal waters in Iceland. The experimental work of

Table 4.1. Equations expressing the temperature dependence of selected geothermometers. C is the concentration of dissolved silica. All concentrations are in mg/kg

Geothermometer	Equation	Restrictions
a. Quartz-no steam loss	$t°C = \dfrac{1309}{5\cdot 19 - \log C} - 273\cdot 15$	$t = 0\text{–}250°C$
b. Quartz-maximum steam loss	$t°C = \dfrac{1522}{5\cdot 75 - \log C} - 273\cdot 15$	$t = 0\text{–}250°C$
c. Chalcedony	$t°C = \dfrac{1032}{4\cdot 69 - \log C} - 273\cdot 15$	$t = 0\text{–}250°C$
d. α-Cristobalite	$t°C = \dfrac{1000}{4\cdot 78 - \log C} - 273\cdot 15$	$t = 0\text{–}250°C$
e. β-Cristabalite	$t°C = \dfrac{781}{4\cdot 51 - \log C} - 273\cdot 15$	$t = 0\text{–}250°C$
f. Amorphous silica	$t°C = \dfrac{731}{4\cdot 52 - \log C} - 273\cdot 15$	$t = 0\text{–}250°C$
g. Na/K (Fournier)	$t°C = \dfrac{1217}{\log(Na/K) + 1\cdot 483} - 273\cdot 15$	$t > 150°C$
h. Na/K (Truesdell)	$t°C = \dfrac{855\cdot 6}{\log(Na/K) + 0\cdot 8573} - 273\cdot 15$	$t > 150°C$
i. Na–K–Ca	$t°C = \dfrac{1647}{\log(Na/K) + \beta[\log(\sqrt{Ca}/Na) + 2\cdot 06] + 2\cdot 47} - 273\cdot 15$	$t < 100°C,\ \beta = 4/3$ $t > 100°C,\ \beta = 1/3$
j. $\Delta^{18}O(SO_4^= - H_2O)$	$1000 \ln \alpha = 2\cdot 88(10^6\ T^{-2}) - 4\cdot 1$ $\alpha = \dfrac{1000 + \delta^{18}O(HSO_4^-)}{1000 + \delta^{18}O(H_2O)}$ and $T = °K$	

Morey et al. (1962) on the solubility of quartz in water at the vapor pressure of the solution, combined with higher temperature results of Kennedy (1950), provided a theoretical basis for the geothermometer (curve A in Figure 4.4). Mahon (1966) showed that the silica concentrations in waters from geothermal wells in New Zealand are

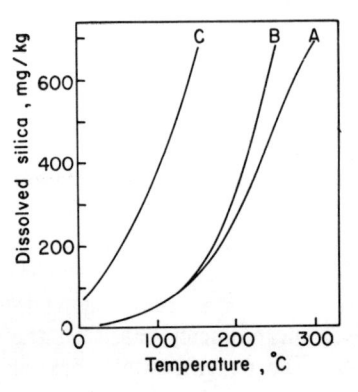

Figure 4.4. Solubility of quartz (curve A) and amorphous silica (Curve C) as a function of temperature at the vapor pressure of the solution. Curve B shows the amount of silica that would be in solution after an initially quartz-saturated solution cooled adiabatically to 100°C without any precipitation of silica (from Fournier and Rowe, 1966, and Truesdell and Fournier, 1976a)

controlled by the solubility of quartz, and Fournier and Rowe (1966) described a method of using the silica concentration in hot-spring and well waters to make quantitative estimates of reservoir temperatures.

The quartz geothermometer works best for well waters where subsurface temperatures are above about 150°C. It has been shown to work well for some hot spring waters (Fournier and Truesdell, 1970), but can give erroneous results when applied indiscriminately. The following factors should be considered when using the quartz geothermometer: (a) the temperature range in which the equations in Table 4.1 are valid, (b) effects of steam separation, (c) possible polymerization and/or precipitation of silica before sample collection, (d) possible polymerization of silica after sample collection owing to improper preservation of the sample, (e) control of aqueous silica by solids other than quartz, (f) the effect of pH upon quartz solubility, and (g) possible dilution of hot water with cold water before the thermal water reaches the surface.

Temperature range of equations. The equations commonly used to describe the solubility of quartz at the vapor pressure of the solution (Morey et al., 1962; Fournier and Rowe, 1966; Truesdell, 1976a; Fournier, 1977) are good to about $\pm 2°C$ over the temperature range 0° to 250°C. Above 250°C the equations depart drastically from the experimentally determined solubility curve (Curve A in Figure 4.4) and should not be used.

Effects of steam separation. As water boils, the silica concentration in the residual liquid increases in proportion to the amount of steam that separates. Thus, when applying the quartz geothermometer to boiling hot springs, it is necessary to correct for the amount of steam formation or the proportion of adiabatic vs. conductive cooling, as outlined by Fournier and Rowe (1966). Vigorously boiling springs with mass flow rates greater than 120 to 130 kg/min can be assumed to have cooled mainly adiabatically (Truesdell et al., 1977), and equation (b) in Table 4.1 (for maximum steam loss at 100 kPa pressure) or curve B in Figure 4.4 can be used to estimate the reservoir temperature. If cooling was partly adiabatic and partly by conduction, the reservoir temperature should be between the temperatures given by equations (a) and (b) in Table 4.1 or between the temperatures indicated by curves A and B in Figure 4. For springs issuing high above sea level, final steam formation can occur at significantly less than 100 kPa, so that equation (b) in Table 4.1 for maximum steam formation and curve B in Figure 4.4 will give a calculated reservoir temperature that is slightly high.

Precipitation of silica prior to sample collection. The quartz geothermometer works because the rate of quartz precipitation decreases drastically as temperature decreases. In the temperature range 200° to 250°C water reaches equilibrium with quartz in a few hours to a few days, depending on the solution temperature, pH, salinity, and degree of initial silica supersaturation or undersaturation. High salinity and pH in the range 5 to 8 favor faster reaction rates. Below 100°C, solutions may remain supersaturated with respect to quartz for years. When waters flow up to the surface from reservoirs with temperatures less than about 225° to 250°C and cool fairly quickly (in less than a few hours), little quartz is likely to precipitate during the upflow. However, where reservoir temperatures are above 250°C, some quartz is likely to precipitate in the deep, hot part of the system as the solution cools. In addition, an ascending solution starting at temperatures above about 225°C will become supersaturated with respect to amorphous silica before reaching the surface owing to the decreased temperature, especially if adiabatic cooling takes place (Figure 4.4). Amorphous silica precipitates much more quickly than quartz at comparable temperatures. Because of the precipitation of quartz at high temperatures and polymerization and precipitation of amorphous silica at low temperatures, the quartz geothermometer applied to hot spring waters seldom indicates temperatures exceeding 225° to 250°C, even where higher temperature reservoirs are known to exist.

Precipitation of silica after sample collection. The solubility of amorphous silica at 25°C is about 115 mg/kg (Morey et al., 1964). The solubility of quartz is 115 mg/kg at about 145°C (Figure 4.4). Therefore, waters coming from reservoirs above 145°C will become supersaturated with respect to amorphous silica upon cooling to 25°C. At high temperatures all of the dissolved silica in equilibrium with quartz is in the monomeric form. When a cooling solution becomes saturated with respect to amorphous silica some of the monomeric silica usually converts to highly polymerized dissolved silica species, with or without the simultaneous precipitation of amorphous silica. Colorimetric methods of analysis used to determine dissolved silica are sensitive to monomeric silica, but fail to detect most highly polymerized forms of dissolved silica. To prevent silica polymerization and precipitation of amorphous silica, a known amount of the sample (5 to 10 ml) should be pipetted into about 50 ml of silica-free water immediately after sample collection.

Control of aqueous silica by solids other than quartz. Quartz is the most stable and least soluble polymorphic form of silica within the temperature and pressure range found in geothermal systems. However, for kinetic reasons other silica phases may form or persist metastably within the stability field of quartz. The solubilities of various silica phases in liquid water at the vapor pressure of the solution are shown in Figure 4.5. Plots of log concentration *vs.* reciprocal of absolute temperature yield essentially straight lines below 250°C that make extrapolation of the data to lower temperatures very simple. Equations relating the solubility, C, in mg SiO_2 per kg water, to temperature for the various silica phases are given in Table 4.1.

In most natural waters at temperatures above 150°C, and in some waters below that temperature, quartz appears to control the dissolved silica concentration. However, under special conditions for short periods of time any of the other silica species may control aqueous silica, even at very high temperatures. This conclusion is based upon (1) compositions of natural waters in drill holes; (2) paragenetic sequences of silica minerals in drill core from hot spring areas; and (3) laboratory investigations in which it has been found that as long as two different silica species contact the solution, e.g., quartz and glass, the more soluble one controls aqueous silica (Fournier, 1973).

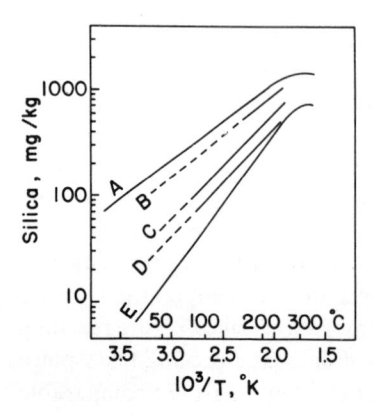

Figure 4.5. Solubilities of various silica phases in water at the vapor pressure of the solution. A = amorphous silica, B = β-cristobalite, C = α-cristobalite, D = chalcedony, and E = quartz (from Fournier, 1973)

In the basaltic terrain of Iceland, chalcedony generally controls aqueous silica at temperatures below 110°C and sometimes at temperatures as high as 180°C (Arnórsson, 1975). In some granite terrains, aqueous silica is controlled by quartz at temperatures above 90°C and by chalcedony at lower temperatures (Christian Fouillac, oral communication, 1977).

Most groundwaters which have not attained temperatures greater than 80° to 90°C have silica concentrations greater than those predicted by the solubility of quartz. Some of these low-temperature waters have equilibrated with chalcedony, as indicated above. However, the silica concentrations in many groundwaters result from non-equilibrium reactions in which silica is released to solution during acid alteration of silicate minerals (for example equations (4.5), (4.7), and (4.9). At low temperatures, the rates of quartz and chalcedony precipitation are very slow so aqueous silica values may become high where acid is continually supplied from an outside source, such as by decay of organic material, oxidation of sulfides, or influx of H_2S or CO_2. According to Ivan Barnes and R. H. Mariner (written commun., 1975) cold waters with high silica concentrations (attaining saturation with respect to amorphous silica) and near neutral pH are particularly prevalent where CO_2 and water react with serpentine. This conclusion is in agreement with results of experimental studies by Wildman et al. (1968) in which serpentine was dissolved at 25°C at various pressures of CO_2.

Effect of pH. The effect of pH upon the solubility of quartz at various temperatures can be calculated from the work of Seward (1974) or Busey and Mesmer (1977), and is shown in Figure 4.6. The dashed line in Figure 4.6 shows pH values at which the solubility of quartz becomes ten percent greater than the solubility in water with pH = 7·0. A ten percent increase in dissolved silica owing to increased pH will cause silica geothermometer temperatures to be about 6° too high at 180° and about 12° too high at 250°C. At 25°C a pH of 8·9 is required to achieve a ten percent increase while at 100°C a pH of just 8·2 is required. The effect of pH upon quartz solubility is most pronounced at about 175°C where a pH of 7·6 will cause an increase in quartz solubility of 10 per cent. However, pH values of solutions in high-temperature geothermal reservoirs are likely to be below 7·5 because of buffering of hydrogen ions by silicate hydrolysis reactions such as

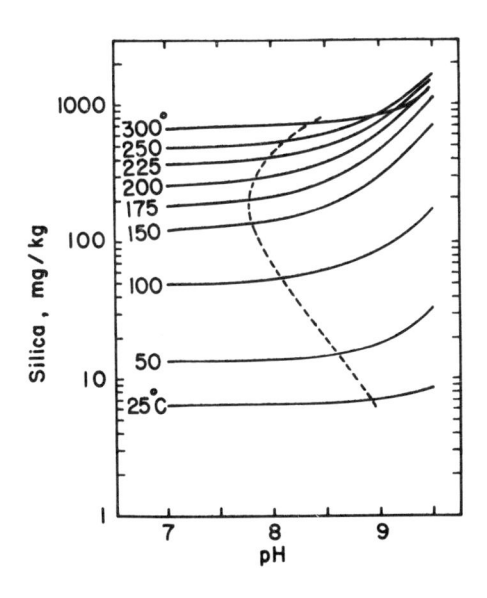

Figure 4.6. Calculated effect of pH upon the solubility of quartz at various temperatures from 25°C to 350°C, using experimental data of Seward (1974). The dashed curve shows the pH required at various temperatures to achieve a 10 per cent increase in quartz solubility compared to the solubility at pH @ 7·0

shown by equations (4.5), (4.7) and (4.9). pH values higher than 7·5 in natural hot-spring waters generally result from the loss of CO_2 after the water leaves the high-temperature reservoir.

Whether or not a correction should be applied to adjust the silica concentration of an alkaline hot spring water for pH effects before applying the silica geothermometer depends upon where and when the solution attained its aqueous silica. If a solution became alkaline and then dissolved additional silica in response to the rise in pH, a pH correction is necessary. In contrast, if a solution attained a high silica concentration owing to high underground temperature and then became alkaline after cooling and loss of CO_2, no pH correction should be applied. In general, if there is other supporting evidence that a thermal water comes from a higher-temperature environment at depth, a pH correction should *not* be applied to the observed silica concentration.

Subsurface dilution of thermal water by cold water. In many places there is evidence that ascending thermal water becomes diluted by cooler water. A new water-rock chemical equilibrium may or may not be attained after mixing. If chemical equilibrium is attained after mixing, the silica geothermometer will give the temperature of that last equilibrium. If chemical equilibrium is not attained after mixing, direct application of the silica geothermometer will give a calculated temperature that is too low. Methods of dealing with diluted thermal waters will be discussed in the section on underground mixing and boiling.

4.3.2. The Na/K Geothermometer

Many investigators have noted the variation of Na^+ and K^+ in natural geothermal waters as a function of temperature (White, 1965, 1970; Ellis and Mahon, 1967; Ellis, 1970; Fournier and Truesdell, 1970, 1973; and Mercado, 1970). Figure 4.7 shows Na^+/K^+ mg/kg concentration ratios in waters from geothermal wells throughout the world,

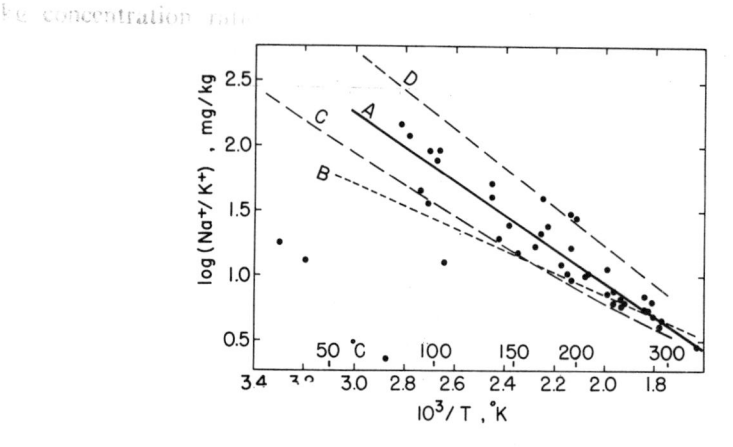

Figure 4.7. Na/K ratios of natural waters plotted at measured downhole temperatures in wells. Curve A is the least-squares fit of the data points above 80°C. Curve B is the combined White (1965) and Ellis (1970) curve used by Truesdell (1976a). Curves C and D show the approximate locations of the low albite-microcline and high albite-sanidine lines from figure 4.3 (from Fournier, 1979a)

plotted at the respective measured reservoir temperatures. In Figure 4.7, curves C and D show the approximate location (assuming activity ratios = 1·7 mg/kg concentration ratios) of the low-albite plus microcline and high-albite plus sanidine curves from Figure 4.3. Note that above 100°C most of the well waters plot between curves C and D. Below 100°C most well waters plot below curve C, indicating that the ratio of dissolved Na^+ to K^+ generally is not controlled by cation exchange between coexisting alkali feldspar pairs at low temperatures.

Line A, Figure 4.7, is the least-squares best fit of the well water data points above 80°C. The data points at temperatures below 80°C were not included in the least squares fit because the Na/K method generally gives unreliable results at low temperatures.

The equation of curve A, Figure 4.7, is labeled g in Table 4.1. Truesdell (1976a) recommended using a combined curve of White (1965) and Ellis (1970) for geothermometry as drawn by Fournier and Truesdell (1973). That curve is shown as line B in Figure 4.7 and its equation is labeled h in Table 4.1.

At temperatures near 270°C it makes little difference whether curve A or curve B in Figure 4.7 (equation (g) or (h)) is used for estimating temperatures. At temperatures between 125° and 200°C, curve B departs significantly from most of the data points. Therefore, curve A is recommended for geothermometry instead of curve B. Curve B does come closer to most data points below 100°C, but it appears that the Na/K method generally fails to give reliable results for waters from environments with temperatures below 100°C. In particular, low-temperature waters rich in calcium give anomalous results by the Na/K method.

Where waters are known to come from high-temperature environments (>180° to 200°C), the Na/K method generally gives excellent results. The main advantage of the Na/K geothermometer is that it is less affected by dilution and steam separation than other commonly used geothermometers, provided there is little Na^+ and K^+ in the diluting water compared to the reservoir water.

4.3.3. The Na–K–Ca Geothermometer

The Na–K–Ca geothermometer of Fournier and Truesdell (1973) was developed specifically to deal with calcium-rich waters that give anomalously high calculated temperatures by the Na/K method. An empirical equation giving the variation of temperature with Na, K, and Ca is listed in Table 4.1, letter i.

When using equation (i) in Table 4.1, first calculate the temperature using a value of $\beta = 4/3$ and cation concentrations expressed either as mg/kg or ppm. If that calculated temperature is <100°C and $[\log (\sqrt{Ca}/Na) + 2 \cdot 06]$ is positive, proceed no further. However, if the $\beta = 4/3$ calculated temperature is >100°C or if $[\log (\sqrt{Ca}/Na) + 2 \cdot 06]$ is negative, use $\beta = 1/3$ to calculate the temperature.

Changes in concentration resulting both from boiling and from mixing with cold, dilute water will affect the Na–K–Ca geothermometer. The main consequence of boiling is loss of CO_2 which can cause $CaCO_3$ to precipitate. The loss of aqueous Ca^{++} generally will result in Na–K–Ca calculated temperatures that are too high.

The effect of dilution on the Na–K–Ca geothermometer is generally negligible if the high-temperature geothermal water is much more saline than the diluting water. However, if a particular water is thought to be a mixture of hot and cold water with less than 20 to 30 per cent hot water component, the effects of mixing on the Na–K–Ca geothermometer should be considered. Figure 4.8 shows $\beta = 1/3$ and $\beta = 4/3$ Na–K–Ca geothermometer temperatures for various mixtures of two hypothetical waters, one starting at 210°C and

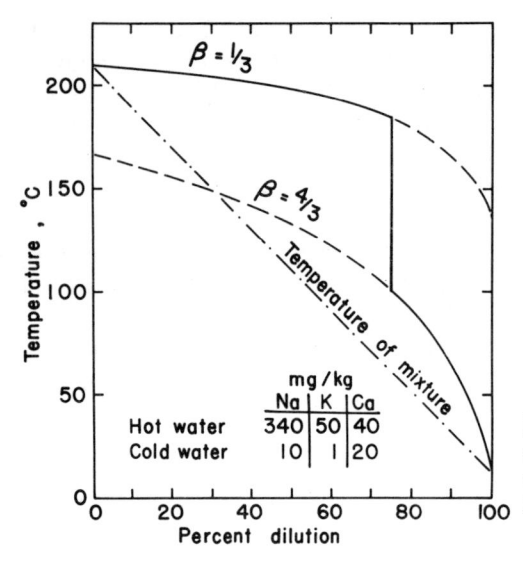

Figure 4.8. For the indicated starting compositions, the effect of dilution of hot water by cold water upon the calculated Na–K–Ca temperature. Different starting compositions will give different results

the other at 10°C. A 50:50 mixture of the two waters would have an actual temperature of about 110°C and would give a $\beta=4/3$ temperature of 143°C and a $\beta=1/3$ temperature of 198°C. By the criteria of Fournier and Truesdell (1973) the $\beta=1/3$ temperature would be selected as most likely. Thus, the Na–K–Ca geothermometer would give a temperature only 12°C below the actual 210°C temperature. However, when the amount of cold water in the mixture becomes greater than 75 per cent, the $\beta=4/3$ geothermometer temperature drops to less than 100°C and a hot spring would emerge at less than 60°C. Because the $\beta=4/3$ temperature is less than 100°C, it normally would be selected as the more likely temperature of water-rock equilibration instead of the $\beta=1/3$ temperature.

Fournier and Potter (1979) showed that the Na–K–Ca geothermometer gives anomalously high results when applied to waters rich in Mg^{++}. Figure 4.9 shows temperature corrections that should be subtracted from the Na–K–Ca calculated temperatures to correct for Mg^{++}. Temperature corrections also can be calculated using the following equations that were derived by Fournier and Potter (1979): For R between 5 and 50

$$\Delta t_{Mg} = 10{\cdot}66 - 4{\cdot}7415R + 325{\cdot}87(\log R)^2 - 1{\cdot}032 \times 10^5 (\log R)^2/T$$
$$- 1{\cdot}968 \times 10^7 (\log R)^2/T^2 + 1{\cdot}605 \times 10^7 (\log R)^3/T^2, \tag{4.11}$$

and for $R<5$

$$\Delta t_{Mg} = -1{\cdot}03 + 59{\cdot}971 \log R + 145{\cdot}05(\log R)^2$$
$$- 36711(\log R)^2/T - 1{\cdot}67 \times 10^7 \log R/T^2, \tag{4.12}$$

where

$R = [Mg/(Mg+Ca+K)] \times 100$, with concentrations expressed in equivalents.

Δt_{Mg} = the temperature correction in °C that should be subtracted from the Na–K–Ca calculated temperature.

T = the Na–K–Ca calculated temperature in °K.

For some conditions, Equations (4.11) and (4.12) may give negative values for Δt_{Mg}. In that event, do not apply a Mg^{++} correction to the Na–K–Ca geothermometer.

As with other geothermometers, the Mg-corrected Na–K–Ca geothermometer is subject to error owing to continued water-rock reaction as an ascending water cools. If the Mg^{++}

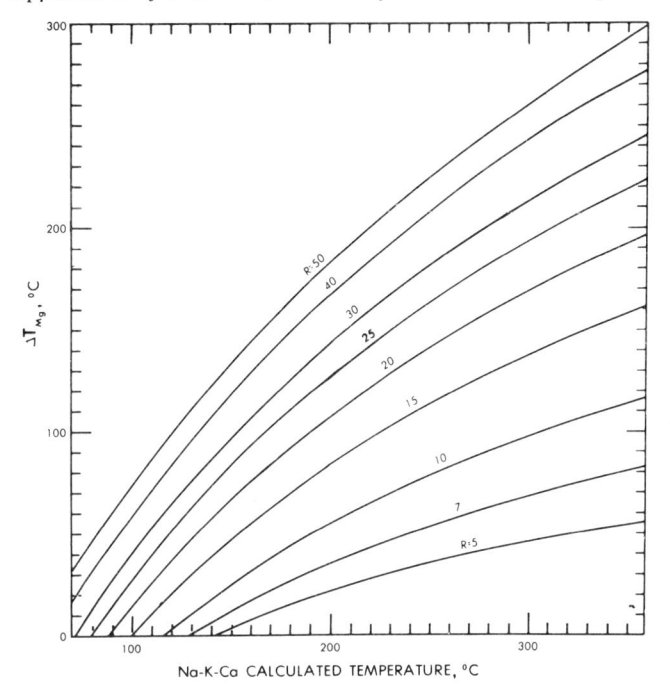

4. Application of Water Geochemistry to Geothermal Exploration

Figure 4.9. Graph for estimating magnesium temperature correction, Δt_{Mg}, to be subtracted from the Na–K–Ca calculated temperature $R = 100 \text{ Mg}/(\text{Mg}+\text{Ca}+\text{K})$, expressed in equivalents. (From Fournier and Potter, 1979)

concentration increases during that upward flow, application of an Mg^{++} correction will lead to an anomalously low calculated reservoir temperature. At this time there is no easy 'rule of thumb' to determine when and when not to apply an Mg^{++} correction to the Na–K–Ca geothermometer. That decision should be based upon the general geologic and hydrologic setting of the particular water that was sampled. However, high magnesium concentrations do indicate that water-rock reactions have occured at relatively low temperatures. Therefore chemical geothermometer results should be used with great caution when applied to Mg-rich waters.

4.3.4. The $\Delta^{18}O(SO_4^= - H_2O)$ Geothermometer

The sulfate oxygen isotope geothermometer is based upon the experimental work of Lloyd (1968), who measured the exchange of ^{16}O and ^{18}O between $SO_4^=$ and water at 350°C, and the work of Mizutani and Rafter (1969) and Mizutani (1972) who measured the exchange of ^{16}O and ^{18}O between H_2O and HSO_4^- at 100° to 200°C. The agreement of the 100° to 200°C HSO_4^- data and the 350°C $SO_4^=$ data and other arguments suggest no fractionation of ^{16}O and ^{18}O between $SO_4^=$ and HSO_4^-. At the pH of most natural fluids SO_4^- is present rather than HSO_4^-.

The rate of equilibration was measured by Lloyd (1968). In the pH range of most deep geothermal waters, the rates of the sulfate oxygen isotope exchange reaction are very slow compared to silica solubility and cation exchange reactions. This can be advantageous for geothermometry, because once equilibrium is attained after prolonged residence time in a reservoir at high temperature, there is little re-equilibration of the oxygen isotopes of sulfate as the water cools during movement to the surface, unless that movement is very slow. Unfortunately, if steam separation occurs during cooling, the oxygen isotopic composition of water will change. The fractionation of ^{16}O and ^{18}O between liquid water and steam is temperature dependent and comes to equilibrium almost immediately at temperatures as low as 100°C. Therefore, if a water cools adiabatically from a high temperature to 100°C, the liquid water remaining at the termination of boiling will have a different isotopic composition depending on whether steam escaped continuously over a range of temperatures or whether all the steam remained in contact with the liquid and separated at the final temperature, 100°C (Truesdell et al., 1977).

Although boiling makes interpretation more complex, it does not preclude use of the $\Delta^{18}O$ ($SO_4^= - H_2O$) geothermometer. McKenzie and Truesdell (1977) showed that sulfate/oxygen isotope geothermometer temperatures could be calculated for three end-member models: (1) conductive cooling, (2) one-step steam loss at any specified temperature, and (3) continuous steam loss. Where water is produced from a well and steam is separated at a known temperature (or pressure), there is no ambiguity in regard to which model to use. Likewise, conductive cooling is assumed for springs that emerge well below boiling.

The validity of temperatures calculated by the $\Delta^{18}O$ ($SO_4^= - H_2O$) method is adversely affected by mixing of different waters (generally hot and cold) unless corrections are made for changes in isotopic composition of both the sulfate and water that result from that mixing. Even if the cold component of the mixture contains no sulfate, calculated sulfate/oxygen isotope temperatures will be in error unless the isotopic composition of the water is corrected back to the composition of the water in the hot component prior to mixing. Examples of calculated corrections for boiling and mixing effects applied to waters from Yellowstone National Park and Long Valley, California, were given by McKenzie and Truesdell (1977) and by Fournier et al. (1979).

The formation of sulfate by oxidation of H_2S at low temperatures is a particularly difficult problem to deal with when applying the sulfate/oxygen isotope geothermometer. A small amount of low-temperature sulfate can cause a large error in the geothermometer result. If analytical data are available for only one or two springs, addition of sulfate by H_2S oxidation may go unnoticed unless pH values are abnormally low. Where analytical data are available for several springs and they all have the same Cl/SO_4 ratio, oxidation of H_2S is probably unimportant. When variations in Cl/SO_4 are found in hot spring waters from a given region, the water with the highest Cl/SO_4 ratio has the best chance of being unaffected by H_2S oxidation.

4.4. Underground Mixing of Hot and Cold Waters

4.4.1. Recognition of Mixed Waters

Mixing of ascending hot water with cold groundwater in shallow parts of hydrothermal systems appears to be common. Mixing can also occur deep in hydrothermal systems, especially at the margins. The effects of mixing upon various geothermometers have been discussed previously.

Where all the thermal waters reaching the surface at a given locality are mixtures of hot and cold water, recognition of that situation can be difficult. The recognition that mixing took place underground is especially difficult where water-rock re-equilibration occurred after mixing. Complete or partial chemical re-equilibration is more likely if the temperature after mixing is well above 110° to 150°C or if mixing takes place in aquifers with long fluid residence times.

Some indications of mixing discussed in Fournier (1979b) are as follows: (1) variations in chloride concentration of boiling springs too great to be explained by steam loss; (2) variations in ratios of relatively conservative elements that do not precipitate from solution during movement of water through rock, such as Cl/B; (3) variations in oxygen and hydrogen isotopes (especially tritium); (4) cool springs with large mass flow rates and much higher temperatures indicated by chemical geothermometers (greater than 50°C); (5) systematic variations of spring compositions and measured temperatures. Generally the colder water will be more dilute than the hotter water. However, in some situations the cold water component could be more concentrated than the hot water, such as where ocean water or closed-basin saline lake water mixes with an ascending hot water. The above indications of mixing may be shown by different compositions of nearby springs or by seasonal variations in a single spring.

4.4.2. The Silica Mixing Model

Under some circumstances the dissolved silica concentration of a mixed water may be used to determine the temperature of the hot water component (Fournier and Truesdell, 1974; Truesdell and Fournier, 1977). The simplest method of calculation uses a plot of dissolved silica vs. enthalpy of liquid water (Figure 4.10). Although temperature is a measured property and enthalpy is a derived property, obtained from steam tables if temperature, pressure and salinity are known (Keenan et al., 1969; Haas, 1976), enthalpy is used as a coordinate rather than temperature. This is because the combined heat contents of two waters at different temperatures are conserved when those waters are mixed (neglecting small heat of dilution effects), but the combined temperatures are not.

For most situations, solutions are sufficiently dilute so that enthalpies of pure water can be used to construct enthalpy-composition diagrams. A straight line drawn from a point representing the non-thermal component of the mixed water (point A, Figure 4.10) through the mixed-water warm spring (point B) to the intersection with the quartz

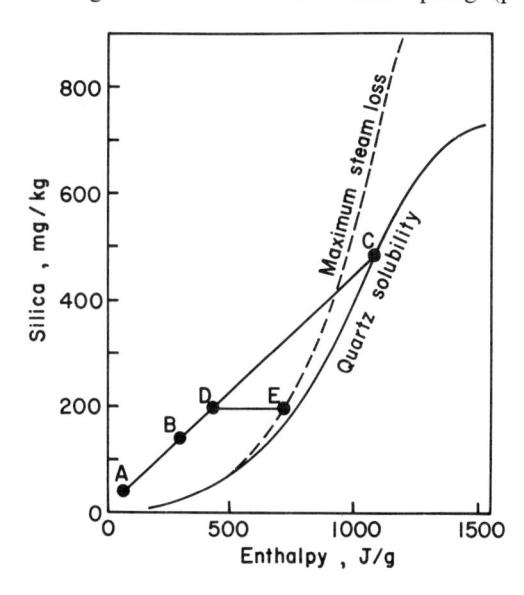

Figure 4.10. Dissolved silica-enthalpy graph showing procedure for calculating the initial enthalpy of a high-temperature water that has mixed with a low temperature water (modified from Truesdell and Fournier, 1977)

solubility curve gives the initial silica concentration and enthalpy of the hot-water component (point C). This procedure assumes that any steam that formed adiabatically as the hot-water component moved up to a more shallow environment did not separate from the residual liquid water before mixing with the cold-water component.

Truesdell and Fournier (1977) discussed the procedure for determining the enthalpy and temperature of the hot water component when steam was lost before mixing took place. As an end-member assumption, consider that steam was lost at atmospheric pressure prior to mixing (point D, Figure 4.10). The horizontal line drawn from point D to the intersection with the maximum steam loss curve gives the initial enthalpy of the hot-water component (point E). If steam had been lost at a higher pressure before mixing, point D would lie above 419 J/g on the extension of line AB, and point E would lie at an appropriate distance between the maximum steam loss and quartz solubility curves.

In order for the above silica mixing model to give accurate results, it is vital that no conductive cooling occurred after mixing. If the mixed water cooled conductively after mixing, the calculated temperature of the hot-water component will be too high. It is also necessary that no silica deposition occurred before or after mixing and that quartz controlled the solubility of silica in the high-temperature water. Even with these restrictions the silica mixing model has been found to give good results in many places. In special circumstances, a silica mixing model could be used in which chalcedony or another silica phase is assumed to control the dissolved silica in the high-temperature component.

4.5. Effects of Underground Boiling

An upflowing hot solution may boil because of decreasing hydrostatic head. If the rate of upflow is fast enough, cooling of the fluid may be approximately adiabatic. Where boiling occurs there is a partitioning of dissolved elements between the steam and residual liquid; dissolved gases and other relatively volatile components concentrate in the steam and non-volatile components become concentrated in the liquid in proportion to the amount of steam that separates. In a boiling process that takes place owing to a decrease in pressure from P_1 to P_2, Truesdell et al. (1977) showed that slightly different amounts of residual liquid water will remain at the end of the process depending upon whether the steam that forms is continuously removed from the system as pressure decreases, or whether the steam maintains contact and re-equilibrates with the cooling water until the steam is removed all at once at P_2 (single-stage steam loss). This distinction between processes involving continuous steam loss and single-stage steam loss is very important when dealing with volatile substances (gases, isotopes), but can be neglected when dealing with non-volatile components.

4.5.1. Calculation of Change in Concentration Resulting from Boiling

For non-volatile components that remain with the residual liquid as steam separates, the final concentration, C_f, after single-stage steam separation at a given temperature, t_f, is given by the formula

$$C_f = \frac{(H_s - H_f)}{(H_s - H_i)} C_i \tag{4.13}$$

where C_i is the initial concentration before boiling, H_i is the enthalpy of the initial liquid before boiling and H_f and H_s are the emthalpies of the final liquid and steam at t_f. For solutions with salinities less than about 10,000 mg/kg, enthalpies of pure water tabulated

4. Application of Water Geochemistry to Geothermal Exploration

in steam tables (Keenan et al., 1969) can be used to solve equation (4.13). For higher salinities, enthalpies of NaCl solutions can be used, such as those tabulated by Haas (1976). Equation (4.13) also can be solved graphically using a plot of enthalpy vs. a non-reactive dissolved constituent (such as chloride), as shown in Figure 4.11. One-step steam loss processes can be represented by straight lines radial to the composition and enthalpy of steam in Figure 4.11. Thus, if the enthalpy and concentrations of constituent x are known prior to boiling (point A in Figure 4.11), the concentrations of x in the residual liquid after boiling can be obtained by extending a straight line from the enthalpy of steam at the final steam separation temperature through point A to the enthalpy of the remaining liquid water after boiling. For example, in Figure 4.11, the initial condition at point A is 300°C and 100 units of x, the final steam separation takes place at 100°C, and point B gives the concentration of x in the residual liquid after steam separation. If final steam separation took place at 200°C, point C would give the concentration of x in the residual liquid. The process can be looked at in reverse. If the final concentration of x is known after steam separation at a given temperature, the initial concentration of x in the non-boiled solution can be determined graphically if the initial enthalpy is known; or the initial enthalpy can be determined if the initial concentration of x is known.

4.5.2. Use of Enthalpy-chloride Diagrams for Estimating Reservoir Temperatures

Where a range in chloride concentration of hot springs appears to result mainly from different amounts of boiling, that range in concentration can give information about the minimum temperature of the reservoir feeding the springs (W. A. J. Mahon in Lloyd, 1972; Truesdell and Fournier, 1976b; Fournier, 1979b). For example, Figure 4.12 shows the chloride range in two chemically distinct types of hot spring waters from Upper Basin in Yellowstone National Park, plotted at the enthalpies corresponding to liquid water at the measured temperature of each spring. The Geyser Hill type waters have ratios of $Cl/(HCO_3 + CO_3)$, expressed in equivalents, greater than four and chloride

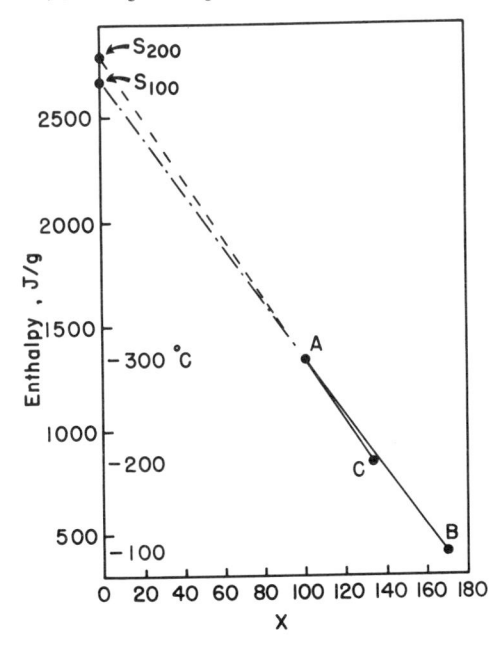

Figure 4.11. Enthalpy-composition diagram showing changes in concentration resulting from adiabatic cooling (boiling) with single-stage steam separation. Points S_{100} and S_{200} show the enthalpies of steam at 100° and 200°C respectively. X is dissolved material (arbitrary units). See text for additional discussion

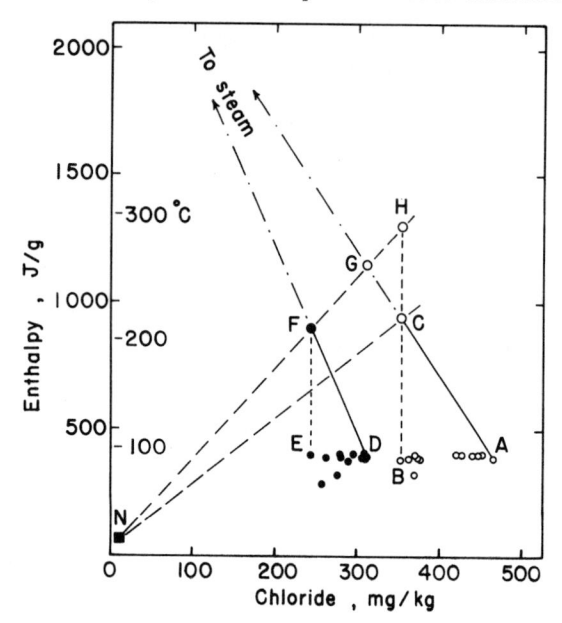

Figure 4.12. Enthalpy-chloride relations for waters from Upper Basin Yellowstone National Park. Small circles indicate Geyser Hill-type waters and small dots indicate Black Sand-type waters

concentrations ranging from 352 to 465 mg/kg. The Black Sand-type waters have Cl/(HCO$_3$ + CO$_3$) ratios close to 0·9 and chlorides ranging from 242 to 312 mg/kg. The minimum temperature of the water in the reservoir feeding the Geyser Hill hot springs can be determined by first drawing a straight line from the spring with maximum chloride (point A) to the enthalpy of steam at 100°C, and then extending a vertical line from the spring with least chloride (point B). The intersection of that vertical line with the previous line, point C, gives the minimum enthalpy of the water in the reservoir, 936 joules, which indicates a temperature of 218°C (Keenan et al., 1969). The silica (quartz) geothermometer applied to water A (assuming maximum steam loss) gave a calculated reservoir temperature of 216°C and applied to water B (assuming no steam loss) gave a temperature of 217°C. The agreement between the calculated reservoir temperatures using silica and chloride relations is strong evidence that the reservoir feeding Geyser Hill has a temperature close to 218°C. The range in chloride concentrations in the Black Sand-type waters, E to D, suggests a reservoir temperature of about 209°C (point F in Figure 4.12), the silica geothermometer gave 205°C.

Although point F in figure 4.12 represents a more dilute and slightly cooler water than point C, water F cannot be derived from water C by simple mixing of hot and cold water (point N) because any mixture would lie on or close to the line CN. Waters C and F are probably both related to a still higher enthalpy water such as G or H. Water F could be related either to water G or H by mixing in different proportions with N. Water C would be related to G by boiling (adiabatic cooling resulting from a decrease in hydrostatic head) and evaporative concentration of chloride in an ascending solution. Water C would be related to H by conductive cooling of a slow moving solution. The route G to C appears more likely because boiling and loss of CO$_2$ into the steam phase can explain the different Cl/(HCO$_3$ + CO$_3$) ratios found in waters C and F. Using additional hot spring

4. Application of Water Geochemistry to Geothermal Exploration 127

data from Midway and Lower Basins, Fournier et al. (1976) concluded that a reservoir at about 270°C underlies the Upper and Lower Basins and that a still hotter reservoir exists at greater depth.

4.5.3. *Effect of Boiling upon Geochemical Thermometers*

The effect of underground boiling owing to decreasing hydrostatic head upon different geothermometers was covered previously as each geothermometer was discussed. In brief, adiabatic cooling affects different geothermometers in various ways: the Na/K geothermometer is not affected; the Na–K–Ca geothermometer is significantly affected only if loss of CO_2 causes $CaCO_3$ to precipitate, resulting in calculated temperatures that are too high; the silica geothermometer must be corrected for separation of steam; and the $\Delta^{18}O(SO_4^= - H_2O)$ geothermometer also must be corrected for steam loss because of the temperature-dependent fractionation of oxygen isotopes between liquid water and steam.

4.6. Vapor-dominated Compared to Hot-Water Systems

The model of a vapor-dominated system formulated by White et al. (1971) is now generally accepted. In that model, relatively impermeable rock and locally derived shallow groundwater provide a cap over a reservoir of considerable vertical extent in which steam is the continuous phase in relatively open channelways and liquid water fills most of the intergranular pore spaces (Figure 4.13). Fluid temperature and pressure increase with increasing depth above the steam zone. Within the steam zone, fluid pressure remains relatively constant because steam weighs very little compared to water, so there is little

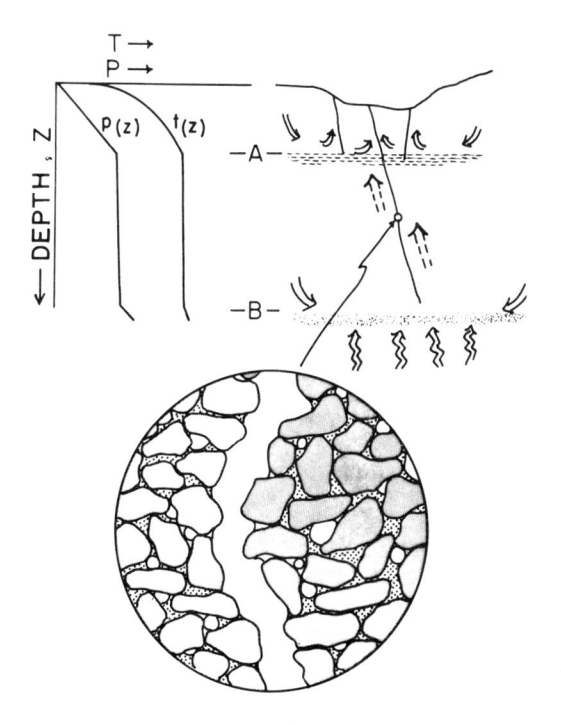

Figure 4.13. Schematic model of conditions in a vapor-dominated geothermal system

128   Geothermal Systems: Principles and Case Histories

change in hydrostatic head. Temperature also is relatively constant throughout the steam zone because the temperature of a mixture of steam and water depends on pressure. Below the steam zone, liquid fills the open channels and pressures and temperatures again increase with increasing depth. In contrast, in a hot-water system, hot water is the continuous phase in the open channels, although bubbles of steam or gas may be present in the water (Figure 4.14). Hydrostatic pressure continuously increases with depth, and the maximum temperature is limited by a boiling point curve (Haas, 1971).

Where a vapor-dominated system is present, it is likely that steam carrying relatively volatile components such as NH_3, CO_2, H_2S, Hg and B will condense in the overlying cap-rock region. The slightly volatile components will redissolve in the condensate, but some of the more volatile components are likely to continue moving up into overlying colder groundwater. Thus, groundwaters and springs over vapor-dominated systems tend to be rich in the above mentioned volatile components of the geothermal fluid or their reaction products and low in non-volatile components, such as chloride, which remain in the residual brine at depth.

Fumaroles, mud pots, acid-sulfate springs with low rates of discharge, sodium bicarbonate spring waters, and acid-altered hot ground are typical surface expressions of vapor-dominated systems. They are not unique to vapor-dominated systems, however, because similar features commonly are present over hot-water systems where underground boiling occurs. Generally, but not always, chloride-rich neutral to alkaline springs emerge above hot-water systems at topographically low places, and acid sulfate springs and mud pots emerge at higher elevations. The Coso, California, geothermal area is an example of a situation where the surface expression of a hot water-dominated system consists only of low-chloride, acid sulfate springs and other features typically equated with vapor-dominated systems. Drilling at Coso Hot Springs showed that a chloride-rich, hot-water system exists at depth with the top of the hot water reaching to within 30 to 45 meters of the ground surface (Fournier et al., 1980; Austin and Pringle, 1970).

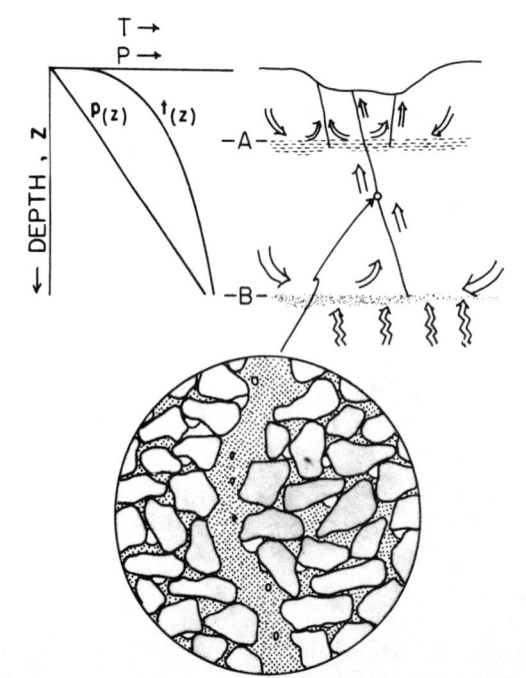

Figure 4.14. Schematic model of conditions in a hot-water-dominated geothermal system where boiling temperatures prevail through a steeply dipping structure filled with water

4.7. Comparison of Thermal Waters from Springs and Wells

Chemical analyses of representative waters from wells and springs at various geothermal localities are listed in Table 4.2.

The enthalpy-chloride diagrams shown in Figure 4.15 were drawn using data from Table 4.2 and assuming cold water components with relatively little chloride and temperatures ranging from 4° to 20°C, depending on latitude and elevation of the geothermal field. Enthalpies of the solutions at known temperatures and salinities were obtained from steam tables (Keenan et al., 1969; Haas, 1976) assuming liquids at the vapor pressure of the solution at the given temperatures and no excess steam. In Figure 4.15a, point 26 (Spring 664) from Orakeikorako, N. Z. lies very close to the straight line extending from the deep hot water component, point 24, to the cold water component. Therefore, it appears there was little conductive cooling after mixing. This is consistent with the relatively high rate of mass flow of Spring 664, 12 L/sec. When additional hot-spring data (W. A. J. Mahon *in* Lloyd, 1972) are employed in the interpretation, it appears that water from a lower temperature aquifer ($\sim 180°$ to $240°C$), such as point P in Figure 4.15a, probably mixes with cold waters to give point 26 (Fournier, 1979b). This conclusion is also supported by the relatively low Na/K temperature of 194°C for water in column 26.

In Figure 4.15b the three Cerro Prieto springs all lie below the line drawn from cold water to the high-temperature deep water, point 13. It is very likely that number 14 cooled by conduction from a temperature near 200°C after mixing. The alkalies in that mixed water have retained the high-temperature imprint of the deep hot-water component but silica has precipitated. Samples 15 and 16 also appear to have cooled by conduction after mixing, but either sample could result by mixing cold water with a hot water from a reservoir with a temperature below 290°C. Alkali geothermometers applied to the mixed waters give temperatures significantly lower than 290°C, indicating considerable water-rock reactions that could have occured in reservoirs with intermediate temperatures.

Hot spring 181 from El Tatio, Chile, point 3 in Figure 4.15c, emerges at the boiling temperature for the elevation. That mixed water could have cooled by conduction, adiabatically, or a combination of these processes. The salinity and maximum enthalpy of the water just after mixing should lie between the points a and b.

The two springs from Ahuachapán, points 7 and 8 in Figure 4.15d, both lie above the mixing line of cold water with deep water. Their position above the line could result (a) from mixing with a high-temperature component having much higher enthalpy than point 6 or (b) from condensation of excess steam derived from boiling water at depth (distillation). Excess steam appears to be the more plausible explanation because the spring waters are relatively rich in volatile components, $SO_4^=$ and HCO_3^-.

In summary, based on the chloride concentrations in spring waters compared to chloride concentrations at depth and in flashed well waters, springs in columns 3, 11, 42, and 45 appear to have cooled mainly adiabatically, springs in columns 22, 25 and 39 partly adiabatically and partly by conduction, and the spring in column 19 by conduction. Based on boron and other dissolved constituents, spring 32 also cooled conductively. The rest of the springs appear to be mixed waters.

The direct application of chemical geothermomcters to the spring waters listed in Table 4.2 gives variable results as shown in Figure 4.16. Chemical geothermometers applied to some of those hot spring waters give accurate indications of reservoir temperatures encountered in drill holes, but most give lower temperatures. However, the results shown in Figure 4.16 are somewhat misleading because in many places reservoirs at intermediate

Table 4.2. Comparison of hot spring and well waters

Column No. Feature	El Tatio, Chile				Ahuachapán, El Salvador				Reykjanes, Iceland		
	1 Well 11 flashed	2 Well 11 at depth[1]	3 Spg. 181	4 Spg. 244	5 Well Ah-1 flashed	6 Well Ah-1 at depth	7 Spg. F	8 Spg. H	9 Well 8 flashed[2]	10 Well 8 at depth[3]	11 Spg. unnamed
Depth, m	—	894	—	—	—	1400	—	—	—	1754	—
Temp °C	85	240	89.5	85	98	232	87	70	100	277	99
pH	6.98	—	6.9	7.30	7.4	—	8.0	6.8	—	6.1	6.2
SiO_2	748	526	122	269	663	490	114	235	988	636	544
Ca	208	146	170	274	416	308	201	29	3480	1530	2260
Mg	0.15	0.11	6	0.4	tr	tr	1	8	25	16	123
Na	4900	3450	2250	4330	6120	4530	768	378	14930	9610	14325
K	825	580	230	525	995	736	18	39	2095	1348	1670
Li	44.9	31.6	—	46	—	—	—	—	—	—	—
HCO_3	41	29	36	27	29	21	52	377	4150[6]	2670[6]	5[6]
SO_4	32	23	36	27	28	21	224	35	47.9	30.8	206
Cl	8716	6130	4009	8126	11046	8172	1528	479	29930	19260	29100
F	—	—	—	—	—	—	—	—	0.15	0.1	0.2
B	202	142	91	183	162	120	20	9.2	—	—	12
T SiO_2	256	257	148	201	247	250	145	191	279	275	232
T Na/K_{RF}	263	262	219	233	259	259	123	220	246	246	229
T Na/K_T	251	251	190	209	246	246	71	191	227	227	205
T Na-K-Ca	260	256	211	231	255	251	90	87	245	240	232
T Na-K-Ca-Mg	260	256	199		255	251	90	87		235	192
Cl/HCO_3	—	364	61	341	—	670	51	2.2	—	124	10013
Cl/SO_4	—	722	302	815	—	1054	18.5	2.2	—	1693	383
Cl/B	—	13.2	13.4	13.5	—	21	23	16	—	—	738
K/Li	—	3.27	—	2.03	—	—	—	—	—	—	—
Cl_{spg}/Cl_{depth}	—	—	0.65	1.33	—	—	0.19	0.06	—	—	1.51
Reference	9	9	9	9	10	10	10	10	11	11	11

4. Application of Water Geochemistry to Geothermal Exploration

Table 4.2. Comparison of hot spring and well waters—continued

Column No. Feature	Cerro Prieto, Mexico					Broadlands, New Zealand			Ngawha, New Zealand		
	12 Well M-26 flashed[4]	13 Well M-26 at depth[1]	14 Spg. 41	15 Spg. 49	16 Spg. 54	17 Well 2 flashed	18 Well 2 at depth[1]	19 Ohaki pool	20 Well 1 flashed	21 Well 1 at depth[1]	22 Jubilee Bath
Depth, m	—	1240	—	—	—	—	1030	—	—	585	—
Temp °C	100	292	40	57	98	99	260	95	99	230	50
pH	8·0	—	2·40	6·50	7·4	8·3		7·05	7·4		6·5
SiO_2	1156	705	247	45	92	805	549	338	460	343	186
Ca	971	592	407	283	492	2·2	1·5				8
Mg	1	0·6	87	20	38	0·08	0·07			—	2·5
Na	10467	6382	4100	1350	3700	1050	716	860	950	708	870
K	2544	1551	1010	233	400	224	143	82	80	60	79
Li	23·8	14·5	12·1	4·4	8	11·7	8·0	7·4	12·2	9·1	11
HCO_3	46	28	0	128	42	128[6]	87	490[6]	61[6]	45	240[6]
SO_4	<6	<3·5	690	960	130	8	5	100	17	13	500
Cl	19548	11918	8410	2930	6700	1743	1188	1060	1625	1211	1336
F	—	—	—	—	—	7·3	5·0	5·2	0·8	0·6	0·3
B	22	13·4	—	—	—	48·4	32·9	32	1200	894	1020
$T\,SiO_2$	203	286	195	97	132	262	261	219	220	220	175
$T\,Na/K_{RF}$	300	300	302	265	223	280	280	213	203	204	209
$T\,Na/K_T$	308	308	310	254	196	277	276	182	170	170	177
$T\,Na$-K-Ca	289	282	277	228	213	294	238				216
$T\,Na$-K-Ca-Mg	—	—	161	178	164	—	—	—	—	—	169
Cl/HCO_3	—	732	—	39	274	—	23	3·7	—	46	9·6
Cl/SO_4	—	—	35	8·3	140	—	643	29	—	252	7·2
Cl/B	—	271	—	—	—	—	11	10	—	—	0·4
K/Li	—	19	15	9·4	8·9	—	3·2	2·0	—	0·4	1·3
Cl_{spg}/Cl_{depth}	—	—	0·71	0·25	0·56	—		0·89	—	1·2	1:10
Reference	12	13	13	13	13	14		15	15		15

Table 4.2. Comparison of hot spring and well waters—continued

	Orakeikorako, New Zealand				Rotokawa, New Zealand			Chingshu, Taiwan		
Column No. Feature	23 Well 2 flashed	24 Well 2 at Depth[1]	25 Spg. 22	26 Spg. 664	27 Well 2 flashed	28 Well 2 at Depth[1]	29 Spg. 6	30 Well 1D-4 flashed	31 Well 1C-4 at Depth[1]	32 Spg. 20
Depth, m	—	1150	—	—	—	880	—	—	1505	—
Temp °C	99	260	99	64	99	220	654	100	195	99
pH	9.1		9.2	8.2	7.8		2.5	8.5		9.7
SiO_2	480	327	280	150	430	329	340	342	280	289
Ca	<1	<1	0.8	4.6	50	38	11	tr	tr	tr
Mg	—	—	—	0.6	—	—	11.3	tr	tr	tr
Na	550	375	370	135	1525	1168	990	1095	896	923
K	54	37	34	10	176	135	102	36	29	39
Li	3.1	2.1	3.4	1.2	10.2	7.8	7.8	7	6	—
HCO_3	290[6]	198	113	201	59[6]	42	144[6]	2807	2296	2586
SO_4	142	97	185	12	120	92	520	32.1	26.2	10.3
Cl	546	372	404	78	2675	2049	1433	18.3	15.0	6
F	5.7	3.9	10.8	2.9	6/6	5.1	—	—	—	—
B	7.7	5.2	3.4	5.4	102	78	45	36	29	30
$T SiO_2$	223	216	187	161	215	217	219	200	204	198
$T Na/K_{RF}$	215	216	210	194	228	229	219	141	140	156
$T Na/K_T$	186	186	179	157	203	204	191	92	91	110
T Na-K-Ca	—	—	223	174	224	221	223	—	—	—
T Na-K-Ca-Mg	—	—	—	138	—	—	88	—	—	—
Cl/HCO_3	—	3.2	6.2	0.69	—	84	17	—	0.01	<.01
Cl/SO_4	—	10.4	5.9	17.6	—	60.4	7.5	—	1.55	1.58
Cl/B	—	22	36.2	4.4	—	8.0	9.7	—	0.16	.06
K/Li	—	3.1	1.8	1.5	—	3.1	2.3	—	0.9	—
Cl_{spg}/Cl_{depth}	—	—	1.09	0.21	—	—	0.70	—	—	0.38
Reference	17	17	17	17	15	15	15	18	19	19

Table 4.2. Comparison of hot spring and well waters—continued

	Tuchang, Taiwan				Kizildere, Turkey		
Column No. Feature	33 Well IT-3 Flashed	34 Well IT-3 at Depth[1]	35 Spg. Unnamed	36 Spg. Unnamed	37 Well KD15 Flashed	38 Well KD15 at Depth[1]	39 Spg. Unnamed
Depth, m	—	445	—	—	—	506	—
Temp °C	100	173	60	96	99	205	100
pH	8·7	—	6·7	8·7	7·8	—	9·0
SiO_2	253	213	25	69	268	214	185
Ca	tr	tr	73	—	6	5	1·2
Mg	tr	tr	17	—	1·2	1·0	0·1
Ma	1100	947	338	710	1172	935	1260
K	15·8	13·6	71	14·7	117	93	125
Li	—	—	—	—	—	—	—
HCO_3	2884	2484	1258	2010	2502	2000	2560
SO_4	36	31	82·3	32	778	621	750
Cl	21	18	14·2	13·4	112	89	103
F	—	—	—	—	16·9	13·5	18
B	30·5	26·3	—	—	25·6	20·4	23
$T\,SiO_2$	183	184	72	118	185	185	174
$T\,Na/K_{RF}$	99	99	117	116	217	216	216
$T\,Na/K_T$	44	44	64	64	187	187	187
$T\,Na-K-Ca$	—	—	73	—	231	228	251[7]
$T\,Na-K-Ca-Mg$	—	—	73	—	—	217	—
Cl/HCO_3	—	0·01	0·02	0·01	0·08	0·08	0·07
Cl/SO_4	—	1·57	0·47	1·13	—	0·39	1·36
Cl/B	—	0·21	—	—	—	1·33	1·36
K/Li	—	—	—	0·74	—	—	—
Cl_{spg}/Cl_{depth}	—	—	0·79	—	20	—	—
Reference	18	20	19	19	20		

Table 4.2. Comparison of hot spring and well waters—*continued*

Yellowstone Park, United States

Column No. Feature	40 Well Y3 Flashed[2]	41 Well Y3 at Depth[3]	42 Ojo Caliente	43 Well Y8 Flashed[2]	44 Well Y8 at Depth[3]	45 Rusty Geyser
Depth, m		88			64	
Temp °C	92	174	95	92	169[8]	92
pH		8·12	8·31		7·9	8·8
SiO_2	—	230	302	258	297	
Ca	1·49	1·26	1·1	1·4	1·2	0·29
Mg	0·02	0·02	0·02	0·05	0·04	<·01
Na	319	270	317	421	360	408
K	13	11	9·2	17·5	15	19
Li	4·1	3·5	4·5	3·0	2·6	2·7
HCO_3	209	177	249	576	493	573
SO_4	22·6	19·1	27	19	16	17
Cl	329	278	331	281	240	292
F	35	30	33	30	26	29
B	4·3	3·6	4·0	3·0	2·6	3·0
$T\,SiO_2$		176	192	198	191	
$T\,Na/K_{RF}$	154	154	134	155	155	162
$T\,Na/K_T$	108	108	84	109	109	118
T Na-K-Ca	167	166	153	174	172	194
T Na-K-Ca-Mg	—	—	—	—	—	—
Cl/HCO_3	—	2·7	2·3	—	0·84	0·88
Cl/SO_4	—	39	33	—	41	47
Cl/B	—	24	25	—	28	30
K/Li	—	0·6	0·4	—	1·0	1·3
Cl_{spg}/Cl_{depth}	—		1·19	—		1·22
Reference	22		22	23		23

[1] Composition calculated from analysis of flashed sample.
[2] Composition calculated from analysis of downhole sample.
[3] Collected using a down-hole sampler, Ave. 11 samples.
[4] Sample collected after flashing at 670 kPa. Analysis recalculated to flashing at 100 kPa.
[5] Total HCO_3^- and $CO_3^=$ recalculated to HCO_3.
[6] Total CO_2 recalculated to HCO_3^-.
[7] Estimated temperature probably is much too high owing to the precipitation of $CaCO_3$.
[8] The well apparently intersected a fracture that taps a deeper reservoir with a temperature of about 190° to 200°C.
[9] Cusicanqui et al. (1976).
[10] Romagnoli et al. (1976).
[11] Björnsson et al. (1972).
[12] Reed (1976).
[13] Mercado (1968).
[14] Mahon and Finlayson (1972).
[15] Ellis and Mahon (1977).
[16] Ellis (1966).
[17] W. A. J. Mahon *in* Lloyd (1972).
[18] Mining Research & Service Organization (1977).
[19] Personal communication, MRSO (1977).
[20] Mineral Research and Exploration Institute, MTA (no date).
[21] Dominco and Samilgil (1970).
[22] Bargar et al. (1973).
[23] Unpublished data, U.S. Geological Survey.

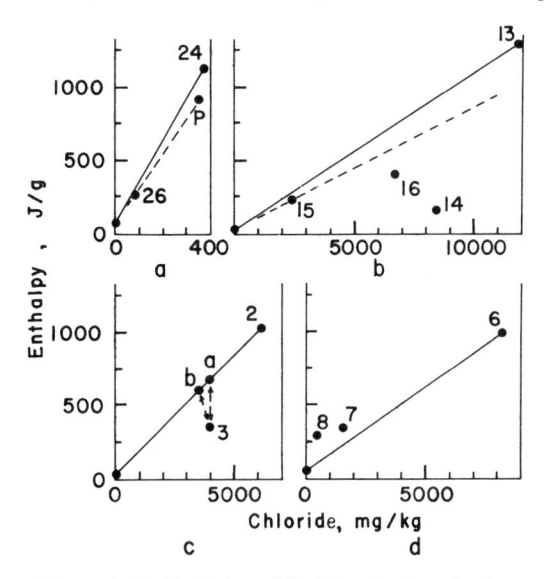

Figure 4.15. Enthalpy-chloride relations for hot-spring and well waters from selected localities listed in Table 4.2. 15a, Orakeikorako; 15b, Cerro Prieto; 15c, El Tatio; and 15d, Ahuachapán. (Numbers in diagrams correspond to columns in Table 4.2)

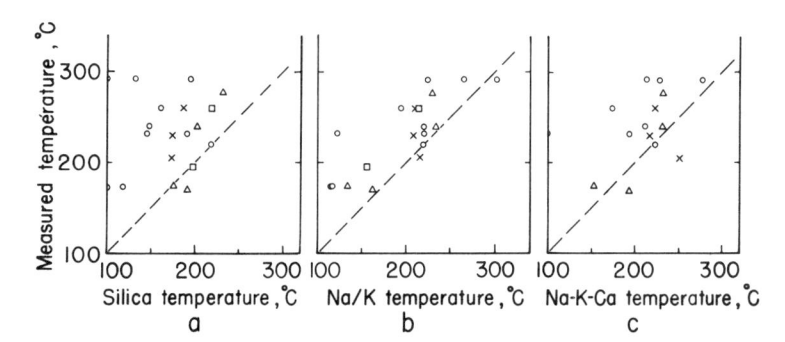

Figure 4.16. Comparison of measured reservoir temperatures in wells and calculated reservoir temperatures using (a) silica, (b) Na/K, and (c) Na–K–Ca geothermometers applied to hot-spring waters listed in Table 4.2. Triangles indicate waters that have cooled adiabatically, squares indicate conductive cooling, X indicates partly conductive and partly adiabatic cooling, and circles are mixtures of hot and cold water

temperatures are present but have been cased off in an attempt to produce only from deeper and higher-temperature parts of the system.

The main generalization that can be drawn from a comparison of the well and spring data shown in Table 4.2 is that waters tend to react chemically with wall rocks after the waters leave deep reservoirs and before emerging at the surface. These reactions may take place in intermediate reservoirs or in the channelways leading to the surface. As temperatures decrease, both HCO_3^- and $SO_4^=$ commonly increase, K^+ decreases relative to Na^+, and Ca^{++} increases relative to Na^+ unless $CaCO_3$ precipitates. The Mg^{++} data are incomplete, but in some places, such as Cerro Prieto and Reykjanes, there are very significant increases in Mg^{++} in the spring waters compared to the deep waters. This increase in Mg^{++} as hot waters cool leads to a dilemma in the application of the Mg correction for the Na–K–Ca geothermometer devised by Fournier and Potter (1979). Waters which were never very hot may give Na–K–Ca temperatures which are much too high unless an Mg^{++} correction is applied. On the other hand, waters which start out hot and react with wall rocks as they cool may give Na–K–Ca temperatures which are closer to the temperature deep in the system than the Mg^{++} corrected temperatures.

Continued water-rock reactions as solutions move up from depth and cool are not always a detriment in regard to obtaining useful information about a given hydrothermal system. The geographic distribution of springs with different water compositions and different gas contents may show directions of underground hot water movement (Truesdell, 1976b) as well as successive chemical re-equilibrations at lower temperatures.

4.8. Applications for Reservoir Engineering

Geochemistry has many applications for reservoir engineering. It provides information about noncondensable gases that decrease turbine efficiency, environmental concerns in regard to waste disposal, and scaling and corrosion associated with production and reinjection. Illustrations of a few other applications are given below.

4.8.1. Early Indication of Aquifer Temperatures in Wells

Geochemical thermometers may be used to estimate aquifer temperatures in wells weeks or months before underground temperatures return to normal after drilling. Flow testing may speed the temperature recovery in the production zone, but interferes with obtaining information about pre-drilling temperatures elsewhere in the well. Also, extensive flow testing immediately after the termination of drilling is not always possible because of limited brine containment or disposal facilities or delayed delivery of test equipment. Production and collection of a small amount of fluid at the wellhead or a downhole water sample, however, may be all that is necessary to provide a good indication of the aquifer temperature.

A 1477 m geothermal exploration well (CGEH No. 1) was drilled near Coso Hot Springs, California, in 1977 (Goranson and Schroeder, 1978). The same day that the well was completed, it was stimulated into production for 1 hour at a rate of 29 to 36 m^3/hr using an air lift. Water samples were collected for chemical analyses every 15 minutes during that production. The Na–K–Ca geothermometer applied to those waters ranged from 191°C at the start to 194°C at the end of the production period, and the Na/K geothermometer gave 195°C to 198°C (Fournier et al., 1980). Before stimulation the maximum temperature in the well was 165°C, and six days after stimulation the maximum temperature was 178°C. After 3 months the maximum temperature in the quiescent well had increased to 187°C, and after 4 months, it was 195°C. It is significant that the first water collected from the well immediately after well completion gave a good indication of the reservoir temperature even though about $1 \cdot 3 \times 10^6$ L of makeup water and $1 \cdot 3 \times 10^5$ kg dry weight of mud were injected into the formation during drilling.

4.8.2. Monitoring Temperature Changes in Production Wells

It is expensive to interrupt production of a geothermal well to run a temperature log. Logging a well is also a relatively slow process, and a temperature survey of a field with several production wells could take many days or weeks. By the time the last well is logged the temperatures in the first wells could have changed. To overcome these disadvantages Mahon (1966) devised a method of monitoring temperatures of waters supplying drillholes at Wairakei using the silica content of that water. Water samples could be collected from all the producing wells the same day without interrupting production. Calculated temperatures were within $\pm 3°C$ of downhole measured temperatures which in turn were good only to $\pm 3°C$. Mahon (1966) also showed that the silica concentration in the water entering wells did decrease in response to decreasing temperatures of the aquifer.

4.8.3. *Confirmation of Production from Multiple Aquifers at Different Temperatures*

Mercado (1970) found that variations in the Na/K ratio in well waters at Cerro Prieto, Mexico were very useful for interpreting well behavior. He discussed well M-20 in detail. Upon completion of that well, water started flowing spontaneously through a small 1·27-cm diameter drain line. Subsequently the rate of discharge was periodically increased by allowing the fluid to escape through larger and larger orifice plates up to 14·6 cm diameter. As the orifice plate diameter was increased, the enthalpy of the discharge decreased and the Na/K ratio in the discharged fluid increased. The variations in the Na/K ratio clearly showed that more than one aquifer supplied water to the well and that the proportion of water from the cooler aquifers increased as the total rate of production from the well was allowed to increase (and downhole pressure decrease).

Variations in the chloride concentration of the produced fluid also can give a good indication of production from different aquifers.

4.8.4. *Flashing in the Reservoir*

Drawdown can cause flashing or boiling in the reservoir before the liquid enters the well. Flashing in the reservoir can result in enthalpies of the discharged fluid that are higher or lower than the initial enthalpy of the liquid in that reservoir.

Where a flashing front moves out into rock away from a well and the fluid pressure in the formation drops below the initial vapor pressure of the solution, the fluid temperature will decrease almost immediately owing to vaporization of liquid. Rock temperatures, however, do not decrease as rapidly as fluid temperatures. Transfer of heat from the reservoir rock to the fluid will cause more steam to form than would form by simple adiabatic expansion of the fluid. If this 'excess' steam enters the well along with the residual liquid, the enthalpy of the discharged fluid will be higher than the enthalpy of the initial fluid in the reservoir. 'Excess' steam has been described in fluids produced from wells at Wairakei and Broadlands, New Zealand (Grindley, 1965; Mahon and Finlayson, 1972) and at Cerro Prieto, Mexico (Truesdell and Mañon, in press).

A relatively low enthalpy of the discharged fluid will result if some or all of the steam fails to enter the well along with the parent water. The steam may escape upward through porous rock or may form a steam cap above the inlet to the well. Where a steam cap forms, enlargement of the steam zone could cause wells to switch eventually from producing fluids with relatively low enthalpy to producing fluids with relatively high enthalpy or even dry steam.

From the foregoing discussion it follows that reservoir temperatures based upon wellhead enthalpy measurements may be too high or too low even where only one aquifer contributes fluid to the well. The application of geochemistry may indicate whether or not flashing is occurring in the reservoir and whether excess or deficient steam accompanies the produced liquid. Truesdell and Mañon (in press) used the silica concentration in flashed (atmospheric pressure) well waters at Cerro Prieto, Mexico to detect excess and deficient steam in the produced fluids. They calculated aquifer temperatures using the silica geothermometer and compared those results with aquifer temperatures calculated from measured enthalpies of the discharged fluids. A higher calculated reservoir temperature, based upon silica, indicates boiling with steam segregation in the formation and liquid water preferentially produced into the well. A calculated reservoir temperature, based upon silica, that is lower than the reservoir temperature based upon enthalpy also suggests boiling in the formation, but excess steam enters the well owing to boiling caused by heat stored in the reservoir rock. The situation is somewhat ambiguous, however, because a relatively low calculated silica temperature also would result if silica precipitated before the water sample was collected for analysis. Even in the absence of silica precipitation in the formation or well, another difficulty is that the silica-enthalpy method might not indicate an outward migrating flashing front unless there is segregation of steam and residual liquid in the formation; without segregation both the silica and enthalpy of the produced fluid will indicate about the same aquifer temperature that is higher than the actual temperature. In some places the above difficulties can be overcome by using the Na/K geothermometer instead of the silica geothermometer (Truesdell et al., 1979). Another approach is to use Na/K geothermometer temperatures in conjunction with silica and enthalpy temperatures to do a better job of evaluating flashing in the formation (Truesdell et al., 1979). Aquifer temperatures calculated independently from silica, Na/K, and enthalpy of the produced fluid can all be in agreement (no indication of flashing in the reservoir), be in partial agreement, or all be different. There are 13 possible combinations of results. Some of the most likely combinations can be interpreted as follows:

Where silica and Na/K temperatures are in near agreement and either higher or lower than the enthalpy temperature, it is likely that the enthalpy temperature is in error owing to flashing in the formation with a disproportionate amount of steam entering the well. Where Na/K and enthalpy agree and silica gives a lower temperature, silica probably precipitated before the sample was collected. Where silica and enthalpy agree and Na/K gives a lower temperature, there may be an outward moving flashing front in the formation without segregation of steam and residual liquid that enter the well.

Where silica and enthalpy agree and Na/K gives a higher temperature, the Na/K value may be residual from a time when the water equilibrated with rock at a higher temperature than that which is present in the immediate aquifer supplying fluid to the well. There are indications at Wairakei and Broadlands, New Zealand, that Na/K does not respond as rapidly as silica to a change in underground temperature (Ellis and Mahon, 1977).

4.8.5. *Evidence of Higher Temperatures Elsewhere within a System*

A comparison of the composition of wellhead and downhole water samples may help verify the existence of a deeper and higher temperature reservoir than that indicated by temperature measurements in a well. An example is at Coso Hot Springs, California, where a shallow 114 m prospect geothermal well was drilled by the U.S. Navy in June 1967. In March 1968, the well was produced by bailing after the bottom hole temperatures had stabilized at 142°C (Austin and Pringle, 1970). A wellhead sample collected toward the end of the bailing activity contained 3042 mg/kg chloride after flashing at atmospheric pressure. Cation geothermometer temperatures applied to that water indicated a reservoir temperature of about 240°C to 250°C. Figure 4.17 shows enthalpy–chloride relations that could yield the wellhead sample, point A. Point B represents the condition of the water at the bottom of the well at 142°C, assuming maximum adiabatic cooling during bailing. The water flowing into the well in response to the bailing could have moved up quickly from the high-temperature reservoir, boiling as it came with little conductive heat loss (route CBA) or it could have cooled entirely conductively before entering the well (route DBA). Line CD shows the probable range in chloride concentration in the deep reservoir, assuming that the reservoir temperature is between 240° and 250°C and cooling is partly adiabatic and partly conductive. In 1978 a downhole sample was obtained and analyzed (Fournier, et al., 1980). The chloride was 2370 mg/kg and geothermometer temperatures were Na–K–Ca = 234°, Na/K = 231°, and $\Delta^{18}O(SO_4^= - H_2O) = 243°C$.

In Figure 4.17 point E shows the condition of the downhole sample at the point of collection, and point F gives the calculated condition according to the goethermometers. The difference in chloride between the wellhead sample after flashing (point A) and without flashing (point E) indicates that flashing occurred in the formation in response to the bailing and that the minimum temperature in the reservoir supplying water to the well should be at least 215°C. The near agreement of all the geothermometer temperatures applied to the wellhead and downhole samples strongly suggests that the actual temperature of the reservoir supplying water to the well is near 240°C and that water from that reservoir cooled partly adiabatically and partly by conduction when the well was bailed.

4.8.6. Geochemical Evidence of Drawdown

In a vapor-dominated system, the movement of a recharge water into the production zone as a result of drawdown may be advantageous for the efficient extraction of heat stored in the rock without significantly lowering the reservoir temperature and pressure. In hot-water systems, however, the rapid influx of surrounding cold water or reinjected waste water may significantly lower the temperature of the produced fluid in a few years. Where there is an influx of cold water into the production zone, Nathenson (1975) showed that chemical changes should precede thermal changes and that the elapsed time between these changes is related to the porosity of the rock. With 0·2 porosity, chemical changes

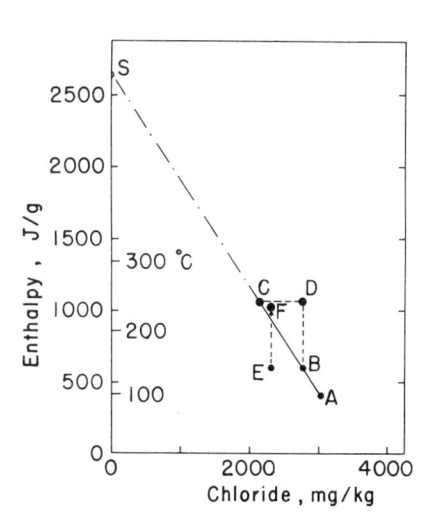

Figure 4.17. Enthalpy-chloride relations for waters from a 114 m well at Coso Hot Springs, California. Modified from Fournier et al. (1980)

that appear X years after the start of production indicate that thermal changes will appear about 3·4 X years later.

At Larderello, Italy, the movement of recharge water into the vapor-dominated system is indicated by the appearance of tritium in steam produced from wells at the margins of the production zone (Celati et al., 1973). In the hot-water system at Cerro Prieto, Mexico, drawdown of fluids from an overlying lower-temperature aquifer is shown in the southeastern part of the field by a decrease in the chloride concentration in well water and an increase in the Na/K ratio (Truesdell et al., 1980).

4.9. Summary

The ratios of the cations in geothermal fluids are controlled by temperature-dependent water-rock reactions with chemical equilibrium attained in high-temperature reservoirs when fluid residence times are relatively long (years).

At present the most useful geochemical thermometers or geothermometers are silica, Na/K, Na–K–Ca, and $\Delta^{18}O(SO_4^= - H_2O)$. Each of these geothermometers requires special consideration in its application. In many places, some or all of these geothermometers applied to hot spring waters give good indications of deep reservoir temperature. In other places, however, these geothermometers give information only about shallow reservoirs containing more dilute and lower-temperature fluids than are present in deeper reservoirs. Under some conditions mixing models may be used to estimate reservoir temperatures and salinities in deeper reservoirs than is otherwise possible.

Geochemistry can be used to estimate reservoir temperatures encountered by newly drilled wells long before temperature logs give a good indication of predrilling conditions. Geochemistry can also be a powerful and sensitive tool for detecting changes in a reservoir during production, particularly when used in conjunction with physical measurements at the wellhead.

References

Arnórsson, S., 1975. 'Application of the silica geothermometer in low temperature hydrothermal areas in Iceland', *Am. J. Sci.*, **275**, 763–784.

Arnórsson, S., 1978. 'Major element chemistry of the geothermal sea-water at Reykjanes and Svartsevgi, Iceland', *Miner. Mag.*, **42**, 209–220.

Austin, C. F., and Pringle, J. K., 1970. 'Geologic investigations at the Coso Thermal Area', *Naval Weapons Center Technical Publication* 4878, 40 pp.

Bargar, K. E., Beeson, M. H., Fournier, R. O., and Muffler, L. J. P., 1973. 'Present-day lepidolite from thermal waters in Yellowstone National Park', *Amer. Mineral.*, **58**, 901–904.

Björnsson, S., Arnórsson, S., and Tómasson, J., 1972. 'Economic evaluation of Reykjanes thermal brine area, Iceland', *Amer. Assoc. Petroleum Geologists Bull.*, **56**, 2380–2391.

Bodvarsson, G., 1960. 'Exploration and exploitation of natural heat in Iceland', *Bull. Volcanol.*, **23**, 241–250.

Browne, P. R. L., 1978. 'Hydrothermal alteration in active geothermal fields', *Annual Review in Earth and Planetary Sciences*, **6**, 229–250.

Browne, P. R. L., and Ellis, A. J. 1970. 'The Ohaki-Broadlands hydrothermal area, New Zealand, mineralogy and related geochemistry', *Amer. J. Sci.*, **269**, 97–131.

Busey, R. H., and Mesmer, R. E., 1977. 'Ionization equilibria of silicic acid and polysilicate formation in aqueous sodium chloride solutions to 300°C', *Inorg. Chem.*, **16**, 2444–2450.

Celati, R., Noto, P., Panichi, C., Squarci, P., and Taffi, L., 1973. 'Interactions between the steam reservoir and surrounding aquifers in the Larderello Geothermal Field', *Geothermics*, **2**(3–4), 174–185.

Cusicanqui, H., Mahon, W. A. J., and Ellis, A. J., 1976. 'The geochemistry of the El Tatio geothermal field, Northern Chile', *Second United Nations Symposium on the Development and Use of Geothermal Resources*, San Francisco, May, 1975, **1**, 703–711.

Dominco, E., and Samilgil, E., 1970. 'The geochemistry of the Kizildere geothermal field, in the framework of the Saraykey–Denizli geothermal area', *United Nations Symposium on the Development and Utilization of Geothermal Resources, Pisa, 1970*, **2**, part 1, 553–560.

Ellis, A. J., 1966. Volcanic hydrothermal areas and the interpretation of thermal water compositions', *Bull. Volcanologique*, **29**, 575–584.

Ellis, A. J., 1969. 'Pres-day hydrothermal systems and mineral deposition', *Proceedings Ninth Commonwealth Mining and Metallurgical Congress, Mining and Petroleum Geology Section*, The Institution of Mining and Metallurgy, London, p. 1–30.

Ellis, A. J., 1970. 'Quantitative interpretation of chemical characteristics of hydrothermal systems', *Geothermics*, Special Issue 2, **2**(1), 516–528.

Ellis, A. J., and Mahon, W. A. J., 1967. 'Natural hydrothermal systems and experimental hot-water/rock interactions (Part 2)', *Geochim. Cosmochim. Acta*, **31**, 519–539.

Ellis, A. J., and Mahon, W. A. J., 1977. *Chemistry and Geothermal Systems*, Academic Press, New York, 392 pp.

Forunier, R. O., 1973. 'Silica in thermal waters: Laboratory and Field investigations', *Proceedings of International Symposium on Hydrogeochemistry and Biogeochemistry, Japan 1970, Volume 1, Hydrogeochemistry*: Washington, D.C., J. W. Clark, 122–139.

Fournier, R. O., 1977. 'Chemical geothermometers and mixing models for geothermal systems', *Geothermics*, **5**, 41–40.

Fournier, R. O., 1979a. 'A revised equation for the Na/K geothermometer', *Geothermal Resources Council Transactions*, **3**, 221–224.

Fournier, R. O., 1979b. 'Geochemical and hydrologic considerations and the use of enthalpy-chloride diagrams in the prediction of underground conditions in hot spring systems', *J. Volcanol. Geotherm. Res.*, **5**, 1–16.

Fournier, R. O. and Potter, R. W. II, 1979. 'Magnesium correction to the Na–K–Ca chemical geothermometer', *Geochim. Cosmochim. Acta*, **43**, 1543–1550.

Fournier, R. O., and Rowe, J. J., 1966. 'Estimation of underground temperatures from the silica content of water from hot springs and wet-steam wells', *Am. J. Sci.*, **264**, 685–697.

Fournier, R. O., Sorey, M. L., Mariner, R. H., and Truesdell, A. H., 1979. 'Chemical and isotopic prediction of aquifer temperatures in the geothermal system at Long Valley, California', *J. Volcanol. Geothermal Res.*, **5**, 17–34.

Fournier, R. O., Thompson, J. M., and Austin, C. ¹ , 1980. 'Interpretation of chemical analyses of waters collected from two geothermal wells at Coso, California', *J. Geophys. Res*, **85**, 2405–2410.

Fournier, R. O., and Truesdell, A. H., 1970. 'Chemical indicators of subsurface temperature applied to hot spring waters of Yellowstone National Park, Wyoming U.S.A.', *Geothermics*, Special Issue 2, **2** (1) 529–535.

Fournier, R. O., and Truesdell, A. H., 1973. 'An empirical Na–K–Ca geothermometer for natural waters', *Geochim. Cosmochim. Acta*, **37**, 1255–1275.

Fournier, R. O., and Truesdell, A. H., 1974. 'Geochemical indicators of subsurface temperature—Part 2, estimation of temperature and fraction of hot water mixed with cold water', *J. Res. U.S. Geol. Survey*, **2**, 263–270.

Fournier, R. O., White, D. E., and Truesdell, A. H., 1976. 'Convective heat flow in Yellowstone National Park', *Second United Nations Symposium on the Development and Use of Geothermal Resources, San Francisco, May, 1975*, **1**, 731–739.

Goranson, C., and Schroeder, R., 1978. 'Static downhole characteristics of well CGEH-1 at Coso Hot Springs, China Lake, California', *Lawrence Berkeley Laboratory Report LBL-7059*, 27 pp.

Grindley, G. W., 1965. 'The geology, structure, and exploitation of the Wairakei geothermal field, Taupo, New Zealand', *New Zealand Geol. Survey Bull.* **75**, 131 pp.

Haas, J. L., Jr., 1971. 'Effect of salinity on the maximum thermal gradient of a hydrothermal system at hydrostatic pressure', *Econ. Geol.*, **66**, 940–946.

Haas, J. L., Jr., 1976. 'Thermodynamic properties of the coexisting phases and thermochemical properties of the NaCl component in boiling NaCl solutions', *U.S. Geol. Survey Bull.*, 1421-B, 71 pp.

Helgeson, H. C., 1969. 'Thermodynamics of hydrothermal systems at elevated temperatures and pressures', *Am. J. Sci.*, **267**, 729–804.

Helgeson, H. C., Brown, T. H., and Leeper, R. H., 1969. *Handbook of Theoretical Activity Diagrams Depicting Chemical Equilibria in Geological Systems Involving an Aqueous Phase at One Atm and 0° to 300°C*, Cooper, San Francisco, 253 pp.

Keenan, J. H., Keyes, F. G., Hill, P. G., and Moore, J. G., 1969. *Steam Tables (International Edition–Metric Units)*, Wiley, New York, 162 pp.

Kennedy, G. C., 1950. 'A portion of the system silica-water', *Econ. Geol.*, **45**, 629–653.

Kharaka, Y. K., and Barnes, Ivan, 1973. 'SOLMNEQ: Solution-mineral equilibrium computations', U.S. Geological Survey Computer Center, U.S. Department of Commerce, National Technical Information Service, Springfield, Virginia 22151, Report PB-215899, 82 pp.

Lloyd, E. F., 1972. 'Geology and hot springs of Orakeikorako', *New Zealand Geol. Survey Bull.*, **85**, 164 pp.

Lloyd, R. M., 1968. 'Oxygen isotope behavior in the sulfate-water system', *J. Geophys. Res.*, **73**, 6099–6110.

Mahon, W. A. J., 1966. 'Silica in hot water discharged from drillholes at Wairakei, New Zealand', *New Zealand J. Sci.*, **9**, 135–144.

Mahon, W. A. J., and Finlayson, J. B., 1972. 'The chemistry of the Broadlands geothermal area, New Zealand', *Am. J. Sci.*, **272**, 48–68.

McKenzie, W. F., and Truesdell, A. H., 1977. 'Geothermal reservoir temperatures estimated from the oxygen isotope compositions of dissolved sulfate and water from hot springs and shallow drillholes', *Geothermics*, **5**, 51–61.

Mercado, S., 1968. 'Localizacion de zonas de maxima actividad hidrotermal por medio de proporciones quimicas. Campo geotermico Cerro Prieto, Baja California, Mexico', *III Congreso Mexicano de Quimica Pura y Aplicada: Comission Federal de Electricidad. Comision de Energia Geotermica*, 32 pp.

Mercado, S., 1970. 'High activity hydrothermal zones detected by Na/K, Cerro Prieto, Mexico', *Geothermics*, Special Issue 2, 1367–1376.

Mineral Research and Exploration Institute (No date). 'Kizildere geothermal field in Turkey', Pamphlet published by the MTA, 12 pp.

Mining Research and Service Organization, 1977. 'Geothermal energy in Taiwan, Republic of China', *MRSO (1977) Report-162*, 10 pp.

Mizutani, Y., 1972. 'Isotopic composition and underground temperature of the Otake geothermal water, Kyushu, Japan', *Geochem. J. (Japan)*, **6**(2), 67–73.

Mizutani, Y. and Rafter, T. A., 1969. 'Oxygen isotopic composition of sulphates—part 3. Oxygen isotopic fractionation in the bisulfate ion-water system', *New Zealand J. Sci.*, **12**(1), 54–59.

Morey, G. W., Fournier, R. O., and Rowe, J. J., 1962. 'The solubility of quartz in water in the temperature interval from 29° to 300°C', *Geochim. Cosmochim Acta*, **26**, 1029–1043.

Morey, G. W., Fournier, R. O., and Rowe, J. J., 1964. 'The solubility of amorphous silica at 25°C', *J. Geophys. Res*, **69**, 1995–2002.

Nathenson, Manuel, 1975. 'Physical factors determining the fraction of stored energy recoverable from hydrothermal convection systems and conduction dominated areas', *U.S. Geological Survey Open-file Report 75–525*, 51 pp.

Reed, M. J., 1976. 'Geology and hydrothermal metamorphism in the Cerro Prieto geothermal field, Mexico', *Second United Nations Symposium on the Development and Use of Geothermal Resources, San Francisco, May, 1975*, **1**, 539–547.

Robie, R. A., Hemingway, B. S., and Fisher, J. R. 1978. 'Thermodynamic properties of minerals and related substances at 298·15 K and one bar pressure and at higher temperatures', *U.S. Geological Survey Bull.*, **1452**, 456 pp.

Romagnoli, P., Cuéllar, F., Jimenez, M., and Ghezzi, G., 1976. 'Aspectos Hidrogeológicos del Campo Geotermico de Ahuachapán, El Salvador', *Second United Nations Symposium on the Development and Use of Geothermal Resources, San Francisco, May, 1975*, **1**, 563–570.

Seward, T. M., 1974. 'Determination of the first ionization constant of silicic acid from quartz solubility in borate buffer solutions to 350°C', *Geochim. Cosmochim. Acta*, **38**, 1651–1664.

Truesdell, A. H., 1976a. 'Summary of section III–geochemical techniques in exploration', *Second United Nations Symposium on the Development and Use of Geothermal Resources, San Francisco, May, 1975*, **1**, liii–lxiii.

Truesdell, A. H., 1976b. 'Chemical evidence of subsurface structure and flow in geothermal system', *Proceedings Int. Symp. on Water-Rock Interaction, Czechoslovakia, 1974*, 250–257.

Truesdell, A. H., and Fournier, R. O., 1976a. 'Calculation of deep temperatures in geothermal systems from the chemistry of boiling spring waters of mixed origin', *Second United Nations Symposium on the Development and Use of Geothermal Resources, San Francisco, May, 1975*, **1**, 837–844, U.S. Gov. Printing Office.

Truesdell, A. H. and Fournier, R. O., 1976b. 'Conditions in the deeper parts of the hot spring systems of Yellowstone National Park, Wyoming', U.S. Geological Survey Open-File Report, 76–428, 22 pp.

Truesdell, A. H. and Fournier, R. O., 1977. 'Procedure for estimating the temperature of a hot water component in a mixed water using a plot of dissolved silica vs enthalpy', *J. Res. U.S. Geological Survey*, 5(1), 49–52.

Truesdell, A. H., and Hulston, J. R. (1980). 'Isotopic evidence on environments of geothermal systems'. Handbook of environmental isotope geochemistry, V.1: The terrestrial environment, P. Fritz and J. C. Fontes, (Eds.), Elsevier Press, p. 179–225.

Truesdell, A. H., and Mañon, A. (in press). 'Geochemical indications of boiling in the aquifer of The Cerro Prieto Geothermal field', *Primera Reunion de Intercambio Tecnico Sobre Geotermia, San Filipe, B.C., 1977*.

Truesdell, A. H., Mañon, A., Jimenez, M., Sanchez, A., and Fausto, J. (1979). 'Geochemical evidence of drawdown in the Cerro Prieto, Mexico geothermal field', *Symposium on the Cerro Prieto, Baja California, Mexico Geothermal Field*, San Diego, September 1978. LBL report 7098, p. 130–138.

Truesdell, A. H., Nathenson, M., and Rye, R. O., 1977. 'The effects of subsurface boiling and dilution on the isotopic compositions of Yellowstone thermal waters', *J. Geophys. Res.*, 82(26), 3694–3703.

Truesdell, A. H., and Singers, W., 1974. 'Computer calculation of down-hole chemistry in geothermal areas', *U.S. Geological Survey J. Res.*, 2(3), 271–278.

Watson, J. C., 1978. 'Sampling and analysis methods for geothermal fluids and gases', Battelle publication PNL-MA-572, UC-66d, available from NTIS, Springfield, Va.

White, D. E., 1965. 'Saline waters of sedimentary rocks', *Amer. Assoc. Petroleum Geologists*, Mem. 4, 352–366.

White, D. E., 1970. 'Geochemistry applied to the discovery, evaluation, and exploitation of geothermal energy resources', *Geothermics*, Special Issue 2, 1, 58–80.

White, D. E., Muffler, L. J. P., and Truesdell, A. H., 1971. 'Vapor-dominated hydrothermal systems compared with hot-water systems', *Econ. Geology*, 66, 75–97.

Wildman, L. D., Jackson, M. L., and Whittig, L. D., 1968. 'Serpentine rock dissolution as a function of carbon dioxide pressure in aqueous solution', *Amer. Miner.*, 53, 1250–1263.

Geothermal Systems: Principles and Case Histories
Edited by L. Rybach and L. J. P. Muffler
© 1981 John Wiley & Sons Ltd.

5. Heat Extraction from Geothermal Reservoirs

Ian G. Donaldson

*Physics & Engineering Laboratory, Department of
Scientific & Industrial Research, Lower Hutt
New Zealand*

Malcolm A. Grant

*Applied Mathematics Division, Department of
Scientific & Industrial Research, Wellington,
New Zealand*

5.1. Introduction

In the practical sense the mechanisms available for the extraction of heat from geothermal reservoirs are rather limited. We may mine the fluids in the reservoir and depend on the natural replenishment of these fluids to sustain the fluid supply, or we may artificially circulate fluid through the reservoir to, in effect, mine the heat alone.

The important questions in the extraction of heat from a geothermal reservoir are how much? for how long? and of what quality? It is to the physical processes relevant to these questions that we will address the majority of our effort in this chapter. We will not consider either the economic or engineering aspects of such extraction.

Geothermal reservoirs are conveniently categorized as either 'vapor-dominated' or 'liquid-dominated'. In each case the name refers to the dominant mobile phase in the reservoir in its undisturbed state. The gradient of pressure with depth is close to that of a static column of that dominant mobile phase. The other phase may also be present and partly mobile. Thus, vapor-dominated systems contain immobile or slightly mobile water, and liquid-dominated systems may either contain liquid water only, or a steam-water mixture. It is also possible for a single reservoir to contain sections of each type. Vapor reservoirs, for example, usually have a liquid-dominated 'condensate layer' near the surface (White et al., 1971). Most reservoirs are liquid-dominated. These latter are further subdivided in this chapter into warm water, hot water and two-phase reservoirs according to how they may react under exploitation. In warm water reservoirs no boiling at all is anticipated under any expected pressure drawdowns; in hot water reservoirs any boiling is only expected to take place in the reservoir at levels above that at which almost all exploitation will take place; while in the two-phase reservoirs at least a significant proportion of the exploitation is expected to take place from within the two-phase (steam-water) zone.

The total heat contained in a geothermal reservoir may be computed if the volume of the reservoir is known and the fluid and rock properties and characteristics can be estimated. The temperature (or enthalpy) of the fluid and the associated rock, together with the reservoir volume, are the primary factors in this heat content estimation. The temperature alone is the dominant factor in the quality estimation—the higher the temperature (or enthalpy) the higher the quality of the heat. When, however, this reservoir is mined, only a fraction of this heat energy may reach the end point of the operation. This may be due to several factors, but we may bracket these into two main categories—the efficiency of the mining operation, and the efficiency of the surface engineering procedures. This latter must depend upon the end use of the energy, as well as on many local factors. The final overall efficiency of utilisation of the geothermal reservoir requires independent study of each field and use situation. It is thus appropriate here to consider only the below ground effects, i.e. the efficiency of the mining operation.

The rate of removal of heat and fluid and the temperature of any reinjected fluid may be controlled by the end use demands. However, the below-ground processes, the actual extraction of the heat energy from the geothermal reservoir, can be studied in general terms as if they were independent of that end use. Each form, or type, of reservoir (warm water, hot water, two-phase, or vapor-dominated, generally permeable or fracture controlled, etc.) may, however, behave differently under exploitation. We therefore discuss the heat extraction from each type of reservoir in order of increasing 'dryness', starting with the 'wet' warm water systems and ending with the vapor-dominated ones. Within each category we also look at the effects of uniform permeability and/or fracture control and of naturally enhanced and artificial recharge, where these are relevant. The field examples offered are those best known to the authors, and hence, there is a strong New Zealand bias in the selection. On the other hand, these fields are all well documented so that data and experience are available for analysis and discussion. The reservoir types that do not conveniently fit our sequence—geopressured, hot-dry-rock and magma—are introduced separately at the end.

There are three themes that we emphasise throughout:

The *first* is the great difference between single-phase (water) and two-phase (water-steam) flow, as it affects pressure transmission and extraction of fluid and heat from the reservoir.

The *second* theme is that, having accepted this one division, there is a continuity in the field response, from warm-water, through liquid-dominated, to vapor-dominated systems. The continuity is provided by a scale along which a field's two-phase portion grows in importance and the two-phase region becomes drier.

The *third* theme is that the extraction of geothermal heat is a mining operation, the heat being removed from a fixed store. This store is replenished to a certain extent by the natural pre-exploitation flow, which continues after exploitation. This flow of hot water is, however, very little enhanced by any effects of the extraction. As will be seen, the pressure drop in the reservoir draws in cooler waters from around the sides, rather than additional hot waters from below.

5.1.1. *Some Background Considerations*

Before we discuss a simple basic model of a geothermal system and the behaviour of the fluid in this system under exploitation we clarify two important concepts on which some of the arguments will be based. Namely, we discuss the importance of the various flow components, and the basic well-to-reservoir link and its effect on pressures and output enthalpies. To be able to generalize our later discussion we also touch briefly on any relevant effects of likely contaminants in the fluids in geothermal systems.

(i) *Horizontal and vertical flows in geothermal systems.* Much of the international geothermal research effort naturally draws upon groundwater and petroleum engineering experience. This experience does, however, lead to a standard model of flow in confined or unconfined horizontal aquifers, and this model can be misleading in many geothermal situations. In the natural state of a geothermal system there is a strong vertical circulation. Vertical permeability must therefore exist, at least in some sections of a field. El Tatio in Chile is an atypical field in that it is apparently horizontally confined (Healy and Hochstein, 1973). It is, however, probably an outflow from a larger system with vertical circulation within the Andean mountain chain.

Although detailed well response analysis generally tends to indicate two-dimensional flows, and there are other features that suggest that some horizontal structuring predominates, these are smaller-scale features of the reservoirs. Both long-term behaviour and overall reservoir response, the indicators of the macroscopic behaviour of the system, show strong three-dimensional trends. We may thus talk of the system in the neighbourhood of individual wells in a manner that implies two-dimensional flow dominance, but in any discussion of field behaviour the three-dimensional character of the system must be taken into account.

This difference between small- and large-scale behaviour is almost certainly due to the fractured nature of most systems. Individual well tests give widely different transmissivities (permeability × aquifer thickness) reflecting the varying nature of the local fissure system from one part of the field to another. At this scale these fissures act as two-dimensional channels. On the reservoir scale, however, these channels interlink to form a full three-dimensional network. On this scale, in the two-phase zone of the reservoir, the dominant physical mechanism may be vertical motion of the two phases under gravity; in the deeper, liquid zone, the flow may be as if to a sink in a general three-dimensional medium.

(ii) *The feed point of a well.* A concept mentioned a number of times in this chapter is the feed point or feed zone of a well. Nearly all geothermal fields, excepting perhaps one such as Heber (California, U.S.A.), which taps a sandy aquifer, have significant fracture permeability. The flow to the well is not supplied by a uniform inflow over the section that penetrates some specific aquifer. Rather, the well intercepts one or a few distinct fractures. As there is usually one of these that dominates, almost all of the well's production will come from an interval of a few metres or less. This feed zone is the point in the well which will then control pressures throughout the remainder of its length. Away from this feed point pressures in the well will simply reflect the weight of the fluid column above or below this junction. This point will also often provide the one point in the well at which the well fluids are in pressure equilibrium with those in the adjacent rock. Alternatively, however, the fissure system may itself become an extension of the well, and such equilibrium may exist at only one point along its own path length.

If there is a water level in the field below which the rock is water-saturated and above which water and steam co-exist, a well will reflect the depth of its feed zone in its behaviour. 'Water-fed' wells, i.e. those feeding from below the water level, produce fluid corresponding in enthalpy to liquid water under the local reservoir conditions. The fluid entering the well is liquid water or fluid that was originally liquid at a distance but boiled on the way to the well; i.e. there may thus be a local two-phase zone surrounding the feed point of the well. Flow through this zone will, however, be isenthalpic so that the enthalpy of the fluid mixture flowing into the well will match that of the original water at a distance. In contrast, 'two-phase' wells, which feed from the established two-phase zone, will produce a fluid of an enthalpy above that of liquid water at the feed zone temperature. Both types of well occur at Wairakei, with the two-phase wells having enthalpies ranging from only a little above that of the appropriate liquid water to that of wet, dry or even super-heated steam.

(iii) *Chemical contaminants and their effects.* Chemical contaminants occuring in geothermal reservoirs are of two types, dissolved solids and gasses. In most situations the dissolved solids are primarily of environmental concern and should have little effect on the heat extraction process. True, high concentrations of materials like calcite, as in Kizildere, Turkey, may cause such rapid clogging of wells that field production is significantly reduced. Nonetheless, if the wells and nearby rock and the associated surface plant could be kept clear, the dissolved solids would have little effect on the response of the reservoir to production. The same applies with other nuisance materials like the salt in Imperial Valley, U.S.A. Provided the fluid can be extracted and the energy utilized, the reservoir will behave much as if the geothermal fluid was of low or even nil contamination.

In contrast, any gaseous contaminants can appreciably alter the reservoir characteristics and hence, the field performance. Roughly, for higher temperature systems, the more gas present the more extensive are the naturally occuring two-phase conditions. But this is really only a shifting of the character of the reservoir along our already defined scale, towards the 'drier' conditions. Thus, provided the two-phase conditions can be defined and the gas content allowed for, the process will still fit within the range of systems considered.

5.2. Idealized Systems

5.2.1. *A Simple Basic Model*

In the introduction to this chapter we defined the primary questions relating to the utilization of energy from a geothermal reservoir as being (i) how long the resource will last under exploitation, (ii) how much heat (or energy) can we extract in that time, (iii) of what quality would this energy be? We can get some insight into the answers to these questions and into the behaviour of the reservoir under exploitation by a study of a simple general model.

The model we have chosen, and illustrated in Figure 5.1, is based on a model first produced by McNabb (1975) for the Wairakei system. The reservoir is pictured as a vertical cylinder, surrounded by cold water. In the illustration this reservoir consists of a lower hot water-saturated zone, overlain by a two-phase (steam-water) zone. Above this is a near-surface zone involving mixing with cold groundwater. This arrangement is currently in use in studies of Wairakei in New Zealand. McNabb (1975) omitted the upper cold mixing layer; removal, or expansion, of other zones might give a better match with other fields. Removal of the two-phase zone would, for example, bring the model into conformity with the warm water system; or the enthalpy or gas content might be high enough that only the two-phase zone need be pictured within the hot zone, thus giving a match with fields like Broadlands ($1\frac{1}{2}$ km to liquid) or Ngawha (>4 km to liquid) in New Zealand. At the high enthalpy limit, we have the vapor-dominated reservoir featuring a two-phase reservoir with counterflowing steam and water, and the deep water surface only a conjecture. Always present, however, is the surrounding cold water. Even the vapor-dominated systems exist in an environment of cold groundwater, although here low

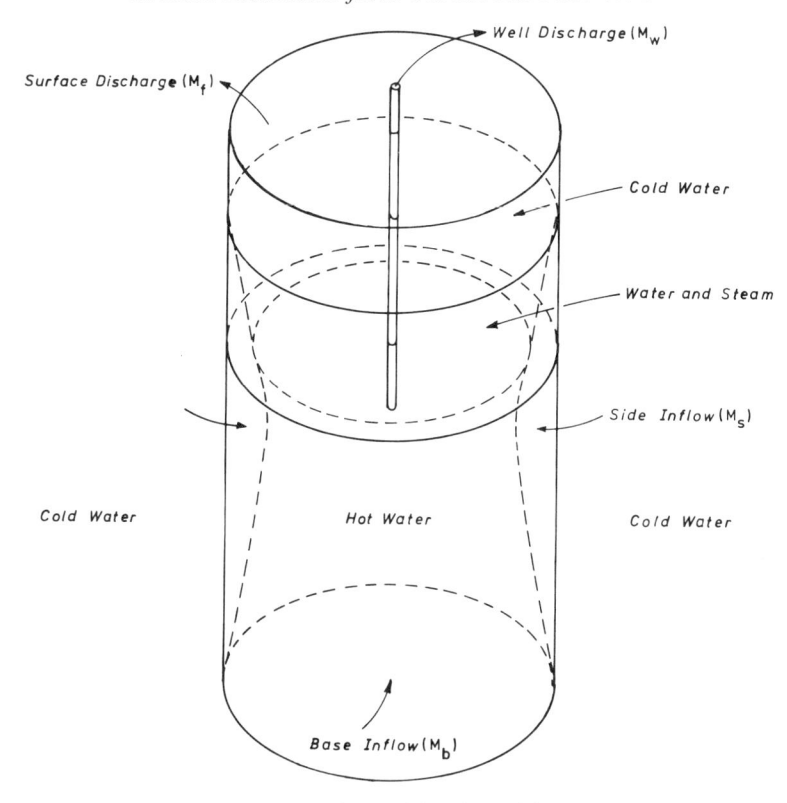

5. Heat Extraction from Geothermal Reservoirs 149

Figure 5.1. The assumed basic model (adapted from McNabb, 1975)

permeability barriers are needed around much of the field boundary to prevent flooding of the deeper, under-pressured part of the vapor system.

We idealize the zone boundaries as simple geometrical forms—planes, straight lines, or cylinders. In reality every such boundary is very diffuse and irregular. Typically at the margins of the field there are intrusions of cold water and outflows of hot water, so that the temperature pattern is quite distorted. The boundary between single-phase and two-phase conditions usually also shows considerable unevenness, with lower temperatures or gas contents giving patches of liquid flow inside the two-phase region, and patches of two-phase beneath the liquid surface. The Wairakei field is quite remarkable in this respect as its water level is very well-defined and smooth. Both the Broadlands and Kawerau fields (see Figure 5.3), in contrast, show quite irregular patterns at this boundary. The upper boundary, where cold water may enter under production, is probably the least regular of all. The main evidence at Wairakei for the entry of surface water is a single 'finger' of cold water that has quenched one well and produced some enthalpy decline in other wells. Many similar quenchings have occurred at Kawerau.

Most of these boundary irregularities are caused by the extreme variability of the permeability of the rock. However, in general, the spatial scale of these variations is usually sufficiently small that for the purposes of field-scale computations we can treat the medium as being of uniform permeability, and the boundaries as smooth and regular. Among the higher temperature fields only Kawerau in New Zealand requires to be considered as a fractured system to explain the effects of exploitation. This field is discussed specifically later. Among the warm water systems, the fractured (karstic) limestones and dolomites at depth in the Carpathian Basin (Boldizsár, 1975) and the dike or planar systems of some Icelandic fields (Bodvarsson, 1962; 1974) also need modified treatment.

The response of the field to production is manifested by the motion of the boundaries and by decompression (boiling) within the two-phase zone. For a hot-water field, where production comes predominantly (or entirely) from the single-phase (water) zone, a good first approximation is to ignore the two-phase zone except for the fact of the fall of the water level. This gives a very simple model that has been applied to Wairakei successfully by several workers (see, for example, McNabb, 1976 or Grant, 1977a). We will therefore start with a brief discussion of the behaviour of the model in a water-only system.

5.2.2. *The Single-phase (Hot-water) Reservoir*

The model visualized is that of Figure 5.1 without the two-phase and upper zones and with all wells tapping the hot water below the (now) free surface. In the natural, undisturbed state the heat in the reservoir has been built up and maintained by a throughflow of hot water from below at the mass flow rate M_b and of enthalpy H_b. Under the assumptions that steady-state initial (pre-exploitation) conditions apply, that there is no flux of mass through the sides and that heat conduction may be neglected, the mass and heat influx at depth is equal to that discharged at the surface. The reservoir is thus isenthalpic and, neglecting the minor pressure effects, isothermal. (It should be noted that if two-phase conditions did exist above this layer the boundary pressure is still fixed at $p_f = p_{sat}(Tb)$, due to the isothermal nature of the water-saturated reservoir.)

If fluid is now withdrawn from wells in this reservoir at the rate M_w, an internal pressure drop will occur which will stimulate a flow M_s in from the sides and may alter the base inflow M_b and surface outflow M_f. For reservoir-scale calculations water may be considered incompressible, and we may write the mass balance as

$$M_b + M_s = M_w + M_f.$$

The heat balance is not so simple. The base flow still brings in the hot water and the outflows, M_w and M_f, carry out water at the reservoir enthalpy. The side flow however brings in cold water. This water does not get into the central reservoir immediately and thus does not lower the overall enthalpy. Rather it removes heat from the hot rock at the side boundaries creating a slow moving cold front that reduces the reservoir's size but not its internal enthalpy characteristics. Thus the interior remains isenthalpic and isothermal. (This means that the 'free' surface pressure, p_f, does not change even in the hot water case). Let us, however, look at these flows in turn.

(i) *The base inflow, M_b*. In most simplistic models of geothermal systems it is assumed that cold meteoric water percolates down to some considerable depth where it is heated and driven, due to the associated density decrease, back up to the surface. This flow establishes and maintains the geothermal anomaly (Wooding, 1963; Donaldson, 1970a,b). The flow path involved is a long one. It is, in fact, so long that the pressure changes occurring under exploitation do not greatly change the flow along it. Thus for Wairakei, taking a circulation depth of 10 km, the natural flux of 400 kg s^{-1} is driven by the 30 MPa difference in pressure between water columns at 20°C and 300°C. The pressure drop under exploitation, 2·6 MPa, can cause an increase of only 37 kg s^{-1}, a negligible contribution to the plant output. For other fields with smaller natural flows, the increase would be correspondingly smaller. Thus for practical purposes, the base inflow does not change under exploitation.

(ii) *The side inflow*, M_s. The side inflow is created by the difference in pressure between the inside and outside of the hot water reservoir, a pressure difference that will be maximal at the well feed level and will tail off both above and below this level. It will, however, still exist some distance below the feed level, and hence, inflows spanning a total depth of twice or more than that of the feed zone must be considered relevant. This inflow will increase with increasing drawdown.

As has been already point out, this 'recharge' of the field with cold water will have the effect of moving in the cold side boundaries of the reservoir, the cold water pushing in the hot water already resident in the pores but being heated itself by the heat stored in the rock. If the volume flux density is $\mathbf{v}$, the velocity of the thermal front is $\lambda \mathbf{v}$ where λ is the volumetric heat capacity of water ($\rho_w c_w$) divided by that of the wet rock (Bodvarsson, 1972; Nathenson, 1975).

As there is no reason for the temperature within the reservoir to change until the cold front arrives, it is the inward movement of this cold front that is mining the heat from the reservoir. (The mining may, however, be limited if short circuiting occurs between the wells and the boundary).

It is important to note that this side flux will not be zero, at least until some time has elapsed, even if the mass withdrawal from the wells is quite small; i.e. we cannot just tap the throughflow of the system without some system adjustment.

(iii) *The well discharge*, M_w. The total well discharge will be controlled by several factors, some man-controlled, such as the number of wells or the valve settings, and some field-controlled, such as the field pressure. Thus additional wells may increase the total discharge while the associated additional drawdown of the field may significantly reduce it. Adding wells may produce only temporary benefits, and the field may have an 'effective' unpumped output limit that it itself sets.

As the enthalpy of the warm water reservoir has not changed, the enthalpy of the discharge will remain h_b. Any downturn in the enthalpy of the discharge must therefore suggest the entry of cooler water.

(iv) *The surface discharge*, M_f. If we increase M_w sufficiently, M_f will reverse in sign, and the water level of the system will fall. This in itself is of no great significance unless the wells are too shallow or if the hot water is overlain with cold. In this latter case a further thermal front is present, and heat mining at the upper boundary will also take place. Unfortunately, with colder water being drawn down above warmer fluid, the potential exists for gravity instability, and 'fingering' of the cold fluid down into the system must be anticipated.

The simple analysis of a hot (or warm) water system illustrates the source of energy and the effects of differing rates and depths of withdrawal. A fuller quantitative analysis could well give estimates of the available energy resource and show the advantages or disadvantages of different rates of withdrawal. This is, however, more pertinent to specific field studies than it is to a general discourse.

There are two further points that we can illustrate with this warm water system—the role of fractures, and the potential effects of reinjection.

Fractures are, in effect, preferred flow paths, and hence a well striking a fracture, or any such freer flow channel, will draw its fluid preferentially from that source. The reduced pressure in this channel will then draw water from smaller cracks and crannies and from the rock pores. The fracture system thus acts as a distribution network for the wells while the porous intermediate beds act as a fluid and energy store. The withdrawal of water and heat from the pores and small cracks has a much longer time scale than has the transfer of fluid along the channels. If a fracture, tapped by a well, reaches a cold water boundary, the cold water will be encouraged along its path into the reservoir. This section of the cold front will move much more quickly than neighbouring areas on account of both the higher velocity and the smaller contact area with the heated rock. High withdrawal rates from wells could therefore result in the amplification of this effect with the possibility of the creation of an effectively dead field with considerable heat still stored but inaccessible to the wells and their feeding fractures. Lower flows could result in a much greater total heat output.

Reinjection of fluid into the system must result in some pressure increase in that system, at least locally. Such a pressure increase, if it affects the reservoir, will assist in maintaining the mass output of the wells. As cooler water will be injected, however, the discharge wells may also be cooled. Obviously this can occur if this fluid is injected within the reservoir and strikes a well feed-channel, as we are likely to maintain a flow path by keeping the pressure up at the reinjection well and down at the withdrawal well. In fact, reinjection at any point within the reservoir will reduce the flow path for the cooler water to the wells. In contrast, reinjection at the boundary may still maintain the field pressures but merely locally enhance the already developed cold front. Naturally the site choice must also depend on other information. Gringarten (1978) shows, for example, the very strong dependence of the reservoir lifetime and the heat recovery factor on the selected development scheme. His study suggests that these may both be increased by alternating injection and production wells; i.e., by *unitizing* geothermal aquifer production as is already done in oil and gas reservoirs, rather than using either isolated withdrawal wells or isolated withdrawal/injection well doublets.

5.2.3. *The Two-Phase System*

In the last section we discussed the behaviour of the single-phase (liquid) system. Under the assumption of incompressibility, the source of the extracted energy is shown to be at the reservoir boundary. The major distinction in geothermal fluid flow is that between the single-phase flow and the two-phase flow. In this latter we include all systems containing two phases even if one phase is immobile. For example, vapor-dominated systems are included in this latter category.

Pressure transmission. In a single-phase flow, in spite of our assumption, some fluid can be made available by decompression. In liquid'dominated geothermal systems, this is generally a negligible amount.

The speed of the diffusion front spreading through compressed liquid is high. It takes only days for a pressure pulse to cross the Wairakei field, for example. Thus the period of production by decompression, when a bore is switched on or off, lasts only this time of a few days. After this a quasi-steady state is established with a falling water level and a continuing cold recharge from the side boundaries of the field. Mass is supplied by this recharge, and energy by the cooling of the field margins. This is a very desirable mode of exploitation. It means that the wells can have a long lifetime, as their region of exploitation is the entire field.

A two-phase mixture of water and steam, on the other hand, is far more compressible than either water or steam. It takes far longer for a pressure pulse to diffuse out to the field boundaries and induce a recharge. The greater compressibility means that more fluid is yielded from storage by a given pressure drop. Also, because two-phase conditions mean that the pressure drop is accompanied by a temperature drop, energy is mined from the rock thus cooled.

The compressibility of a two-phase steam-water mixture has been determined by a simple approximate approach by Grant and Sorey (1979). They suppose that the pressure of the fluid in a unit volume of porous rock is decreased by Δp. This causes a corresponding temperature drop (along the saturation curve $p = p_{sat}(T)$).

$$\Delta T = -\Delta p/(\mathrm{d}p_{sat}/\mathrm{d}T)$$

Heat is thus liberated from the rock, and this boils some of the water to steam. As this steam occupies more volume than the water it comes from, there is a volume increase ΔV. This may be related to the pressure drop to obtain an effective compressibility, i.e.

$$c_2 = \frac{1}{V}\frac{\Delta V}{\Delta p}$$

$$= \frac{1}{\phi}\frac{1}{H_s - H_w}\frac{\rho_w - \rho_s}{\rho_s\rho_w}\frac{\overline{\rho c}}{(\mathrm{d}p_{sat}/\mathrm{d}T)}.$$

Here we use the subscript 2 on the compressibility to emphasize that this is only valid under two-phase conditions. In the full expression for this compressibility, ϕ is the porosity of the medium, H_s and H_w are the enthalpies of the steam and water phases, ρ_s and ρ_w are the associated steam and water densities and $\overline{\rho c}$ is the heat capacity of the wetted rock. Full derivations of this compressibility are given by Garg (1978), Grant (1978) and Moench and Atkinson (1978). This compressibility is orders of magnitude greater than either the steam or water compressibility. From the form of derivation of c_2, it can be seen that it is not compression of each or either phase that matters, but the transfer of matter between the two. The fluid in the pores responds to a pressure drop by yielding fluid to flow and replaces the lost volume by boiling some water to steam. The mechanism is the same whether the flowing fluid is primarily water, or steam. If a cubic meter of fluid is removed from the pore space in rock at 250°C, a further 0·025 m³ boils to fill the 1·025 m³ with steam. The amount of boiling, and hence of cooling or decompression, is the same whether the fluid removed is water or steam.

The greater compressibility of the two-phase mixture means that for a given pressure drop we get a proportionately greater yield of fluid volume from each rock volume. We also take heat from the rock, by the very process of boiling. If we begin with a two-phase mixture in the rock, there is some amount of steam initially in place. When we decompress, and force boiling to take place, all the additional steam that flows is supplied from water. The heat to boil it is supplied by the rock.

The source of the extracted energy. We have already indicated that for the single-phase system, due to the rapid transmission of any pressure change, all the energy is mined from the perimeter of the reservoir after the first few days of operation or change. In the two-phase situation, however, it takes time, probably of the order of years, before this condition eventuates. It is thus of interest to see just where this energy comes from in the interim. A study of a single well penetrating and extracting its fluid from the two-phase zone is useful in this regard.

If we consider the area around the well as being initially at a uniform pressure p and a uniform water saturation (the ratio of water volume to total pore space) S, and then allow only small perturbations from this state, a diffusion equation for pressure may be derived (Grant, 1978; Grant and Sorey, 1979).

We may thus write

$$\frac{1}{\kappa}\frac{\partial p}{\partial t} = \nabla^2 p$$

where κ is the hydraulic diffusivity $(=k/\phi\mu_t c_2)$
k is the permeability of the medium
and μ_t is the total dynamic viscosity of the fluid

$$(=F_w(S)/\mu_w + F_s(S)/\mu_s)$$

In this latter expression μ_w and μ_s are the dynamic viscosities of the independent phases (water and steam) and $F_w(S)$ and $F_s(S)$ are the water saturation dependent relative permeabilities of the water and steam phases in the mixed flow system. The relative permeability to either phase is the fraction of the actual permeability that has to be taken if Darcy's Law is to be applied to each phase independently (see, for example, Scheidegger, 1957, p. 155).

As we pointed out earlier, the analysis of single well behaviour tends to indicate a dominance of horizontal structure. We will therefore restrict this consideration to the problem of transient flow around the well in a uniform horizontal aquifer containing the two-phase mixture. The conclusions reached apply equally to other geometries, but this arrangement would appear to be most compatible with both the field behaviour and the assumptions made in the derivation of the pressure equation.

Solutions to this equation are well known (see, for example, Carslaw and Jaeger, 1959). For the case of a well opened at time $t=0$ and discharging at the constant mass rate W we have, for example,

$$p = (Wv_t/4\pi kh)(-Ei(-r^2/\kappa t))$$

where the total kinematic viscosity v_t is related to the kinematic viscosities of water and steam, v_w and v_s, by the expression

$$\frac{1}{v_t} = \frac{F_w(S)}{v_w} + \frac{F_s(S)}{v_s}$$

The mass of fluid stored per unit volume of rock is

$$\phi[\rho_w S + \rho_s(1-S)]$$

The rate at which fluid is lost from storage, or yielded to the flow in the reservoir, is the rate of change of this with time. This is proportional to

$$c_2 \frac{dp}{dt} \propto c_2 \frac{1}{t} e^{-(r^2/\kappa t)}$$

Figure 5.2 illustrates the pressure profile at time t with distance r out from the well, the rate of loss of fluid from storage with radial distance (at a particular time t), and this rate of loss with time (at a particular radius r).

After time t the pressure front has diffused out a distance of the order of $(\kappa t)^{1/2}$, the diffusion radius (Figure 5.2(a)). If we look at the pattern of fluid yield at this time, Figure 5.2(b), we see that fluid is being taken from storage (at a rate that declines as t increases) within this same diffusion radius. Little is taken beyond this circle. As Figure 5.2(b)

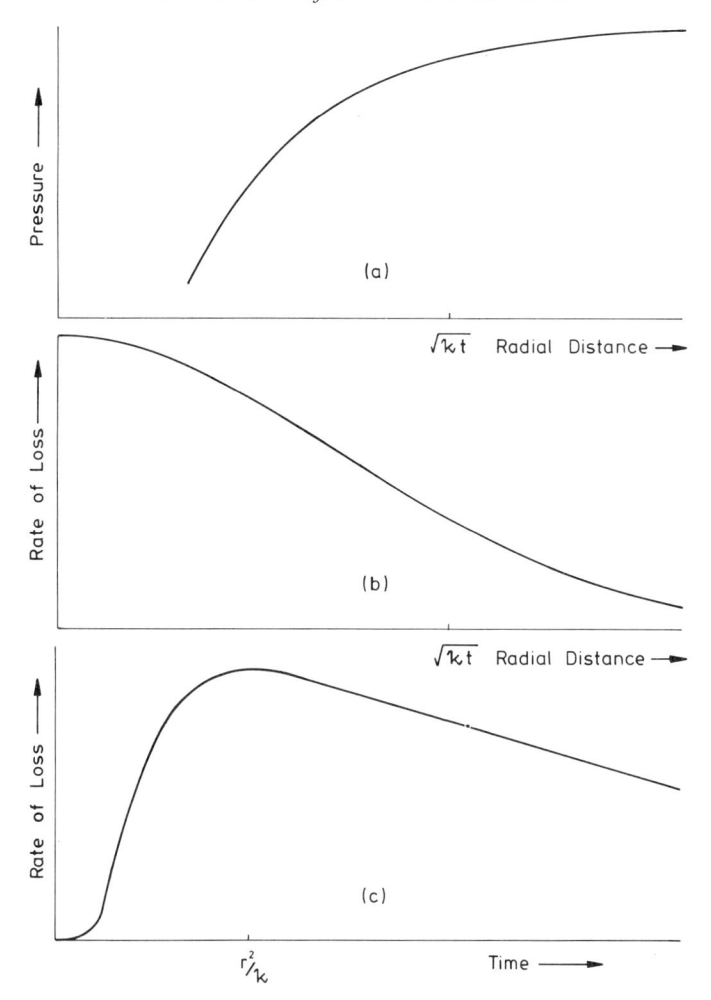

Figure 5.2. The role of the diffusion radius $(\kappa t)^{1/2}$ at any given time t after the commencement of a constant mass withdrawal from a well. (a) The profile of pressure drawdown. After this time t the pressure drop has diffused out a distance $r \approx (\kappa t)^{1/2}$. (b) The profile of rate of loss of fluid mass (kg m^{-3} s^{-1}) from the rock matrix. (c) The variation of this rate of loss of fluid mass with time at some fixed radial distance r from the well. Note the peak loss at $t \approx r^2/\kappa$

illustrates only the rate of fluid yield per unit volume, to obtain the rate of yield per unit area at the radial distance r this function must be multiplied by $2\pi r$. This product has a maximum at $r = (\kappa t)^{1/2}$ and is small for both small and large r. Thus the bulk of the fluid entering the well is supplied from a broad band around this diffusive radius.

If we look at a particular point to see the pattern of fluid yield with time, Figure 5.2(c) results. At first little fluid is yielded, as the point is outside the diffusion radius. The yield rate rises to a maximum at $t = r^2/\kappa$ as the diffusion front goes past, then slowly declines. Note that the rate of fluid yield depends on the rate at which the pressure is changing with *time*. Thus, although the pressure increases markedly with *distance* near the well, this does not mean correspondingly large amounts of fluid yield. The pattern of fluid yield from storage is very smooth, out to the diffusion radius $(\kappa t)^{1/2}$ (Figure 5.2b).

Along with the yield of mass from storage in the pore space, there is a yield of energy from the rock. The transmitted fluid is a mixture of water and steam, with an enthalpy above that of liquid water. The excess of enthalpy is supplied by the rock, and the profile of this yield of heat is proportional to that of the yield of mass. This means that the pattern of heat yield, the extraction of heat from the rock, is evenly spread out to the diffusion radius. The gradient of pressure, and hence temperature, with distance becomes very large near the well. Little heat is however transferred to the flowing fluid as the temperature of the rock is falling very little *with time*. Although there are large temperature changes near the well, the flow is nearly isenthalpic, i.e. the enthalpy of the flowing steam/water mixture stays almost static. As its pressure and temperature fall, the steam/water ratio (and hence the steam saturation) increases to maintain this condition. It is not possible under normal conditions for a quasi-steady flow to consist only of moving water at some distance, and only of moving steam at the well.

At this point it is of interest to look at two examples of two-phase systems by using variations of our simple basic model.

(i) *The liquid-dominated two-phase system.* If the basic model is taken, more or less in the form illustrated (Figure 5.1), with a two-phase zone overlying the single-phase water-saturated reservoir and with all exploitation wells tapping the fluid in the two-phase zone, we have a good example of the liquid-dominated two-phase system. An example of such a field is Broadlands, New Zealand, described in more detail later in this chapter.

In such a field it would take one or more years for the diffusion front to reach the edge of the field. Thus, for time less than this after turning on a well, heat and mass are withdrawn from an ever-increasing region around the well. For longer times, a quasi-steady state is established, similar to that for a liquid field in which the cold surroundings gradually encroach upon the hot zone. Again we have the situation that mass flow from the field is balanced by cold inflow, so that the ultimate mass source is outside the field. The heat flow out of the well is sustained by the heat supplied to the incoming fluid by rock cooled on the field boundary, so that again the well mines heat from the field boundary. There is also a continuing contribution of heat and mass from the two-phase zone if there is a continuing pressure decline with time.

Along with the sideways propagation of the pressure wave and of the mass and heat withdrawal front, there will also be some vertical propagation effects. In time the pressure wave will also reach the base of the steam/water zone and thence propagate relatively quickly through the water-saturated section. Water will thus boil at the interface, the interface level will drop, and energy will be extracted from the rock in the immediate vicinity. Some energy will also be mined at the sides of this hot, liquid section, even though no wells directly tap it.

(ii) *The vapor-dominated system.* We can use another variation of the model for a steam well in a vapor-dominated system. In this case the reservoir will contain steam and immobile water and the all-liquid zone, if it exists at all, will be assumed to be at such great depth that it does not appear in the section of the model of interest. Pressure transmission here is a particularly simple case of two-phase propagation. Only steam moves, and with the pressure drop some of the water boils to contribute more steam to that flow. In so doing, it takes heat from the rock, and this process thus mines both mass and heat from this local material.

The diffusive pressure front in these conditions can take years to travel a kilometre. Thus we can dispense with the sides of the model and think of the well as being in an infinite medium. This ultra-simplification is compensated by a new complication—there may be a dry steam zone in the system. The water saturation changes in the two-phase zone are linearly related to the temperature changes. Hence, depending on the initial conditions, the pressure and temperature may decay enough to drop the water saturation, S, to zero. If this occurs, there will be a dry steam zone around the well. Within this dry steam zone there is a little further yield of mass by expansion of the steam, and a little further yield of energy as the steam pressure drops, creating a slight superheat of the steam. The profile of drawdown is, however, little affected by this additional zone. The steam well thus has an expanding zone of drawdown with the diffusion radius $(\kappa t)^{1/2}$. The outer parts of this are two-phase, but there may be an inner zone containing dry steam. The well draws mass and energy almost entirely from the two-phase zone—the dry zone being effectively a 'dead core' of the well's exploitation area.

If cold water is reinjected into a hot water reservoir, the idealized picture suggests establishment with time of a zone of cool water around the reinjection well. This will in turn be surrounded by a warm water zone, created by the extraction of heat from the neighbouring rock by the cooler reinjection water; beyond that will be the natural uncooled geothermal waters (Bodvarsson, 1972). A similar picture arises with reinjection into two-phase conditions: cold water by the well, then hot water, and finally the natural two-phase conditions. The hot liquid acts as a buffer between the cold liquid and the two-phase fluid.

Under most normal permeabilities and injection rates the reinjected cold water spreads radially out from the reinjection well, the pattern being distorted only to a minor extent by the bouyancy effects. The movement will naturally be enhanced along any lower permeability channels. In a vapor-dominated system, in which the vertical pressures relate much more closely to the steam conditions, the downflow of water will be much more significant.

With our limited experience with reinjection this picture can only be an idealized one. The actual performance of reinjection wells shows, however, the very strong influence of permeability variations, and the fluid motion is thus far from that of uniform radial spreading. A reinjection test has, for example, been carried out at well BR7 at Broadlands, New Zealand, for the past two years. This well has three feed zones. One of these has been accepting reinjection fluid at 90–150°C. Another, only 50 metres deeper, still contains 280°C fluid (P. F. Bixley, personal communication). Such details are difficult to allow for and, at this time, analysts can only hope that the idealized picture has some validity on the large scale.

5.3. Heat Extraction from some Typical Existing Fields

To illustrate the above idealised behaviour in some real systems we have selected some typical existing fields for discussion. As several New Zealand fields are mentioned, the locations of these are indicated in Figure 5.3. We start with the warm water reservoir and work our way through to the vapor-dominated one. The continuous nature of the change in field response may be seen as we move through the sequence of reservoir types.

5.3.1. *The Warm Water Reservoir*

For this discussion we define a warm water reservoir as one containing water at a sufficiently low temperature, under the prevailing reservoir conditions, that boiling does not occur within the reservoir environs either under natural conditions or under

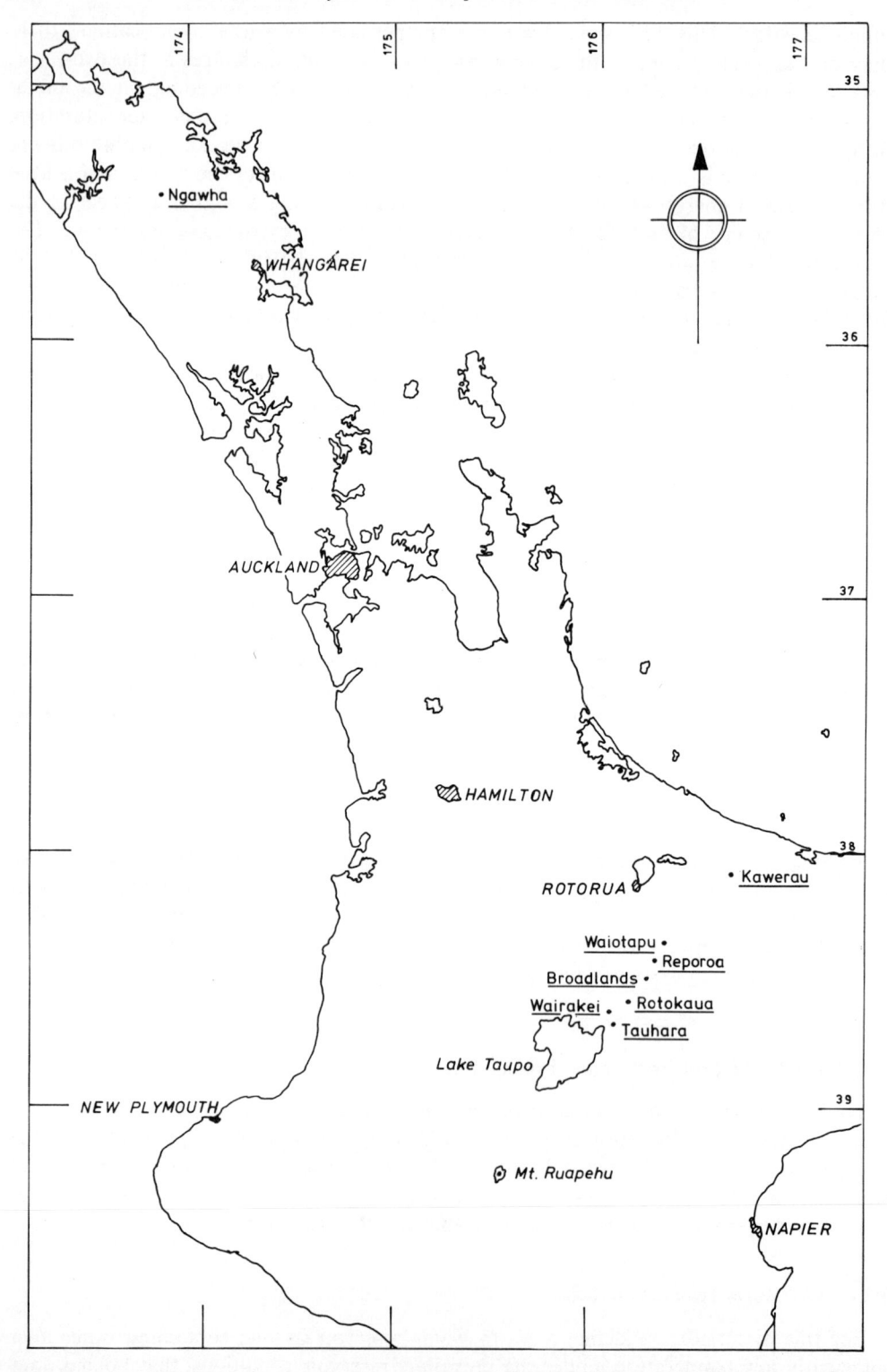

158 *Geothermal Systems: Principles and Case Histories*

Figure 5.3. The location of the New Zealand geothermal fields discussed in this chapter

5. Heat Extraction from Geothermal Reservoirs

exploitation. It thus includes systems with temperatures of up to 100°C, commonly only exploited for home heating, agricultural purposes and balneological use, and the Icelandic 'low temperature systems'. It may also include some systems with recorded temperatures as high as 200°C.

A system that fits into this category, although at the high-temperature end of the scale, is that at Heber, California, U.S.A. We consider this in some detail as the geology alters the conceptual picture of the field but leaves the mechanics of heat removal much as it is in other warm water fields. The following description is based on the study of Tansev and Wasserman (1977). Figure 5.4 is Figure 1 of their paper.

The field is roughly circular, with an area of 30 km² inside a 120°C contour. The maximum temperature is 190° and a plant is planned that will use fluid at 160°. The field consists of interbedded sand and shale layers overlain by some 600 metres of shale alone. The sand is permeable, and its volume is the fluid reservoir.

Although the layering may be expected to impose a horizontal flow on the reservoir under exploitation, several features indicate the presence of vertical permeability and the absence of any boundaries. Pressure tests have, for example, investigated out to a radius of more than 6 km, i.e. beyond the boundaries of the useful reservoir.

The natural gradient of pressure with depth for this field corresponds to the static

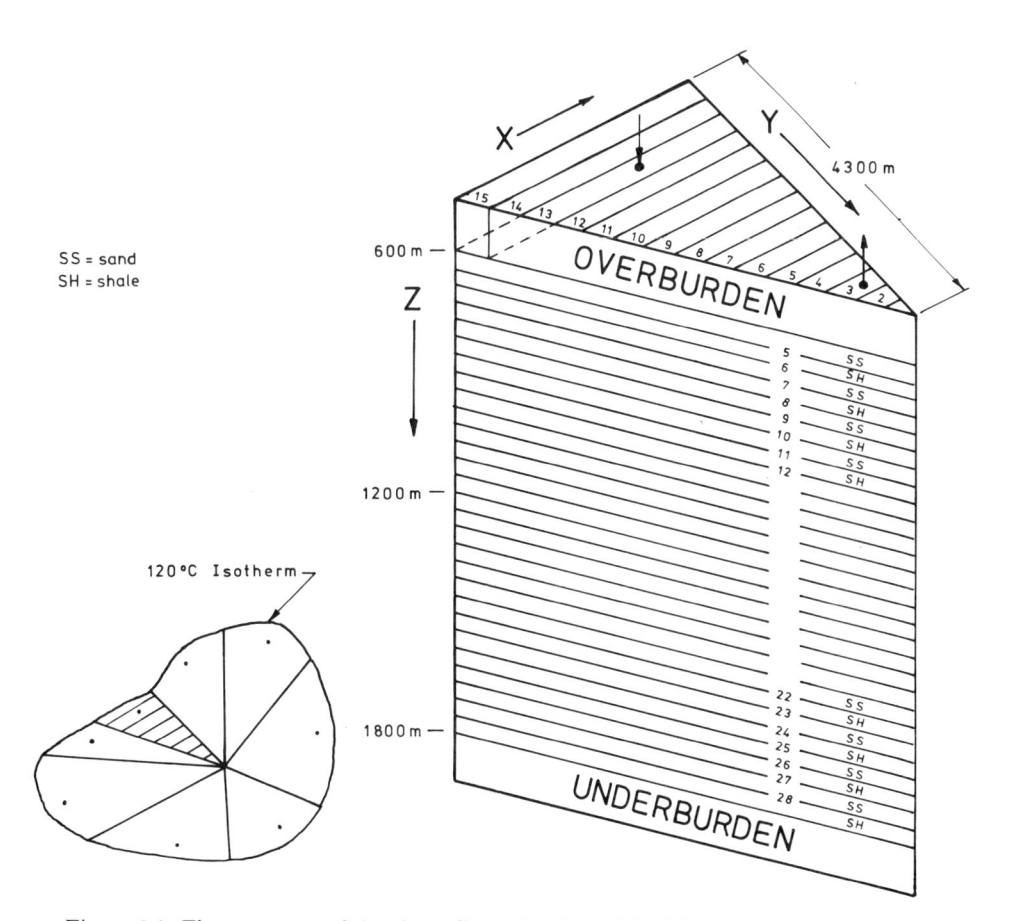

Figure 5.4. The geometry of the three-dimensional model of the Heber geothermal field (from Figure 1, Tansev and Wasserman, 1977, reproduced with permission)

pressure of water at 85°C. The temperature distribution in the reservoir is uniform vertically, but with temperatures diminishing with horizontal distance from the centre of the reservoir. This combination indicates a natural vertical circulation with the warmer central water rising while that further out, and cooler than 85°C, will be falling. This again implies permeability extending into the cooler environs of this field.

Tansev and Wasserman modelled the Heber field under exploitation assuming horizontal flow along the sand layers, with heat conduction from the interbedded shale. The flow is driven by the discharge from the producing wells and by fluid injection at the thermal boundary. Any deficit in fluid is assumed to be made up by inflow across the boundaries. Because pressure diffuses so rapidly in liquid, the liquid is assumed to be incompressible. The flow field is then determined geometrically from its sources and sinks. The main interest in the model calculations is not in the pressure field, but in the temperature, and hence the heat extraction. The pressure field does, however, vary passively as the fluid viscosity and density are allowed to vary with temperatures, and hence the fluid's resistance to flow depends on the temperature. As the fluid moves inwards, the isotherms contract. In this model, the simple sweeping out of energy along each layer is modified by conduction from the bounding shale layers. Thus at any given time the temperature profiles will be pulled in somewhat more in the sand than in the shale. It is estimated that about 30 per cent of the heat in place is recoverable.

5.3.2. Hot Water Reservoirs

In contrast with the warm water reservoir, boiling can occur in the hot water reservoir in both the natural and exploited states. All withdrawal will, however, take place from the water-saturated zones. Unless drilling is therefore particularly deep this implies a somewhat shallow two-phase region. Two examples are considered here.

(i) *Wairakei, New Zealand.* The classical example of such a hot water system is probably the Wairakei field in New Zealand. Although some wells do extract water and steam from the upper two-phase zone, the majority are sufficiently deep and so cased that they draw only from the deeper water section. Most of the field's behaviour may thus be explained by treating it as only a hot water system.

A typical geological section across the Wairakei field is illustrated in Figure 5.5 (from Grindley, 1965). This shows considerable horizontal structure, but again, as with the Heber field, Salifornia, the indications are that vertical connections, and hence permeability, exist at all observable levels and that vertical circulation of both fluid and heat therefore takes place.

In its natural (undisturbed) state, the Wairakei reservoir had an input temperature of 260–270°C; i.e. water at this temperature rose from depth under natural convection and heated the reservoir fluid and rocks. At a depth of about 400 m the fluid pressure and temperature were such that boiling could take place. Thus from that level to the surface, both steam and water moved upwards, the temperature and water saturation of the mixture being set by the pressure at the level concerned and by the natural heat flux. At the shallower levels, for example, the lower pressure means a lower temperature ($T = T_{\text{sat}}(p)$), and, to maintain the heat flow, a higher proportion of steam (a higher steam saturation) is required. At the surface, i.e. at 100 kPa and 100°C, under isenthalpic flow conditions, one sixth of the mass flux must be steam. This is equivalent to about 99·7 per cent of the volume flux. An estimate of the temperature, pressure and water saturation variations with depth in this natural situation is illustrated in Figure 5.6. A fuller

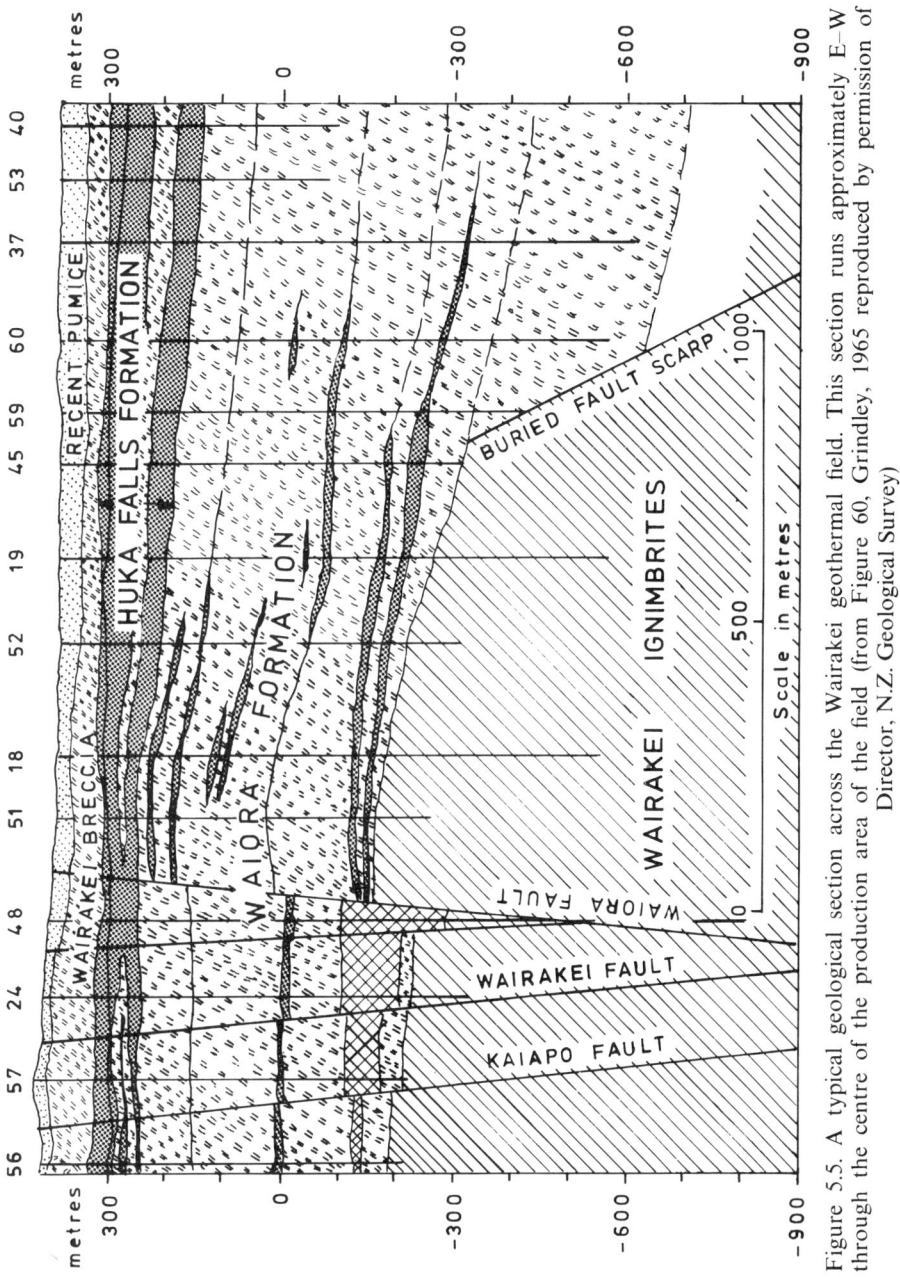

5. *Heat Extraction from Geothermal Reservoirs*   161

Figure 5.5. A typical geological section across the Wairakei geothermal field. This section runs approximately E–W through the centre of the production area of the field (from Figure 60, Grindley, 1965 reproduced by permission of Director, N.Z. Geological Survey)

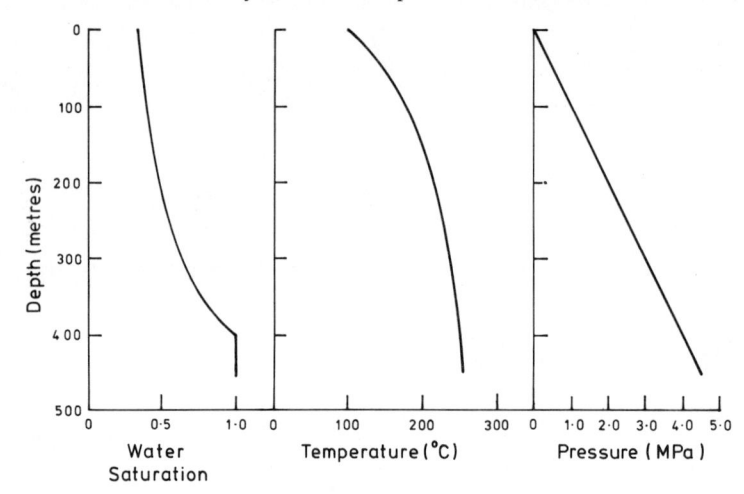

Figure 5.6. The estimated temperature, pressure and water saturation variation with depth for the Wairakei geothermal field in its natural, undisturbed state

discussion of an undisturbed system of this nature has previously been presented by one of the authors (Donaldson, 1968).

The two-phase zone is often referred to as the steam zone or steam cap. It should be remembered, however, that it contains water as well as steam, and that the pressure gradient with depth is nearly hydrostatic. It is not a static bubble of steam overlying water.

Production holes feed both from this two-phase zone and from the deeper water zone. The bulk of the wells feed deep enough to produce only liquid water (at the enthalpy of the field water). The few shallower wells feeding in the two-phase zone, in contrast, produce both liquid water and steam, and the enthalpy of their feed, essentially the proportion of steam, increases towards the surface.

Most of the production comes from beneath the water level (the water/two-phase interface). The published curves of aquifer pressures (Bolton, 1970) in fact refer to pressures averaged over the well readings taken in any given calendar month. They are adjusted to a particular level in the water-saturated section of the system.

These pressures, recently reassessed by L. J. Fradkin (personal communication) for the period 1958–68 and by one of the authors (MAG) and A. McNabb (personal communication) for earlier and later periods, are plotted for a 580 m depth in Figure 5.7. Monthly mass discharges and annual average mean discharge enthalpies (from New Zealand Ministry of Works and Development data) are also illustrated on this figure.

In the liquid zone, the outstanding feature about the Wairakei reservoir is its uniformity. The pressure drop in the field is spread nearly uniformly, with only about 100 kPa difference from the production zone to the boundaries. This is not true in the two-phase zone. There, on account of the much slower rate of pressure diffusion, there is a localized pressure defect in the production area which is apparently not present at the edge of the field.

With continuing production the pressures in this water zone have fallen steadily (Bolton, 1970), although the decline is now small (Figure 5.7). The total pressure drop is 2·6 MPa. With this pressure drop, the water level has fallen correspondingly and is now

5. Heat Extraction from Geothermal Reservoirs

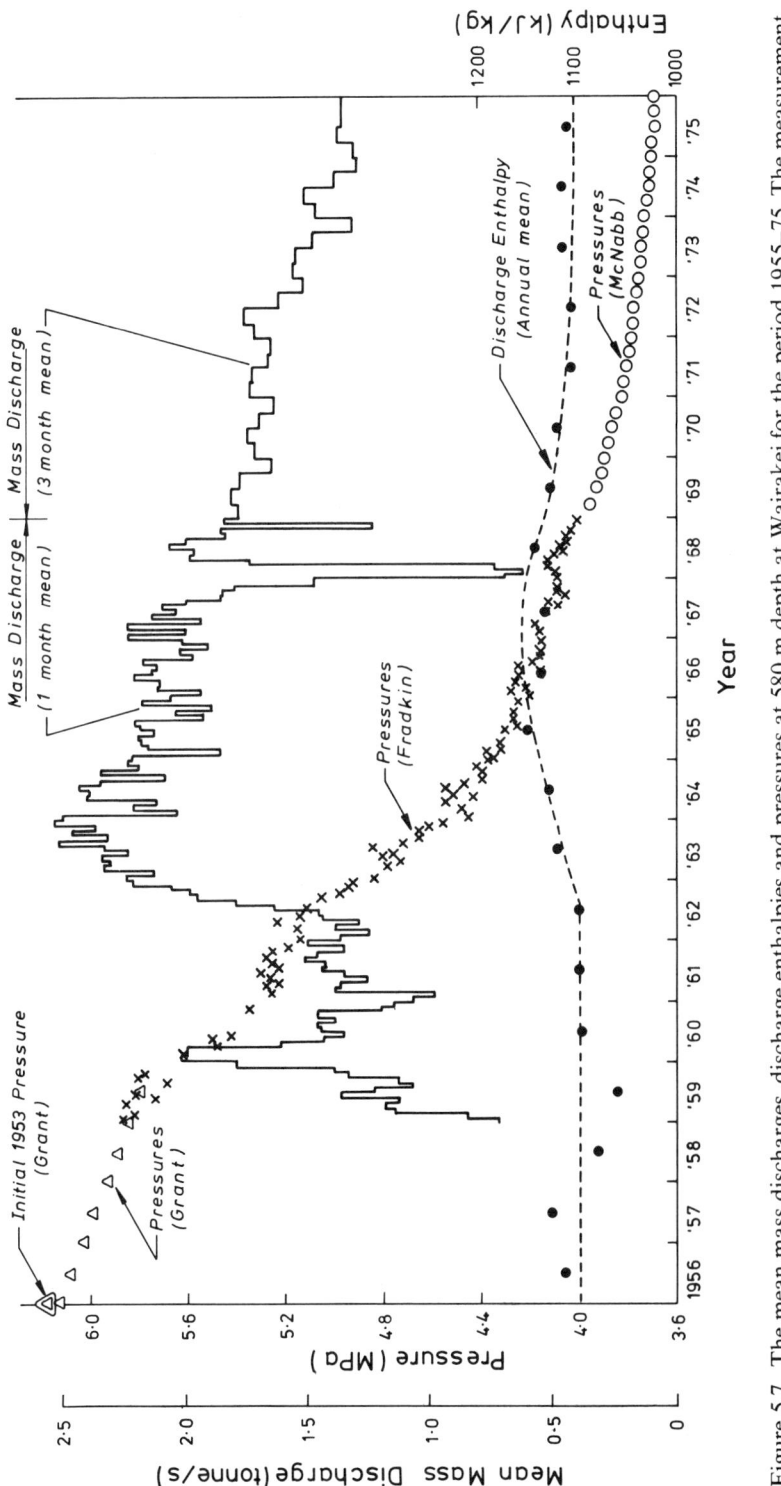

Figure 5.7. The mean mass discharges, discharge enthalpies and pressures at 580 m depth at Wairakei for the period 1955–75. The measurement data are all from New Zealand Ministry of Works and Development records. The mean monthly pressures have been estimated from these data by M. A. Grant (1953–59), L. J. Fradkin (1958–68) and A. McNabb (1969–75) (personal communications). Reproduced by permission of the International Institute for Geothermal Research, Pisa, Italy

at a depth of 600 m. There has also been an apparent fall in the temperature of the water feeding these wells. The cause of this is not yet fully understood.

With the fall in water level, the feed zones of some wells have moved from the water zone to the two-phase section. Wells feeding from this zone are isolated. Recharge is effectively a long way off, and they thus produce their fluid by boiling in their own neighbourhood. This results in a steady and continuing local decline in pressure and temperature in the two-phase zone. With this, the performance of the two-phase wells also goes down.

In the early stages of production, most wells fed from the liquid zone, and the overall production initially had an average enthalpy of 1100 kJ/kg, i.e. near that of liquid water at 253°C. With exploitation, beginning in 1953, the two-phase zone has grown, and within it conditions have tended to become drier. Thus by 1967 the enthalpy of the two-phase wells had risen sufficiently to raise the average output enthalpy to 1160 kJ/kg. Thereafter the enthalpy fell for two reasons: the two-phase wells declined in performance more rapidly than did the water-fed wells, and there has been a slight drop off in the enthalpy of the water wells due to the temperature decrease. The enthalpy history of the discharged fluid may thus be taken as steady at around that of liquid water at 260°C with a perturbation of the order of 10 per cent over a period of a few years in the 1960's. Overall it has deviated little from that of liquid water. This, together with the pressure decay effects and the relatively minor temperature drop, suggests that the overall field response can be modelled as though all withdrawal came from below the water level, i.e. below 600 m.

There have now been several attempts to model the Wairakei field, presumably on account of the amount of data available throughout its exploitation lifespan. Whiting and Ramey (1969), for example, considered it as a compressed, sealed unit, similar in most respects to a standard oil reservoir. Marshall (1966, 1970, 1975a, 1975b), in contrast, set up a one-dimensional vertical model and allowed for the vertical throughflow and some inflow (recharge) from the sides. Detailed numerical models have now also been produced by several research groups. Mercer et al. (1975) and Mercer and Faust (1976, 1979) have developed two-dimensional horizontal models of the system, using integrated averages over the vertical dimension in the Waiora formation to give them meaningful average values of their variables for their program, while Pritchett et al. (1976a,b) have concentrated their major effort on a vertical section. To date, however, the simple drainage model (McNabb, 1975) and its more recent variations do appear to best describe and match most features of Wairakei's response to exploitation.

This model is virtually that described earlier in the section on the single-phase (hot water) reservoir. With the two-phase layer above there are only minor differences. In this case the natural state is an upflow of hot water which boils at about 400 m depth. Above this level both water and steam rise to the surface to give the natural output.

Under exploitation the water level in the model falls, and pressures in the liquid zone fall with it. The natural flow is now mainly drawn into the wells, but a surface discharge continues as a transient feature of the two-phase zone; i.e. even though the pressure gradient decreases the steam is still driven upwards. At greater depth, as was illustrated above, the natural upflow is only minimally increased by the exploitation. The pressure drop in the field thus mainly stimulates the inflow of cold water from the sides.

Associated with the fall in water level there is also drainage from the two-phase zone. The additional volume of rock exposed by the falling water level is not dry—it does in fact retain some water. This water drains downwards, and as the pressure drop slowly diffuses upwards, this drainage extends into the pre-existing two-phase zone. There may also now be some entry of cold water at the surface.

Selecting reasonable numbers for the parameters in the equations that apply for this model gives on solution the correct form for the pressure decay curve and a reasonable match of the field behaviour. With adjustment of the more doubtful parameters, good fits of the data are readily obtained. These good fits should not, however, be taken as the ultimate check on the model or the parameter values, as various factors can introduce a bias into these parameter estimates. Only the successful forecasting of field behaviour will be the indicator of real success.

It should be noted that no geological boundaries are indicated, or included even implicitly, in this model, At this level of aggregation they do not appear to exist; i.e. the fluid flows as though all strata are equivalent. There are numerous geological details in the field (see, for example, Figure 5.5 and Grindley, 1965), and these affect the performance of individual wells and can be seen in other features of the response to exploitation. The most obvious is that horizontal permeability in the field must be very high in order to produce the observed pattern of uniform drawdown right across the field. It has also been suggested that the Huka Mudstones form a cap. The initial pressure distribution and the pre-exploitation surface manifestations contradict this as these both suggest no significantly greater impediment to flow at this or any other level. These mudstones do, however, appear to affect the horizontal transmission of pressure to the adjoining Tauhara field (Wooding, 1980). For the field response, the existence of capping or otherwise is irrelevant. The field responds as an unconfined system, because of the motion of the free surface. A similar openness has been recently observed for the Momotombo reservoir in Nicaragua (Dykstra and Adams, 1977).

This simple model suggests that cold water inflow from the sides is one mechanism limiting the lifetime of a field. For ideal conditions one might therefore be able to estimate a 'recovery factor' (the fraction of the total heat in a defined reservoir that could be extracted). Such exercises have been carried out elsewhere (Nathenson, 1975; Nathenson and Muffler, 1975). Their recovery factor of 50 per cent for ideally permeable conditions corresponds with that obtained for Wairakei with this ideal model. To allow for the structural inhomogeneity, only a qualitative correction factor is possible, and Nathenson and Muffler (1975), in fact, halve their ideal figure. Current experience at Wairakei is that cooler water now enters from above in local production areas, and as this has not yet been allowed for in any analysis even this 25 per cent factor may be too high.

(ii) *Kawerau, New Zealand.* In marked contrast with Wairakei, where the geological structure appears to play only a very minor role in the control of the transfer of mass and heat within the reservoir, the Kawerau field appears to show a very strong lack of homogeneity of its structures in its behaviour. In particular, it shows the phenomenon dreaded in all discussions of reinjection but so far, fortunately, seldom seen—preferential paths for the rapid entry of cold water.

When first drilled, most Kawerau wells tap two-phase conditions, and hence Kawerau perhaps should be categorised as a two-phase field. The cooling discussed, however, quickly reduces the feed zones to the liquid state, and hence it was thought more appropriate to discuss the behaviour of this field here.

The Kawerau field is located in the New Zealand thermal belt, about 80 km north-east of Wairakei (see Figure 5.3). It has been exploited at a relatively low level for over twenty years. Drilling began in the early 1950's (Studt, 1958) to obtain steam to supply a pulp mill. Seven holes were drilled initially to a depth of about 600 m, and sufficient steam was obtained to satisfy the requirements of the time. After a few years these wells failed and some of them were deepened (Dench, 1962). This operations was successful, and the deepened wells produced reliably for some years, the last of them only now failing. Three more holes were drilled in 1967, and a slow programme of exploration drilling recommenced in 1974 with the intent of obtaining enough steam for electric power production as well as processing in the mill. The average discharge rate over the past twenty years has thus been small, equivalent to about 20 MW (electrical).

The field was not fully defined until quite late in its development when a resistivity survey (Macdonald and Muffler, 1972) established that it covered an area of 6–8 km^2. The initial drilling had been concentrated in a very small part of this area, and even now the drilling is very uneven, with half the field having three holes and the remainder seventeen. Current drilling aims at a more even spread.

The history of development is a steady pattern of drilling deeper and deeper, and over a wider and wider area. At the same time the wells already on line were steadily declining in output. Each new sequence of drilling was to a greater depth and produced wells with higher outputs and longer life. This does not seem startling until the amount of withdrawal is considered. The withdrawal has been at an average rate of 3–6 million tonnes per year—a cumulative withdrawal to 1976 of some 80 million tonnes. By comparison, 50 million tonnes are discharged from Wairakei each year, and in an equivalent twenty years this has not produced any rundown in well performance of a similar nature.

Unfortunately, in comparison with Wairakei, the wells at Kawerau have been poorly monitored, but, from the measurements available, it is apparent that there is little, if any, drawdown in pressure in the immediate vicinity of the wells. This is consistent with the low rate of withdrawal and the good permeability indicated by well performance.

What has damaged, and eventually destroyed, well performance is a decline in temperature of the fluid tapped by the wells. Figure 5.8 illustrates this, showing temperature profiles in well KA8 at different times. KA8 was one of the original wells drilled in 1956, and at that time it was very productive from a feed at 600 m. By 1960 it had failed so it was then deepened to 900 m. A new feed was encountered at 800 m, and it is the performance of this feed that is of interest.

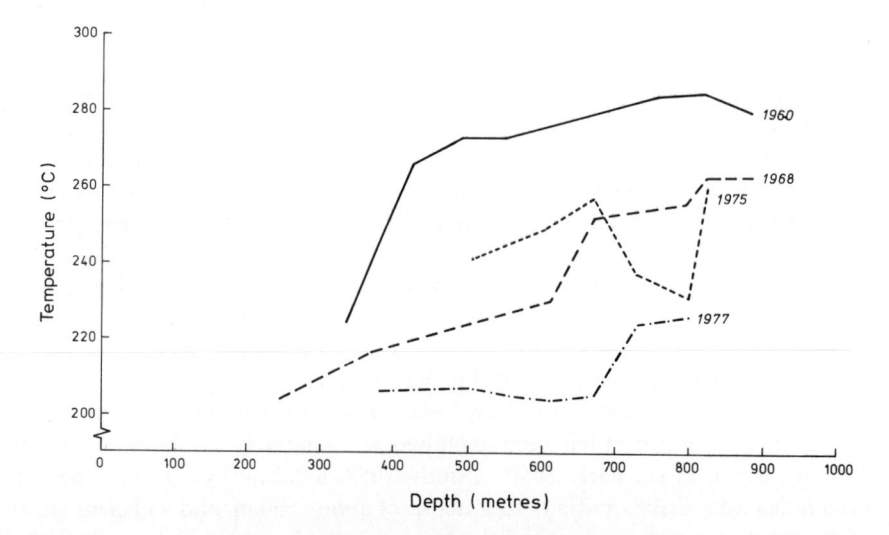

Figure 5.8. Temperature profiles down well KA8 at Kawerau, New Zealand at various times from 1960 to the present. (Data supplied by Ministry of Works and Development, New Zealand)

The first temperature profile, in 1960, indicates a maximum temperature of 285°C at the feed zone. The remaining temperatures in the well at this time are uninteresting as they simply reflect a boiling column of water above this point. The bore then produced a high enthalpy fluid—around 1400 kJ/kg. This is well above the enthalpy of liquid water at this temperature (1260 kJ/kg) indicating that the well was drawing on two-phase conditions. The bore was at this time the best in New Zealand, providing about half of the output of the entire Kawerau field. The next profile, in 1968, shows a maximum temperature of only 265°C. Pressures were also measured at that time, and these were well above saturation, indicating that the well was now drawing on liquid water at around that temperature. The enthalpy of the discharge confirmed this. Subsequent temperatures were progressively lower, with a temperature inversion eventually developing near the feed. Most recently the feed temperature was down to 205°C, and the well nearly dead.

The interpretation placed on this is that cold water is now getting right to the feed zone of the well KA8. Initially the well fed from two-phase conditions. With the withdrawal of fluid by the well, hot fluid will move in from further out, and some cooler, denser fluid may move down from above. In any system it will be the particular balance of these that will control what actually happens at the well. If the cold water inflow is sufficient it will condense some of the steam, so that the water saturation will rise. This is what in fact did happen at Kawerau, and by about 1964–65 this bore was feeding from liquid conditions; i.e. all the steam in its feed zone neighbourhood had condensed, and the well discharge enthalpy had fallen to that of liquid water. The progressive temperature decline thereafter is indicative of increasing encroachment of the cold water. Thus in this field the cold water appears to have preferential access to the permeable feed zones.

During its lifetime KA8 discharged about half of the total Kawerau field discharge, or 30 million tonnes. With a porosity of 15 per cent, this would fill a sphere of rock less than 200 metres in radius. The nearest cold water to the feed zone is the ground water some 800 m away, just below the ground surface. This water must therefore flow along permeable paths and thus only drain a very small segment of the rock of its stored heat. The well, as a means of extracting heat from the storage in the rock, failed when only the fraction of the rock containing the fissure system had cooled. The remainder of the heat not in close contact with the fissures is at present inaccessible.

Fractured reservoirs of this nature may be assessed with the simple planar model developed by Bodvarsson (1974). This model has recently been re-evaluated by Muffler and Cataldi (1978) to determine appropriate (but still theoretical) geothermal recovery factors for fractured systems. Although the fracture spacing is restricted (a 338 m spacing is assumed in the Bodvarsson and in the Muffler and Cataldi analysis), and the idealised recovery factors are too high, the study does show the significant potential increase in recoverability with lower rates of withdrawal. The estimated recoverability in fact doubles if the withdrawal is adjusted for a 100 year rather than a 25-year lifetime.

5.3.3. *Two-Phase Systems*

In their response to exploitation, and in their mode of heat removal, there is a continuous gradation between liquid- and vapor-dominated systems. The natural gradient of pressure with depth provides a clear-cut distinction between these extremal systems, but this distinction is not particularly relevant to the heat extraction. There are, for example, shallow wells at Wairakei that produce dry steam from conditions of low water saturation and behave very much like wells in a vapor system, and yet they produce from an environment that initially had a hydrostatic gradient of pressure with depth.

This continuous gradation is characterised by the increasing size of the two-phase zone, the decreasing water saturation of this zone, and the decreasing hydraulic diffusivity. All of these go along with an increasing proportion of the heat being extracted from within the reservoir by boiling, and an equivalent decreasing proportion being mined at the margin of the field by the encroachment of cold water. This inflow of water, which appears to the reservoir as a recharge of hot fluid due to the exchange of heat between the rock and the cold fluid at the field margins, thus decreases as we move from system to system along this path from liquid- to vapor-domination.

Along this gradation we can see examples. One wet two-phase field is Broadlands, New Zealand. Another, a bit hotter and drier, is Krafla, Iceland (described by Stefánsson in this volume). Drier still, but with near hydrostatic pressure gradients, are Ngawha, New Zealand, a very gassy field, and Rotokaua, New Zealand. The shallow Wairakei wells also fit into the sequence at about this point. Finally there are the classical vapor-dominated systems, such as the The Geysers, United States, Kawah Kamojang, Indonesia, and Larderello, Italy. The geological or geophysical controls that give rise to any fields's specific location in this gradation are yet to be established. At the present time there is little forecasting of any field's characteristics. Rather, these are discovered during drilling.

In this section we will discuss two of the two-phase systems—Broadlands and Ngawha, both in New Zealand.

(i) *Broadlands, New Zealand.* This field is the wettest of the listed two-phase examples. At sufficient depth it is a conventional liquid-dominated field, i.e. similar to Wairakei, with a temperature of 310°C and a gas content of 2–3 per cent by mass. If all wells were drilled into this sector, the field would behave as a hot water system. The boiling water interface is, however, here at about 1 km depth, and most of the production takes place in the two-phase region above. These wells produce from a zone in which the temperatures are in the 250–280°C range.

A few wells, e.g. BR13, do produce from liquid conditions, but the best permeability and temperature are found in the two-phase zone. The sections of Figure 5.9 illustrate schematically a fluid state profile across the productive zone and out into the fringes of the field, and temperature and permeability profiles along this section at a depth of 600 m. The permeability profile is only indicative as actual values of the permeability and its variation are not known. It does, however, appear to decrease as we move out from the two-phase section.

Wells tapping the two-phase zone produce fluid with an average enthalpy of around 1300 kJ/kg, some 110 kJ/kg above that of liquid water at 270°C, the average production temperature. It has therefore been estimated that the liquid saturation in this zone was probably initially of the order of 80–85 per cent. It should be noted that the higher enthalpy here is a valuable asset as this heat can be converted to electrical energy at an efficiency of about 20 per cent (cf. 8–10 per cent for the liquid enthalpy).

The response to exploitation is interesting. It takes about one to two years for significant pressure disturbances to cross the two-phase zone horizontally and a longer time for them to propagate downwards to the boiling water interface. The permeability is anisotropic, being higher horizontally than vertically. Thus when a well is discharged, for the first year or so it will draw on an expanding zone in the two-phase region. Only then does it begin to feel the liquid surroundings, either at the sides or below. In this field, the behaviour at the side boundaries is complicated by the lower permeability (cf. Figure 5.9(c)), but nonetheless, just as in Wairakei, the exterior water will be pulled into the field extracting heat from the margins on the way. At the same time, the decrease in pressure in

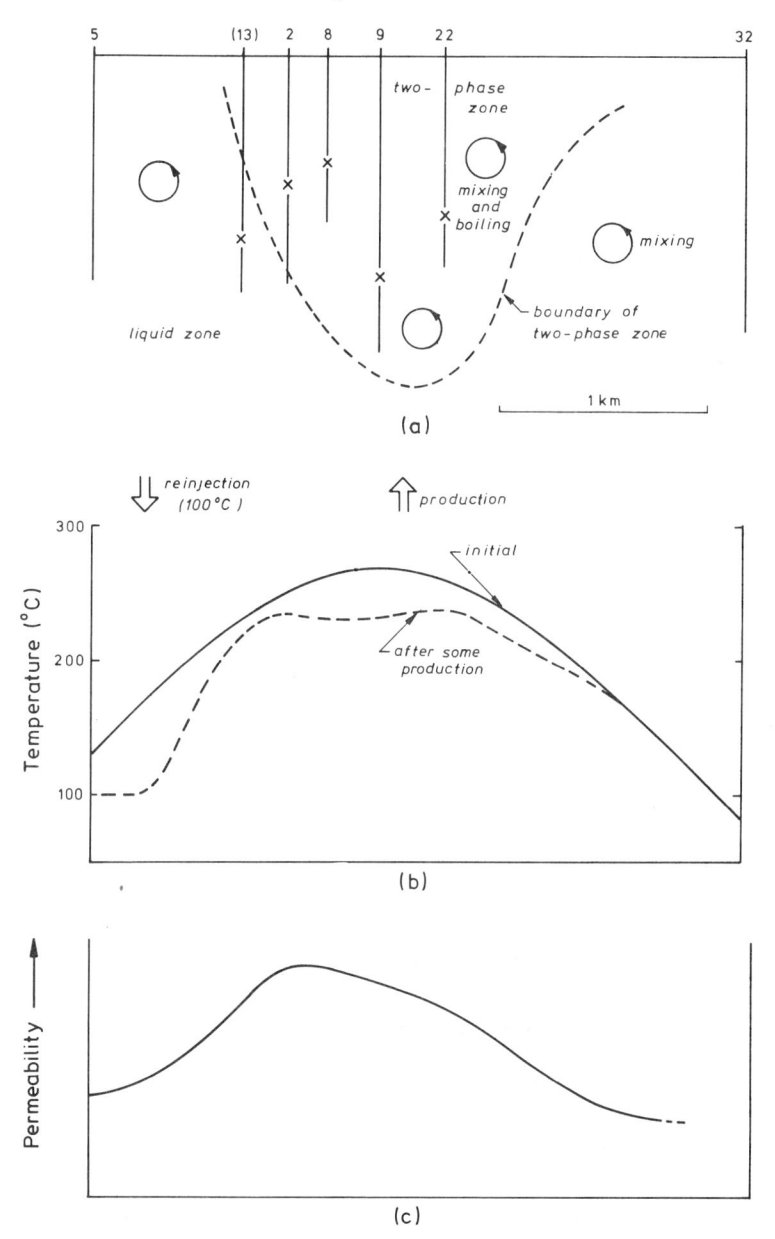

Figure 5.9. Variations across the Broadlands geothermal field, New Zealand. (a) The liquid and two-phase zones. (b) Temperature profiles at 600 m. (c) Permeability variation (qualitative only)

the centre of the field will drop the temperatures in the two-phase zone itself and lower the water level by boiling water off at the single-phase/two-phase interface. The temperature profile will thus become, after some years, that illustrated by the dashed curve in Figure 5.9(b). During this period heat will be mined from the production zone, from the neighbourhood of the dropping water/two-phase interface, and from the margins of the field. With the development of a quasi-steady field condition, after a few years' exploitation, the temperature change in the production zone and the drop in water level will virtually cease. The bulk of the heat will then be mined from the side boundary regions.

Under exploitation, some rise in the enthalpy of the well discharge is expected. As fluid is first withdrawn and pressure falls, the saturation also falls. This increases the steam mobility and decreases water mobility, so that discharge enthalpy rises. However, the recharge fluid is principally or entirely liquid. As the field later approaches its quasi-steady exploited state, the discharge enthalpy falls again, towards the recharge enthalpy. Thus part of the response to exploitation is a transient enthalpy rise. For a suitably sized power station, the peak enthalpy is at most 50 per cent higher than the initial value (Grant 1977b). The field does not 'dry out' under exploitation—there is no way this basically liquid-dominated field will discharge dry steam.

If, as is proposed, reinjection takes place at the field boundary, an additional pressure pulse will be generated in this area, this time a positive one. This pulse will propagate through the two-phase zone at the same rate as the drawdown pulse. It will still, however, take the same order of time for the two pressure waves to interact, i.e. one to two years. In the area of recharge, pressures will also be sustained, and this will tend to reduce the drop in pressure, and hence temperature, within the field reservoir. The water level will also drop less. This will mean that less heat will be mined from within the field, either in the two-phase zone or at the water/two-phase interface. More will therefore be mined from the field margins, especially in the area of maximum reinjection. Reinjection increases the speed of advance of the cold front and hence the rate at which rock volume is being cooled.

(ii) *Ngawha, New Zealand.* Ngawha and similar fields, e.g. Matsukawa in Japan (Baba et al., 1970) and Rotokaua, in New Zealand, have one feature in common with vapor-dominated fields—the wells discharge dry or nearly dry steam—and one feature that is in contrast—pressures are hydrostatic rather than vapor-controlled. They are thus liquid-dominated fields, but with a comparatively high vapor mobility.

As this natural vapor flow is not large even though it is driven by the hydrostatic gradient, the rock permeability must be low, at least in the vertical direction. This low permeability ($\ll 10^{-15}$ m^2) has so far been confirmed by the few well tests carried out at Ngawha and Rotokaua. It is almost certainly an inherent feature of the Ngawha-type systems, and shows a second contrast with the vapor-dominated ones.

In combination with the two-phase conditions, the low permeability makes the hydraulic diffusivity extremely small. Each well will thus mine its own independent volume of rock, and the pressure wave, in effect, will never reach the side boundaries. Steam will be extracted from rock pores containing steam and immobile, or only slightly mobile, water, and this steam will be replaced by the boiling of some of that water. With continuing withdrawal, there will thus be a local decline in both temperature and pressure in both the rock pores and the well discharge. Each local 'mine' will last until the resistance to flow in the increasing volume being tapped drops the well bottom pressure and temperature to such a level that the operation can no longer be satisfactorily sustained. The volume so mined will thus depend directly on the local permeability. For maximum extraction, these local 'mines' will need to be sited quite close together, both horizontally and vertically.

As Ngawha complements its steam discharge with a high gas content (mainly CO_2), it is also essential to consider the likely chemical effects of this extraction. The chemical behaviour here is probably passive, and hence it is only the gas in or out of solution with which we will be concerned. Thus, as the pressures and temperatures decrease, gas transfer from the liquid to the vapor will take place, and gas will continue to be discharged from the wells. This will affect the heat discharge of the system.

The initial gas content of the discharge reflects the equilibrium conditions present in the field. For well Ng1, with initial gas partial pressures of 3·0 MPa (at 230–235°C), this means 50–70 per cent gas (by mass) in the vapor phase. This gas content, however, rapidly declined, reflecting a process of boiling within the confined volume that the well exploited. This decay in gas content may thus not indicate a long term trend. This will be difficult to estimate as any recharge would also be very gassy.

In this Ngawha-type of system we also have to develop a different outlook with regard to reinjection. This will here only be a means of waste disposal. The cool liquid injected is most likely to move only slowly away from the injection well and hence will not rapidly affect the mass balance. Any pressure, and temperature, perturbation will also be restrained to the region immediately adjacent to the well input zone.

The total heat extraction potential of this system will thus be primarily restricted by the number of wells that can be drilled and discharged economically rather then by any size limitation or temperature restraint.

5.3.4. *Vapor-Dominated Systems*

The term 'vapor-dominated' was coined by White et al. (1971). The essential characteristics of the system are: (a) a discharge of steam, and (b) that this discharge comes from a region where the pressure is nearly constant with depth. The steam discharge may initially be wet, dry saturated, or superheated. There is, however, a trend for the discharge to become dry, or increasingly superheated, with exploitation. The pressure of the production zone is often around 3·3 MPa, equivalent to a saturation pressure for water and steam at 235°C. Extremal values of 2·4 MPa at Bagnore, Italy (Atkinson et al., 1978a, b), and 7·0 MPa at Travale, Italy (Atkinson et al., 1978c), have, however, been measured. In most systems there appears to be little leakage to the surface, and capping structures are generally assumed.

There are several vapor-dominated systems now known and under development or exploitation around the world. These include Larderello, Bagnore and Travale in Italy, The Geysers in the U.S.A., and Kawah Kamojang in Indonesia. These all fit in the same niche in our graded sequence and behave similarly in most respects of interest. Only a general discussion is therefore given.

In all these fields, the vapor-dominated zone occurs in a general region of water-saturated rock. There may, in fact, be geothermal liquid-dominated or cold water regions around or intermingled with the vapor region (Petracco and Squarci, 1975, Lipman et al., 1978; Cataldi et al., 1978). In the natural state there is a flow of mass and energy between the liquid and vapor zones. The exploitation behaviour is also intimately linked with that of the water within and surrounding the reservoir.

We follow White et al. (1971), and the bulk of work on vapor-dominated systems, in assuming that the vapor reservoir contains, in its natural state, immobile or only slightly mobile water. This water is the mass supply for the reservoir steam. Some authors (Brigham and Morrow, 1977; Atkinson et al., 1978a) use a 'falling water level' model in which a zone of dry steam overlies a liquid layer. In either case, a supply of liquid is needed as the amount of steam produced from these vapor fields far exceeds what could be stored in them as vapor alone (James, 1968). As it is doubtful whether these falling water level models truly model all the processes involved, we concentrate on the White model (White et al., 1971, and Truesdell and White, 1973) for this discussion.

Measurements at Kawah Kamojang in Indonesia are consistent with the White model and the presence of a dispersed water phase. The water saturation has been measured at 35 per cent (Grant, 1979). The residual saturation is not known, but it must be somewhat less than this as the steam is slightly wet, with a representative enthalpy of 2775 kJ/kg.

In the natural state of these reservoirs, we therefore anticipate an upward flow of steam and a downward flow of water. This latter is the condensate of the steam, very little of which is able to escape through the capping structures to the surface, although heat can be lost to the surface by conduction. The down flow is probably predominantly in various cracks and fractures in the system with the remaining water virtually immobile in the rock pores. (In contrast, freely available water in a relatively permeable reservoir would continue to flow to the wells.) These cracks and fractures are the fine geological structure of the reservoir, and thus the preferred paths for some flows. Water bearing strata have, for example, been encountered in The Geysers (Ramey, 1970), and wells at Kawah Kamojang frequently discharge wet steam (Grant, 1979). Corresponding with this pattern of patchiness in the flow distribution of water and steam there will also be some variation in the water saturation; i.e. some rocks will be wetter than others. There will also be some interchange of water and steam at the sides of the system, i.e. steam condensing and water boiling.

A well extracting fluid and heat from a vapor reservoir discharges either saturated or superheated steam (Truesdell and White, 1973). If the steam is superheated there is a dry zone around the well, through which the steam flows isothermally. This dry zone would have originally contained some water, but, as lowering the pressure in the vapor system reduces the water saturation, this water will have completely boiled off.

Beyond the dry zone is a zone containing immobile water. A well producing saturated steam contacts this zone directly; i.e. in this area the dry zone is not yet established. In either case, saturated or superheated, the mass and energy discharged from the well is supplied from the wetted zone. The water boils to form steam, and this flows to the well, taking the mass and the latent heat with it.

Outside the vapor-dominated section of the reservoir we may have a liquid-dominated region and rocks saturated with cooler ground water (White et al., 1971). At Larderello, for example, external cooler water supplies part of the recharge (Petracco and Squarci, 1975). As such cold water enters, it first crosses into the hot liquid region, mining heat from that boundary, and then boils, mining its latent heat from the rock, and enters the vapor reservoir. The mass supply to the well comes from the cold ground water, and the energy from the boundaries where the fluid gains enthalpy. Other recharge may come from the hyothesized deep brine layer (White et al., 1971).

In general, as the pressures in vapor-dominated geothermal reservoirs are largely well below hydrostatic, to prevent flooding by cold exterior water there must be low permeaability defining boundaries on all sides. This is the case at The Geysers (Bredehoeft, J. D., Ramey H. R., Jr., personal communications) and at Kawah Kamojang. These boundaries may reduce external recharge to negligible amounts. In this event all the steam supplied to the wells comes from boiling in the vapor-dominated zone. The water in this zone supplies the mass, and the rock supplies the heat to convert this to the steam discharged by the wells.

An estimate of recoverability from a vapor-dominated reservoir, based on the model of White et al. (1971) and Truesdell and White (1973), has been made by Nathenson (1975). In their application of this approach to The Geysers, Nathenson and Muffler (1975), assuming 5 per cent for the volume percentage of water and a 240°C reservoir, estimate an ideal recovery factor of 19·4 per cent. A qualitative allowance (of 50 per cent) for structural inhomogeneity and its effects reduces this factor to 9·7 per cent. The assignment

5. Heat Extraction from Geothermal Reservoirs

of a recovery factor is, however, subjective, but even for a favourable vapor-dominated geothermal reservoir like Larderello–Travale with an assumed 215°C reservoir temperature the figures are low (11–19 per cent) (Muffler and Cataldi, 1978). Their intermediate estimate of 15 per cent compares with that (> 13·3 per cent) estimated from intergranular vaporization based on the model and data of Barelli et al (1975) for that area.

5.4. Other Systems

There are three other types of geothermal systems that are of interest to scientists and engineers at the present time—the geopressured system, the hot dry rock system and the magma chamber. Although these do not fit in with our sequence, the first two of these may be regarded as relatively simple variations on the basic theme. The magma chamber may perhaps be a little different.

5.4.1. *The Geopressured System*

The reservoir mechanics of a geopressured system have so far only been discussed to a very limited extent (see, for example, Pritchett et al., 1977, Garg and Pritchett, 1977, 1978, and Garg et al., 1977, 1978). These authors indicate that there are four driving mechanisms which tend to expel fluid from a geopressured stratum—water compressibility, pore collapse, evolution of methane gas, and clay dehydration or shale dewatering. The decrease in permeability accompanying pore collapse and the relative permeability effects due to the evolution of the gas tend to impede the fluid flow. This would, however, be outweighed in general by the positive driving effects.

In its simplest form a geopressured reservoir may just be regarded as a high pressure sealed container filled with porous material saturated with compressed water and dissolved gas. Pull out the plug and the water and gas flow out, most of the boiling and separation taking place only in the wells or above ground on account of the great reservoir depths of these fields (4800 to 5000 metres in the Brazoria geothermal fairway, Texas Gulf Coast, U.S.A.; Bebout et al., 1977a,b). The majority of the water thus comes available purely from decompression in the strata. Left to run in this mode, this process would in effect give us all that water held in compression above the natural pressure. This will, however, only be a small fraction of the fluid that might be available. The fraction of the heat mined would be even smaller as the heat contained in the rock would not be touched.

Recirculation of fluid through the system is thus essential, both to maintain the pressures and to extract the heat in the rock. In this case, as we have no natural inflow (recharge) from the sides, the only cold water front will be that associated with the water from the injection wells, and if these and the extraction wells are well sited, there are good prospects of a high percentage return from this reservoir.

5.4.2. *The Hot Dry Rock System*

In various areas of the world hot but dry and impermeable rock is known to exist at relatively shallow depth. By drilling wells into this rock and fracturing the material between these wells it is possible to set up a circulation path for injected fluid to extract heat from such a system. Only one system of this nature has so far been attacked, that in the Jemez Mountains, New Mexico, U.S.A. (Laughlin, this volume). The full process and progress with the research program to date is described in papers like Cummings et al. (1979) and Tester and Albright (1979). This discussion is therefore somewhat superficial and is aimed only at relating the process of heat extraction here to that in hydrothermal systems.

From the heat extraction point of view, the process is in fact little different from that of systems already discussed. Although there is no natural fluid, in principle we are again only mining the heat from the rock with a cold fluid, just as we are presently doing at the margins of fields like Wairakei. In this case the 'margin' is the injection well, since the cooler fluid flowing from the injection well extracts heat by conduction from the rock bounding the fissure. The success or otherwise of the venture will thus depend on the quality of the inter-well fracturing. A few dominant connectors will short-circuit the rock mass so that only a limited hot-rock volume will be accessible. This could soon be depleted of its heat. In contrast, a high-density fracture structure should access a wealth of heat.

5.4.3. Magma Chambers

Magma chambers are mentioned here for completeness. They appear to have considerable potential as a source of energy, but their exploitation will probably require techniques well outside the range of technologies used in exploiting the other fields and systems described in this chapter. A discussion of magma as a geothermal resource is given by Stoller and Colp (1978).

5.5. Summary

Detailed geothermal reservoir analysis is an extremely complex process. Structural complexity, three-dimensional behaviour, mixed mass and energy transport, phase changes, dissolved gasses, the likely necessity of reinjection and a welter of unknown parameters or parametric relationships all must be taken into account. For the ultimate management of geothermal resources such detailed studies may be essential, but for resource valuation the much simpler approach that has been discussed in this chapter may be all that is necessary (see, for example, Donaldson and Grant, 1978).

Heat extraction from a porous medium, and hence a geothermal reservoir, requires a generous amount of fluid to carry the heat conveniently to the exterior of the system and efficient heat transfer mechanisms to move the bulk of the stored heat from the rock to this fluid carrier. The fluid may be 'generated' within the medium, by expansion of that already compressed in the system or by expansion due to a change of phase, it may be sucked in from outside the heated area, or it may be deliberately injected by the field operators.

In this study we have shown the outward diffusion of the pressure depression from the production zone to be the stimulus both of fluid 'generation' and of the inflow of external liquid (the 'recharge'). In liquid-dominated systems, the pressure wave moves out rapidly to the field boundaries, and the recharge thus becomes the dominant mass source within only a day or so of first production. The expansion effects, except perhaps in the case of the geopressured fluid, are virtually negligible. As we move along the sequence to two-phase systems, pressure diffusion becomes slower, and boiling effects result in the replacement of some of the water discharge with steam, a definite advantage from the heat extraction point of view. For a typical system, such as Broadlands, New Zealand, however, the pressure wave still reaches the reservoir boundaries—it just takes longer, in this case the order of one to two years.

In low permeability reservoirs, such as Ngawha, New Zealand, the situation is a little different, and the wells may there tap only the local water. Again boiling will, however, sustain the discharge, at least for a time. Artificial fracturing may be of benefit in these systems.

At one extreme of our sequence, the vapor-dominated reservoir, we return to a situation akin to that of the other two-phase cases, with here the potential of boiling of all the water in the immediate vicinity of the wells.

In time the fluid thus comes from further and further away, by boiling of water, and ultimately again by an inflow at the field margins.

To all intents and purposes injection or reinjection of cooler or used fluid just replaces the cooler side inflow. The choice of location of this inflow, however, rests with the field operators.

The heat going up the well comes in part from the stored fluid. The bulk, however, comes from the rock through which it flows. It is this latter heat that we need to extract for greatest benefit. There are two main ways in which this may be done—by boiling of water in adjacent pores and cracks, or by passing cooler water through the structures. The former normally takes place right inside the reservoir, while the latter will occur at the margins of the field or around injection wells. Either is effective, although the boiling process will result in some drop in temperature at the well. The margin mining process, in contrast, is one of frontal movement with little decay in temperature on the forward side of the front except where reservoir inhomogeneities create problems.

Both vertical and horizontal flows will be important in this heat extraction process, but, for many fields, the horizontal effects appear to dominate. Pressure drops in the two-phase zone will, however, stimulate boiling at any two-phase (steam-water)/liquid interface and cause a drop in its level. Colder water layers overlying such two-phase zones may also be sucked down into the system. Since these cold layers may be closer to some wells than the water at side boundaries and they have an additional density drive, this water could ultimately be the limiting control for these wells, quenching them before they have extracted more than a fraction of the heat from the rock around them.

Fractured systems can also cause problems as fronts can move in much more quickly along such preferred paths than through the rock mass. Quenching of the well system may thus take place while the majority of the heat, stored in the rock between the various fractures, remains virtually untouched. Kawerau, New Zealand, described earlier in this chapter, is a prime example of such a fractured field.

At the start of this chapter we posed the three questions that we believe are the most crucial in geothermal reservoir development—how much energy can we extract? For how long may it be extracted? And, of what quality is the energy extracted? Answers to these questions are now available.

To answer the third question first, the quality of the extracted energy, as defined by the enthalpy of the fluid discharged, depends directly on the temperature of the reservoir and on the type of reservoir system involved. The quality increases as the reservoir temperature goes up and also as we move along the sequence of field types from the warm water to the vapor-dominated systems. In most fields the mean discharge enthalpy, and hence the energy quality, does not change greatly with exploitation until the field commences to collapse due to the cold water inflow reaching the production area. Only in the liquid-dominated two-phase systems are any significant changes anticipated, and these may only occur for a limited time. An enthalpy increase of about 50 per cent, as estimated for Broadlands, New Zealand, is as large as is to be expected anywhere.

It is the cold water inflow that defines how much heat we can extract from any reservoir. If the reservoir is homogeneous this inflow is uniform, and much of the heat stored in the rock outside of the production zone will be swept into this zone and extracted. In this case the rate of extraction has little effect on the cumulative amount of heat extracted as exactly the same volume will be swept out with only the time-scale changed. The 'how long' thus purely depends on the rate of extraction. This rate is, however, controlled in the reservoir itself by the drawdown that can be tolerated.

If there are marked inhomogeneities in the reservoir, as have shown up at Kawerau, New Zealand, the cold front advances preferentially along the higher permeability channels, and wells are quenched while heat remains stored in the surrounding rock. Here the amount of heat extracted depends on a balance between the rate of heat transfer from the rock to the fluid in these channels and the rate of advance of the local cold fronts. The slower the advance of the cold front, the more heat is transferred, and the greater is the efficiency of the mining operation. The mining efficiency decreases significantly as the rate of extraction goes up (Bodvarsson, 1974; Muffler and Cataldi, 1978).

Provided that we can tolerate the lower rate of fluid extraction in this latter situation, the end point of both exploitations will be much the same. The mining efficiency (or recovery factor) is controlled by the geometry and nature of the reservoir. For ideal uniformly permeable liquid-dominated reservoirs this recovery factor has been estimated at about 50 per cent (Nathenson, 1975), and a similar recovery factor has been obtained with the idealised model for Wairakei. To allow for the inhomogeneous nature of the structures in these reservoirs this figure must be reduced. A subjective halving of the ideal recovery factor has been suggested by Nathenson and Muffler (1975).

For vapor-dominated reservoirs the analysis is even more complex, but even for favourable reservoirs like Larderello–Travale the recovery factors appear lower than those for their liquid-dominated counterparts. Muffler and Cataldi (1978), in their overview suggest a recovery factor of about 15 per cent for such fields, comparable with the >13.3 per cent based on the model and data of Barelli et al. (1975). For The Geysers, Nathenson and Muffler (1975) give an ideal 19·4 per cent and a real (subjective) 9·7 per cent.

Acknowledgements

The ideas expressed in this chapter can in no way be attributed to any one source, nor did they just come overnight. They come rather from many discussions over many years with many colleagues. The authors cannot hope to name more than a few, but we wish to thank all those colleagues, both in New Zealand and overseas, with whom we have reacted over the years.

Our specific thanks must go to our direct associates at this time, Drs Lara Fradkin and Michael L. Sorey, presently at Physics and Engineering Laboratory, Lower Hutt, New Zealand, and Drs Alex McNabb and Robin Wooding of Applied Mathematics Division, DSIR, Wellington, New Zealand, with whom we have had an almost continuous dialogue during our thinking and writing up of this work. We should also like to thank Paul Bixley, Ministry of Works and Development, Wairakei, New Zealand, and Sabodh Garg, Systems, Science and Software, La Jolla, California, U.S.A., for their comments and ready assistance.

All field data used in the studies of New Zealand fields discussed in this chapter have been made available by the New Zealand Ministry of Works and Development. The authors would like to thank the officers of that Department for making this information and data so readily available.

References

Atkinson, P., Celati, R., Corsi, R., and Kucuk, F., 1978a. 'Behaviour of two-component vapor-dominated geothermal reservoirs' presented at the 1978 California Regional Meeting of the Society of Petroleum Engineers of AIME, San Francisco, California, April 12–14, 1978. SPE Paper 7132.

Atkinson, P., Celati, R., Corsi, R., Kucuk, F., and Ramey, H. J., Jr., 1978b. 'Thermodynamic behaviour of the Bagnore Geothermal Field', presented at the Larderello Workshop on Geothermal Resource Assessment and Reservoir Engineering, Sept. 12–16, 1977. *Geothermics 7* (2–4), 185–208.

Atkinson, P., Barelli, A., Brigham, W., Celati, R., Manetti, G., Miller, F., Meri, G., and Ramey, H. J., Jr., 1978c. 'Well testing in Travale-Radicondoli Field', presented at the Larderello Workshop on Geothermal Resource Assessment and Reservoir Engineering, Sept. 12–16, 1977. *Geothermics 7* (2–4), 145–184.

Baba, K., Takahi, S., and Matsuo, G., 1970. 'On the reservoir at Matsukawa Geothermal Field', United Nations Symposium on the Development and Utilisation of Geothermal Resources, Pisa, Italy, *Geothermics* (Special Issue 2), **2** (2), 1440–1447.

Barelli, A., Calamai, A., and Cataldi, R., 1975. 'Stima del potenziale geotermico della fascia preappenninica centro-meridionale', *ENEL Centro di Ricerca Geotermica, Relazione T3/149 Z*, 33 pp.

Bebout, D. G., Loucks, R. G., and Gregory, A. R., 1977a. 'Study looks at Gulf Coast Geothermal Potential', *Oil and Gas Journal*, Sept. 26, 1977.

Bebout, D. G., Loucks, R. G., and Gregory, A. R., 1977b. 'Texas geothermal prospect slated to begin operations at Martin Ranch', *Oil and Gas Journal*, Oct. 3, 1977.

Bodvarsson, G., 1962. 'An appraisal of the potentialities of the geothermal resources in Iceland', *Sixth World Power Conference, Melbourne, Australia*, Paper 206–111.

Bodvarsson, G., 1972, 'Thermal problems in the siting of reinjection wells', *Geothermics*, **1**(2), 63–6.

Bodvarsson, G., 1974. 'Geothermal resource energetics', Geothermics, **3**(3), 83–92.

Boldizsár, T., 1975. 'Research and development of geothermal energy production in Hungary', *Geothermics*, **4**, 44–56.

Bolton, R. S., 1970. 'The behaviour of the Wairakei geothermal field during exploitation', United Nations Symposium on the Development and Utilisation of Geothermal Resources, Pisa, Italy. *Geothermics* (Special Issue 2), 1426–1439.

Brigham, W. E., and Morrow, W. P., 1977. 'P/Z behaviour of geothermal steam reservoirs', *SPE Journal* (Dec. 1977), 407–412.

Carslaw, H. S., and Jaeger, J. C., (1959). *Conduction of Heat in Solids*, Clarendon Press, Oxford (2nd Edition), 386 pp.

Cataldi, R., Lazzarotto, A., Muffler, P., Squarci, P., and Steffani, G., 1978. 'Assessment of geothermal potential of Central and Southern Tuscany', presented at the Larderello Workshop on Geothermal Resource Assessment and Reservoir Engineering, Sept 12–16, 1977. *Geothermics 7* (2–4), 91–132.

Cummings, R. G., Morris, G. E., Tester, J. W., and Bivins, R. L., 1979. 'Mining earth's heat: hot dry rock geothermal energy', *Technology Review*, February 1979, 59–78.

Dench, N. D., 1962. 'Reconditioning of steam bores at Kawerau', *N.Z. Engineering*, **17**(10), 1–8.

Donaldson, I. G., 1968. 'The flow of steam water mixtures through permeable beds: a simple simulation of a natural undisturbed hydrothermal region', *N.Z. Journal of Science*, **11**(1), 3–23.

Donaldson, I. G., 1970a. 'The simulation of geothermal systems with a simple convective model', United Nations Symposium on the Development and Utilisation of Geothermal Resources, Pisa, Italy, *Geothermics* (Special Issue 2), 649–654.

Donaldson, I. G., 1970b. 'A possible model for hydrothermal systems and methods of studying such a model', *Proceedings of the Third Australasian Conference on Hydraulics and Fluid Mechanics, Sydney, Australia*, 25–29 Nov., 1970.

Donaldson, I. G., and Grant, M. A., 1978. 'An estimate of the resource potential of New Zealand geothermal fields for power production', presented at the Larderello Workshop on Geothermal Resource Assessment and Reservoir Engineering, Sept. 12–16, 1977, *Geothermics 7* (2–4), 243–252.

Dykstra, H., and Adams, R. H., 1977. 'Momotombo Geothermal Reservoir', *Summaries, Third Workshop Geothermal Reservoir Engineering*, Dec. 14–16, 1977, Stanford University, Stanford, California, 96–106.

Garg, S. K., 1978. 'Pressure transient analysis for two-phase (liquid water/steam) geothermal reservoirs', *Paper SPE 7479*, presented at the 53rd Annual Fall Technical Conference SPE–AIME, October 1–3, 1978, Houston. Texas.

Garg, S. K., and Pritchett, J. W., 1977. 'Simulation of drive mechanisms in geopressured reservoirs', *Proceedings 18th U.S. Symposium on Rock Mechanics, Keystone, Colorado*, 1B5-1 to 1B5-4.

Garg, S. K., and Pritchett, J. W., 1978. 'Two phase flow in geopressured geothermal wells'. Third Geopressured Geothermal Energy Conference, University of Southern Louisiana, Lafayette, Louisiana, Nov. 1977. *Energy Conversion*, **18**, 45–51.

Garg, S. K., Pritchett, J. W., Brownell, D. H., Jr., and Riney, T. D., 1978. 'Geopressured geothermal reservoir and wellbore simulation', *Systems, Science and Software, La Jolla, California, Report SSS-R-78-3639*.

Garg, S. K., Pritchett, J. W., Rice, M. H., and Riney, T. D., 1977. 'U.S. Gulf Coast geopressured geothermal reservoir simulation', *Systems, Science and Software, La Jolla, California, Report SSS-R-77-3147*.

Grant, M. A., 1977a. 'Approximate calculations based on a simple one-phase model of a geothermal field', *N.Z. Journal of Science*, **20**, 19–25.

Grant, M. A., 1977b. 'Broadlands—a gas-dominated geothermal field', *Geothermics*, **6**, 9–29.

Grant, M. A., 1978. 'Two-phase linear geothermal pressure transients—a comparison with single-phase transients', *N.Z. Journal of Science*, **21**, 355–364.

Grant, M. A., 1979. 'The water content of the Kawah Kamojang geothermal reservoir', *Geothermics*, **8**, 21–30.

Grant, M. A., and Sorey, M. L., 1979. 'The compressibility and hydraulic diffusivity of a water-steam flow', *Water Resources Research* **15**(3), 684–686.

Grindley, G. W., 1965. 'The geology, structure, and exploitation of the Wairakei Geothermal Field, Taupo, New Zealand', *N.Z. Geological Survey Bulletin n.s. 75*.

Gringarten, A., 1978. 'Reservoir lifetime and heat recovery factor in geothermal aquifers used for urban heating', *Pure and Applied Geophysics*, **117**, 297–308.

Healy, J., and Hochstein, M. P., 1973. 'Horizontal flow in hydrothermal systems', *Journal of Hydrology (N.Z.)*, **12**(2), 71–82.

James, C. R., 1968. 'Wairakei and Larderello; geothermal power systems compared', *N.Z. Journal of Science*, **11**, 706–719.

Lipman, S. C., Strobel, C. J., and Gulati, M. S., 1978. 'Reservoir performance of The Geysers Field', presented at the Larderello Workshop on Geothermal Resource Assessment and Reservoir Engineering, Sept. 12–16, 1977. *Geothermics* **7** (2–4), 209–220.

Macdonald, W. J. P., and Muffler, L. J. P., 1972. 'Recent geophysical exploration of the Kawerau geothermal field, North Island, New Zealand', *N.Z. Journal of Geology*, **15**, 303–317.

McNabb, A., 1975. 'A model of the Wairakei geothermal field', unpublished report, Applied Mathematics Division, Department of Scientific and Industrial Research, Wellington, New Zealand.

Marshall, D. C., 1966. 'Preliminary theory of the Wairakei Geothermal Field', *N.Z. Journal of Science*, **9**(3), 651–673.

Marshall, D. C., 1970. 'Development of a theory of the Wairakei Geothermal Field by the "simplest cases first" technique', United Nations Symposium of the Development and Utilisation of Geothermal Resources, Pisa, Italy. *Geothermics* (Special Issue 2), 669–676.

Marshall, D. C., 1975a. 'Theory of the Wairakei geothermal field. 1. Single-phase, drawdown model', *N.Z. Journal of Science*, **18**, 453–463.

Marshall, D. C, 1975b. 'Theory of the Wairakei geothermal field. 2. Permeability and diffusivity variations', *N.Z. Journal of Science*, **18**, 591–603.

Mercer, J. W., and Faust, C. R., 1976. 'Status of modelling efforts for the Wairakei Geothermal Field', *Summaries, Second Workshop Geothermal Reservoir Engineering*, Dec. 1–3, 1976, Stanford University, Stanford, California, 308–309.

Mercer, J. W., and Faust, C. R., 1979. 'Geothermal Reservoir Simulation III: Application of liquid- and vapor-dominated hydrothermal modeling techniques to Wairakei, New Zealand' *Water Resources Research*, **15**(3), 653–671.

Mercer, J. W., Pinder, G. F., and Donaldson, I. G., 1975. 'A Galerkin finite-element analysis of the hydrothermal system at Wairakei, New Zealand', *Journal of Geophysical Research*, **80**(17), 2608–2621.

Moench, A. P., and Atkinson, P., 1978. 'Transient-pressure analysis in geothermal steam reservoirs with an immobile vaporizing phase', *Geothermics* **7**(2–4), 253–264.

Muffler, L. J. P., and Cataldi, R., 1978. 'Methods for regional assessment of geothermal resources', presented at the Larderello Workshop on Geothermal Resources Assessment and Reservoir Engineering, Sept. 12–16, 1977. *Geothermics* 7(2–4), 53–90.

Nathenson, M., 1975. 'Physical factors determining the fraction of stored energy recoverable from hydrothermal convection systems and conduction-dominated areas', *U.S. Geological Survey Open-File Report 75-525*, 35 pp.

Nathenson, M., and Muffler, L. J. P., 1975. 'Geothermal resources in hydrothermal convection systems and conduction-dominated areas'. In *Assessment of Geothermal Resources of the United States—1975*, D. E. White and D. L. Williams (Eds.), U.S. Geological Survey Circular 726, 104–121.

Petracco, C., and Squarci, P., 1975. 'Hydrological balance of Larderello Geothermal Region', *Proceedings Second United Nations Symposium on the Development and Use of Geothermal Resources, San Francisco, California, 22–29 May, 1975*, 521–530.

Pritchett, J. W., Garg, S. K., and Brownell, D. H., Jr., 1976a. 'Numerical simulation of production and subsidence at Wairakei, New Zealand', *Summaries Second Workshop Geothermal Reservoir Engineering*, Dec. 1–3, 1976, Stanford University, Stanford, California, 310–23.

Pritchett, J. W., Garg, S. K., Brownell, D. H., Jr., Rice, L. F., Rice, M. H., Riney, T. D., and Hendrickson, R. R., 1976b. 'Geohydrological environmental effects of geothermal power production phase IIA', *Systems, Science and Software, La Jolla, California, Report SSS-R-77-2998*.

Pritchett, J. W., Garg, S. K., and Riney, T. D., 1977. 'Numerical simulation of the effects of reinjection upon the performance of a geopressured geothermal reservoir', *Geothermal: State of the Art*, Transactions Annual Meeting of the Geothermal Resources Council, San Diego, California, p. 245.

Ramey, H. J., Jr., 1970. 'A reservoir engineering study of The Geysers Geothermal Field' submitted as evidence, Reich and Reich, Petitioner *v.* Commissioner of Internal Revenue, 1969 Tax Court of the United States, 52, T.C. No. 74, 1970.

Scheidegger, A. E., 1957. *The Physics of Flow through Porous Media*, University of Toronto Press, Toronto 236 pp.

Stoller, H. M., and Colp, J. L., 1978. 'Magma as a geothermal resource—a summary', *Geothermal Resources Council Transactions*, **2**, 613–615.

Studt, F. E., 1958. 'Geophysical reconnaissance at Kawerau, New Zealand', *N.Z. Journal of Geology and Geophysics*, **1**(2), 219–246.

Tansev, E. O., and Wasserman, M. L., 1977. 'Modeling the Heber Geothermal Reservoir', *Summaries, Third Workshop Geothermal Reservoir Engineering*, Dec. 14–16, 1977, Stanford University, Stanford, California.

Tester, J. W., and Albright, J. N., 1979. 'Hot dry rock energy extraction field test: 75 days of operation of a prototype reservoir at Fenton Hill, Segment 2 of Phase 1', *Los Alamos Scientific Laboratory, LASL-7771-MS*, 104 pp.

Truesdell, A. H., and White, D. E., 1973. 'Production of superheated steam from vapor-dominated geothermal reservoirs', *Geothermics*, **2**, 154–173.

White, D. E., Muffler, L. J. P. and Truesdell, A. H., 1971. 'Vapor-dominated hydrothermal systems compared with hot water systems', *Economic. Geology*, **66**, 75–97.

Whiting, R. L., and Ramey, H. J., Jr., 1969. 'Application of material and energy balances to geothermal steam production', *Journal of Petroleum Technology*, 893–900 (July, 1969).

Wooding, R. A., 1963. 'Convection in a saturated porous medium at large Rayleigh number or Peclet number', *Journal of Fluid Mechanics*, **15**, 527–544.

Wooding, R. A., 1980. 'Analysis of pressure drawdown measurements in the Wairakei–Tauhara geothermal area' (submitted to *Water Resources Research*).

Geothermal Systems: Principles and Case Histories
Edited by L. Rybach and L. J. P. Muffler
© 1981 John Wiley & Sons Ltd.

6. Geothermal Resource Assessment

L. J. Patrick Muffler

U.S. Geological Survey, Menlo Park, CA 94025 U.S.A.

6.1. Introduction

Resource assessment can be defined as the broad-based estimation of future supplies of minerals and fuels. This assessment requires not only the estimation of the amount of a given material in a specified part of the Earth's crust, but also the fraction of that material that might be recovered and used under certain assumed economic, legal, and technological conditions. Furthermore, resource assessment includes not only the quantities that could be produced under present economic conditions, but also the quantities not yet discovered or that might be produced with improved technology or under different economic conditions.

Geothermal resources consist primarily of thermal energy, and thus geothermal resource assessment is the estimation of the thermal energy in the ground, referenced to mean annual temperature, coupled with an estimation of the amount of this energy that might be extracted economically and legally at some reasonable future time. Geothermal resource estimation also includes estimates of the amount of byproducts that might be produced and used economically along with the thermal energy. These byproducts can be metals or salts dissolved in saline geothermal fluids (e.g., the fluids of the Salton Sea geothermal system; White, 1968; Helgeson, 1968) or gases such as methane dissolved in geopressured fluids (Wallace et al., 1979).

A resource assessment is a statement made at a given time using a given data set and a given set of assumptions concerning economics, technology, etc. With respect to most commodities, both the data and the assumptions can change rapidly, the former primarily in response to exploration activities, the latter in response to technology development, economics, political vagaries, environmental constraints, social policy, etc. Consequently, a resource assessment is of only transitory value and must be updated periodically. This is particularly so for a resource like geothermal energy, for which exploration, development, and use are increasing rapidly (see Figure 6.1 for an example) and where the world-wide energy picture is in a state of flux as countries try to come to grips with finite fossil fuel resources, environmental pollution, nuclear waste disposal, etc..

6.2. Resource Terminology

The need for explicit terminology concerning geothermal energy was emphasized by Muffler and Cataldi (1977; 1978). This terminology must be uniform and comparable from country to country and also must be compatible with the resource terminology used for

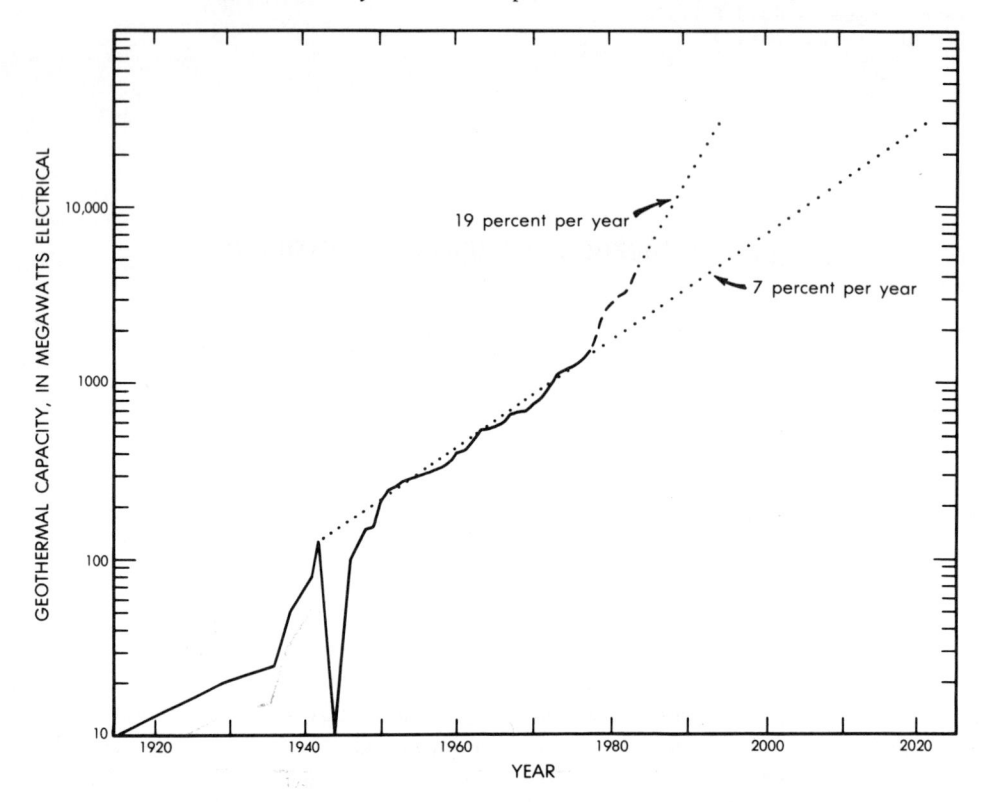

Figure 6.1. Graph showing worldwide installed geothermal electrical capacity as a function of time. Dashed line indicates plants under construction or committed up to 1983. The dotted extrapolations can be interpreted as upper and lower limits of expected growth. From Muffler and Guffanti (1979), based on data collated in great part by Donald E. White

other energy sources. A primary goal of energy resource estimation is the comparison of different energy sources, and to do this, the terminology used in the various disciplines must be coherent. An internally consistent geothermal terminology is of little value if it is incompatible with petroleum, uranium, and coal terminology. For a general description of mineral and fuel resource assessment and a trenchant critique of terminology, see Schanz (1975a; 1975b).

Resource base was defined by Netschert (1958) and by Schurr and Netschert (1960, p. 297) as '... the sum total of a mineral raw material present in the earth's crust within a given geographic area ... whether its existence is known or unknown and regardless of cost considerations and of technological feasibility of extraction.' *Resource* was defined by Netschert (1958) and by Schurr and Netschert (1960, p. 297) as '... that part of the resource base (including reserves) which seems likely to become available given certain technologic and economic conditions.' And finally, *reserve* was defined by Flawn (1966, p. 10) as '... quantities of minerals ... that can be reasonably assumed to exist and which are producible with existing technology and under present economic conditions.' The term resource base refers to material in the ground. The terms resource and reserve, however, refer only to the fraction of the material that can be recovered.

When dealing with a specific concentration of material such as an ore deposit, a petroleum reservoir, or a geothermal reservoir, the ratio of the material (or energy) that can be recovered to the material (or energy) in place is termed the *recovery factor*. This factor is termed the *reserve recovery factor* or the *resource recovery factor*, depending on whether it is used in the determination of reserves or resources (e.g., Cataldi et al., 1977; 1978).

The recovery factor can range up to one for a few metallic deposits where almost all the material can be mined and brought to the surface. As pointed out by Schanz (1975b), in such a situation it makes little difference whether one considers the resource to be the quantity in place or the quantity brought to the surface; the uncertainties in resource estimation are commonly greater than the quantities left behind in the mine or lost in processing.

Considering mineral deposits with recovery factors substantially less than one, however, it is important to distinguish between the amount of material in place in the ore deposit and the amount of material that can be extracted. The latter quantity is of primary interest to man, both in estimating the absolute quantities of a given commodity (e.g., do we have sufficient chromium resources to support the steel industry?), in comparing two different sources of the same element, or in comparing one commodity with another.

The discrepancy between material in place and the recoverable fraction becomes even more striking when one considers sources of energy. For example, the petroleum resource calculations of Miller et al. (1975) assume that the recoverable petroleum ultimately could be as much as 60 per cent of the petroleum in the ground. Similarly, their reserve calculations assume that the petroleum recoverable under current economics is 32 per cent of the petroleum in the ground. For coal, the recoverability depends on depth and thickness of the coal bed and is currently about 50 per cent for deep-mined coal (Schanz, 1975b). Clearly, the important quantity is the amount of the fuel that can be extracted, particularly when one considers that fuels are converted to energy and that the comparison between fuels has to be on the basis of energy. It does little good to compare the energy equivalent of petroleum in place with the energy equivalent of natural gas in place when the recovery factors are so strikingly different (currently 32 per cent for petroleum *vs.* 80 per cent for natural gas; Schanz, 1975b, p. 28). As discussed below, the situation becomes even more acute for geothermal energy, for which recovery factors are commonly 25 per cent or less.

Consequently, Muffler and Cataldi (1977; 1978) recommended that geothermal energy terminology be made compatible with the terminology for other energy sources and that the term 'geothermal resource' be restricted to that part of the thermal energy in the ground (referenced to mean annual temperature) that can be extracted economically and legally at some specified future time. This figure, expressed in terms of energy, is the only figure that can be compared meaningfully to the thermal energy equivalent of barrels of recoverable oil, cubic meters of gas, tons of coal, or kilograms of uranium. It is the *recoverable* geothermal energy that is useful and meaningful to man and therefore should be termed the geothermal resource.

In recent years, it has become common, at least in North America, to depict resources, reserves, and similar terms on rectangular diagrams that depict the degree of economic feasibility on the vertical axis and the degree of geologic assurance on the horizontal axis. These diagrams were introduced by McKelvey (1968) and have come to be termed McKelvey diagrams. As adopted by the U.S. Bureau of Mines and the U.S. Geological Survey (1976), the McKelvey diagram encompasses 'total resources', with the major vertical subdivisions being 'economic' and 'subeconomic' and the major horizontal subdivisions being 'identified' and 'undiscovered' (Figure 6.2).

It has been pointed out by Schanz (1975b) that this McKelvey diagram of the U.S. Geological Survey is open-ended and does not encompass all of a given material. Schanz

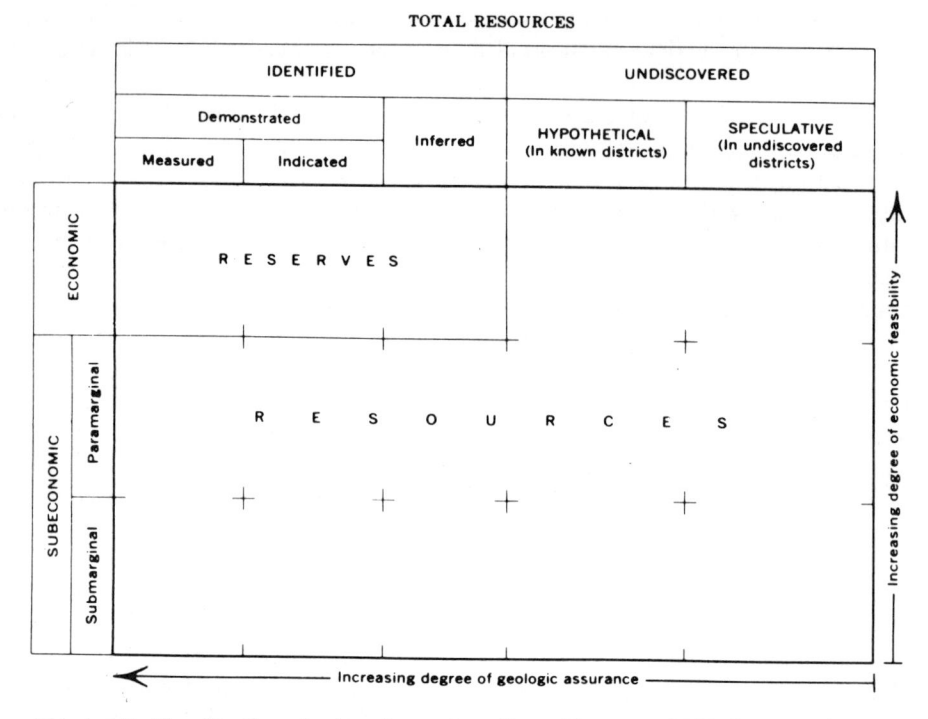

Figure 6.2. Classification of mineral resources (from Figure 1 of U.S. Bureau of Mines and U.S. Geological Survey, 1976)

further emphasizes that, in using this McKelvey diagram, one must continually keep in mind the physical existence of materials and energy beyond resources (i.e., materials and energy which we may not use until far into the future; Schanz, 1975b, p. 11). As an example, there is no question that there is an immense quantity of aluminium in almost all types of crustal rocks. But aluminum *resources* are not considered to include all this aluminum, but only that aluminum in specific, restricted rock types where the aluminum occurs in such a concentration and such a form that it can be extracted under reasonable economics and technology (i.e., bauxite or laterite). Numbers such as the aluminum in the crust '... can be found intermingling indiscriminately among our statistics of reserves and resources. Since we cannot say categorically that they will never have any value at some point in the future, there must be a place for them in our terminology; and we must make every effort to relegate them to where they properly belong.' (Schanz, 1975b, p. 11.)

6.3. Geothermal Resource Terminology

Building on this general background, Muffler and Cataldi (1977; 1978) attempted to develop a logical resource terminology specifically for geothermal energy. Adapting the general definition for resource base given by Netschert (1958) and by Schurr and Netschert (1960), *geothermal resource base* is defined as 'all the heat in the Earth's crust beneath a specific area, measured from local mean annual temperature.' The *accessible resource base* is the thermal energy at depths shallow enough to be tapped by drilling in the forseeable future, whereas that fraction of the accessible resource base that might be extracted economically and legally at some reasonable future time is the *geothermal resource*. Both the accessible resource base and resource include *identified* and *undiscovered* components. Finally, the *geothermal reserve* is identified geothermal energy that can be extracted legally today at a cost competitive with other energy sources.

These terms can be depicted simply on a McKelvey diagram whose vertical axis is extended to include the residual accessible resource base and the inaccessible resource base (Figure 6.3). This diagram depicts all the important categories of geothermal energy, illustrates the cumulative nature of reserves, resources, and resource base (Schanz, 1975b, p. 19), emphasizes that resources are only a part of the accessible resource base, and restricts reserves to only those resources that are identified and economic today. It should be emphasized that all categories are measured in units of energy and referenced to mean annual temperature.

6.4. Methodology for Geothermal Resource Assessment

Methodologies used for geothermal resource assessment prior to 1977 were reviewed by Muffler and Cataldi (1977; 1978) and divided into four main categories: (1) surface thermal flux, (2) volume, (3) planar fracture, and (4) magmatic heat budget.

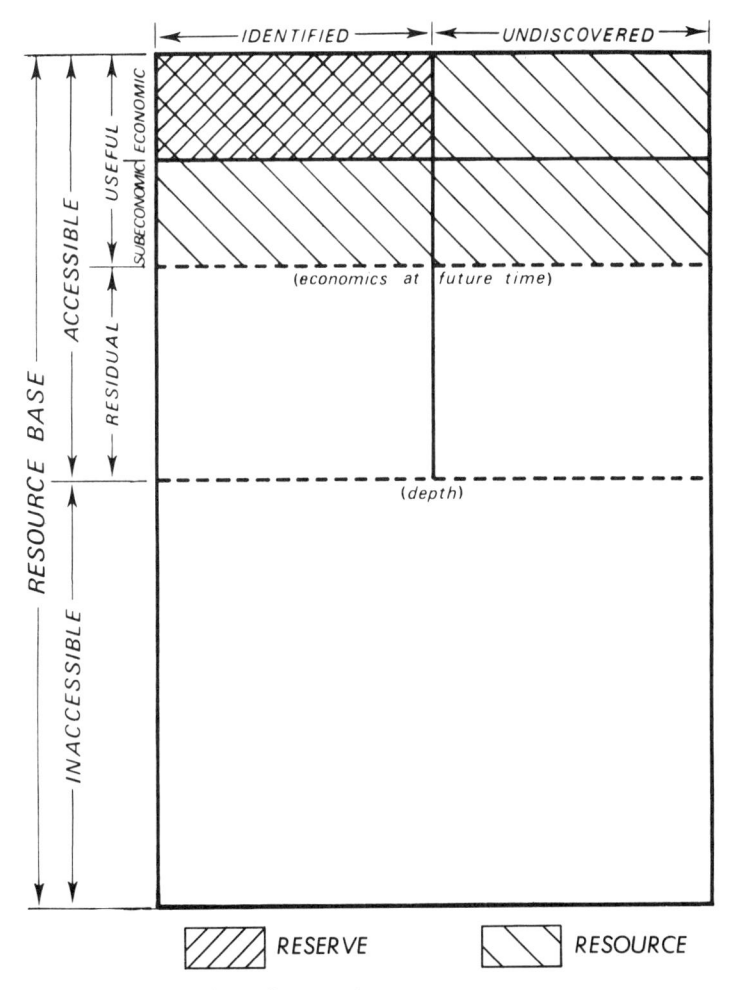

Figure 6.3. McKelvey diagram for geothermal energy, showing derivation of the terms resource and reserve (from Muffler and Cataldi, 1977; 1978). Vertical axis is degree of economic feasibility; horizontal axis is degree of geologic assurance

The method of surface thermal flux consists of measuring the rate of thermal energy loss at the ground surface by conduction, steaming ground, hot springs, fumaroles, and discharge of thermal fluids directly into streams. Experience from already developed geothermal fields is then used to relate this rate of energy loss to the rate at which thermal energy might be produced through drillholes (e.g., White, 1965; Kenzo Baba in Suyama et al., 1975).

The volume method involves the calculation of the thermal energy contained in a given volume of rock and water and then the estimation of how much of this energy might be recoverable. The thermal energy in the ground can readily be calculated as the product of the volume of a geothermal reservoir, the mean temperature, the porosity, and the specific heats of rock and water. Alternatively, one can calculate the thermal energy approximately as the product of just volume, temperature, and an assumed volumetric specific heat (White, 1965; Renner et al., 1975; Brook et al., 1979). Calculation of the amount of recoverable thermal energy is more complex, however, and requires a knowledge of reservoir properties such as permeability. In most cases, the recovery factor can be specified only approximately (Nathenson, 1975).

The planar fracture method (Bodvarsson, 1970; 1972; 1974) involves a model wherein thermal energy is extracted from impermeable rock by flow of water along a planar fracture. The calculations are based on conductive transfer of heat to the fracture and require estimation of fracture area, fracture spacing, initial rock temperature, minimum acceptable outflow temperature, and the thermal conductivity of the rock. The method does allow the direct calculation of recoverable thermal energy without going through the intermediate step of calculating thermal energy in place. However, the method is strictly applicable only to terrains such as the unfolded flood basalts of Iceland and is of questionable use in the complex, three-dimensional fracture systems that characterize most hydrothermal convection systems.

The method of magmatic heat budget involves calculating the thermal energy still remaining in young igneous intrusions and adjacent country rock, as a function of emplacement temperature, size, age, and cooling mechanism (e.g., Smith and Shaw, 1975). Although the method is very useful for giving an indication of the order of magnitude of geothermal energy to be expected in young volcanic terrains, it is not an inventory of geothermal energy, and the estimates can not be translated directly into geothermal resources.

Muffler and Cataldi (1977; 1978) concluded that the volume method appeared to be the most useful because (1) it was applicable to virtually any geologic environment, (2) the required parameters could in principle be measured or estimated, (3) the inevitable errors were in part compensating, and (4) the major uncertainties (the recovery factor and the resupply of heat) were amenable to resolution in the foreseeable future. The recovery factor was deemed to be a major weakness that required substantial investigation, both in terms of modeling and in terms of field reservoir engineering. Several simple models suggested that resupply of heat is significant only for hot-water systems of high natural discharge, as indicated previously by Nathenson (1975).

6.5. Selected Recent Geothermal Resource Assessments

The last five years have seen a great increase in attention paid to systematic geothermal resource assessment. This effort began in 1975 with an assessment of the Preapennine belt of central Italy (Barelli et al., 1975a; 1975b) and with the first comprehensive geothermal resource assessment of the United States (White and Williams, 1975). Various aspects of geothermal resource assessment were also treated in a joint workshop of the Japan and United States Geological Surveys (see Suyama et al., 1975). Subsequently, the National Electrical Energy Agency of Italy (ENEL) and the U.S. Department of Energy implemented a joint research project on geothermal resource assessment, in cooperation with the U.S. Geological Survey and the International Institute for Geothermal Research (Pisa, Italy). The results of these investigations were presented in September 1977 at the Larderello Workshop on Geothermal Resource Assessment and Reservoir Engineering (Muffler and Cataldi, 1977; Cataldi et al., 1977) and are published in Geothermics (Muffler and Cataldi, 1978; Cataldi et al., 1978). In additon, a geothermal resource assessment of New Zealand was prepared for the Larderello Workshop (Donaldson and Grant, 1977; 1978). In 1978 the assessment of Cataldi et al. (1977) for central and southern Tuscany was extrapolated to the entire Preapennine belt (Cataldi and Squarci, 1978), and the U.S. Geological Survey prepared a revised geothermal resource assessment of the United States (Muffler, 1979a).

6.5.1. Central and Southern Tuscany Italy

The volume method was used by Cataldi et al. (1977; 1978) in the geothermal resource assessment of central and southern Tuscany. An area of 8661 km² in the Preapennine belt of central Italy was divided into 31 zones of reasonably homogeneous geology and thermal regime. The upper 3 km of each zone was then divided by horizontal surfaces into three volumes: an upper impermeable complex, a reservoir complex, and a lower basement complex. Each volume was then assigned a temperature and a total porosity, allowing calculation of the thermal energy in each of the 93 volumes. For volumes likely to be produced by intergranular vaporization, the recovery factor was determined from Figure 4 of Nathenson (1975) using effective porosity rather than total porosity. For volumes likely to be produced by intergranular flow and having mean temperature greater than 60°C, the resource was calculated by multiplying the thermal energy in the ground by a resource recovery factor, scaled from 0·5 at an effective porosity of 20 per cent to zero at an effective porosity of 0 per cent. Reserves were calculated similarly, with the resource recovery factor being reduced by a depth factor that ranged from 1·0 at the land surface to zero at 3 km. Geothermal resources for electrical production (from volumes of mean temperature greater than 130°C) were estimated to be 53×10^{18} J, and the electricity producible from these resources was estimated at 134×10^9 watt-years (electrical) or 4500 MWe for 30 years. Geothermal resources for other uses (from volumes of mean temperature less than 130°C) were estimated to be 93×10^{18} J.

6.5.2. New Zealand

An entirely different method (the method of analogy) was used by Donaldson and Grant (1977; 1978) to estimate the electrical capacity that might be supported by the geothermal resources of New Zealand. Observing that the temperatures of most identified fields in New Zealand are bracketed by Wairakei and Broadlands (260°C and 300°C maximum temperatures, respectively), Donaldson and Grant (1977; 1978) conclude that the electrical capacity of these fields per unit area should also be bracketed by the Broadlands and Wairakei values (10–11 MWe/km² and 13–14 MWe/km²). Assuming adequate permeability, they then estimate the power potential of each identified undeveloped field from an estimate of the area (primarily from resistivity surveys) and the subsurface temperature (from direct measurements or geothermometry). Donaldson and

188   *Geothermal Systems: Principles and Case Histories*

Grant estimate proven, inferred, and speculative electrical capacities for 13 geothermal fields of The North Island, concluding that New Zealand has some 1100–2500 MWe available from its geothermal resources, presumably for a long if not infinite time.

6.5.3. *United States—1975*

The first geothermal resource assessment of the entire United States (White and Williams, 1975) evaluated geothermal energy in four major categories:

(a) Regional conduction-dominated environments;
(b) Geopressured-geothermal systems;
(c) Igneous-related systems;
(d) Hydrothermal convection systems with temperatures $\geq 90°C$.

For each category, the assessment used a two-step process: (1) calculation of thermal energy in the ground (termed the 'heat content'), and (2) calculation of the recoverable thermal energy (termed the 'resource').

For regional conduction-dominated environments, the thermal energy in the ground was calculated to a depth of 10 km using 'best estimate' thermal gradients derived from regional heat-flow data (Diment et al., 1975). Each heat-flow province of the United States was assigned to one of three types (Sierra Nevada type of low heat flow, Eastern type of intermediate heat flow, and Basin and Range type of high heat flow); calculations assumed uniform thermal conductivity and exponential decrease of radioactive heat generation with depth. Because of the immense volumes of rock involved, the resultant value $(33,000,000 \times 10^{18}$ J) is huge and serves primarily as a background value or upper limit for any discussion of geothermal energy of the United States. The geopressured-geothermal 'fluid resource base' of the northern Gulf of Mexico Basin is included in the regional conduction-dominated calculations. On the other hand, the anomalous heat of the hydrothermal convection and the igneous-related systems are superimposed on the regional conduction-dominated environments (White and Williams, 1975), with the total thus representing the 'resource base' (to 10 km).

For igneous-related systems, the thermal energy still remaining in silicic intrusions and adjacent country rock was calculated by conductive cooling models using estimates of the size and age of the intrusions (Smith and Shaw, 1975). Calculations assumed an initial temperature of 850°C, a final temperature of 300°C, and a magma chamber ranging in thickness from 2·5 to 10 km with its top at 4 km. Cooling of the igneous body by hydrothermal convection was assumed to be offset by the effects of magmatic preheating and additions of magma after the assumed time of emplacement. The resulting figures do not represent an inventory of measured thermal energy, but instead are estimates based on a model. The sum of the thermal energy in evaluated igneous-related systems and the thermal energy estimated for unevaluated igneous-related systems is $420,000 \times 10^{18}$ J, approximately 1·2 per cent of the thermal energy in regional conduction-dominated environments (White and Williams, 1975, p. 149).

Thermal energy in reservoirs of identified hydrothermal convection systems was calculated by Renner et al. (1975) using the volume method. Reservoir tops were either estimated from drillhole data or assumed (at 1·5 km), and reservoir bottoms were assumed to be at a depth of 3 km. Areas were estimated from geological, geophysical, and available drillhole data, and reservoir temperatures were estimated from drillhole measurements and chemical geothermometers (quartz and Na–K–Ca). Thermal energies were tabulated separately for high-temperature systems (>150°C) and intermediate-temperature systems (90–150°C).

Renner et al. (1975) also estimated the thermal energy yet to be discovered in hydrothermal convection systems. This undiscovered thermal energy consists of (1) additional thermal energy due to upward revisions of the volumes of identified systems, (2) additional thermal energy due to upward revisions of temperature estimates, and (3) thermal energy in systems that have not yet been identified (Renner et al., 1975, p. 54). For high-temperature hydrothermal convection systems, the undiscovered thermal energy was estimated for the United States as a whole to be five times the identified (excluding National Parks). Similarly, the undiscovered thermal energy in intermediate-temperature hydrothermal convection systems was estimated for the United States as a whole to be three times the identified. The total identified and undiscovered thermal energy in hydrothermal convection systems (to 3 km) was calculated to be $12,800 \times 10^{18}$ J, only 0·04 per cent of the thermal energy in regional conduction-dominated regimes to 10 km (White and Williams, 1975, p. 149).

The average recovery of thermal energy from hydrothermal convection systems was estimated by Nathenson and Muffler (1975) to be 25 per cent. Using this factor, the total identified and undiscovered geothermal resource of hydrothermal convection systems >150°C in the United States was estimated to be 1520×10^{18} J, and the electricity producible from this resource to be 153,000 MWe for 30 years. The total geothermal resource in hydrothermal convection systems at 90–150°C was estimated to be 1450×10^{18} J.

For geopressured-geothermal systems, only thermal energy in the pore water of onshore Tertiary rocks of the northern Gulf of Mexico Basin was calculated (Papadopulos et al., 1975). Approximately 250 wells were used to provide temperature data and estimates to the top of the geopressured zone for 21 subareas. These data then allowed calculation of the volume of water in storage (in both shales and sandstones), the thermal energy in the water, the energy of methane dissolved in the water (assuming saturation at the mean temperatures and pressures of each subarea), and the mechanical energy due to the high pressures. Calculations were made to a depth of 6 km in Texas and 7 km in Louisiana. The identified geopressured-geothermal energy consisted of $46,000 \times 10^{18}$ J of thermal energy, $25,000 \times 10^{18}$ J of energy in dissolved methane, and 230×10^{18} J of mechanical energy.

The thermal energy in geopressured waters in unassessed parts of the northern Gulf of Mexico Basin (offshore, deeper Tertiary, and Cretaceous rocks) to a depth of 10 km was estimated to be $1\frac{1}{2}$ to twice that in the assessed parts (Papadopulos et al., 1979, p. 125), and the thermal energy in other geopressured basins of the United States was estimated to be approximately the same as that in the assessed parts of the northern Gulf of Mexico Basin (White and Williams, 1979, p. 151–152). The total identified and undiscovered thermal energy of geopressured-geothermal fluids of the northern Gulf of Mexico Basin thus totalled $115,000–161,000 \times 10^{18}$ J, with perhaps another $46,000 \times 10^{18}$ J in other unevaluated geopressured basins of the United States (White and Williams, 1979, p. 152).

Recovery of geopressured-geothermal energy in the northern Gulf of Mexico Basin was estimated by Papadopulos et al. (1975) according to three development plans: (1) restriction of wellhead pressure to 14 MPa, (2) unrestricted wellhead pressure, and (3) wellhead pressure maintained sufficiently high to restrict ground subsidence to 1 meter. Assuming that the fluids were saturated with methane at the reservoir conditions, the recoverable geopressed-geothermal energy (thermal, dissolved methane, and mechanical) was 1520×10^{18} J for plan 1, 2300×10^{18} J for plan 2, and 360×10^{18} J for plan 3. In terms of percentage of the accessible fluid resource base, the recoverable energy was 2·1 per cent for plan 1, 3·3 per cent for plan 2, and 0·5 per cent for plan 3.

6.5.4. United States—1978

A refined and updated geothermal resource assessment of the United States was prepared by the U.S. Geological Survey based on data available on July 1, 1978. Once again, the assessment consists of a USGS Circular (Muffler, 1979a) with component chapters (Muffler and Guffanti, 1979; Sass and Lachenbruch. 1979; Smith and Shaw, 1979; Brook et al., 1979; Sammel, 1979; Wallace et al., 1979; Muffler, 1979b). The Circular is supported by four USGS Open-File Reports (Mariner et al., 1978; Nathenson, 1978; Smith et al., 1978; Haas, 1978) giving data and methodologies too voluminous to be included in Circular 790 itself.

Parts of Circular 726 that are still valid are not repeated in the 1978 assessment. Among these parts are:

1. The calculations of thermal energy in regional conductive environments (Diment et al., 1975).
2. The methodology by which thermal energies are calculated for young silicic igneous systems (Smith and Shaw, 1975).
3. The analysis of the recovery of energy from magma (Peck, 1975), which concludes that the large quantity of thermal energy in magmas at depth is not presently recoverable and may never be so.
4. The analysis of recovery of thermal energy from hydrothermal convection systems given by Nathenson (1975).
5. The three recovery plans presented by Papadopulos et al. (1975) for geopressured-geothermal energy of the northern Gulf of Mexico Basin.

On the other hand, two major items were added to the 1978 assessment:

1. A report describing and depicting areas favorable for the discovery and development of low-temperature geothermal waters from depths less than 1 km (Sammel, 1979).
2. Three colored maps depicting a variety of geothermal data: (a) the conterminous western United States (at 1:2,500,000), (b) Alaska (at 1:5,000,000) and Hawaii (at 1:2,500,000), and (c) the northern Gulf of Mexico basin (at 1:1,000,000).

The thermal energy still remaining in silicic intrusions and adjacent hot country rock is estimated by Smith and Shaw (1979) using the refined data base of Smith et al. (1978). As in 1975, this estimate assumes that cooling of the igneous body by hydrothermal convection was offset by the effects of magmatic preheating and additions of magma after the assumed time of emplacement. The total energy in evaluated systems ($101{,}000 \times 10^{18}$ J) is not significantly changed from 1975, but the estimate of Smith and Shaw (1979) for unevaluated systems (i.e., systems for which adequate age and volume data are not available) is significantly greater ($>900{,}000 \times 10^{18}$ J vs. $310{,}000 \times 10^{18}$ J indicated in table 36 of White and Williams, 1975).

The igneous-related thermal energy estimated by Smith and Shaw (1975; 1979) occurs as magma, as low-permeability rock (both solidified igneous rock and hot country rock), and as associated hydrothermal systems. Smith and Shaw (1975, p. 76) and White and Williams (1975, table 26) estimate that about 50 per cent of the igneous-related geothermal energy exists as molten or partly molten magma. Muffler (1979, p. 162) concludes that the hydrothermal energy at depths greater than 3 km is unlikely to be more than 5 times greater than the hydrothermal energy at depths less than 3 km (i.e., a total of $65,000 \times 10^{18}$ J or 7 per cent of the total igneous-related geothermal energy). By differences, the low-permeability rock would thus make up approximately 43 per cent of the total igneous-related thermal energy. Even considering that these estimates are little more than guesses, it is clear that approximately equal amounts of the igneous-related geothermal energy exist as magma and low-permeability rock, with only a few percent represented by hydrothermal convection systems.

The methodology used to estimate the thermal energy of hydrothermal convection systems in 1978 is essentially similar to that used in 1975, with the addition of statistical methodology that allows expression of the temperatures, volumes, and thermal energies as mean values with standard deviations. This methodology, treated in detail by Nathenson (1978), requires that three values be estimated for each variable (temperature, area, and thickness) of each system; these values are the minimum, most likely, and maximum estimates of a triangular probability density (Figure 6.4) that most nearly approximates the considered estimate of the real probability density of that variable. The means and standard deviations of these triangular probability densities can then be summed analytically to give the means and standard deviations of temperature, volume, and energy for all identified hydrothermal convection systems (Brook et al., Tables 4–6). Monte Carlo methods can be used to give the resultant probability distribution for the total energy (Brook et al., 1979, Figure 7).

The calculation of recoverable energy from hydrothermal convection systems is essentially the same as Circular 726 (Nathenson and Muffler, 1975), with both analyses based on Nathenson (1975). A recovery factor of 25 per cent is used for all identified hot-water geothermal systems, and a recovery factor of 9·3 per cent for vapor-dominated geothermal systems.

A quantity used frequently in thermodynamic calculations is *available work*, the maximum amount of work that can ideally be obtained from a given amount of thermal energy. Although available work was not tabulated in Circular 726, the concept was involved in the calculations of electrical energy and is explicitly addressed by Nathenson (1975). Tables 4 and 5 of Brook et al. (1979) do contain an explicit tabulation of available work for each identified high-temperature hydrothermal convection system.

In a real conversion cycle, however, the amount of electricity produced is always less than the available work and is related to available work by a *utilization factor*, which is a function of the wellhead temperature and the particular conversion cycle chosen (Brook et al., 1979, Figure 6). In Circular 790, a representative utilization factor of 0·4 was used for all hot-water systems, and a factor of 0·5 for The Geysers.

The undiscovered thermal energy in hydrothermal convection systems was estimated by

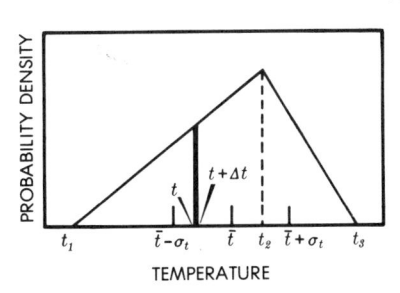

Figure 6.4. Example of a triangular probability density. The parameters t_1, t_2 and t_3 are the minimum, most likely, and maximum characteristic reservoir temperatures, respectively. The mean, $\bar{t}$, and the mean plus or minus one standard deviation, $\bar{t} \pm \sigma_t$, are also shown. The area of the solid vertical band gives the probability that the characteristic reservoir temperature is between the values t and $t + \Delta t$, where Δt is a small number. The total area of the triangle is the probability of all events and equals one. From Brook et al. (1979), Figure 4

Brook et al (1979) for each major geologic province of the United States. In most cases, the undiscovered component was estimated as a multiple of the identified component, but in two cases (the Island Park area of Idaho and the Aleutian Volcanic Chain of Alaska) the undiscovered component was estimated as 1 per cent of the thermal energy contained in the corresponding igneous-related system of Smith and Shaw (1979). The allocation of the undiscovered thermal energy into intermediate-temperature and high-temperature categories was based on the frequency of identified hydrothermal convection systems in 20°C temperature intervals and on two limiting cases for the relation between reservoir volume and temperature (Brook et al., 1979, pp. 37–39). The ranges given in Table 11 of Brook et al. (1979) for the undiscovered accessible resource base and resource in hydrothermal convection systems reflect these two limiting assumptions. It should be emphasized that these ranges and the ranges for undiscovered potential for electricity and beneficial heat are complementary; the higher value for electricity cannot occur simultaneously with the higher value for beneficial heat.

The total identified and undiscovered geothermal resource of hydrothermal convection systems estimated in 1978 (2400×10^{18} J) is 20% less than the figure estimated in 1975 (3000×10^{18} J). This decrease is due primarily to the decrease in estimated area of some of the biggest identified systems, particularly Bruneau–Grand View (2250 km^2 to 1480 km^2), Klamath Falls (240 km^2 to 69 km^2), Long Valley (225 km^2 to 82 km^2), and Coso (168 km^2 to 27 km^2). In 1975, these four giant systems alone represented 65 per cent of the thermal energy estimated for identified hydrothermal convection systems excluding National Parks; in 1978, the corresponding figure is 35 per cent. The large decreases in estimated energy for these systems thus have a strong effect on the totals, not only of the identified hydrothermal convection systems, but also the undiscovered component, since in most cases it is calculated as a multiple of the identified component. A secondary reason for the lower 1978 estimates is the use of refined geothermometers. This use resulted in lower estimated temperatures for many systems in the 110–150°C range (compare Figures 10a and 10b of Brook et al., 1979) and a decrease in the total number of systems of temperature ≥ 90°C from 283 to 215.

The 1978 assessment of geopressured-geothermal energy of the northern Gulf of Mexico (Wallace et al., 1979) is based on data from over 3500 wells, located both onshore and offshore, and essentially completes the preliminary inventory of approximately 250 onshore wells presented in the 1975 assessment (Papadopulos et al., 1975). The total thermal energy and dissolved methane energy in the accessible fluid resource base to 6·86 km is $170{,}000 \times 10^{18}$ J, approximately equal to the upper bound of the estimate of $115{,}000$–$161{,}000 \times 10^{18}$ J given by Papadopulos et al. (1975) for both identified and undiscovered thermal energy alone in the northern Gulf of Mexico Basin. White and Williams (1975, Table 28) estimate that an additional $46{,}000 \times 10^{18}$ J of thermal energy are likely to be in other, unevaluated geopressued basins of the United States (Wallace et al., 1979, Figure 26), along with an uncertain but appreciable amount of dissolved methane.

The few production tests carried out to date in the northern Gulf of Mexico basin have not significantly modified the recovery analysis presented by Papadopulos et al. (1975). However, Wallace et al. (1979) note that only the sandstones of the northern Gulf of Mexico Basin have aquifer properties favorable for production and conclude that only the $17{,}000 \times 10^{18}$ J in fluids of these sandstones represents the accessible fluid resource base from which initial production will be drawn. Applying plan 3 (controlled development with limited pressure reduction and subsidence) and plan 2 (depletion of reservoir pressure) of Papadopulos et al. (1975) to the energy in fluids of only the sandstones,

Wallace et al. (1979) calculate that the recoverable thermal and dissolved methane energy in the northern Gulf of Mexico Basin (i.e., the geopressured-geothermal resource) is between 430 and 4400×10^{18} J.

6.6. Uncertainties and Topics that Need Investigation

6.6.1. *Undiscovered Hydrothermal Convection Systems*

A major uncertainty in the estimates of thermal energy in hydrothermal convection systems $\geq 90°C$ is the thermal energy in the undiscovered component, which represents 83 per cent of the total of the 1978 assessment. The philosophy behind the 1978 assessment of the undiscovered component of hydrothermal convection systems in the United States is the same as that expressed in 1975 and clearly is one of optimism that the many unevaluated tracts of land in the western United States do contain undiscovered hydrothermal convection systems and that significant numbers of the identified systems will turn out to have areas significantly greater than the nominal area of 2 km² assumed for 63 per cent of the identified systems. In particular, the Aleutian Volcanic Chain, the Cascade Range, the eastern Snake River Plain, and the Basin and Range province are particularly promising for exploration and discovery of new hydrothermal convection systems. In the Cascades and the eastern Snake River Plain, extensive near-surface cold aquifers may well mask higher-temperature reservoirs associated with these major crustal volcanic belts.

Evaluation of the undiscovered component is clearly dependent on continued exploration by private industry, such as that responsible for the recent discovery in Dixie Valley, Nevada (not known at the time of preparation of Circular 790). In succeeding assessments, one would expect to see the balance between the identified and undiscovered components shift from the present dominance of the undiscovered component to a future dominance of the identified component.

6.6.2. *Recovery of Thermal Energy from Hydrothermal Convection Systems*

One of the major uncertainties in the assessment of the geothermal resources of hydrothermal convection systems is the estimation of the recovery factor for hot-water hydrothermal convection systems. There seems to be some agreement that the recovery factor for a reservoir of ideal permeability is approximately 50 per cent (Nathenson, 1975) or even more (Donaldson and Grant, this volume). But an ideally permeable reservoir occurs rarely if at all in nature, and one is left with the question of how much the ideal recovery factor should be reduced to account for permeability variations, blocks of the reservoir that can not be tapped, etc. In Brook et al. (1979, pp. 27–28), the recovery factor (Rg) was separated into a recovery factor for an ideal reservoir (Rg_i) and a correction factor (k). The uncertainty in Rg was then expressed by assigning k a triangular probability density with a minimum of 0, a most likely value of 0·5, and a maximum of 1. This procedure is admittedly unsatisfactory, but is the best available at the present for estimating the mean recovery of a number of systems of varying character, particularly when reservoir properties are unknown for most of the systems. As recommended by Muffler and Cataldi (1978), intensive research should be directed toward the question of recoverability of hydrothermal reservoirs, under conditions of both natural and induced permeability. This research should not be limited to mathematical modeling, but must include appropriate reservoir engineering studies and case histories of producing fields.

6.6.3. Evaluation of Igneous-related Geothermal Energy

Smith and Shaw (1979) estimate that approximately $101,000 \times 10^{18}$ J of thermal energy still remain in young silicic systems for which data are adequate to make thermal estimates. This estimate is based on a model of conductive cooling (Smith and Shaw, 1975) and is critically dependent on the assumption that cooling of an igneous body by hydrothermal convection is offset by the effects of magmatic preheating and additions of magma after the assumed time of emplacement (Smith and Shaw, 1979, p. 17). Smith and Shaw recognize, however, that the nature, effects and importance of hydrothermal cooling of intrusions are the subject of considerable debate and that their assumption needs further evaluation by quantitative modeling of hydrothermal effects and by geologic evaluation of hydrothermal and volcanic longevity of specific igneous systems.

Smith and Shaw (1979, p. 15) estimate that the total igneous-related energy is at least an order of magnitude greater than their calculation of $101,000 \times 10^{18}$ J for evaluated systems. This estimate is recognized to be unavoidably speculative and subjective and clearly needs substantiation through geologic, geophysical, and geochronologic studies of numerous young volcanic systems. Such substantiation is critically important because $1,000,000 \times 10^{18}$ J of igneous-related energy represents a geothermal target approximately 100 times greater than the total energy in hydrothermal convection systems.

In a previous section (p. 190–191), it was speculated that roughly 50 per cent of the total igneous-related energy of Smith and Shaw (1979) is likely to occur as molten or partly molten magmas, 43 per cent as hot rock of low permeability, and perhaps 7 per cent as hydrothermal convection systems. These figures, however, are little more than educated guesses, in the absence of direct information about the geometry and hydrologic characteristics of hydrothermal convection systems at depths greater than 3 km. Direct sampling and measurement in drillholes at depths of 3 to 10 km in igneous-related systems are critical to the evaluation of the thermal budget of igneous-related systems and to the estimation of thermal energies of various types. Such a deep research drilling effort is currently being considered as part of a proposed program of continental drilling for scientific purposes in the United States (U.S. Geodynamics Committee, 1979).

The analysis on p. 190–191 suggests that approximately $430,000 \times 10^{18}$ J of thermal energy occurs in low-permeability rock associated with igneous systems. This huge amount of energy at less than 10 km represents the prime target for extraction of energy by 'hot dry rock' technology (Laughlin, this volume). Clearly, temperatures in igneous-related systems are substantially higher than at equivalent depths in regional conduction-dominated environments. It remains to be demonstrated, however, just how much of this energy can be extracted under reasonable assumptions of technology and economics (Muffler, 1979b, p. 162).

6.6.4. Recovery of Energy from Geopressured-geothermal Fluids

Although the inventory of geopressured-geothermal energy of the northern Gulf of Mexico Basin in essentially complete (to a depth of 6·86 km), there remains great uncertainty as to how much of the energy contained in the accessible fluid resource base can be recovered. The few production tests carried out since 1975 have not significantly modified the recoverability analysis of Papadopulos et al. (1975), and the ranges in energy given by Wallace et al. (1979) for the geopressured-geothermal resource reflect the extremes of the three analyses of Papadopulos et al. (1975). A number of aquifer tests are being carried out or are planned by the Department of Energy (see Wallace et al., 1979, p. 145), and these tests should yield data that will permit recoverable geopressured-geothermal energy to be estimated in the future with a much greater degree of certainty. In addition, these tests should supply field data on the content of dissolved methane in geopressured waters and thus allow evaluation of the assumption of Papadopulos et al. (1975) and Wallace et al. (1979) that the geopressured-geothermal fluids are saturated with methane.

6.6.5. *Low-temperature Inventory*

In the 1978 assessment of geothermal resources of the United States, no quantitative estimate was made of the thermal energy of geothermal waters at temperatures less than 90°C. This deliberate omission reflected the lack of sufficient reliable data upon which an estimate could be based. In particular, the arbitrary assignment of many of the low-temperature boundaries on map 1 of Circular 790 and the limited understanding of the nature of low-temperature systems precluded estimation of reservoir volumes for nearly all areas. Furthermore, estimating reservoir temperatures in these low-temperature mixed-water systems by geochemical methods was deemed to be especially unreliable. The net evaluation of these factors in 1978 was that numerical estimates of thermal energy in geothermal waters at temperatures less than 90°C would have been wholly unreliable and meaningless and would have done more harm than good.

Estimation of the thermal energy in low-temperature thermal waters should be possible in the future, however, and is a major thrust of the Federal geothermal program. It has been demonstrated throughout the world that thermal waters at temperatures of 55°C and even lower can be used for direct applications such as domestic and agricultural heating, for example in the Paris Basin (Varet, 1979), in Hungary (Ottlik et al., this volume), and in Iceland. Consequently, low-temperature geothermal energy provides a potential alternative for many present uses of fossil fuel and electricity, and quantitative estimates should therefore be made of the low-temperature geothermal waters available in the United States. Current programs leading toward this evaluation include (1) the Western States Cooperative Direct Heat Geothermal Program of the Department of Energy (Wright et al., 1978) and (2) a program being carried out on the Atlantic Coastal Plain by Virginia Polytechnic Institute and State University to evaluate possible low-temperature geothermal waters in sedimentary rocks underlain by granitic basement containing anomalously higher quantities of the heat-producing elements uranium and thorium (Costain, 1979).

Acknowledgements

In preparing this paper I have made extensive use of data and ideas developed during my association with three major resource assessments (White and Williams, 1979; Cataldi et al., 1977; 1979; and Muffler, 1979a). These assessments all were cooperative efforts of many persons, and I am deeply indebted to all of them for the many contributions that I have tried to summarize here. In particular, I wish to acknowledge my debt to the ideas and insight provided by my colleagues Raffaele Cataldi, Manuel Nathenson, and Donald E. White. I also thank Nathenson, White, and Marianne Guffanti for their very helpful reviews of this paper.

References

Barelli, A., Calamai, A., and Cataldi, R., 1975a. 'Stima del potenziale geotermico della fascia preappenninica centro-meridionale, *ENEL Centro di Ricerca Geothermica, Relazione T3/149 Z*, 33 pp.

Barelli, A., Calamai, A., and Cataldi, R., 1975b. 'Estimation of the geothermal potential of the pre-Apennine belt of central-southern Italy (abs.)', *2nd United Nations Symposium on the Development and Use of Geothermal Resources*, San Francisco, Abstract I–3.

Bodvarsson, G., 1970. 'An estimate of the natural heat resources in a thermal area in Iceland,' *Geothermics*, Special Issue 2, **2** (2) 1289–1293.

Bodvarsson, G., 1972. 'Thermal problems in the siting of reinjection wells,' *Geothermics*, **1**, 63–66.

Bodvarsson, G., 1974. 'Geothermal resource energetics', *Geothermics*, **3**, 83–92.

Brook, C. A., Mariner, R. H., Mabey, D. R., Swanson, J. R., Guffanti, Marianne, and Muffler, L. J. P., 1979. 'Hydrothermal convection systems with reservoir temperatures $\geq 90°C$.' In *Assessment of Geothermal Resources of the United States—1978*, U.S. Geol. Survey Circ. 790, Muffler, L. J. P. (Ed.), 18–85.

Cataldi, R., and Squarci, P., 1978. 'Valutazione del potenziale geotermico in Italia con particulare riguardo alla Toscana centrale e meridionale', *Rendiconti della LXXIX Riunione Annuale della Associazione Elettrotecnica Italiana*, Fasc. I, Nota I. 32, 8 pp.

Cataldi, R., Lazzarotto, A., Muffler, P., Squarci, P., and Stefani, G., 1978. 'Assessment of geothermal potential of central and southern Tuscany', *Proceedings Larderello Workshop on Geothermal Resource Assessment and Reservoir Engineering (Sept. 12–16, 1977, Larderello, Italy), ENEL Studi e Ricerche*, 351–412.

Cataldi, R., Lazzarotto, A., Muffler, P., Suarci, P., and Stefani, G., 1978. 'Assessment of geothermal potential of central and southern Tuscany', *Geothermics*, 7, 2–4, 91–131.

Costain, John, 1979. 'Geothermal exploration methods and results—Atlantic Coastal Plan', *Geothermal Resources Council Special Rept. No. 5*, 13–22.

Diment, W. H., Urban, T. C., Sass, H. H., Marshall, B. V., Munroe, R. J., and Lachenbruch, A. H., 1975. 'Temperatures and heat contents based on conductive transport of heat.' In *Assessment of Geothermal Resources of the United States—1975*, U.S. Geol. Survey Circular 726, White, D. E., and Williams, D. L. (Eds.). 1975, p. 84–103.

Donaldson, I. G., and Grant, M. A., 1977. 'An estimate of the resource potential of New Zealand geothermal fields for power generation', *Proceedings Larderello Workshop on Geothermal Resource Assessment and Reservoir Engineering (Sept. 12–16, 1977, Larderello, Italy): ENEL Studi e Ricerche*, 413–428.

Donaldson, I. G., and Grant, M. A., 1978. 'An estimate of the resource potential of New Zealand geothermal fields for power generation', *Geothermics*, 7, no. 2–4, 243–252.

Flawn, P. T., 1966. *Mineral Resources*, Chicago, Rand McNally & Co., 406 pp.

Haas, J. L., Jr., 1978. 'An empirical equation with tables of smoothed solubilities of methane in water and aqueous sodium-chloride solutions up to 25 weight per cent. 360°C, and 183 MPa', *U.S. Geol. Survey Open-File Rept. 78–1004*, 41 pp.

Helgeson, H. C., 1968. 'Thermodynamic and geochemical characteristics of the Salton Sea geothermal system', *Am. Jour. Sci*, **266**, 129–166.

Mariner, R. H., Brook, C. A., Swanson, J. R., and Mabey, D. R., 1978. Selected data for hydrothermal convection systems in the United States with estimated temperatures $\geq 90°C$', *U.S. Geol. Survey Open-File Rept. 78–858*, 493 pp.

McKelvey, V. E., 1968. 'Mineral potential of the submerged parts of the continents', *Proceedings of a symposium on mineral resources of the world ocean, University of Rhode Island, Naragansett Marine Laboratory, Occasional Publication No. 4*, 31–38.

Miller, B. M., Thomsen, H. L., Dolton, G. L., Coury, A. B., Hendricks, T. A., Lennartz, F. E., Powers, R. B., Sable, E. G., and Varnes, K. L., 1975. 'Geological estimates of undiscovered oil and gas resources in the United States', *U.S. Geol. Survey Circ. 725*, 78 pp.

Muffler, L. J. P. (Ed.), 1979a. 'Assessment of geothermal resources of the United States—1978', *U.S. Geol. Survey Circ. 790*, 163 pp.

Muffler, L. J. P., 1979. 'Summary'. In *Assessment of Geothermal Resources of the United States—1978, U.S. Geol. Survey Circ. 790*, Muffler, L. J. P. (Ed.), 156–163.

Muffler, L. J. P., and Cataldi, Raffaele, 1977. 'Methods for regional assessment of geothermal resources', *Proceedings Larderello Workshop on Geothermal Resource Assessment and Reservoir Engineering (Sept. 12–16, 1977, Larderello, Italy), ENEL Studi e Ricerche,* 131–207.

Muffler, L. J. P., and Cataldi, R., 1978. 'Methods for regional assessment of geothermal resources', *Geothermics,* 7, 2–4, 53–89.

Muffler, L. J. P., and Guffanti, Marianne, 1979. 'Introduction.' In *Assessment of Geothermal Resources of the United States—1978, U.S. Geol. Survey Circ. 790,* Muffler, L. J. P. (Ed.), 1–7.

Nathenson, Manuel, 1975. 'Physical factors determining the fraction of stored energy recoverable from hydrothermal convection systems and conduction-dominated areas', *U.S. Geol. Survey Open-File Rept. 75-525,* 35 pp.

Nathenson, Manuel, 1978. 'Methodology of determining the uncertainty in the accessible resource base of identified hydrothermal convection systems', *U.S. Geol. Survey Open-File Rept. 78-1003,* 50 pp.

Nathenson, Manuel, and Muffler, L. J. P., 1975. 'Geothermal resources in hydrothermal convection systems and conduction-dominated areas.' In *Assessment of Geothermal Resources of the United States—1975, U.S. Geol. Survey Circular 726,* White, D. E., and Williams, D. L. (Eds.), 1975, 104–121.

Netschert, B. C., 1958. *The Future Supply of Oil and Gas,* Baltimore, Johns Hopkins University Press, 134 pp.

Papadopulos, S. S., Wallace, R. H., Jr., Wesselman, J. B., and Taylor, R. E., 1975. 'Assessment of geopressured-geothermal resources in the northern Gulf of Mexico basin', In *Assessment of Geothermal Resources of the United States—1975, U.S. Geol. Survey Circular 726,* White, D. E., and Williams, D. L. (Eds.), 1975. 125–146.

Peck, D. L., 1975. 'Recoverability of geothermal energy directly from molten igneous systems.' In *Assessment of Geothermal Resources of the United States—1975, U.S. Geol. Survey Circular 726,* White, D. E., and Williams, D. L. (Eds.), 122–124.

Renner, J. L., White, D. E., and Williams, D. L., 1975. 'Hydrothermal convection systems.' In *Assessment of Geothermal Resources of the United States—1975, U.S. Geol. Survey Circular 726,* White, D. E., and Williams, D. L. (Eds.), 5–57.

Sammel, E. A., 1979. 'Occurrence of low-temperature geothermal waters in the United States.' In *Assessment of Geothermal Resources of the United States—1978, U.S. Geol. Survey Circ. 790,* Muffler, L. J. P. (Ed.), 86–131.

Sass, J. H., and Lachenbruch, A. H., 1979. 'Heat flow and conduction-dominated thermal regimes.' In *Assessment of Geothermal Resources of the United States—1978, U.S. Geol. Survey Circ. 790,* Muffler, L. J. P. (Ed.), 8–11.

Schanz, J. J., Jr., 1975a. 'Problems and opportunities in adapting U.S. Geological Survey terminology to energy resources.' In *First IIASA Conference on Energy Resources (May, 1975),* Grenon, M. (Ed.), International Institute for Applied Systems Analysis, 2361 Laxenburg, Austria, 85–101.

Schanz, J. J., Jr., 1975b. 'Resource terminology: an examination of concepts and terms and recommendations for improvement', *Palo Alto, Calif., Electric Power Research Institute, Research Project 336,* August 1975, 116 pp.

Schurr, S. H., and Netschert, B. C., 1960. *Energy in the American Economy, 1850–1975,* Baltimore, Johns Hopkins Press, 774 pp.

Smith, R. L., and Shaw, H. R., 1975. 'Igneous-related geothermal systems.' In *Assessment of Geothermal Resources of the United States—1975, U.S. Geol. Survey Circular 726,* White, D. E., and Williams, D. L. (Eds.), 1975, 58–83.

Smith, R. L., and Shaw, H. R., 1979. 'Igneous-related geothermal systems.' In *Assessment of Geothermal Resources of the United States—1978, U.S. Geol. Survey Circ. 790,* Muffler, L. J. P. (Ed.), 12–17.

Smith, R. L., Shaw, H. R., Luedke, R. G., and Russell, S. L., 1978. 'Comprehensive tables giving physical data and energy estimates for young igneous systems of the United States' *U.S. Geol. Survey Open-File Rept. 78-925,* 28 pp.

Suyama, J., Sumi, K., Baba, K., Takashima, I., and Yuhara, K., 1975. 'Assessment of geothermal resources of Japan', *Proc. United States–Japan Geological Surveys Panel Discussion on the Assessment of Geothermal Resources, Tokyo, Japan, Oct. 17, 1975, Geological Survey of Japan,* 63–119.

U.S. Geodynamics Committee, 1979. 'Continental scientific drilling program', *U.S. National Academy of Sciences,* 192 pp.

U.S. Bureau of Mines and U.S. Geological Survey, 1976. 'Principles of the mineral resource classification system of the U.S. Bureau of Mines and U.S. Geological Survey', *U.S. Geol. Survey Bull.* **1450-A**, 5 pp.

Varet, J., 1979. 'Low enthalpy geothermal fields, with reference to geothermal energy in France', *Jour. Japan Geothermal Energy Assn.*, **15**(4), 3–16.

Wallace, R. H., Jr., Kraemer, T. F., Taylor, R. E., and Wesselman, J. B., 1979. 'Assessment of geopressured-geothermal resources in the northern Gulf of Mexico basin.' In *Assessment of Geothermal Resources of the United States—1978*, Muffler, L. J. P. (Ed.), *U.S. Geol. Survey Circ. 790*, 132–155.

White, D. E., 1965. 'Geothermal energy', *U.S. Geol. Survey Circ. 519*, 17 pp.

White, D. E., 1968. 'Environments of generation of some base-metal ore deposits, *Econ. Geol.*, **63**, 301–335.

White, D. E., and Williams, D. L. (Eds.), 1975. 'Assessment of geothermal resources of the United States—1975, *U.S. Geol. Survey Circular 726*, 155 pp.

Wright, P. M., Foley, Duncan, Nichols, C. R., and Grim, P. J., 1978. 'Western States cooperative direct heat geothermal program of DOE', *Geothermal Resources Council Transactions*, **2**, 739–742.

Geothermal Systems: Principles and Case Histories
Edited by L. Rybach and L. J. P. Muffler
© 1981 John Wiley & Sons Ltd.

7. Environmental Aspects of Geothermal Development

MAX D. CRITTENDEN, JR.

U.S. Geological Survey, Menlo Park, CA 94025 U.S.A.

7.1. Introduction

The development and utilization of geothermal energy result in an array of environmental consequences, both onsite and off, that depend in detail on

- the nature of the resource;
- the way it is used;
- the character of the geologic and ecologic setting.

High-temperature resources are those most likely to be utilized onsite to produce electric power and those most likely to contain significant quantities of undesirable chemical constituents. Low-temperature resources are the most amenable to direct utilization for space heating or process heat but require either that the source be close to some urban or industrial site or that the materials being processed be readily transportable. In general, the severity of environmental problems tends to diminish with decreasing reservoir temperature, whereas the area of surface disturbance tends to increase.

Because the general impact of energy production has been dealt with extensively elsewhere, for example, in Inhaber (1979), the discussion here concerns only those environmental aspects that are unique to geothermal energy and are primarily geologic in nature. Most discussion on environmental impact centers around those effects that are perceived to be deleterious. To achieve a more realistic view of the consequences of undertaking a particular project, negative impacts should be balanced against those that are positive.

7.2. Evolution of Environmental Concern

Early development of many geothermal areas preceded today's awareness of the need to examine the impact of any major new activity on the human and natural environment. As a consequence, some aspects of operations at several of the world's largest geothermal sites would not be acceptable in today's climate of environmental concern. Moreover, many geothermal areas were first developed on an experimental basis and involved some economic risk. As a result, investment in environmental engineering tended to be minimal at early stages.

One such operation is the plant at Wairakei, New Zealand, originally designed in the mid-1950's to produce heavy water by distillation together with a small amount of power

(Haldane and Armstead, 1962). Axtmann (1974), after observing the plant for 4 months during 1974 summarized the principal negative impacts as:

'(1) Concentrations of hydrogen sulphide in the gaseous effluents are far in excess of levels acceptable in a new plant. (2) While producing 143 MWe, the plant discharges approximately 850 MWt to the Waikato River. (3) During periods of drastically reduced river flow, such as during April 1974, arsenic effluents from the bore field produce concentrations considerably in excess of those recommended by the World Health Organization for potable water. (4) The plant may make some contribution to the mercury contamination of the Waikato River.'

Axtmann also reports the following positive results of the plant's operation: (1) It accounts for 7 per cent of the national total of electrical power produced. It has had the highest load factor and nearly the lowest cost. (2) The waste water added to the Waikato River produced an additional 2·4 MWe in the hydroelectric plants downstream. (3) It is a prime tourist attraction in North Island. As Axtmann (1974) points out, some of the negative impacts can be minimized by reinjection of waste water from the wells, by treatment and/or recovery of H_2S from the gas extractors, or by redesign to incorporate cooling towers, rather than the 'run of the river' cooling scheme now in use.

At the Cerro Prieto field in northern Mexico, an area of desert terrain, it has been possible to dispose of both liquid and gaseous effluent from a 75 MWe power plant at the surface, though neither would be acceptable in many other environmental settings. Mercado (1976) reports that waste water from steam separators, together with blow-down from cooling towers, is piped to an 8-km^2 evaporation pond. The waste water contains on the order of 20,000 mg/kg of total dissolved solids (TDS), including a variety of chemical constitutents (Table 7.1). In 1974 the pond contained a primarily chloride brine with a TDS of more than 75,000 mg/kg (Mercado, 1976). Sea water typically contains about 35,000 mg/kg TDS. Original plans considered the construction of a waste canal leading to the Gulf of California, or to a nearby playa, the Laguna Salada. The present system of containment and evaporation is more environmentally acceptable than either of the earlier proposals and has the added advantage of potential for eventual recovery of the potassium and lithium contained in the brines (Mercado, 1977).

Table 7.1. Composition of geothermal fluids in typical hot-water systems (mg/kg)

	Cerro Prieto Mexico[1]	Ahuachapán El Salvador[2]	Wairakei New Zealand[3]	East Mesa, Calif. U.S.A.[4]
Na	6956	4554	1190	548
K	1670	792	185	28
Li	17	—	11	—
Ca	439	326	23	2·1
Cl	13,190	8253	2100	450
B	12	112	28	2·1
As	—	7·8	4·3	0·11
CO_3	10	—	—	0
HCO_3	95	27	14	530
SiO_2	770	483	560	148

[1] Average of 27 wells, fluid composition *prior* to flashing (Mercado, 1976, p. 1386).
[2] A typical well, *prior* to flashing (Einarsson *et al.*, 1976, p. 1350).
[3] Mixed wells, cooled fluid *after* flashing (Rothbaum and Anderton, 1976, p. 1418).
[4] Well RGI 38–30, prior to flashing (Republic Geothermal Inc., unpublished data, 1978).

Cerro Prieto steam contains unusually large amounts of non-condensible gases. Mercado (1977) reports that the steam entering the turbine contains 14,000 mg/kg of CO_2, 1500 mg/kg of H_2S, and 110 mg/kg NH_4. At full capacity (75 MWe), about 1100 kg/hr of H_2S enters the plant; of this, approximately 90 per cent, (990 kg/hr) is emitted to the atmosphere. This amounts to about 8600 metric tons per year. Plans to enlarge the plant call for treatment of all stack gases by the Stretford process; this should make it possible to double the size of the plant without significantly increasing the output of H_2S.

The Geysers area in northern California, currently (May 1979) producing approximately 600 MWe—began operation in 1960 using a turbine discarded from an earlier fossil-fuel plant (Allen and McCluer, 1976). At first, excess condensate was allowed to flow into Big Sulphur Creek. Since 1969, however, all excess condensate has been reinjected (Chasteen, 1975). The primary environmental problem in this area is that of H_2S emission. Because the steam in The Geysers area averages on the order of 225 mg/kg H_2S, the 4,790,000 kg of steam per hour required to produce the 608 MWe currently on-line would result in the production of some 25,000 kg per day of H_2S if abatement procedures were not in effect. Since 1971, Pacific Gas & Electric Company (PG&E) has been conducting extensive tests of several abatement techniques, and new plants will be equipped with surface (heat exchanger type) condensers in order to increase the proportion of H_2S removed by the gas ejectors. That gas will be treated by the Stretford process. But because the amount of H_2S that will dissolve in the condensate is still uncertain, PG & E is investigating various other chemical techniques for treating cooling water and upstream abatement systems that would conrol not only emissions from the power plant but emissions from the wells themselves when the plants are closed down.

It is evident from the history of operations at these fields that the awareness of environmental problems and the engineering sophistication that is being put into their solution have increased enormously during the brief lifetime of many geothermal projects. Some of the more critical problems will be examined as they impact various components of the environment: air quality, water quality, landslides, seismicity, and subsidence.

7.3. Air Quality

Gaseous components of geothermal effluents range widely in both composition and concentration, and their impact on the environment varies accordingly. From typical concentrations of the major noncondensible gases in major geothermal areas shown in Table 7.2, it is evident that H_2S, the gaseous component creating the most serious environmental problems, ranges among these fields by almost four orders of magnitude, and the total emissions (Table 7.3) vary proportionately. Even these figures, however, fail to reflect the actual environmental impact, which depends to a high degree on the factor of sensitivity. The emission of more than 8600 metric tons/yr of H_2S at Cerro Prieto apparently does not result in significant environmental complaint, largely because the field is in an arid setting. But the emission of only 4500 to 6000 metric tons/yr over a much larger area of production at The Geysers has resulted in widespread objections, partly because the surrounding woodland and grasslands are widely used for recreation and partly because the operation is close to valuable agricultural crops such as wine grapes. The current air quality standards in California, U.S.A., require that the maximum levels of H_2S in ambient air not exceed 42 g/m^3 (0·03 mg/kg) averaged over 1 hour. Measurements conducted at The Geysers by SRI International in 1977 (Ruff et al., 1978) at sites 2 to 10 km from the center of development showed that this standard was exceeded in 0·05 to 4 per cent of the total hours sampled. Although the monthly averages of H_2S

Table 7.2. Concentration of noncondensible gases in typical geothermal areas (wt%)

Gas	The Geysers, Calif., U.S.A.[1]	Cerro Prieto, Mexico[2]	Wairakei, New Zealand[3]	East Mesa, Calif., U.S.A.[4]
CO_2	0·326	0·689–0·897	0·1465	0·0893
H_2S	0·0222	0·017–0·035*	0·004	0·000005
CH_4	0·0194	0·015–0·032	0·00044	0·0011
NH_3	0·0052	0·012–0·019	0·00075	—
N_2	0·0056	0·0015–0·0041	0·00046	0·0037
C_2H_6	—	—	0·00018	0·0001
H_2	—	0·016–0·041	0·0001	0·000005
H_3BO_3	—	—	0·00003	—
HF	—	—	0·00178	—
Ar	—	0·00004–0·0001	—	0·0001

[1] Allen and McClure, 1976, p. 1313.
[2] Nehring and Fausto, 1979.
[3] Axtmann, 1974, p. 12, converted to wt.%
[4] Maximum values observed, assuming total noncondensible gas content of 0·094 wt.% (Republic Geothermal Inc., unpublished data, 1978).
* Mercado, 1976, reports that the total steam flow into the plant averages 0·157% H_2S.

Table 7.3. Approximate quantities of chemicals emitted per 100 MWe of power generated by some developed fields (in metric tons per year) (After Ellis, 1978)

Constituent	Cerro Prieto[2]	Wairakei[2]	Broadlands[2]	Larderello[2]	The Geysers[2]
Li	320	350	300	—	—
Na	1.5×10^5	3.5×10^4	3×10^4	—	—
K	4×10^4	5500	5000	—	—
Ca	1×10^4	500	80	—	—
F	40	200	200	—	—
Cl	3×10^5	6×10^4	5×10^4	—	—
Br	500	150	150	—	—
SO_4	170	700	200	—	—
NH_4	800	50	500	1300	1700
B	400	750	1100	200	200
SiO_2	4×10^4	2×10^4	2×10^4	—	—
As	30	100	100	—	—
Hg	—	0.025	0.035	—	0.04
CO_2	1.5×10^5	8000	4×10^5	4×10^5	3×10^4
H_2S	11,500[1]	300	6000	5000	1200(1976)[3]
					900(1977)[3]

[1] Mercado, 1977; [2] Ellis, 1978; [3] Dorighi, 1978, includes effects of abatement.

content are only 0·001 to 0·01 mg/kg, further reduction of H_2S emission will be required. It is expected that the heat-exchanger type condensers being installed by PG & E on its new plants will reduce emissions of H_2S to the atmosphere by about 50 per cent. In addition, PG & E has conducted extensive experiments designed to reduce the H_2S output from existing plants that use direct-contact condensers. Studies begun several years ago showed experimentally that the amount of H_2S emitted from the cooling tower to the atmosphere could be greatly reduced by addition of an iron salt to the circulating water (Allen and McCluer, 1976), thereby causing oxidation of the sulfide to sulfur, which can be removed by filtration. In full-scale operation, PG & E estimates that this process achieves a reduction of approximately 60 per cent in the amount of H_2S emitted (Dorighi, 1978). The resulting sludge, consisting of native sulfur, iron compounds, and fine rock dust, accumulates at the rate of several thousand liters per day, and because of its composition, must be trucked to a distant disposal site. These difficulties, together with less-than-expected reduction in H_2S emissions in full-scale operations, have led PG & E to explore other techniques such as removal of H_2S from the steam before it enters the power plants. Although it has not been determined which techniques will be most efficient and economical, some combination of available methods appears capable of greatly reducing the environmental impact from this source.

Other possibly deleterious elements present in geothermal steam at The Geysers include traces of boron, arsenic, mercury, and radon (Table 7.4). The precise partitioning of these elements between the condensate reinjected, the atmosphere, and the liquid-spray drift from power plant cooling towers has not been determined. Altshuler (1978) notes that the boron content of soils 500 to 800 m downwind from the older plants is 10 to 1000 times that of the surrounding region, and the vegetation is showing significant levels of stress. Mercury content is 5 to 10 times greater than normal. The radon content, though detectable, is reported to be comparable to the normal emission from the soils of the area. Emission by drift of fluid particles from the cooling towers of the newer plants (Table 7.4)

Table 7.4. Minor elements in steam at The Geysers, California, U.S.A., and projected output from new type plants (Unit 17) (from Altshuler, 1978)

	Concentration in steam in mg/kg			Projected cooling-tower emission in g/MWhr	
	Low	High	Average	Gas or vapor	Liquid drift
Arsenic	0·002	0·05	0·019t	0·15	0·001
Boron	2·1	39·0	16·0	—	0·7
Mercury	0·00031	0·0018	0·005	0·039	0·00009
Radon	5·0*	30·5*	16·2*	0·13†	—

* In nanocuries/kg.
† In microcuries/MW hr.

is expected to be one-third to one-tenth that of the earlier plants as a result of the redesign of the cooling towers.

'Radioactive' thermal waters have long been credited with a variety of exotic curative powers (Chiostri, 1976). Measurements of radioactive components in springs (Wollenberg, 1976) have shown that uranium and members of its decay series tend to be highest in low-temperature waters high in $CaCO_3$ and relatively low in high-temperature systems in which SiO_2 predominates. Wollenberg (1976) found radiation levels of 0·25 to 0·5 milliroentgens per hour over enclosed pools at Kyle Hot Springs in northern Nevada, U.S.A. Well-defined radioactive anomalies were detected downwind from the pools, indicating that the source was the gas ^{222}Rn. Most hot-spring areas, even calcareous ones, yielded values only one-tenth as high. Most siliceous springs showed no anomalous radioactivity, or at most, levels two orders of magnitude lower than at Kyle Hot Springs.

7.4. Water Quality

Protecting the quality of surface and ground water is a major objective of recent environmental legislation. The character of geothermal fluids (Table 1) ranges from waters that are actually potable to fluids so concentrated that they are regarded as virtual ore-depositing fluids (White et al., 1963). As a consequence, the problems associated with their utilization vary widely (Ellis, 1978). In general, environmental problems increase with increasing temperature.

For any geothermal area in the United States whose water effluent is lower in quality than existing surface waters, the rule is *no surface discharge*. Although few countries adhere to such rigorous standards, in practice this means that most geothermal effluent will have to be reinjected. Exceptions may exist in areas where thermal springs already flowing at the surface are merely diverted to utilize their heat content, and prior water rights may require that surface flow be maintained. As low-temperature sources are increasingly utilized for direct-heat applications, the use of 'waste' water for agricultural purposes (stock watering, irrigation, mariculture) may be expected to increase.

In addition to the general limitations imposed by total salinity and the typical array of chemical constituents, the presence of certain trace elements may limit the way in which fluids otherwise of high quality may be used. The sensitivity of citrus and other fruits to boron in quantities as small as 0·5 mg/L is well documented (Eaton and Wilcox, 1939).

Table 7.5. Trace-element tolerances for irrigation waters. (From U.S. National Technical Advisory Committee on Water Quality Criteria, 1968)

	For water used continuously on all soils	For short-term use on fine-textured soils only
	mg/L	mg/L
Aluminum	1·0	20·0
Arsenic	1·0	10·0
Beryllium	0·5	1.0
Boron	0·75	2·0
Cadmium	0·005	0·05
Chromium	5·0	20·0
Cobalt	0·2	10·0
Copper	0·2	5·0
Fluorine	1	1
Iron	1	1
Lead	5·0	20·0
Lithium	5·0	5·0
Manganese	2·0	20·0
Molybdenum	0·005	0·05
Nickel	0·5	2·0
Selenium	0·05	0·05
Tin	1	1
Tungsten	1	1
Vanadium	10·0	10·0
Zinc	5·0	10·0

[1] Tolerance not determined.

Similar limitations may exist for arsenic and selenium if water is to be used for agriculture (see Table 7.5).

Although reinjection offers a solution for many of the environmental problems associated with the development of high-temperature fields, careful engineering may be required to prevent plugging of wells or of the receiving formations. In steam fields such as The Geysers, whose reservoir pressure is well below the equivalent hydrostatic head, reinjection takes place by gravity. Moreover, because the only fluids being injected are excess condensate and blowdown from cooling towers, they are relatively clean, and have low salinity. Under these conditions, only settling is required (Chasteen, 1976). In other situations, both chemical pretreatment and filtration may be necessary. Under certain conditions, it appears that even relatively saline waters can be reinjected successfully for short periods at least, as demonstrated at Ahuachapán (Einarsson et al., 1976). Nevertheless, reinjection of hypersaline brines, which produce scale and chemical precipitate with any drop in temperature, may prove to be a major obstacle to extensive utilization despite their high heat content (Muffler, 1979) (White et al., 1963).

In addition to the direct benefits that result from reinjection of fluids that cannot be disposed of on the surface, return of some part of the produced fluids to the reservoir should have indirect benefits. Axtmann (1974) has pointed out that reinjection of waste now flowing to the Waikato River at Wairakei could conserve the equivalent of some 450 MWe of heat now presumably being used to heat meteoric water. If reservoir conditions permit, conservation of these quantities of heat could materially extend the lifetime of a geothermal field. And using reinjection to reduce the total volume of fluid withdrawn should reduce the amount of subsidence, as it does in oil or ground-water reservoirs (Poland et al., 1975). At Wairakei, for example, if it is assumed that subsidence is proportional to the amount of fluid withdrawn, reinjection throughout the project to date might theoretically have reduced the observed 4·5 m of subsidence to only a meter or so. In practice, this might merely have delayed the attainment of equilibrium between withdrawal and recharge, which gravity data suggest is now about 90 per cent complete. Yet subsidence appears to be continuing (Hunt, 1975). Reinjection for subsidence control alone may not be economically or environmentally sound except where small amounts of subsidence cannot be tolerated, as in Imperial Valley, U.S.A., should geothermal reservoirs beneath irrigated fields be developed. The network of irrigation and drainage canals might be seriously disrupted by as little as a meter of localized subsidence.

7.5. Landslides

Landslides and similar geologic hazards have received little attention in the development of geothermal resources. In certain kinds of terrain, they place severe constraints on the placement of both wells and power plants or other facilities. The Geysers field is located in an area of steep terrain, formed by erosion of structurally incoherent rocks (Figure 7.1). Recent geologic mapping (McLaughlin, 1978) indicates that as much as 50 per cent of the surface in some parts of the geothermal area is underlain by landslides

Figure 7.1. Steep landslide terrain in the northeast side of Big Sulphur Creek, The Geysers geothermal area, California, U.S.A. Power plants 3 and 4 (lower left) and 7 and 8 (on ridge top) are located on bedrock outcrops. Photograph by courtesy of Pacific Gas and Electric Company

7. Environmental Aspects of Geothermal Development

(Figure 7.2). Virtually all of these slides have been active during the Quaternary, and many are still active today. Bacon (1976) reported that more than half of the wells at The Geysers are sited on such landslides, and two of the three well blowouts that have occurred in the area are ascribed to casing failure as a result of landslide movement. In each case, the well that failed was drilled during the early years of development before the full potential for such damage was recognized. As a result of these experiences, developers are now locating well pads on islands of stable bedrock and drilling one or more slant holes from a single site. This has made it possible to maintain the bottom hole spacing in the producing horizons needed to adequately develop the resource while locating well heads on stable ground. The costs of necessary onsite geologic and soil stability studies, though high, are small compared with loss of a well as the result of a blowout. The cost of

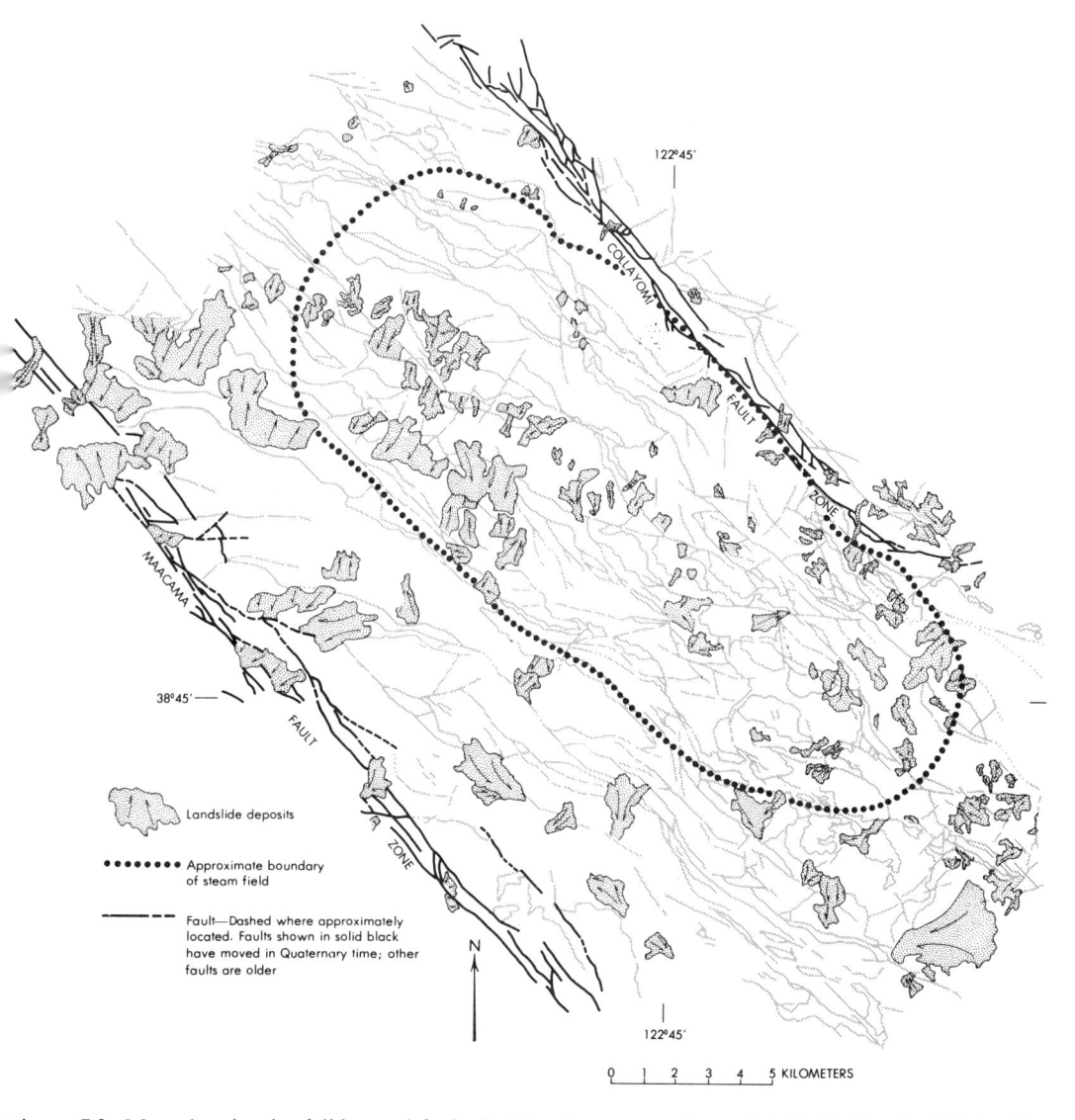

Figure 7.2. Map showing landslides and faults in The Geysers geothermal field, California, U.S.A. After McLaughlin (1978)

controlling, plugging, and abandoning a well that blew out in 1975 was reported to nearly equal that of a new well (Bacon, 1976).

Similar but more stringent considerations control the siting of power plants. The costs of slope failure either upslope from or beneath a power plant would involve much more serious losses. Limitations on the distance steam can be carried by pipeline place severe constraints on the selection of available sites. The need to select sites at The Geysers with maximum air circulation to minimize the impact of H_2S emission from cooling towers has favored sites on ridge tops wherever aesthetic considerations permit.

Although the situation at The Geysers may be extreme, many other areas underlain by steep terrain and weak rocks, such as some parts of Japan, the Philippines, and Central America, may experience similar difficulties as large-scale development takes place.

7.6. Seismicity

The worldwide association between geothermal resources and recent silicic volcanism has long been recognized (Smith and Shaw, 1975). More recently the concepts of plate tectonics have not only provided a rationale for this association but also demonstrated an inevitable worldwide localization of volcanic, tectonic, and thermal phenomena along plate boundaries (See Rybach, this volume). An inescapable consequence of this association is that a large part of the world's geothermal development will take place in areas of high regional seismicity. The most critical environmental constraints are requirements for:

- earthquake resistant construction;
- avoidance of active faults at plant sites;
- avoiding injection of fluids into active faults.

7.6.1. Regional Seismicity

The consequence of the intimate association of geothermal areas and active plate boundaries is that geothermal areas are subject to numerous earthquakes, some of large magnitude. In the Imperial Valley–Mexicali area at the head of the Gulf of California, on the border between the United States and Mexico (Figure 7.3), at least five major geothermal sites—Salton Sea, Brawley, Heber, East Mesa, and Cerro Prieto—lie near the projection of the San Andreas fault system close to where it intersects the East Pacific Rise. In this area, strain appears to be distributed across five or more identified active faults in a zone about 100 km wide. Geothermal areas are localized both near the center of the zone and along some of the bounding faults. Predictably, this is one of the most seismically active areas in the United States. Earthquakes greater than magnitude 6 were recorded in and near this area in 1899, 1903, 1915, 1918, 1934, 1948, and 1968 (Hamilton, 1976). Similar earthquakes must be expected to continue at 10- to 20-year intervals, and quakes of magnitude 4 to 6 will probably occur every few years.

Adequate protection against these expectable earthquake hazards clearly involves design and construction of geothermal facilities to resist the acceleration and vibration frequencies associated with earthquakes likely to occur within the plant's lifetime. The experience of the San Fernando earthquake revealed that ground acceleration locally can exceed 0·5 g. This is much higher than provided for by many engineering designs in the past (Maley and Cloud, 1971). Design requirements are particularly stringent for facilities utilizing inflammable secondary fluids such as propane or isobutane under pressure as a means of recovering heat from medium-temperature or saline geothermal solutions. In

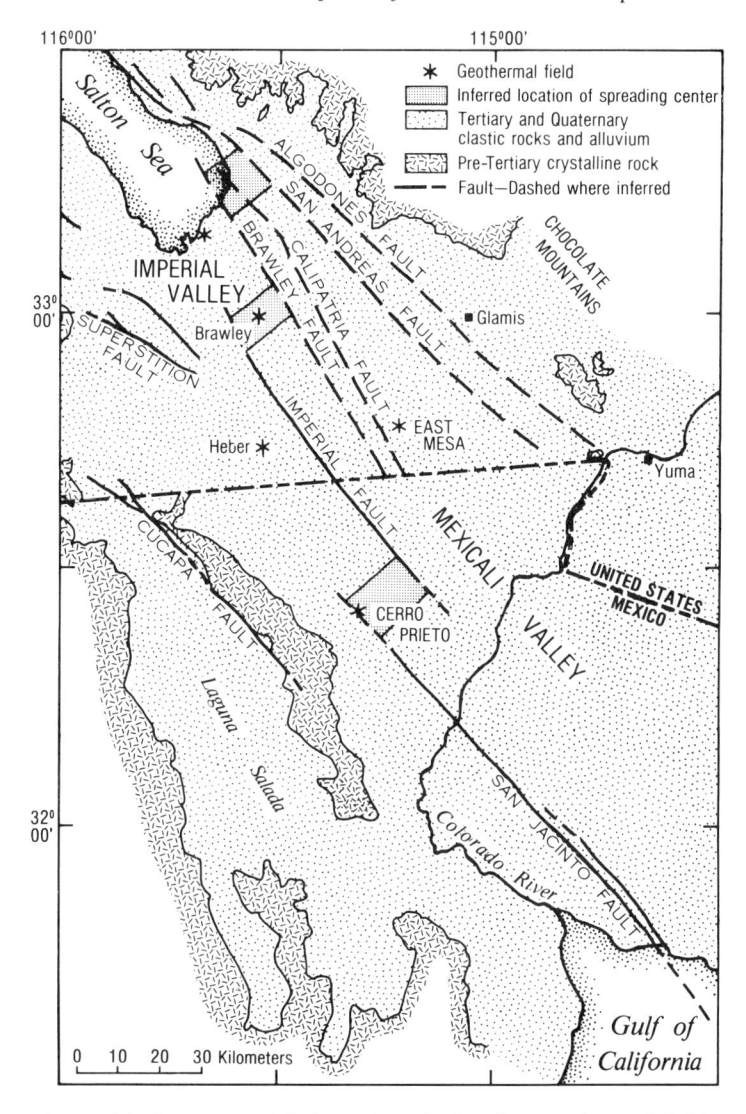

Figure 7.3. Structure of Salton Trough showing geothermal fields and inferred location of spreading centers after Elders et al. (1972), reproduced by permission of the American Association for the Advancement of Science

this respect, the hazards are similar to those associated with design and operation of petroleum refineries.

7.6.2. Surface Rupture

The possibility that geothermal facilities may inadvertently be located directly across an active fault and subsequently be damaged by surface rupture can be eliminated by careful advance geologic and engineering studies. Well-publicized difficulties of locating safe sites for nuclear reactors anywhere in California provide abundant evidence that the problem is far from trivial. Fortunately, unlike the hazards associated with nuclear power plants, even serious damage to a geothermal plant is not likely to affect anyone outside the area of the plant itself.

7.6.3. Induced Seismicity

The discovery in 1966 that injection of liquid waste at moderate pressure into a deep well was responsible for triggering earthquakes near Denver, Colorado (Evans, 1966) has led to widespread concern that reinjection of excess geothermal fluids is capable of inducing earthquakes elsewhere (Swanberg, 1976). Recent careful analysis (Yerkes and Castle, 1976) has shown that a low level of seismicity is also associated with certain cases of fluid withdrawal, most notably that of the Wilmington oil field, U.S.A. Experience in geothermal areas in many parts of the world during the past 10 years does not indicate that this is a serious hazard, although sensitive instrumentation will probably detect induced microearthquakes in any area of extensive production.

The record at The Geysers, U.S.A., is perhaps the best available. Chasteen (1976) reports that since reinjection began in 1969, condensate amounting to more than 1.6×10^{10} L has been returned to the reservoir at a rate as high as 1.7×10^8 L per day. At the present time, earthquakes large enough to feel occur monthly within the area of production, and microearthquake activity is high. Monitoring was carried out briefly in 1971 by means of a net of temporary stations (Hamilton and Muffler, 1972). The results showed that activity was centered along mapped faults in the area of steam production but did not clearly distinguish whether the microquakes recorded were the direct result of steam production and fluid reinjection or represented release of accumulated natural regional stress. Continuous monitoring since 1975 by means of an improved network of permanent stations (Bufe et al., in press) shows an unusually large number of small quakes sharply localized in both area and depth. A concentration of epicenters, mainly of quakes less than magnitude 2, is centered within the producing area and consequently the area of maximum pressure reduction (Lipman et al., 1978). The hypocenters of these quakes are shallow, mainly 1 to 4 km, which places them directly within the areas of steam production. The largest earthquakes (magnitudes between 3·5 and 3·8) occur near injection wells. This pattern of localization leads Bufe et al. (in press) to conclude that these effects are a direct result of the geothermal operations. Yet first-motion studies of quakes within the field and those in the surrounding area indicate that both are responding to the same regional stress pattern. Earthquakes outside the producing field that appear to be part of the normal regional (non-geothermal) tectonic regime are both deeper (4 to 12 km) and less frequent. These events lie along a major right-lateral shear zone (Herd, 1978) that is subparallel to the San Andreas fault and passes within a few kilometers of the steam field.

Carefully monitored injection tests have been conducted in a few other areas with negative results. Cameli and Carabelli (1976) report that in the Viterbo region of Italy, gravity flow into a well at a rate of 3×10^6 L/d produced no seismic or microseismic activity clearly attributable to the injection operations. Similar negative results were reported from the Otaki area, Japan, by Kubota and Aosaki (1976). Flow by gravity into three wells at rates up to 1.3×10^7 L/d per well began in 1972 and is continuing. To May 1975, no seismic events had been detected.

Injection tests, but without onsite monitoring of microseismic activity, have been carried out in several areas. At Ahuachapán, El Salvador, Einarsson et al. (1976) describe injection tests at rates as high as 1.4×10^7 L/d. No significant seismic effects are noted. At East Mesa, California, injection tests at rates as high as 1.3×10^6 L/d have been conducted by the U.S. Bureau of Reclamation (Mathias, 1976) and by private companies without producing seismic events sufficient to register on the regional net (sensitivity threshold about magnitude 1·75) (U.S. Geological Survey, unpublished data, 1978). Similar results were obtained at Roosevelt Hot Springs, Utah U.S.A.

As a result of concern about the possibility of inducing earthquakes by fluid injection, controlled experiments were begun by the U.S. Geological Survey in 1969 to determine the conditions under which earthquakes are both induced and prevented (Raleigh et al., 1976). These experiments were carried out at the Rangely oil field in western Colorado where injection of water to stimulate oil production appeared to be inducing earthquakes. The results showed the validity of the Hubbert-Rubey principle which relates the fluid pressure necessary to initiate fracture to the stress normal to a preexisting fault plane and shear stress parallel to it. First, measurements were made to determine the existing stress values. These values were then used to calculate the critical value above which fluid pressure would be expected to result in fault movement. Finally, during cyclic pumping tests, it was demonstrated repeatedly that earthquakes were *initiated* when the fluid pressure was raised above the critical value and were *turned off* when the pressure was reduced.

On the basis of these experiments, it is possible to predict with considerable confidence the conditions under which fluid injection is likely to generate earthquakes. These are:

- existence of a fault across which a significant level of shear stress has already been built up;
- injection of fluids into the fault zone at pressure exceeding the critical value required to initiate faulting.

As a generalization, this suggests (1) that unless a potentially active fault exists in the area penetrated by wells, neither withdrawal nor injection of fluids is likely to induce major earthquake activity, (2) if wells penetrate an active fault on which the shear stress is already high and an earthquake is about to occur naturally, injection of even a small amount of fluid could trigger the earthquake. Conversely, the withdrawal of fluid by pumping would increase intergranular friction along the fault zone and delay the natural occurrence of the quake until the shear stress rose sufficiently to overcome the increased friction. Note that for a fault on which stress was accumulating continuously, pumping would delay but not *prevent* the fault movement and accompanying earthquake.

In summary, the evidence available indicates that significant seismicity induced by either withdrawal or reinjection of geothermal fluids is comparatively rare and where it does occur is generally of small magnitude. The most cost-effective means of insuring that it does not present a hazard to people or facilities is to monitor each producing area; except for those in immediate proximity to settled areas, existing regional seismic nets are likely to be adequate. If seismicity is detected, for example by earthquakes large enough to feel, their cause can be determined by the installation of portable arrays of seismometers, and appropriate ameliorative measures can be devised if any are needed.

7.7. Subsidence

Prolonged withdrawal of fluid from almost any reservoir is likely to result in detectable subsidence. The actual amount observed will depend on both the quantity withdrawn and the strength of the reservoir rocks. Moreover, the significance of the resulting deformation depends on the land use.

The best-documented example of subsidence is at Wairakei, New Zealand (Stilwell et al., 1975), where measurements were begun in 1956 soon after well-testing began. Maximum subsidence from 1964 to 1974, about 4·5 m (Figure 7.4), occurred about 1500 m east of the well field. Horizontal surveys indicate simultaneous inward movement toward the center of vertical subsidence of as much as 0·5 meter. Despite the large amount of subsidence, damage to production facilities has been small because the pipelines and power plant are peripheral to the center of subsidence. The most obvious effects were observed in concrete drains designed to carry hot water from well-head steam separators to the Waikato River; these have been telescoped as much as 0·8 m. Stilwell et al. (1976) relate the area of maximum subsidence at Wairakei to the thickness of the underlying Waiora Formation, a rhyolite pumice having porosity between 20 and 40 per cent (Grindley and Browne, 1975).

A similar pattern of centralized subsidence has been observed at The Geysers,

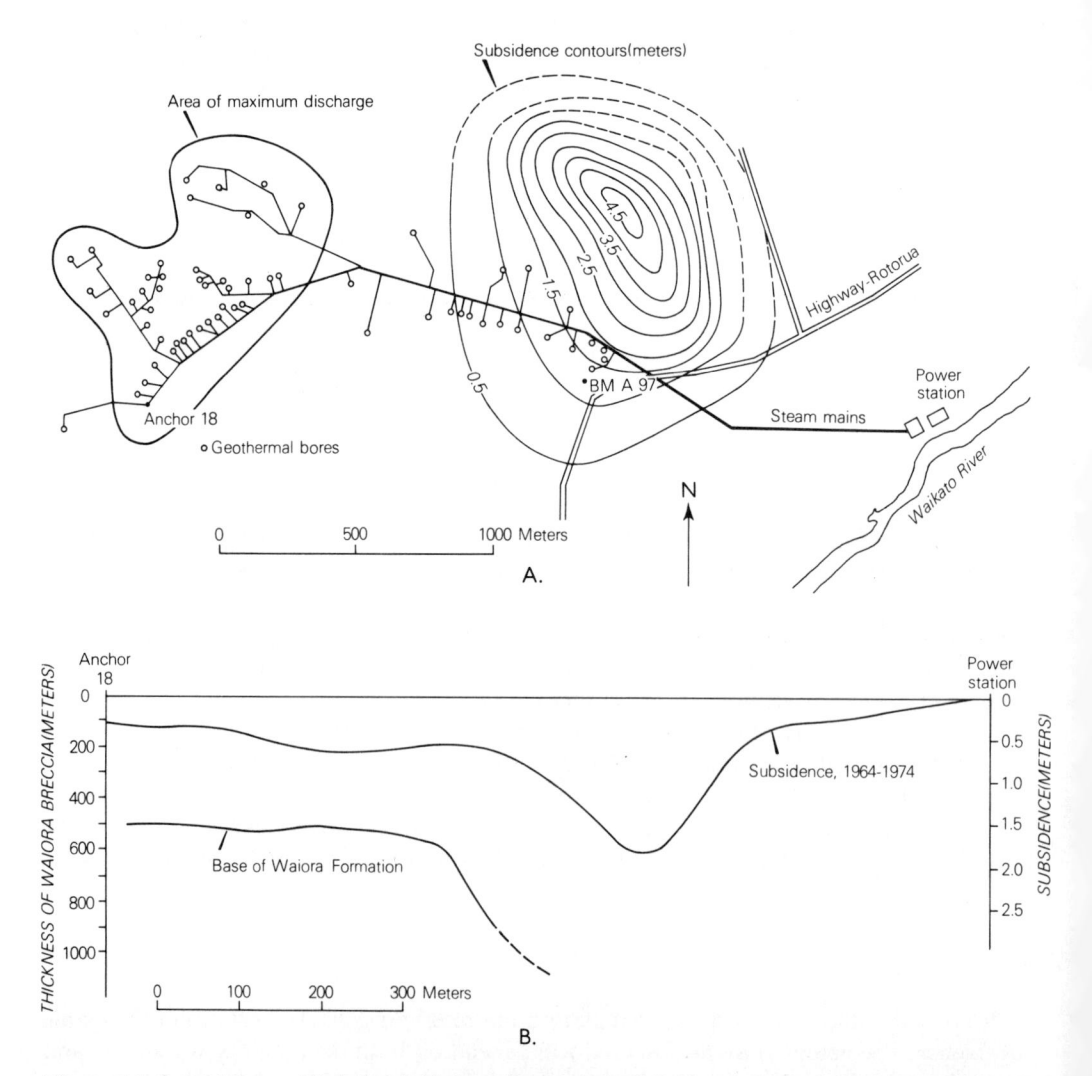

Figure 7.4. Subsidence at Wairakei field, New Zealand. A. Subsidence 1964 to 1974, contour interval 0·5 m. B. Profile along line of steam mains, showing relation of subsidence to thickness of Waiora Formation. After Stilwell et al., (1976)

California (Lofgren, 1978). Here, however, despite vastly greater production, the maximum subsidence over a period of 4 years (1973–1977) was only about 14 cm because the reservoir is in rocks of high strength and low compressibility. Although the amount of subsidence is small, it is closely associated with the area of greatest pressure drop and appears to be a direct result of steam production (Figure 7.5). Local horizontal movements are of about the same amplitude as the vertical and suggest that the reservoir is contracting both horizontally and vertically as the internal pressure decreases.

In both the Wairakei field and The Geysers area, regional tectonic strains are small compared with the local effects of fluid removal, and in neither field are the impacts producing serious environmental effects. An area where the reverse situation will probably prevail is in the Imperial Valley, California, U.S.A. (Fig. 7.3) where geothermal resources at the Salton Sea, Heber, and Brawley geothermal areas are being developed. Irrigation water from the Colorado River is distributed to approximately 400,000 hectares served by the Imperial Irrigation District through a 2700-km network of concrete-lined canals. In such areas, local subsidence of as much as a meter could seriously disrupt operation. Regional subsidence of as much as 5 cm per year occurred near the edge of the Salton Sea between 1972 and 1977 (Imperial County, Department of Public Works). This takes the form of a slight northerly or northeasterly tilt toward the Salton Sea. These large-scale components should not be difficult to distinguish from local effects that may be attributable to withdrawal of geothermal fluids. Government regulations for this area require precise (1st or 2nd order) surveys of wellhead bench marks at least once a year to provide adequate warning of local subsidence. Possible remedial measures include an increase in the quantity of fluid injected, or a decrease in the rate of fluid removal, or both.

7.8. Conclusions

Rigorous analysis of any form of energy production requires assessment of the entire use cycle; production of raw materials, transportation, construction and operation and operation of facilities, and waste disposal and deactivation if required. Because the geothermal industry is relatively new, data adequate for a statistical comparison with other energy forms probably are not available. For this reason, the summary here will address only parts of the energy cycle and will be largely qualitiative.

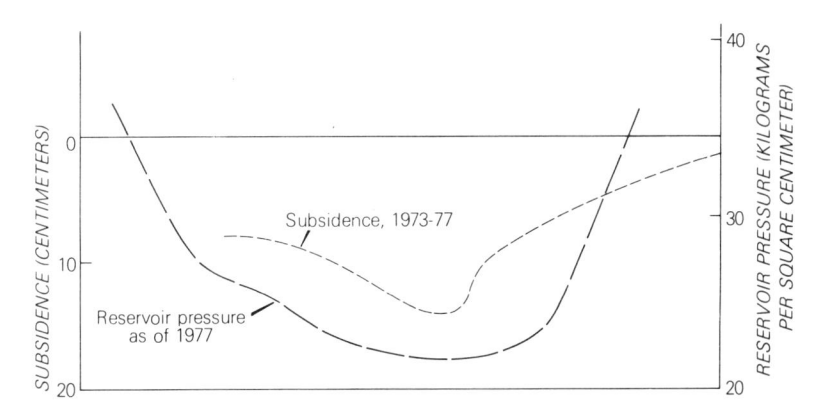

Figure 7.5. Relation of subsidence to reservoir pressure through the central part of The Geysers field, California, U.S.A. Subsidence data and interpretation from Lofgren (1978). Pressure data from Lipman et al. (1978), reproduced by permission of Ben E. Lofgren

The most obvious difference between geothermal energy and the more common fossil-fuel sources is that it is generally produced and utilized onsite, thus eliminating the more remote impacts associated with mining and transportation. The drilling of an individual geothermal well requires the disturbance of about 2 hectares of ground surface, but this can generally be reduced to 100 m^2 for the remainder of the operation. About 15 such wells are required to produce 100 MWe. In most areas the network of production pipelines produces the greatest visual impact, but this can be minimized by appropriate selection of camouflage colors. Where appropriate, pipelines can be elevated so that normal uses of the surface for farming or grazing can be continued, as it has for decades at Lardarello, Italy. The strongest objections to the development of geothermal energy arise where it appears to threaten scenic values in areas used primarily for recreation or other outdoor activities. Even there, careful siting of power plants or other facilities to take advantage of terrain concealment, and attention to routing of pipelines or electrical-distribution lines to minimize the impact of unbroken visual corridors can go far toward making this form of energy production compatible with many natural settings. Moreover, the operation at Wairakei, New Zealand demonstrates that geothermal energy itself may become a major tourist attraction.

The impact of geothermal energy on air quality ranges from essentially zero for systems that utilize closed-loop circulation or systems with extremely low gas content (East Mesa, U.S.A.) to moderate where the content of noncondensible gases is large (Cerro Prieto, Mexico, The Geysers, U.S.A., Lardarello, Italy, or Wairakei, New Zealand). For comparison, the output of major noncondensible gases from The Geysers geothermal area and a typical coal-fired plant are shown in Table 7.6. This geothermal source produces less than 5 per cent as much CO_2 and only 36 per cent as much sulfur as the equivalent steam plant burning coal. As the chemical impact of geothermal development on surface and underground water can be virtually eliminated by fluid reinjection, the principal remaining impact is that of thermal loading. Where cooling towers are used, this will affect the atmosphere; in areas like New Zealand, where run-of-the-river, rather than evaporative cooling is used, it will affect surface waters. If the additional heat results in increased eutrophication or plant growth, as it appears to in New Zealand (Axtmann, 1974), cooling towers may be more advantageous.

In addition to the effects on air, land, and water discussed here, the development of geothermal energy may involve factors such as noise, whose significance will vary with the human environment. Drilling in dry steam fields may result in noise levels of 80 dB (decibels) (equivalent to that of jet plane takeoff). Although the noise level can be reduced by specially designed mufflers, such sounds may carry for many kilometers. In already settled areas or areas used primarily for recreation, noise may present an unacceptable intrusion. Fortunately, once production has begun, direct venting of wells is reduced to a few days per year, at most.

With the engineering technology known at this time, geothermal energy is capable of

Table 7.6. Comparison of noncondensible gases produced by typical fossil-fuel and geothermal plants (mg/kg/day/MWe) (after Bowen, 1973)

	CO_2	H_2S (abated)	SO_2	S (content)
Coal (1% S)	18,000	—	127	64
The Geysers	780	25	—	23·5

fulfilling its widely accepted role of an energy source whose environmental effects are small compared with many other conventional sources.

References

Allen, G. W., and McCluer, H. K., 1976. 'Abatement of hydrogen sulphide emissions from The Geysers geothermal power plant.' In *Proc. Second United Nations Symposium on the Development and Use of Geothermal Resources*, San Francisco, Calif., U.S.A., 1975, 1313–1315.

Altshuler, S. L., 1978, 'Studies of cooling tower emissions at The Geysers power plant', *Report 420—78.104*, Pacific Gas and Electric Co., San Francisco, California.

Axtmann, R. C., 1974. 'An environmental study of the Wairakei Power Plant', *Report No 445*, Physics and Engineering Laboratory, New Zealand Department of Scientific and Industrial Research, Lower Hutt, N.Z., 38 pp.

Axtmann, R. C., 1975. 'Environmental impact of a geothermal power plant', *Science*, **187** (4179), 795–803.

Babcock, E. A., 1971. 'Detection of active faulting using oblique infra-red aerial photography in the Imperial Valley, California', *Geological Society of America Bulletin*, **82**, 3189–3196.

Bacon, C. F., 1976. 'Blowout of a geothermal well', *California Geology*, **29**, 13–17.

Biehler, S., Kovach, R. L., and Allen, C. R., 1964. 'Geophysical features of northern end of Gulf of California structural province.' In *Marine Geology of the Gulf of California*, American Association of Petroleum Geologists Memoir 3, 126–143.

Bowen, R. G., 1973, 'Environmental impact of geothermal development.' In *Geothermal Energy, Resources, Production, Stimulation*, Kruger, P. and Otte, C. (Eds.) Stanford Univ. Press, Stanford, California, 197–215.

Bufe, C. G., Marks, S. M., Lester, F. W., Ludwin, R. S., and Stickney, M. C. (in press). 'Seismicity of The Geysers-Clear Lake region.' In *U.S. Geological Survey Professional Paper 1141*.

Cameli, G. M., and Carabelli, E. (1976). 'Seismic control during a reinjection experiment in the Viterbo Region (Central Italy).', In *Proc. Second United Nations Symposium on the Development and Use of Geothermal Resources*, San Francisco, Calif., U.S.A., 1975, 1329–1334.

Chasteen, A. J., 1976. 'Geothermal steam condensate reinjection.', In *Proc. Second United Nations Symposium on the Development and Use of Geothermal Resources*, San Francisco, Calif., U.S.A., 1975, 1335–1336.

Chiostri, E., 1975. 'Geothermal Resources for Heat Treatments.' In *Proc. Second United Nations Symposium on the Development and Use of Geothermal Resources*, San Francisco, Calif., U.S.A., 1975, 2094–2097.

Combs and Hadley, 1977. 'Microearthquake investigations of the Mesa geothermal anomaly, Imperial Valley, California', *Geophysics*, **42**, 17–32.

Dorighi, G. P., 1978. 'Hydrogen sulfide emissions inventory for The Geysers power plant January, 1976–December, 1977', *Report 420-77.119*, Pacific Gas and Electric Co., San Francisco, California.

Eaton, F. M., and Wilcox, L. V., 1939. 'The behavior of boron in soils', *U.S. Department of Agriculture, Technical Bulletin 696*.

Einarsson, S. S., Vides, R. A., and Cuéllar, G., 1976. 'Disposal of geothermal waste water by reinjection.', In *Proc. Second United Nations Symposium on the Development and Use of Geothermal Resources*, San Francisco, Calif., U.S.A., 1975, 1349–1363.

Elders, W. A., Rex, R. W., Meidav, Tsvi, Robinson, P. T., and Biehler, Shawn 1972. 'Crustal spreading in southern California', *Science*, **178**, (4056), 15–24.

Ellis, A. J., 1978, 'Geothermal fluid chemistry and human health', *Geothermics*, **6**, 175–182.

Evans, D., 1966. 'The Denver area earthquakes and the Rocky Mountain Arsenal disposal well', *Mountain Geologist*, **3**, 23–26.

Grindley, G. W., and Browne, P. R. L., 1976. 'Structural and hydrologic factors controlling the permeabilities of some hot-water geothermal fields.' In *Proc. Second United Nations Symposium on the Development and Use of Geothermal Reousrces*, San Francisco, Calif., U.S.A., 1975, 377–386.

Haldane, T. G. N., and Armstead, H. C. H. (1962). 'Proceedings of a joint meeting of the Institutions of Civil Engineers, Mechanical Engineers, and Electrical Engineers', London, **176**(23), 603–634. Reprinted in *Hearings before Subcommittee on Energy of Committee on Science and Astronautics, House of Representatives, 93rd Congress, Sept. 1973*, U.S. Govt. Printing Office, 1973.

Hamilton, R. M., and Muffler, L. J. P., 1972. 'Microearthquakes at The Geysers geothermal area, California', *Journal of Geophysical Research*, 77, 2081–2086.

Hamilton, Warren, 1976. 'Plate tectonics and man', *U.S. Geological Survey, Annual Report for Fiscal Year 1976*, 39–53.

Herd, D. G., 1978. 'Intracontinental plate boundary east of Cape Mendocino, California', *Geology*, 6, 721–725.

Hunt, T. M., 1975. 'Repeat Gravity Measurements at Wairakei, 1961–74', *New Zealand Department of Scientific and Industrial Research, Geophysics Division Report 111*, 27 pp.

Inhaber, H., 1979. 'Risk with energy from nonconventional sources', *Science*, 203, February 23, 718–723.

Kubota, K., and Aosaki, K., 1976. 'Reinjection of geothermal hot water at the Otake Geothermal Field.' In *Proc. Second United Nations Symposium on the Development and Use of Geothermal Resources*, San Francisco, Calif., U.S.A., 1975, 1379–1383.

Lipman, S. C., Strobel, C. J., and Gulati, M. S., 1978. 'Reservoir performance at The Geysers Field' *Geothermics*, 7(2–4), 209–219.

Lofgren, B. E., 1978. 'Monitoring crustal deformation in The Geysers-Clear Lake Geothermal area, California', *U.S. Geological Survey Open-File Report*, 78–597, 99 pp.

Maley, R. P., and Cloud, W. K. 1971. 'Preliminary strong motion results from the San Fernando earthquake of February 9, 1971', *U.S. Geological Survey Professional Paper 733*, 163–176.

Mathias, K. E., 1976, 'The Mesa geothermal field a preliminary evaluation of five geothermal wells! In *Proc. Second United Nations Symposium on the Development and Use of Geothermal Resources*, San Francisco, Calif., U.S.A., 1975, 1741–1747.

Mercado, Sergio, 1976. 'Cerro Prieto Geothermoelectric Project: Pollution and Basic Protection.' In *Proc. Second United Nations Symposium on the Development and Use of Geothermal Resources*, San Francisco, Calif., U.S.A., 1975, 1385–1398.

Mercado, Sergio, 1977. 'Disposiciones de desechos geotermicos.' In *Simposio International sobre energia geothermica en America Latina*, Guatamala City, 1976; Instituto Italo-Latino Americano, Rome, Italy, 777–799.

McLaughlin, R. J., 1978. 'Preliminary geologic map and structural section of the central Mayacamas Mountains and The Geysers Steam Field, Sonoma, Lake, and Mendocino Counties, California', *U.S. Geological Survey Open-File Report 78–389* (2 sheets), scale 1 : 24,000.

Muffler, L. J. P., and Doe, B. R., 1968. 'Composition and mean age of detritus of the Colorado River delta in the Salton Trough, Southeastern California', *Journal of Sedimentary Petrology*, 38, 384–399.

Muffler, L. J. P., (Ed.), 1979. 'Assessment of geothermal resources of the United States–1978', *U.S. Geological Survey Circular 790*, 163 pp.

Nehring, N. L., and Fausto, J., 1979. 'Gases in steam from Cerro Prieto, Mexico, geothermal wells with a discussion of steam/gas ratio measurements', *Proc. Symposium on Cerro Prieto*, San Diego, Calif., U.S.A., Lawrence Berkeley Laboratory, LBL-7098, 127–129.

Poland, J. F., Lofgren. B. E., Ireland, R. L., and Pugh, R. G., 1975. 'Land subsidence in the San Joaquin Valley as of 1972', *U.S. Geological Survey Professional Paper, 437-H*, 75 pp.

Raleigh, C. B., Healy, J. H., and Broedehoeft, J. D., 1976. 'An experiment in earthquake control at Rangely, Colorado', *Science*, 191, 1230–1237.

Reed, M. J., and Campbell, G. E., 1975. 'Environmental Impact of Development in The Geysers Geothermal Field, U.S.A.' In *Proc. Second United Nations Symposium on the Development and Use of Geothermal Resources*, San Francisco, Calif., U.S.A., 1975, 1399–1415.

Rex, R. W., 1970. 'Investigation of geothermal resources in the Imperial Valley and their potential value for desalination of water and electricity production', *Institute of Geophysics and Planetary Physics*, University of California, Riverside, California, 14 pp.

Rothbaum, H. P., and Anderton, B. H., 1976. 'Removal of silica and arsenic from geothermal discharge water by precipitation of useful calcium silicates.' In *Proc. Second United Nations Symposium on the Development and Use of Geothermal Resources*, San Francisco, Calif., U.S.A., 1975, 1417–1425.

Ruff, R. E., Cavanaugh, L. A., and Carr, J. D., 1978. '1977 Executive Summary, Specialized Monitoring Services', *S.R.I. International*, Menlo Park, California, U.S.A.

Smith, R. L., and Shaw, H. R., 1975. 'Igneous-related geothermal systems', In *Assessment of Geothermal Resources of the United States—1975*', White, D. E., and Williams D. L. (Eds.), *U.S. Geological Survey Circular 726*, 58–83.

Stilwell, W. B., Hall, W. K., and Tawhai, J., 1976. 'Ground movement in New Zealand Geothermal

Fields.' In *Proc. Second United Nations Symposium on the Development and Use of Geothermal Resources*, San Francisco, Calif., U.S.A., 1975, 1427–1434.

Swanberg, C. A., 1976. 'Physical aspects of pollution related to geothermal energy development.' In *Proc. Second United Nations Symposium on the Development and Use of Geothermal Resources*, San Francisco, Calif., U.S.A., 1975, 1435–1443.

U.S. National Technical Advisory Committee on Water Quality Criteria, 1968. 'Report on water quality criteria', *Federal Water Pollution Control Administration*, 234 pp.

White, D. E., Anderson, E. T., and Grubbs, D. K., 1963. 'Geothermal brine well: Mile-deep drill hole may tap ore-bearing magmatic water and rocks undergoing metamorphism', *Science*, **139**(3558), 919–922.

Wollenberg, H. A., 1976. 'Radioactivity of geothermal systems.' In *Proc. Second United Nations Symposium on the Development and Use of Geothermal Resources*, San Francisco, Calif., U.S.A., 1975, 1283–1292.

Yerkes, R. F., and Castle, R. O., 1976. 'Seismicity and faulting attributable to fluid extraction', *Engineering Geology*, **10**, 151–167.

Case Histories

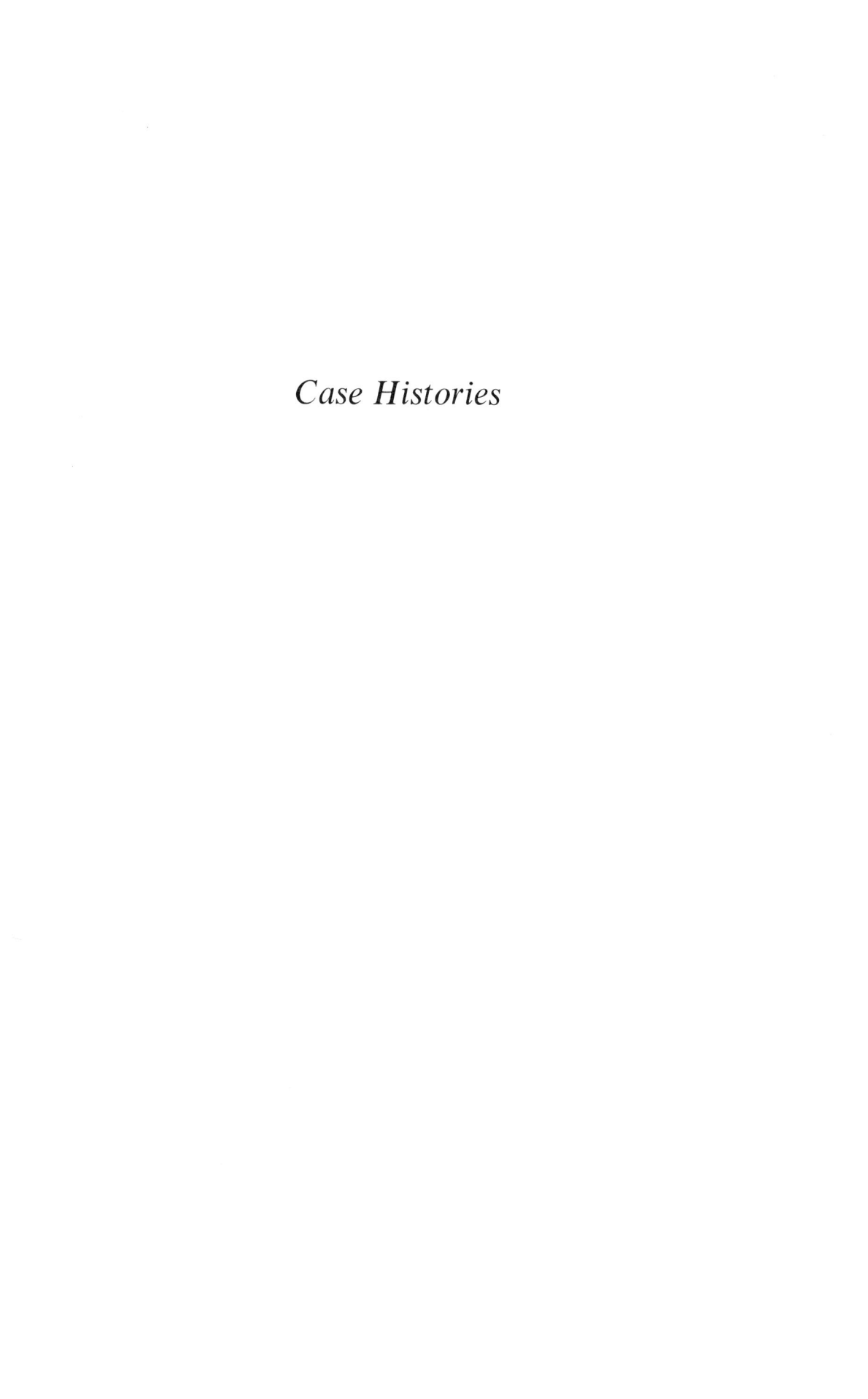

8. The Low Enthalpy Geothermal Resource of the Pannonian Basin, Hungary

PÉTER OTTLIK

Hungarian Water Authority, 1011 Budapest, Fö U.44.

JÁNOS GÁLFI

Research Institute for Water Management, 1095 Budapest, Kvassay J.U.1.

FERENC HORVÁTH

Eötvös University, 1083 Budapest, Kun Béla Tér 2.

KÁROLY KORIM

Water Exploration Enterprise, 1051 Budapest, Zrinyi U.1.

LAJOS STEGENA

Eötvös University, 1083 Budapest, Kun Béla Tér 2.

8.1. Introduction

Geothermal conditions of Hungary deviating from the average state have been indicated by numerous thermal springs known since antiquity. First of all, two areas became known through the occurrence of thermal springs: (1) the Western part of Budapest and (2) the Héviz area lying at the south-west end of Lake Balaton (Figure 8.1). The thermal springs had been used for centuries in natural conditions for curative purposes.

Since the nineteenth century the quantity of thermal water appearing as a natural discharge has been increased by artificial exploration (Table 8.1). The first such activity is connected with the name of V. Zsigmondy, the famous Hungarian engineer. Data of the three wells deepened by him in the nineteenth century and that of other three wells of historical importance are collected in Table 8.1.

In the decade after World War I, prospecting for oil started in Hungary. In the course of this, reserves of thermal waters stored in the upper Pannonian sandy strata of Pliocene age were discovered and exploited in the NE and SW parts of the country as well as in the other parts of the Great Hungarian Plain.

Exploration speeded up after World War II, and, up to 1953, the country acquired about 80 wells giving thermal water. Between 1953 and 1965, 110 new thermal wells were drilled. As of 1977, 525 thermal wells were registered, all of them giving water with a

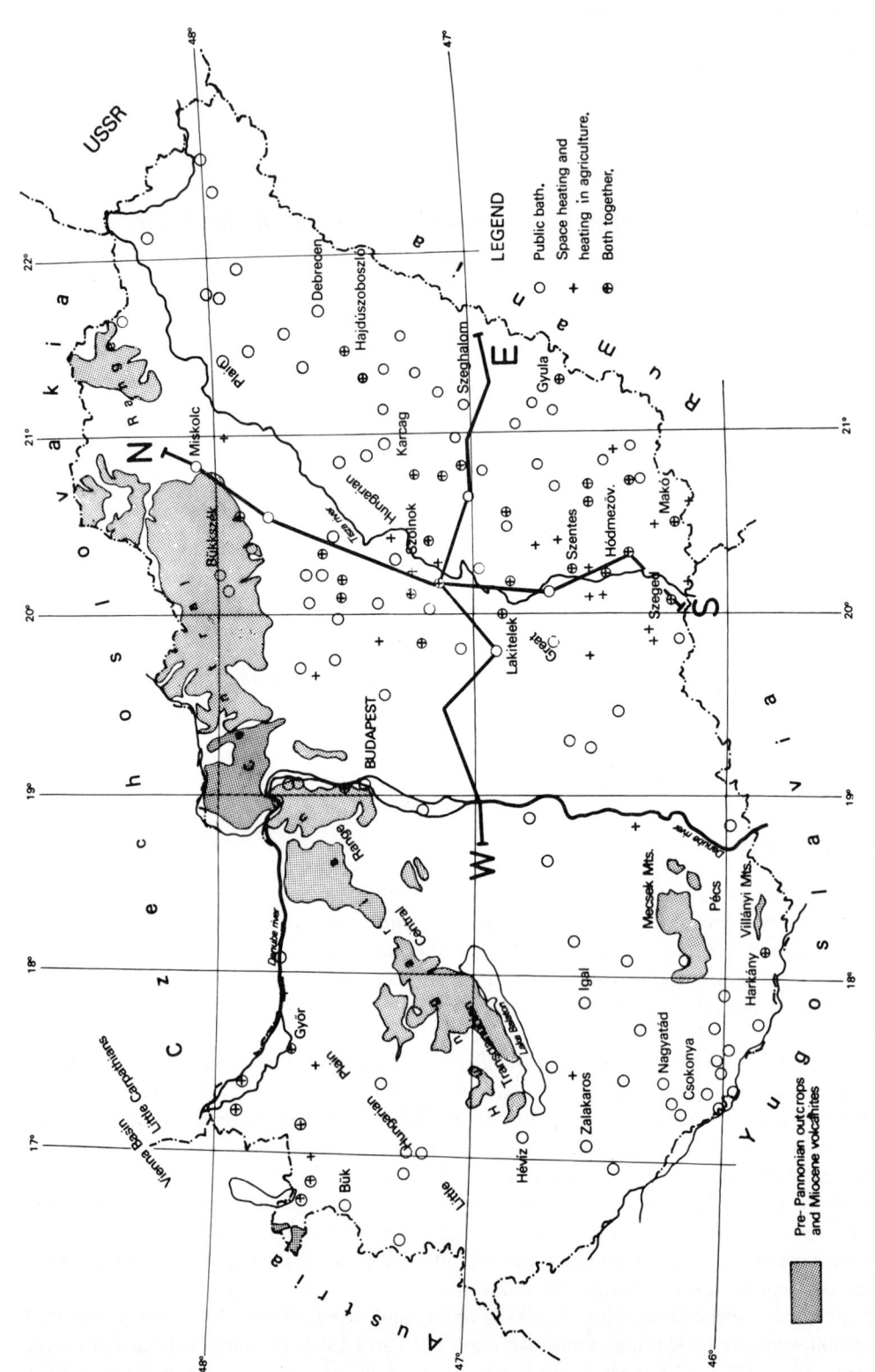

Figure 8.1. Sketch map showing the utilization of geothermal water in Hungary in 1977–78 (Korim, 1978)

8. *The Low Enthalpy Geothermal Resource of Hungary*

Table 8.1. Historical geothermal wells in Hungary

Location	Year of drilling	Depth (m)	Producing formation	Temperature (°C)	Discharge (L/min)
Harkány	1866	37·7	Triassic limestone	61	—
Budapest, Margaret Island of the Danube, No. 1.	1866	118·5	Eocene marl	44	6200
Budapest, Town Park, No. 1.	1868/77	970	Triassic dolomite	74	350–500
Nagyatád, south from Lake Balaton	1911	413	Pannonian sediments	32	900
Budapest, Margaret Island of the Danube, No. 2.	1920/30	310	Eocene marl	71	3500
Budapest, Town Park, No. 2.	1920/30	1257	Triassic limestone	75	3570

temperature higher than 35°C with a total output of 438,000 L/min. 370 wells have temperatures of 35–59°C, 125 wells have temperatures of 60–89°C, and 30 wells have temperatures higher than 90°C.

Parallel to the drilling activity or even prior to it, the scientific observation of geothermal phenomena has been going on. First of all, the thermal waters coming to the surface in Budapest were investigated; later on, studies were extended to the whole Pannonian basin.

After World War II, the utilization of thermal waters, the development of baths, and the exploration of geothermal resources became a national task.

8.2. Geological Framework

The region encircled by the Eastern Alps, the Carpathian arc and the Dinarides is usually called the Carpathian basin. However, this region is genetically not uniform and it should be divided into three sub-basins. The central and largest part is the *Pannonian basin*, whereas in the northwest and southeast parts, the Little Carpathians and the Apuseni Mountains separate the Vienna basin and the Transsylvanian basin respectively from the Pannonian basin (Figures 8.1 and 8.8). The Pannonian basin proper can further be divided into two parts; the outcropping pre-Neogene rocks of the Transdanubian Central Range separate the Little Hungarian Plain on the NW from the Great Hungarian Plain on the SE.

The Pannonian basin can be classified tectonically as a back-arc basin encircled by the Carpathian arc. Stegena et al. (1975) suggested a plate tectonic model for the evolution of the Pannonian basin (Figure 8.2). The subduction of the 'oceanic' lithosphere of the Carpathian flysch trough directed towards the Pannonian area led to thermal mantle diapirism under this region. The rising mantle material caused the thinning out of the lithosphere and crust by subcrustal erosion; that means that the flowing mantle material below the crust eroded and melted its lower part. The thinned-out crust, having an average thickness of 25 km, sunk isostatically, leading thus to the formation of the Pannonian basin.

Figure 8.3 shows the crustal structure of the Pannonian basin along a NE-SW profile (see Figure 8.5 for profile trace).

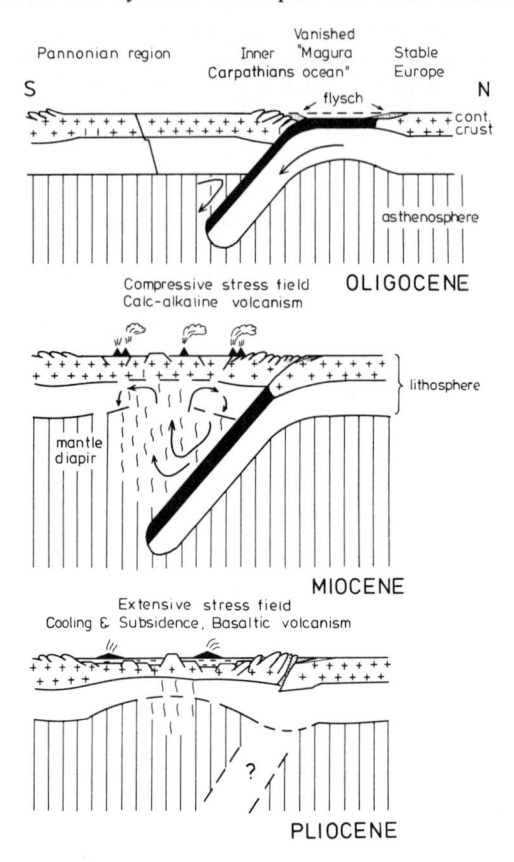

Figure 8.2. Evolutional scheme of the Pannonian basin in the Tertiary Period

8.2.1. Geological Overview

The Pannonian basin is filled with sediments of Late Cretaceous–Paleogene age and mainly Neogene–Quaternary age. The complex of all the rocks and consolidated formations forming the floor of these sediments will be referred as 'basement' in the following discussion. Thus, in this sense basement means not only the crystalline or metamorphic rocks of Precambrian age, but also formations of Paleozoic and Mesozoic (pre-upper Cretaceous) age. In a large part of the basin, crystalline basement is overlain directly by Miocene sediments; however, in the surroundings of the Hungarian Central Range, Eocene sediments are underlain directly by Triassic rocks, which will be called also as basement, according to the definition given above. The Precambrian and Paleozoic basement consists of metamorphic crystalline rocks (granites, phyllites, schists, gneisses, sandstones, dolomites and limestones). They are usually impermeable formations, but the carbonates locally include fractured-fissured aquifers.

Figure 8.4 displays a geologic profile along two sections. The Miocene sediments of the basin are deposited mostly directly on Mesozoic and older rocks forming the basement of the Neogene basin. The Mesozoic basement is made up of limestone, dolomite, sandstone, marl and silicic to mafic magmatic rocks. The main mass of the Mesozoic rocks is made up of Triassic limestone and dolomite. They have wide areal extent, and their thickness

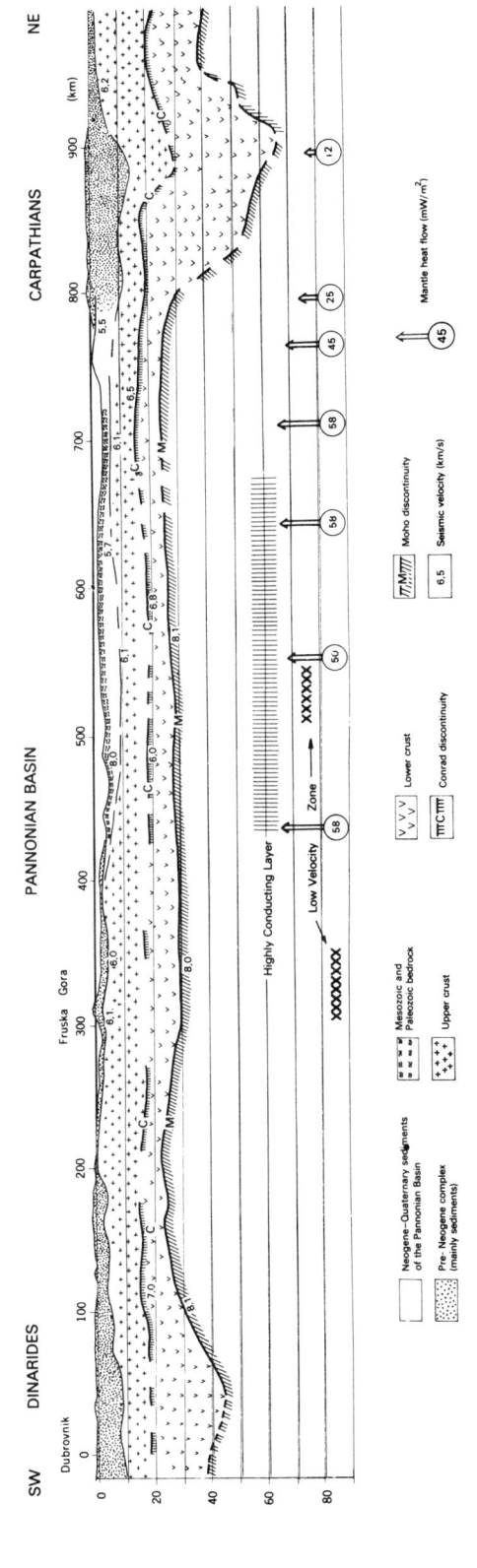

Figure 8.3. Crustal profile from the Carpathians to the Dinarides, with heat flow values calculated in mW/m² for the upper mantle (Buntebarth, 1976), with the position of the high conductivity layer (Ádám, 1976) and of the low velocity zone (Bisztricsány and Egyed, 1973). Geographical position of the profile see AA′A″ in Figure 8.5. After Stegena et al. (1975)

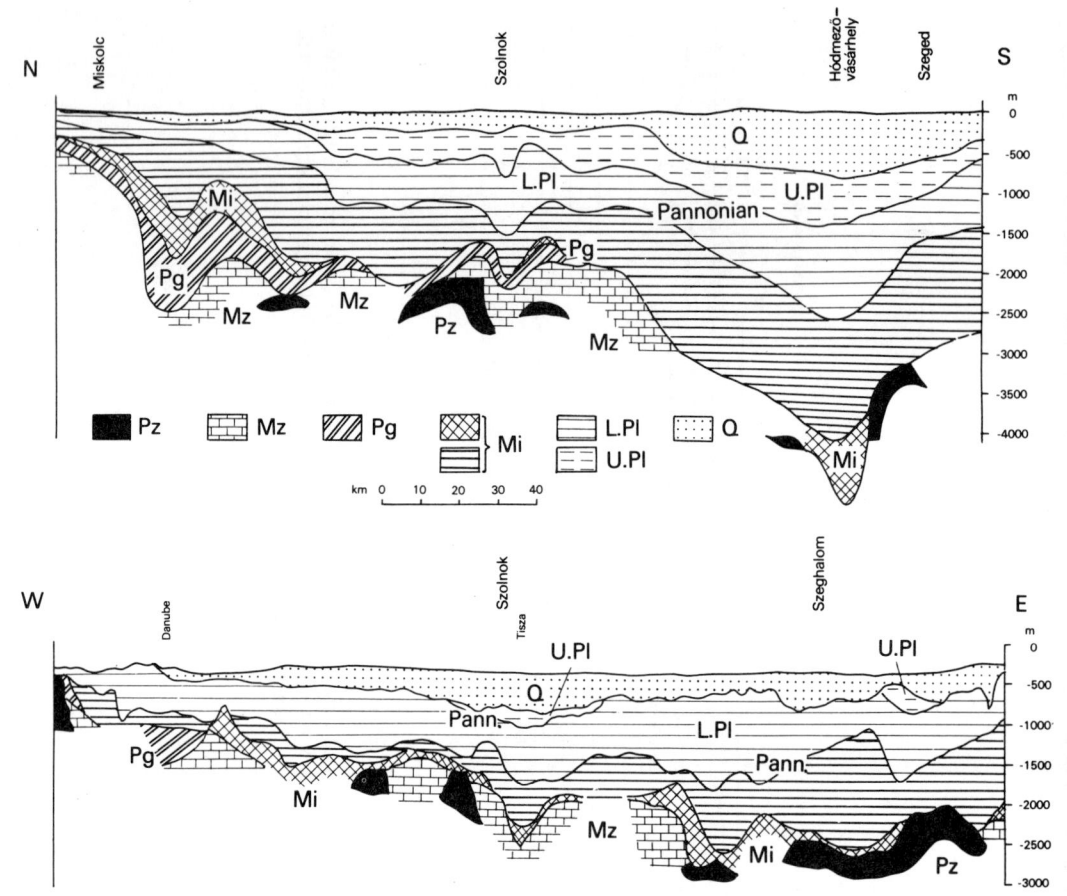

Figure 8.4. Geological profiles of strike directions N–S and E–W across the Great Hungarian Plain (after Ronai, 1978). Geographical position of the profiles see in Figure 8.1. Pz: Paleozoic. Mz: Mesozoic. Pg: Paleogene. Mi: Miocene. Pl: Pliocene. Q: Quaternary

can reach 4000–5000 meters. This often carstic and highly fractured-fissured carbonate complex constitutes a regional thermal water reservoir.

The basement is overlain mainly in the NE part of the Great Hungarian Plain by a series of sediments consisting of flysch, sandstone, marl, clay, limestone and andesitic volcanites. The series, which can not always be divided exactly, represents rocks of Upper Cretaceous–Paleogene age. As the bulk of the series is poorly permeable or impermeable, there is no water circulation.

The stratigraphy of the Neogene–Quaternary rocks is shown in Table 8.2. The thickness of the Tertiary sediments is represented by cross sections (Figure 8.4). This series consists of shallow marine, lacustrine, and terrestrial sediments and of volcanogenic rocks.

The Pannonian area was emergent during the most of the lower Miocene. Transgression started in the Karpatian stage (Table 8.2), became nearly general during the Badenian stage, and was accompanied by intense andesitic-dacitic-rhyolitic volcanism. During the Sarmatian, as a consequence of vertical movements, the Pannonian area became separated from the Paratethys sea which extended during the Oligocene-Miocene through the southern part of Europe. The Upper Miocene–Pliocene time of subsidence is characterized by lacustrine sedimentation.

It should be noted that the chronostratigraphical correlation of the geological sections

8. The Low Enthalpy Geothermal Resource

Table 8.2. Chronostratigraphy and lithostratigraphy of the Pannonian basin (modified after Rögl et al., 1978)

Million years	Periods	Epochs			Pannonian basin stages	Mediterranean European stages	
0	Quaternary	Holocene Pleistocene			Holocene Pleistocene		
5	Neogene	Pliocene	Lower	Upper	Pannonian	Upper	Piacenzian
							Zanclian
						Lower	Messinian
10		Miocene	Upper				Tortonian
					Sarmatian		
15			Middle		Badenian		Serravallian
							Langhian
					Karpatian		Burdigalian
					Ottnangian		
20			Lower		Eggenburgian		Aquitanian
					Egerian ↓		
25		Oligocene ↓					Chattian ↓

between the separated basins is yet not solved perfectly, and details of the chronostratigraphy shown in Table 8.2 are yet discussed. The thickness of the Karpatian-Sarmatian series varies from 0 to 500 m. In the NE–SW striking belt, however, pre-Pannonian Miocene volcanites were accumulated up to a 2000 m thickness (Figure 8.5). The Pannonian sediments have a great thickness (up to 4000 m). They constitute a sedimentary sandwich, composed of sandy, clayey, and silty beds of highly variable horizontal and vertical dimensions. In the lower Pannonian, the impermeable beds dominate, but the upper Pannonian and Quaternary formations include a vast porous bed system. The majority of thermal water resources of the Pannonian basin occurs within the Pliocene sand and sandstone beds. Locally the hot water reservoirs can be discharged (or recharged) through lithological 'chimneys'.

8.2.2. Tectonism and Volcanism

The pre-Neogene basement is characterized by normal faults, part of which is synsedimentary. The rate of subsidence varied in space and time resulting in a basin-and-range topography ('tectonique cassant'). For the large-scale tectonics of the basin, the strike direction is mainly NE–SW; nevertheless, one can encounter transverse tectonic lines nearly perpendicular to the main ones.

The young sediments are loose or plastic and, owing to the compaction, their bedding follows the more significant forms of the brittle basement with a certain smoothing. Over the elevated parts of the basement horsts, slightly bent forms of atectonic character may be observed within strata series of great thickness. The value of compaction at a depth of 3 km may reach 20–30 per cent, reducing strongly the porosity. However, seismic measurements have demonstrated that quite young tectonic movements have also taken place within the basin-filling sedimentary rocks, with faults in the brittle basement being reflected by mild flexures in the overlying sedimentary rocks.

In the inner part of the Carpathian arc calc-alkaline volcanism took place in the Eocene, but the most important period of volcanism falls into the Miocene and Pliocene (Figure 8.5). During that period, activity of volcanoes and also tuff eruption from these is widespread, especially in the northern parts of the Great Hungarian Plain, where a tuff layer more than a hundred meters thick can be found under the Pannonian strata. Another period of volcanism is at the end of Pliocene–early Pleistocene, although with a significantly weaker activity. In that period, basaltic lava came to the surface and formed isolated domes at several places (Figure 8.5). Today's cold CO_2-rich wells/springs are considered as a last phase of postvolcanic action.

It is worth mentioning that in Hungary there is no significant correlation between the Miocene volcanic activity and heat flow. To the north and east, however, where the volcanoes were active during the Pliocene, these areas are characterized by heat flow highs (Fig. 8.8).

8.3. Geothermics

The occurence of several thermal springs and some natural hot water ponds has long ago called the attention to the high thermal activity of the Pannonian basin. Increased underground temperatures were also very early observed in mines. Reports on the downward increase of temperature in deep Hungarian mines initiated the first systematic discussion of underground temperature by Boyle in 1671 (see Bullard, 1965).

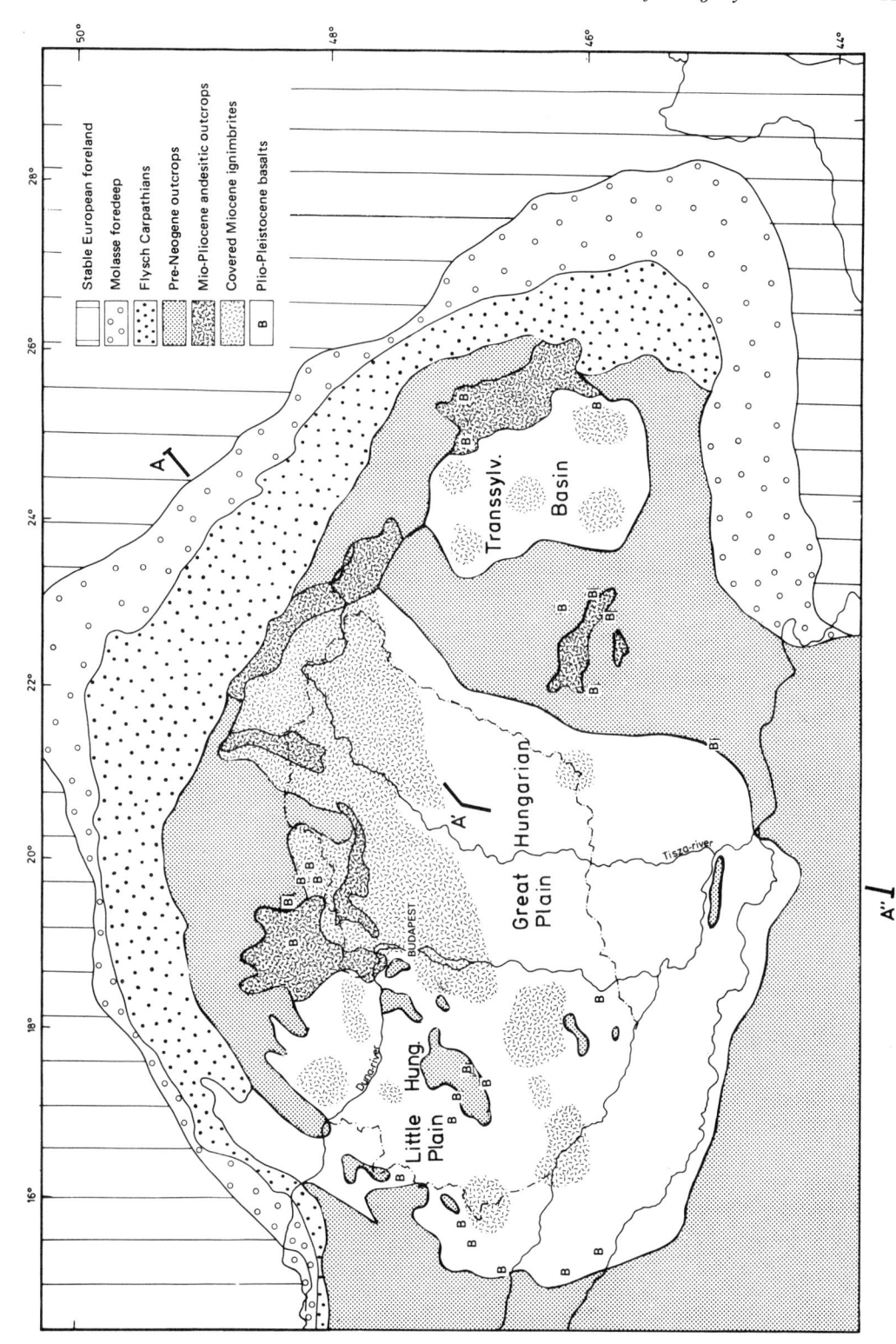

Figure 8.5. Neogene–Quaternary volcanism of the Carpatho-Pannonian region

8.3.1. Rock Temperature Measurements

In recent years the routine well-logging of drillholes has supplied several thousand temperature data in Hungary. It is, however, a very inhomogeneous set of data because of the different measuring techniques. The majority of the data are bottom-hole temperatures measured by mercury thermometers soon (1–48 hours) after the cessation of drilling operation and of mud circulation. Normally the temperature is measured at 2–4 different times and this set of data is available for 3–5 depth points (which were the one-time bottoms of the hole). Apparently, these values are lower (at least by 10–30 per cent according to measurements) than the equilibrium rock temperature, which is defined as the hole temperature being constant, within the error of measurement, during long-term repetitions of the bottom-hole measurements.

Many temperature data have been obtained in oil and water exploration boreholes by measuring the temperature of the fluid during flow-rate measurements. These values can also differ from the equilibrium reservoir temperatures. Moreover there are some hundred outflowing-water temperature data. These data give estimates for the temperature at the depth-range of the aquifer using some theoretical or empirical relationship for the cooling of the water during ascent. Equilibrium or quasi-equilibrium temperatures are available only in some particular wells (e.g. those used for heat flow determination).

A temperature map prepared for the depth of 1000 m using corrected outflowing water temperatures is presented in Figure 8.6.

8.3.2. Heat Flow Measurements

Heat flow data measured in Hungary have recently been reanalysed and recalculated (Horváth et al., 1979), from which 13 heat flow determinations are now available (Figure 8.7). Four heat flow data determined in the Mecsek Mountains suggest an unequivocally high value (103–139 mW/m^2) for this area.

Heat flow measurements performed in the region of the Hungarian Central Range in or above the Mesozoic carbonate complex (Figure 8.7) meet certain difficulties since in the heavily faulted and fissured Mesozoic carbonates and marls convective heat transport due to water migration may occur. Measured temperatures sometimes show large lateral temperature variations, suggesting upward migration of hot water along faults, or downflow of cold meteoric water in the loose sedimentary complex.

The rock matrix of clayey and marly sedimentary samples will be destroyed after a short time when saturated by water. Therefore their thermal conductivity in a water-saturated state cannot be measured directly in the laboratory by the widely used divided bar instrument. This caused difficulties and inaccuracies during the early flow determinations. Therefore, a new apparatus for thermal conductivity determination has recently been constructed based on the differentiated line source method (Scott et al., 1973; Cull, 1974). In this method, only a thin surface layer of the sample is saturated by water, and the measurements take only a few minutes. During such a short time no significant alteration occurs in the clayey rocks, and therefore the measured thermal conductivity approximates well the in-situ conductivity. The variation of clay content with depth, determined from electrical and nuclear well-logs, was also taken into account for the determination of the thermal conductivity of the sedimentary column.

All the 13 heat-flow determinations carried out in boreholes are reliable (that is with an error less than 25 per cent) mainly due to the good location of wells. The difficulty of precise thermal conductivity determination of clayey sediments and/or the lack of stationary temperatures has hindered obtaining heat-flow values of high accuracy (that is with an error less than 15 per cent) in the basinal part of Hungary.

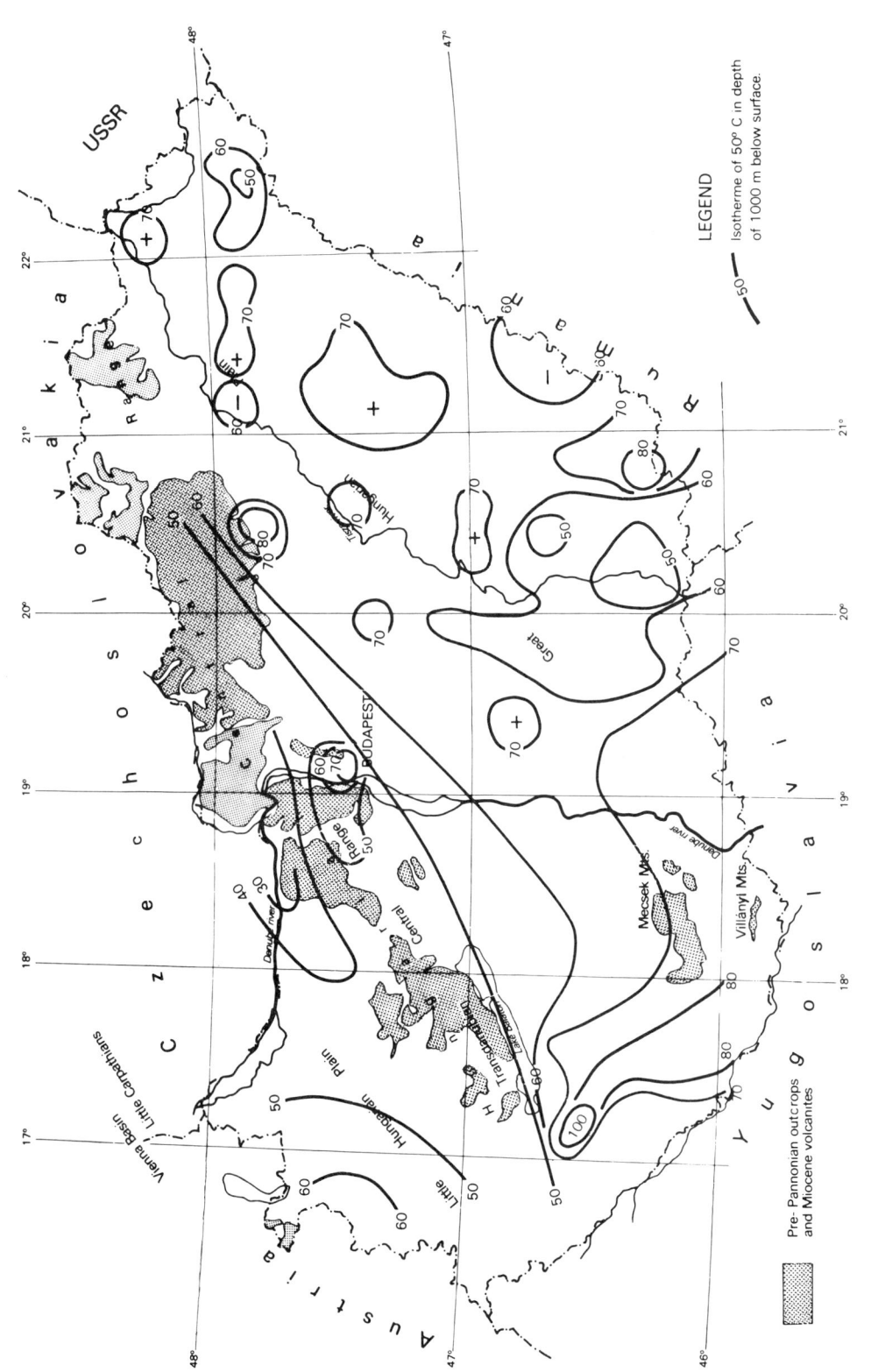

8. The Low Enthalpy Geothermal Resource of Hungary

Figure 8.6. Geoisotherms for Hungary at 1000 m depth. Temperatures in °C

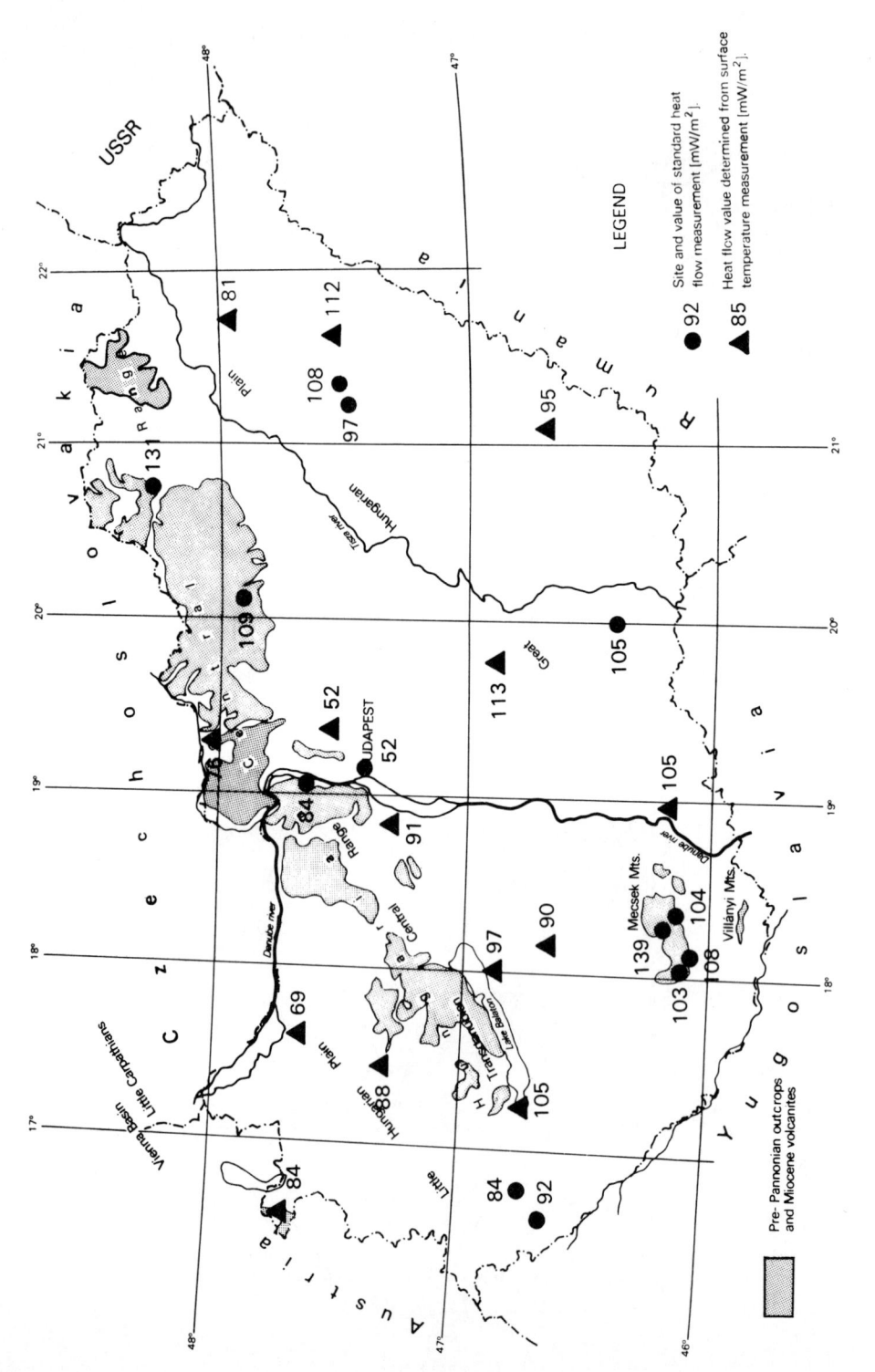

Figure 8.7. Heat flow data for Hungary, in units of mW/m^2

Recently 14 heat flow estimates have been made by L. Bodri (Horváth et al., 1979) using shallow-depth (20–200 cm) temperature records over long periods (5–15 years) (Figure 8.7). For the construction of the heat flow map of Hungary both sets of data were used, and the temperature maps of Stegena (1976) were also taken into consideration (Horváth et al., 1979). Several heat flow determinations have been made for the Pannonian basin outside Hungary and for the surrounding regions (see in Čermák and Rybach, 1979). The heat flow map of the carpatho-Pannonian region, slightly modified after the European heat flow map of Čermák and Hurtig 1979), is shown in Figure 8.8. In Table 8.3 are given statistics of heat-flow data for different parts of the Intra-Carpathian region. Both the map and the table show that the Pannonian basin is characterized by high heat flow: The mean of 38 heat flow values gives 90·4 mW/m^2. Good negative correlation with the thickness of basin-fill can be observed, suggesting that positive thermal anomalies are caused probably by the elevated position of the well conducting basement. Regions encircled by the 100 mW/m^2 isoline (Figure 8.8) are associated with the updoming of the basement. Both the Vienna basin and the Transsylvanian basin are characterized by subnormal heat flow (49·7 and 47·0 mW/m^2 mean values, respectively). Although their geologic evolutions are thought to be similar to that of the Pannonian basin, the strong heat-flow contrast indicates that their thermal history must have been different.

Geothermal model calculation and other geophysical data indicate that high heat flow in the Pannonian basin is caused by the updoming of the asthenosphere due to the thermal mantle diapir (Figure 8.2). Conductive cooling of the diapir led to the formation of the Pannonian basin by subsidence and extension (Stegena et al., 1975; Horváth et al., 1979; Ádám et al., 1979).

8.4. Hydrogeology

Among the Paleozoic rocks, the dolomite of Devonian age is water bearing. One occurrence is known where water is exploited with yield of 139 L/min with outflow temperature of 58°C.

South of the Lake Balaton, limestone of a supposed Carboniferous age occuring at a depth of 600–800 m was opened by drilling, and from the fissures water of 70°C temperature was exploited with a slight overpressure. Recently, in the Permian limestone, water of 100°C was found at a depth of 1000 m, south of Budapest.

Carbonate rocks of Triassic age are widespread in the basement of the basin and represent one of its main aquifer systems. This group of formations is significant for both its prevalence and its great thickness, estimated to be more than 3000 m at some places. Its hydrogeological significance is due to the great dimensions and to the secondary porosity and permeability acquired during the geological history of the basin. The secondary porosity is composed of joint porosity, since the brittle rock of great thickness became fractured during Alpine tectonic movements, and of vug porosity, since a significant part of the Pannonian area was emerged for a long time (till the end of the Lower Miocene), and so these rocks, remaining at the surface, became karstified. At present in the Transdanubian Central Range 30–35 per cent of the precipitation falling on the uncovered and karstified Triassic rocks at the surface penetrates downwards, thus assuring the continuous recharge of deep karstic water. The faulted structure enables the downward percolation of water and the development of large-scale convection. The existence of such circulation is indicated by thermal springs coming to the surface along fault lines. The hydrogeological characteristics of the Triassic and younger aquifers are summarized in Table 8.4.

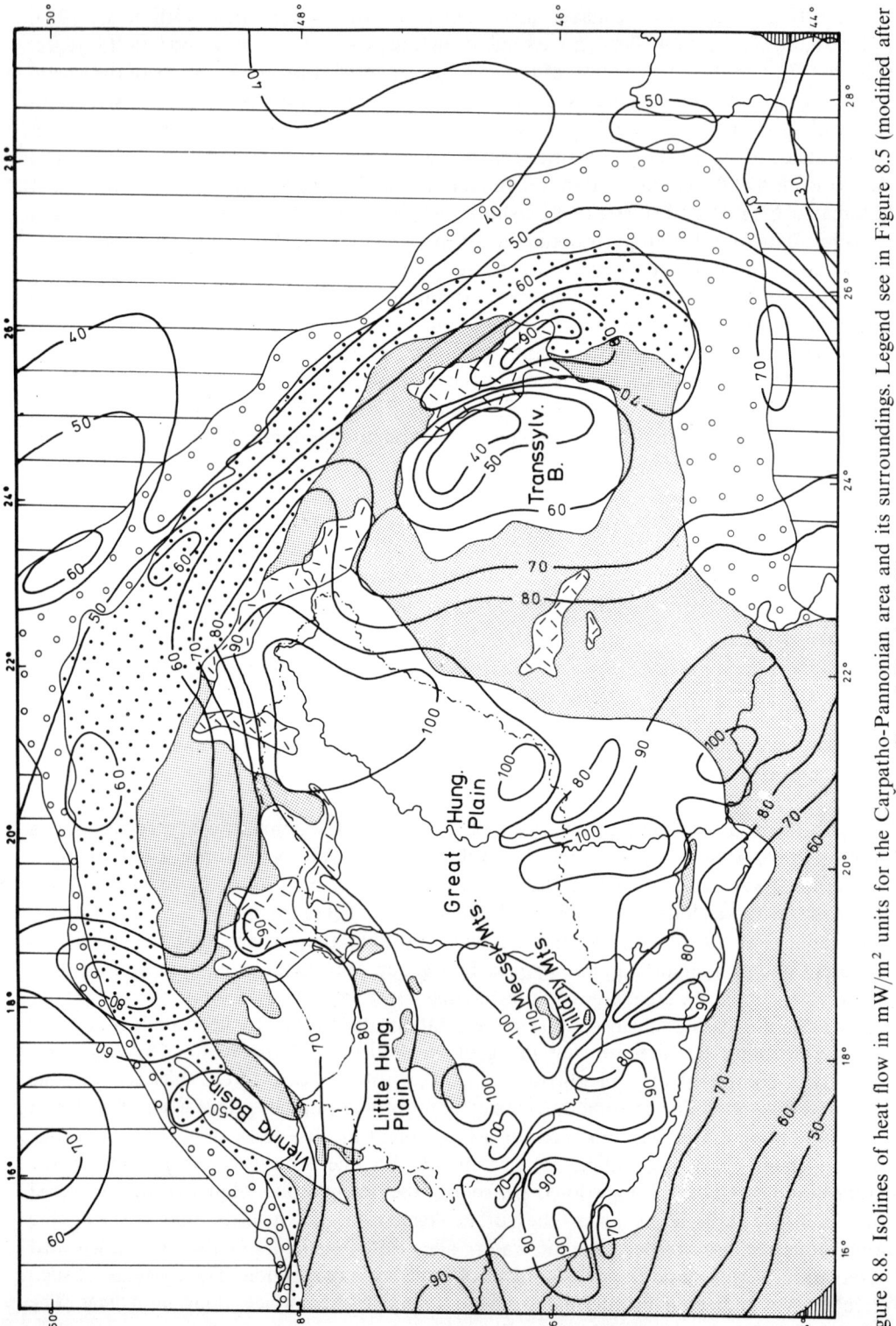

Figure 8.8. Isolines of heat flow in mW/m² units for the Carpatho-Pannonian area and its surroundings. Legend see in Figure 8.5 (modified after Čermák and Hurtig, 1979)

Table 8.3. Mean heat flow data for the Intra-Carpathian region

Area	Mean heat flow (mW/m^2)	Standard deviation (mW/m^2)	Number of data
Pannonian Basin	90·4	17·6	38
Great Hungarian Plain	94·5	16·5	31
Little Hungarian Plain	72·3	8·8	7
Vienna basin	49·7	7·0	8
Transsylvanian basin	47·0	16·2	6

Table 8.4. Porosity and hydraulic conductivity of thermal water-bearing formations in the Pannonian basin

Epochs and stages	Type of rocks	Bulk porosity (per cent)	Hydraulic conductivity (10^{-6} m/s)
Quaternary and Pleistocene	Gravel and sand	30–40	10–50
Pliocene (Upper and Middle)	Sand	25–30	10–20
Upper Pannonian	Sand, sandstone	20–30	5–10
Lower Pannonian	sand, sandstone, conglomerate	10–20	2–5
Miocene (Middle and Lower)	Sandstone, limestone conglomerate	5–15	2–5 10–200
Oligocene	Sand, sandstone		0·5–10
Eocene	Dolomite, limestone breccia	0–5	3–100 20–80
Triassic	Dolomite, limestone	0–3	3–30

Jurassic rocks occur only to a limited extent, mostly developed in a limestone facies in sedimentary continuity with the Triassic. The Jurassic formations form a common hydrologic system with the Triassic.

Cretaceous rocks are mainly represented by series of flysch facies in the basement of the basin. Where Cretaceous limestones have been recognized they are not independent hydrologically from the underlying limestones.

The basin fill proper starts at many places with Eocene formations transgressing on Triassic. These consist in their main mass of terrestrial, littoral or shallow marine marly/limestone formations. From the hydrogeological point of view, the Eocene and Triassic limestone series form a single karstic water system in many areas. Thermal waters have been encountered in Eocene limestone in many places.

The extension of Oligocene formations is limited. These are represented by conglomerates, sandstones, thick clay and sand strata transgressing on the Eocene or immediately on the Triassic. The hydrogeological role of the Oligocene formations is not significant; nevertheless they give thermal water at a few places.

The extension of Miocene strata is greater than that of the proceeding ones. In a few places the starting member is a transgressive conglomerate deposited on the basement. The series shows variable facies. There are in the series littoral limestones, clays and marls, volcanic tuffs and flows. The sedimentary features indicate a transition in fresh-water character toward the upper parts of the series. The conglomerates and limestones are good water storing rocks. The formations are not very extensive owing to their littoral character and so they do not furnish a significant quantity of water even in the case of a favourable porosity or of promising hydrogeological conditions.

The Pannonian series is the most significant one among the sediments of the basin as regards both their thickness and their extension. The entire series can stratigraphically be divided into two parts. The lower part, the Lower Pannonian series, which belongs to the Upper Miocene (Table 8.2), consists predominantly of clay and sandy clay with interbedded more porous lenses at places. In the upper part of the Lower Pannonian series, the fine sandy fraction becomes more conspicuous. In the Upper Pannonian series, the proportion of sandy strata increases to as much as 50 per cent. The thickness of individual sand layers varies between 2 and 25 m. Although these layers usually are not of great lateral extension, they are frequently interconnected, and, accordingly, large water-storage systems are often encountered. The total volume of porous water-bearing sediments lying between 500–2500 m is estimated to be about 9000 km^3, giving pore space of 1800–2700 km^3 (see Table 8.4).

During the Pleistocene, sand and gravel predominantly of fluvial origin were deposited in young depressions. The thickness of these clastic layers amounts to 400–700 m in some areas. The porosity and permeability of the detrital layers is very favourable (Table 8.4).

8.4.1. *Water Migration Systems*

As mentioned in the previous section, the subsurface waters of the Pannonian basin are not in hydrostatic equilibrium, in general. A little surplus pressure exists; as a consequence of this a certain water migration exists in the sediments, and the wells are of artesian character. Considering the movement of water from the surface to depth in large, closed sedimentary basins, three zones are to be distinguished generally. In the Pannonian basin, these may be characterized as follows:

1. Zone of infiltration: the uppermost 300–700 m of sediments, approximately to the bottom of the Upper Pannonian sediments. In this zone, there is a free communication between meteoric and subsurface waters. As a consequence, fresh waters and approximately hydrostatic pressures predominate in this zone; slight hydraulic pressures are controlled by topography.
2. Zone of expellation: below the zone of infiltration to a depth of 1–3 km, approximately to the bottom of Neogene. In this zone the water migration is controlled by the compression and consolidation of sediments; the water is remnant old sea-water expelled from the marine sediments by compaction and by consolidation and moves upwards and sidewards. Mineralized waters and local or regional overpressures characterize this zone.
3. Zone of reinfiltration: At greater depths, the compaction of the rocks comes to end, and the consolidated rocks are capable of a secondary infiltration. Beneath the Pannonian basin, this zone exists (if at all) in the Mesozoic and older basement of the basin. No sufficient experience exists to characterize this zone; hydrostatic pressures and mixed waters are supposed to be present.

The three zones mentioned above give but a generalized picture for the water regime of the Pannonian basin; the boundaries between the three zones are very probably quite diffuse and changed by local effects such as faults and geometry of various rocks.

The most obvious method to determine the subsurface water movements is the use of pressure measurements to determine the hydraulic gradients, either directly (that is, to measure pressure in boreholes) or indirectly (that is, to measure the height of the

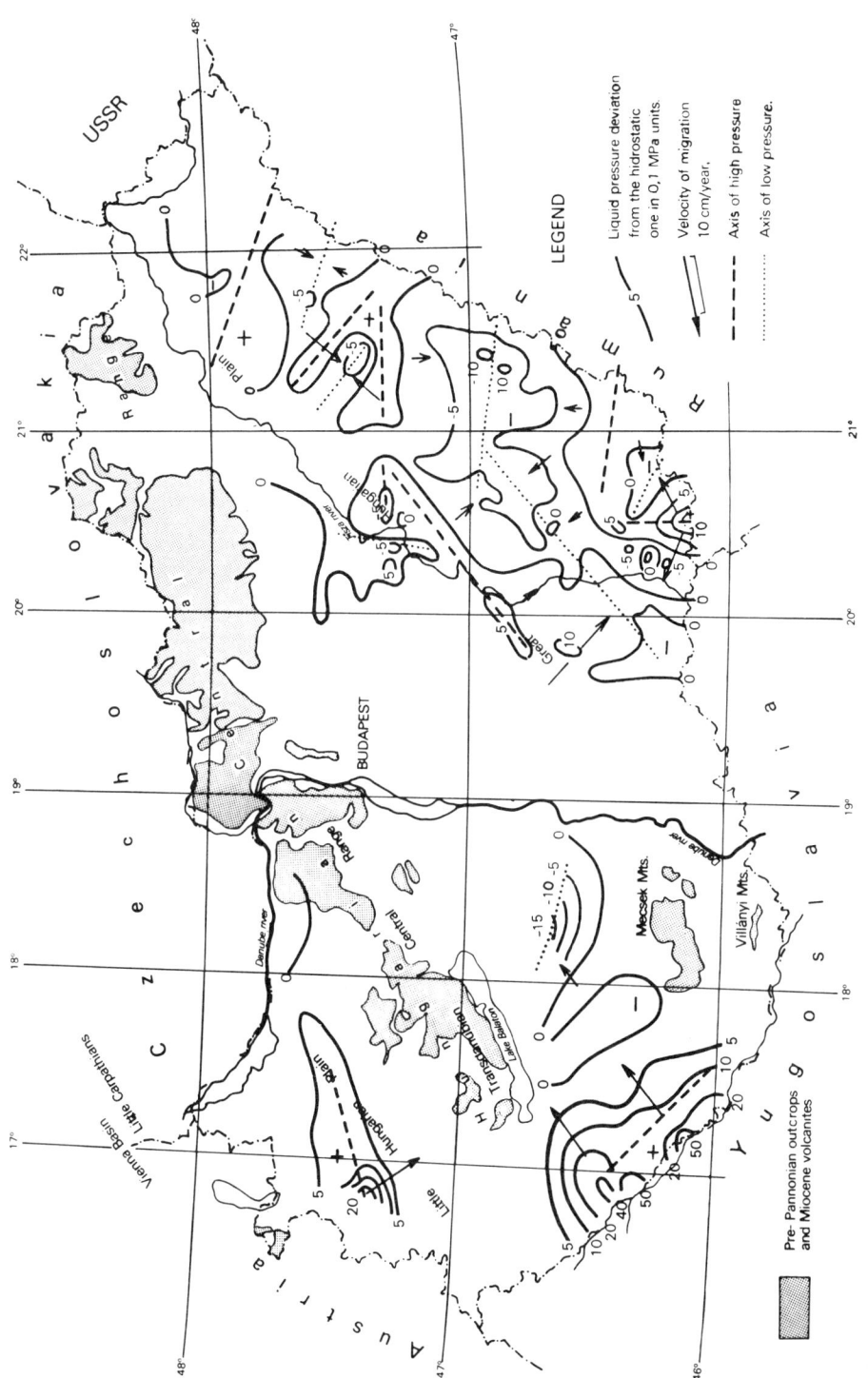

8. The Low Enthalpy Geothermal Resource of Hungary 237

Figure 8.9. Dynamic pressures and migration velocities at the bottom of the Lower Pannonian sediments. Velocities are calculated using hydraulic conductivity of 10^{-6} m/s

equilibrium water level). Both methods are more or less uncertain, because of disturbing factors like water salinity, temperature, mud-content etc., and because of the relatively little differences between hydrostatic and piezometric pressures.

Collecting all the data of pressure measurements, an attempt was made to give an overview of the hydraulic pressure conditions and, based on this, to estimate the water-migration directions for the central part of the Pannonian basin (Fig. 8.9).

8.4.2. Hydrochemistry

The chemical composition of the deep waters in the Pannonian Basin corresponds to the world-wide phenomenon that mineralization increases with increasing depth.

The ground-water originating from precipitation starts as bicarbonate water and changes, when migrating downwards, through a sequence of bicarbonate+chloride to chloride+bicarbonate to chloride+sulphate or sulphate+chloride and finally to a predominantly chloride water approaching the composition of the sea-water.

In the Pannonian Basin, the total dissolved solid content of ground-water increases from the Upper Pannonian (4–6 g/L) downwards to the Miocene (10–15 g/L). The average value both of the NaCl content and of ratio Na/Ca versus depth is represented in Fig. 8.10.

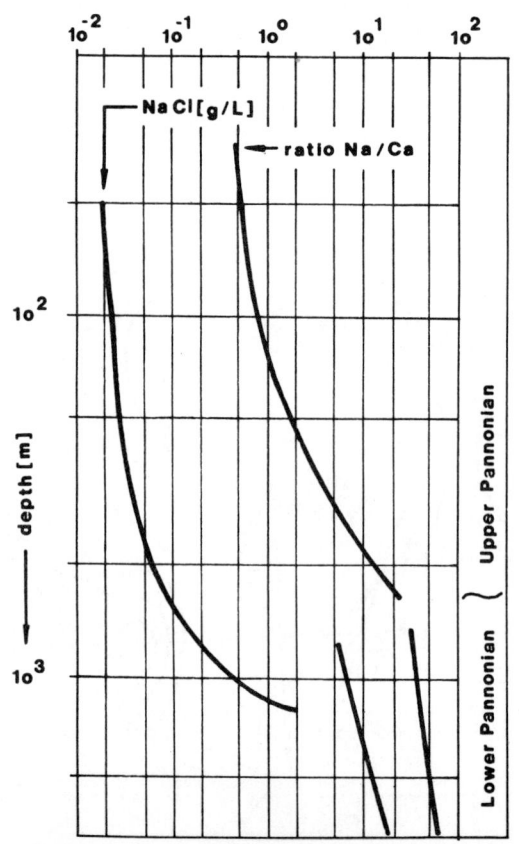

Figure 8.10. The variation of the average NaCl content (in g/L) and of the Na/Ca ratio vs. depth of the groundwater in the Pannonian basin. The components are calculated in equivalent weight (Kleb, 1971)

8. *The Low Enthalpy Geothermal Resource of Hungary* 239

The vertical hydrochemical zonality exists only where no movement of water occurs. The sequence of composition change depends on the magnitude of flow of water. Consequently, the flow-path of the groundwater may be traced when the chemical composition of ground-water is measured. The areas of deep ground-water discharge can be delineated, therefore, as local maxima of chloride content in an environment of shallow groundwater (Fig. 8.11).

8.5. The Recovery and Utilization of the Geothermal Energy

The geothermal aquifers within the Pannonian Basin are divided into two parts, according to their lithological pattern, hydrogeological characteristics and areal extent: on one hand to great regional systems, and on the other to small local systems.

The dominant type is the Upper Pannonian reservoir system (Fig. 8.12), which extends nearly through the entire Hungary, even beyond its border region to the neighbouring countries (Austria, Czechoslovakia, Rumania, Yugoslavia). It includes a vast geothermal energy resource in the form of low enthalpy waters. About 65 per cent of the total thermal

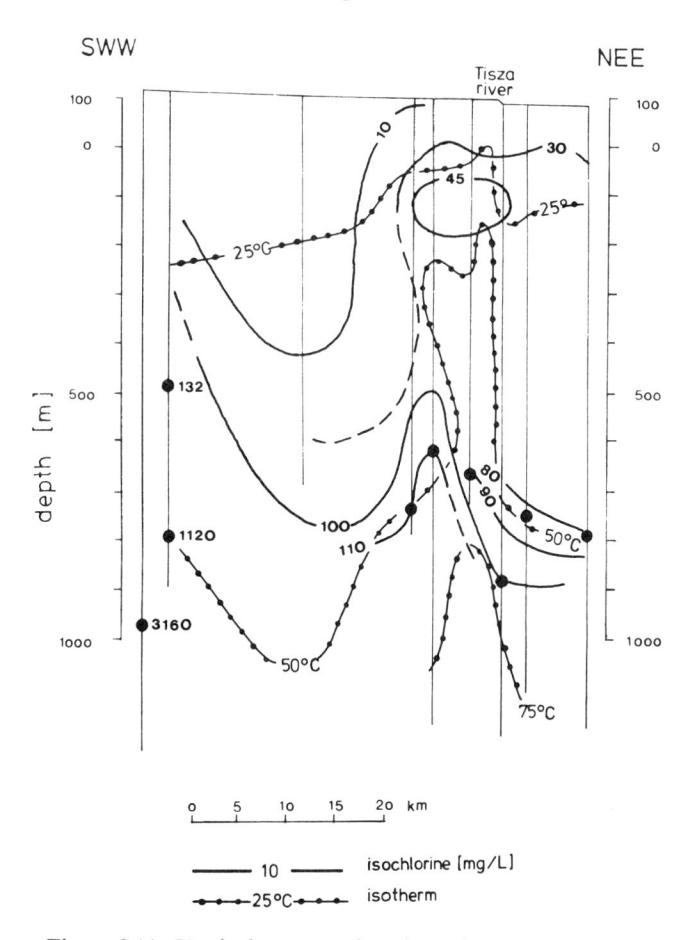

Figure 8.11. Vertical cross-section through a geothermal maximum showing the variation of chloride content (mg/L) and of temperature (°C) vs. depth in eight thermal wells. A local maximum of 45 mg/L chloride content indicates the cross-section of the path of upward migrating thermal water in the Lakitelek (see Figure 8.1) region (after Erdélyi, 1979)

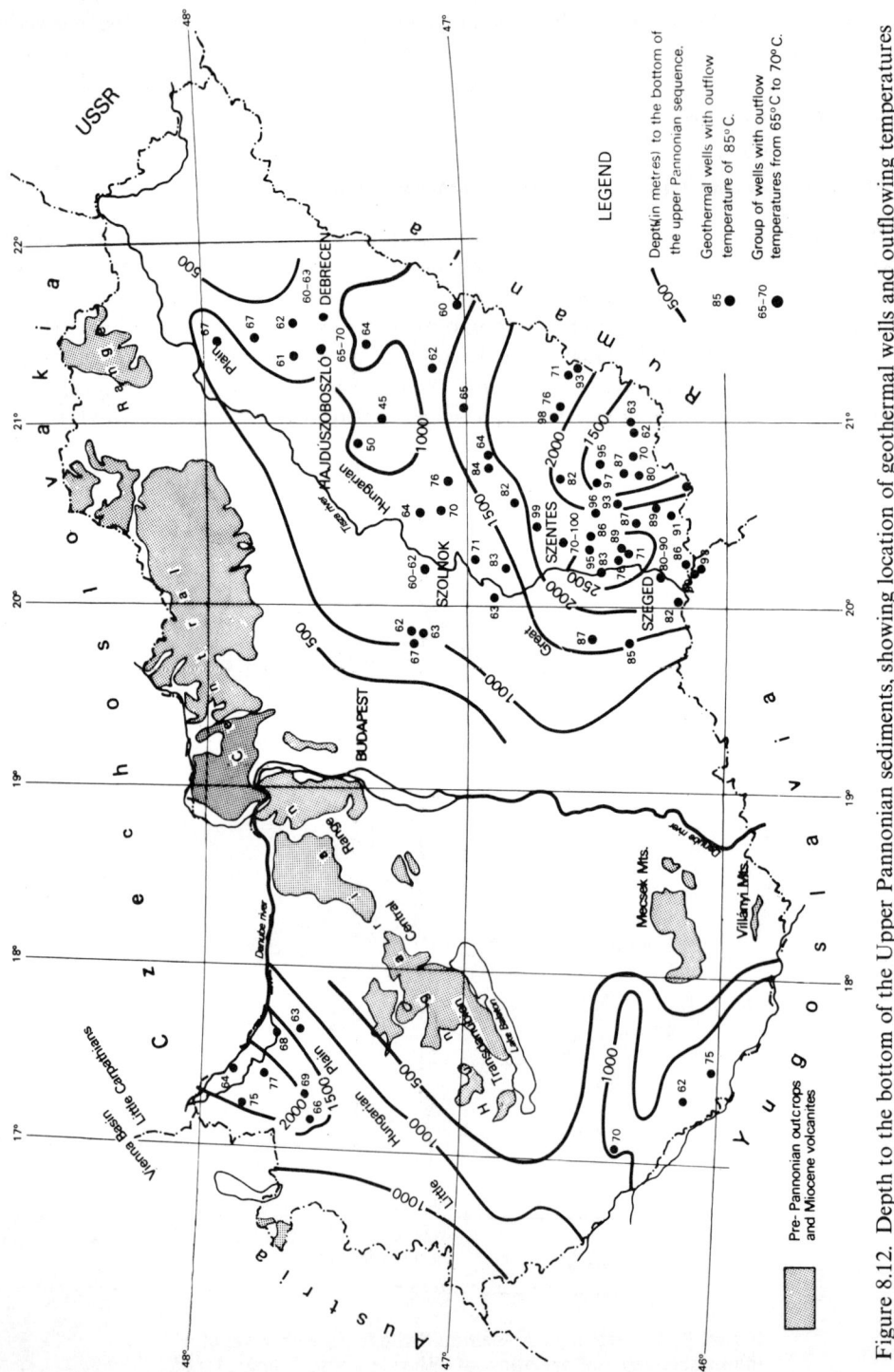

Figure 8.12. Depth to the bottom of the Upper Pannonian sediments, showing location of geothermal wells and outflowing temperatures (in °C)

wells are producing from these formations. For the quantity of thermal water (in km^3) and for the thermal energy (in Joule) stored in the upper three kilometers, Boldizsár (1978) gives the following figures:

Tertiary: 1·60 10^5 km^3 sediment stores 4·4 × 10^{22} J energy
Pannonian, depth < 1 km: 0·38 10^5 km^3 sediment stores 0·8 × 10^{22} J energy
Pannonian, depth > 1 km: 0·18 10^5 km^3 sediment stores 0·3 × 10^{22} J energy

The average yield of the Upper Pannonian thermal wells is about 1500 L/min. The total dissolved solid content of the water ranges from 2000 to 5000 mg/L.

The other main thermal water reservoir system occurs within the fractured-fissured carbonate rock complex of Triassic age. The upper part of this carbonate rock formation is intensively karstified including numerous paleokarst horizons. This highly heterogeneous and anisotropic thermal water reservoir system with its mainly vertical or quasi-vertical flow-paths has a less extension and minor significance in comparison to the Upper Pannonian system. About 20 per cent of the thermal wells in Hungary are producing from Triassic carbonate rock formations. The yield of these wells is highly varied and ranges from a few hundred liters to many thousands of liters per minute. Their total dissolved solid contents are also varied and represent a Ca–Mg bicarbonate water type. The Triassic carbonate thermal water reservoirs also represent low-enthalpy thermal water resources.

The small local thermal water reservoir systems are rather isolated and limited in extension. They comprise many geologic and hydrostratigraphic horizons from Devonian dolomite formations to Quaternary sediments.

There is a wide range of utilization according to the areal distribution and hydrogeological conditions. The thermal wells are selectively suitable for all kinds of thermal water utilization from balneology to district heating. The most remarkable and most significant branch is the agricultural utilization highly developed especially in the southern part of the Great Hungarian Plain where good geologic, hydrogeologic and hydrodynamic conditions are prevailing. As a consequence, the highest density of thermal water wells can be found here.

Thermal water exploited from the Triassic carbonate system is utilized mainly for balneological purposes in the well known bathing resorts, spas and therapeutical centres, for example, at Budapest, Héviz, Harkány (Fig. 8.1). Similarly, the small local thermal reservoirs are the sites of balneological use, and in some parts of the country, very important bathing resorts occur (e.g. Bük, Fig. 8.1).

The distribution of different kinds of thermal water utilization is shown in Table 8.5.

8.5.1. The Main Branches of the Thermal Water Utilization

Balneological and therapeutical use. There are a total of 214 thermal wells for all kinds of balneological purpose (see Table 8.5). Due to the abundance of thermal water resources and the great number of thermal wells, a highly developed balneological network was established in the whole country. The most famous bathing centers and resorts are Budapest, Héviz, Harkány, Hajduszoboszló, Gyula, Bük, Debrecen and Zalakaros (Figure 8.1). In Budapest, for example, there are 28 thermal wells and nearly 100 individual natural thermal springs. The majority of bathing facilities (indoor baths as well as outdoor swimming pools) occur in the Great Hungarian Plain. At present there are 145 bathing resorts in Hungary.

District and space heating. The first apartment heating projects were the buildings

Table 8.5. Utilization of thermal water by wells as by January 1978 (see Figures 8.1 and 8.12)

Kind of utilization	Number of wells with temperatures	
	From 35 to 100°C	Higher than 60°C
Balneology	214	50
Drinking water	143	2
Agriculture (greenhouses)	84	63
Industrial use	15	2
For secondary oil recovery (repressuring)	10	4
District heating	9	8
Observation wells (for hydrologic studies)	11	—
Shut-down or abandoned	74	21
Total	560	150

around bathing centers in Budapest, Harkány, Hajduszoboszló, Debrecen, Gyula, Szolnok and Györ (Figure 8.1). The primary use of the thermal water was here for balneological purposes, and only the secondary use was for building heating related to the bathing complex.

In 1959, the municipal hospital of Szentes was supplied by thermal water of outflowing temperature of 79°C for heating the wards and other rooms, and later in the sixties a program for hospital and district heating with geothermal energy was carried out.

A district heating project comprising 1000 flats in Szeged was started in 1962 by the use of thermal water of 89°C coming from a nearby well drilled for this special purpose to 1900 meters. This initiative was followed at the university in Szeged, then in the hospitals of Hódmezövásárhely and Makó as well as in a ceramic plant in Hódmezövásárhely (Figure 8.1).

The estimated calorific value of the geothermal energy used for district and apartment heating is about 1.25×10^{14} J/year ~ 4 MWt.

Agricultural use of geothermal energy is the most important branch of the thermal water utilization in Hungary, especially for horticultural purposes. At present, the majority of horticultural facilities heated by geothermal energy in the world occurs in Hungary. Greenhouses totalling about 500,000 m^2, as well as plastic tents, tunnels, and soil heating of about 1,200,000 m^2, are supplied by thermal water. About 80 per cent of all existing greenhouses in the southern part of the Great Hungarian Plain are heated by geothermal energy.

In the horticultural use of thermal water, the greenhouses are first heated with thermal water of a temperature ranging from 60 to 100°C. Then the outgoing thermal water of lower temperature is introduced to the plastic tents and tunnels. Finally the soil heating phase follows.

In addition, there are about 50 facilities for animal husbandry, mainly in southern Hungary (chicken, pig, calf, etc.), which are heated by thermal waters, among them fishponds which are supplied partly by thermal water, enabling more rapid growth. More recently a few plants for agricultural drying process were established in south Hungary.

In exceptional cases some thermal water, where the total dissolved solid content is less than 2000 mg/L, is used for irrigation.

There are 11 thermal wells in the Great Hungarian Plain which supply water of 40–50°C for hemp-processing plants.

The estimated total heat production referred to $+15°C$ in the field of agricultural utilization is about 1.55×10^{16} J/year ~ 500 MWt.

Thermal water used as domestic water is supplied mainly in the same districts and apartments as well as hospitals which are heated at the same time by geothermal energy. At present, 3000 flats use 2700 m^3/day thermal water as domestic water, whose energy per year equals about 1.25×10^{14} J/year $\sim$ 4 MWt.

Multipurpose utilization of the thermal water is a highly desirable objective which was realized in many parts of Hungary. Earlier, in the period between the two World Wars, in Hajduszoboszló (Figure 8.1), beside the balneological as primary use, thermal water was utilized for horticulture and for bottling mineral water. In addition, the water had initially a high dissolved methane gas content (gas–water ratio was 1.8 m^3/m^3), which was separated and used for a small power plant. Later the gas content was depleted, especially after the discovery and production of a nearby gas-field.

In Budapest, two wells supplied thermal water not only for balneological/therapeutical facilities but also domestic water for the neighbouring apartments, hospitals and for the zoo.

A remarkable example of the multipurpose utilization of thermal water is the municipal hospital of Szentes (Figure 8.1). The hospital was heated and supplied simultaneously with domestic water, while the nearby municipal bath and swimming pool as well as the neighbouring greenhouse complex were supplied by the same thermal water.

Another example for the multipurpose utilization is in the Little Hungarian Plain, where thermal water of 64°C supplies greenhouses, plastic tents, bathing facilities, fishponds and an irrigation system.

The most favorable region for multipurpose utilization is the southern part of Hungary. Near Szeged, two thermal wells supply the heating of apartments, greenhouses, plastic tunnels, domestic water and dry processing plants.

In Szeged (Figure 8.1), at the above mentioned site, where 1000 flats are heated and 400 flats are supplied by domestic water through the geothermal energy, the nearby swimming pool is also supplied by this thermal water.

The multipurpose utilization of the thermal water in cascades is the best way to eliminate any wasting of the available geothermal energy.

Thermal water as mineral water for bottling: Many thermal waters are suitable for drinking or for certain therapeutical purposes due to their chemical composition. Well known mineral water occurs in Budapest, Harkány, Bük, Szeged, Szolnok, Hajduszoboszló, Debrecen and Csokonyavisonta. Well known medicinal water is the 'Salvus' in Bükkszék (Figure 8.1).

8.5.2. Problems Related to the Operating of Thermal Wells

The majority of the thermal wells have outflowing water due to sufficient reservoir energy especially at the beginning of the production. Later, during the production, the yield of the wells will decrease, and later (after years) the water-level will fall below the ground surface. This calls for artificial production by air lifting or pumping. In some exceptional cases gas-lift is also applied.

The decreasing yield is caused not only by the depletion of the reservoir energy but also by other factors. The aging of the well structure can cause such problems. The hazard of sanding is also a common phenomenon since the deep wells (more than 1700 to 1800 m deep) were completed by perforations without screens. The lower portion of the section yielding thermal water is often filled by the intruding fine sand, thus causing a decline in the thermal water production. In this case a workover job is needed.

Another frequently encountered problem of thermal wells is scaling. The rate of precipitation and deposition of $CaCO_3$ dissolved in thermal waters is highly different. In certain thermal wells the rate of scaling is so high (1 to 2 mm per day) that the thermal water production will be greatly reduced or even stopped after a continuous production over one or two weeks. Among the more than 500 operating thermal wells in Hungary, there are 47 wells where scaling occurs and causes production problems.

Scaling appears usually in the upper portion of the well to a depth of 20 to 80 meters, exceptionally to 200 meters. In some thermal waters of corrosive character, the dissolution of the metallic surface of the casing results in corrosion products which will be mixed with the precipitation and will produce ultimately a protective coating against further corrosion.

Tests of scaling control carried out in the last decade and a half showed that every well presents a unique, individual problem and there is no general solution. The most reliable scaling control is the chemical treatment using hydrochloric acid and inhibitors. Another effective method for scaling control is the stabilization of the production of the thermal water by pressure control, which maintains the carbonate equilibrium in the thermal water.

The previous methods for scaling control, the mechanical one by drilling or replacing of the filled tubings, is nowadays progressively abandoned.

8.5.3. High Enthalpy Reservoir Possibilities in the Pannonian Basin

As a result of the mantle diapirism and the associated Miocene and Pliocene volcanism, the sediments of the basin, the crust, and the upper mantle are still hot today. This fact provides the theoretical basis to assume that at greater depth, reservoirs of high enthalpy might be present. The mean temperature over the area of the Pannonian basin at 1 km depth is about 70°C, at 2 km about 120°C and at 3 km it can be taken—mainly by extrapolation—as 165°C. So at a depth of 4–5 km temperatures might be reached, that can be used for producing electric energy (the lower limit for this can be assumed at present to be 180°C).

At present, studies are going on with the task of detecting reservoirs of high enthalpy. Methods applied are as follows: (a) resurveying of the geothermal temperature observations and carrying out of new measurements, (b) regular accomplishment of heat-flow measurements, (c) carrying out geoelectric measurements of high peneration; reinterpretation of the measurements made to date, (d) widespread application of SiO_2 and Na–K–Ca geochemical thermometers, (e) more exact delimitations of deep-seated volcanites (by means of magnetic interpretation) and more exact age determinations, and (f) reassessing of borings of great depth (4–6 km).

References

Ádám, A., 1976. 'Results of deep electromagnetic investigations.' In *Geoelectric and Geothermal Studies*, A. Ádám, (Ed.), Academic Press, Budapest. 547–560.

Ádám, A., Horváth, F., and Stegena, L., 1979. 'Geodynamics of the Pannonian basin: Geothermal and electromagnetic aspects', *Acta Geol., Acad. Sci., Hung.*, Budapest.

Bisztricsány, E., and Egyed, L., 1973. 'The determination of LVL depth from data of spaced seismological stations', *Geophys., Trans., Hung. Geophys. Inst. R. Eötvös*, 21, 1–4.

Boldizsár, T., 1978. 'Geothermal energy from hot rocks.' In *Proc. Nordic Symp. on Geothermal Energy*, Ch. Svensson and S. A. Larson (Ed.), 46–51, Göteborg.

Bullard, E. C., 1965. 'Historical introduction to terrestrial heat flow'. In *Terrestrial heat flow*, W. H. K. Lee (Ed.), American Geophysical Union Monograph No. 8, 1–6.

Buntebarth, G., 1976. 'Temperature calculations on the Hungarian seismic profile section NP-2,' In *Geoelectric and Geothermal Studies*, A. Ádám, (Ed.), Academic Press, Budapest, 561–566.

Čermák, V., and Hurtig, E., 1979. 'Heat flow map of Europe'. Annex to Čermák, V. and Rybach, L. (Eds.), *Terrestrial Heat Flow in Europe*, Springer-Verlag, Berlin, 328 p.

Čermák, V., and Rybach, L., (Eds.), 1979. *Terrestrial Heat Flow in Europe*, Springer-Verlag, Berlin, 328 p.

Cull, J. P., 1974. 'Thermal conductivity probes for rapid measurements in rock', *Journ. E. Sci. Intr.*, 7, 771–774.

Erdélyi, M., 1979. 'Hydrodynamics of the Hungarian basin', *Proceedings of the Research Institute for Water Resources Development*, Budapest.

Horváth, F., and Stegena, L., 1977. 'The Pannonian basin, a Mediterranean interarc basin.' In *Structural History of the Mediterranean Basin*, B. Biju-Duval and L. Montadert, (Eds.), Ed. Technip, Paris.

Horváth, F., Bodri, L., and Ottlik, P., 1979. 'Geothermics of Hungary and the tectonophysics of the Pannonian basin "red spot".' In *Terrestrial Heat Flow in Europe*, V. Čermák and L. Rybach (Eds.), Springer-Verlag, Berlin, p. 206–217.

Kleb, B., 1971. 'Stratigraphical and geochemical study of the Pannonian sediments in respect to the change towards fresh-water character.' In *Investigations of the Formations of Pannonian age in Hungary*, Budapest (in Hungarian).

Korim, K., 1978: 'The thermal wells of Hungary III', Ed. VITUKI, Budapest (in Hungarian).

Rögl, F., Steinger, F. F., and Müller, C., 1978. 'Middle Miocene Salinity crisis and Paleogeography of the Paratethys (Middle and Eastern Europe).' In *Initial Reports of the 6. Deep Sea Drilling Project*, K. Hsü, Montadert et al. (Eds.), 42(1), 987–990.

Rónai, A., 1978. 'Hydrogeological Peculiarities of the Great Hungarian Plain.' In *Hydrogeology of Great Sedimentary Basins*, Annales of the Hungarian Geological Institute, LIX. 1–4 (in French).

Scott, R. W., Fountain, J. A., and West, E. A., 1973. 'A comparison of two transient methods of measuring thermal conductivity of particulate samples', *Rev. Sci. Instrum.*, 44, 1058–1063.

Stegena, L., Géczy, B., and Horváth, F., 1975. 'Late Cenozoic evolution of the Pannonian basin', *Tectonophysics*, 26, 71–90.

Stegena, L., 1976. 'Geothermal temperature map of Central and Eastern Europe.' In *Geoelectric and Geothermal Studies*, A. Ádám, (Ed.), Academic Press, Budapest, 381–383.

Geothermal Systems: Principles and Case Histories
Edited by L. Rybach and L. J. P. Muffler
© 1981 John Wiley & Sons Ltd.

9. Exploration and Development at Takinoue, Japan

HISAYOSHI NAKAMURA
Japan Metals and Chemicals Co., Ltd.

KIYOSHI SUMI
Geological Survey of Japan

9.1. Introduction

The Takinoue area is located in northeastern Japan (Figure 9.1) about 50 km northwest of the city of Morioka. Known as a hot-spring bath area from olden times, Takinoue is situated along the Kakkonda River and occupies the southern corner of the Towada–Hachimantai National Park at an altitude of 650 m above sea level. Takinoue is 8.4 km southwest of the Matsukawa geothermal power plant, which is the first of its kind in Japan and was constructed by the Japan Metals and Chemicals Co., Ltd. (JMC) in 1969. Unfortunately, however, there are no roads connecting the Matsukawa area with Takinoue over the mountains.

Geothermal exploration by JMC in the Takinoue area was begun in 1958 and carried out with the cooperation of the Geological Survey of Japan (G.S.) for about 10 years. The purpose of this exploration was to collect information about the relation between geological structure and occurrence of geothermal fluid in the Takinoue area. Judging from the data from an underground temperature survey, an electric resistivity survey, and a geochemical survey of the fumarolic gases and thermal waters, it was concluded that geothermal development in this area would be feasible.

In accordance with the conclusion obtained from the G.S.–JMC cooperative investigations, a more detailed survey was undertaken by JMC from 1969 to 1972. Based on the results obtained from this detailed survey, drilling of six exploratory wells was begun in 1972 and completed in 1974. During this period, negotiations between JMC and Tohoku Electric Power Co., Ltd. resulted in a program of cooperative development in which JMC would supply geothermal steam for the electric power company, who would construct a power plant of 50 MWe.

Since drilling of exploratory wells had confirmed that it would be possible to tap geothermal steam for power generation of 50 MWe, drilling of production and reinjection wells was started in August of 1974. In August, 1977, drilling of 11 wells for production and 15 for injection had been completed. Construction of steam collection pipelines (3.5 km in total length) together with the power plant began in May, 1974 and was completed in November, 1977.

Successful test operation of the power plant began in December, 1977 using a method of gradually increasing output to 50 MWe. The power plant has been in full operation since the end of May, 1978.

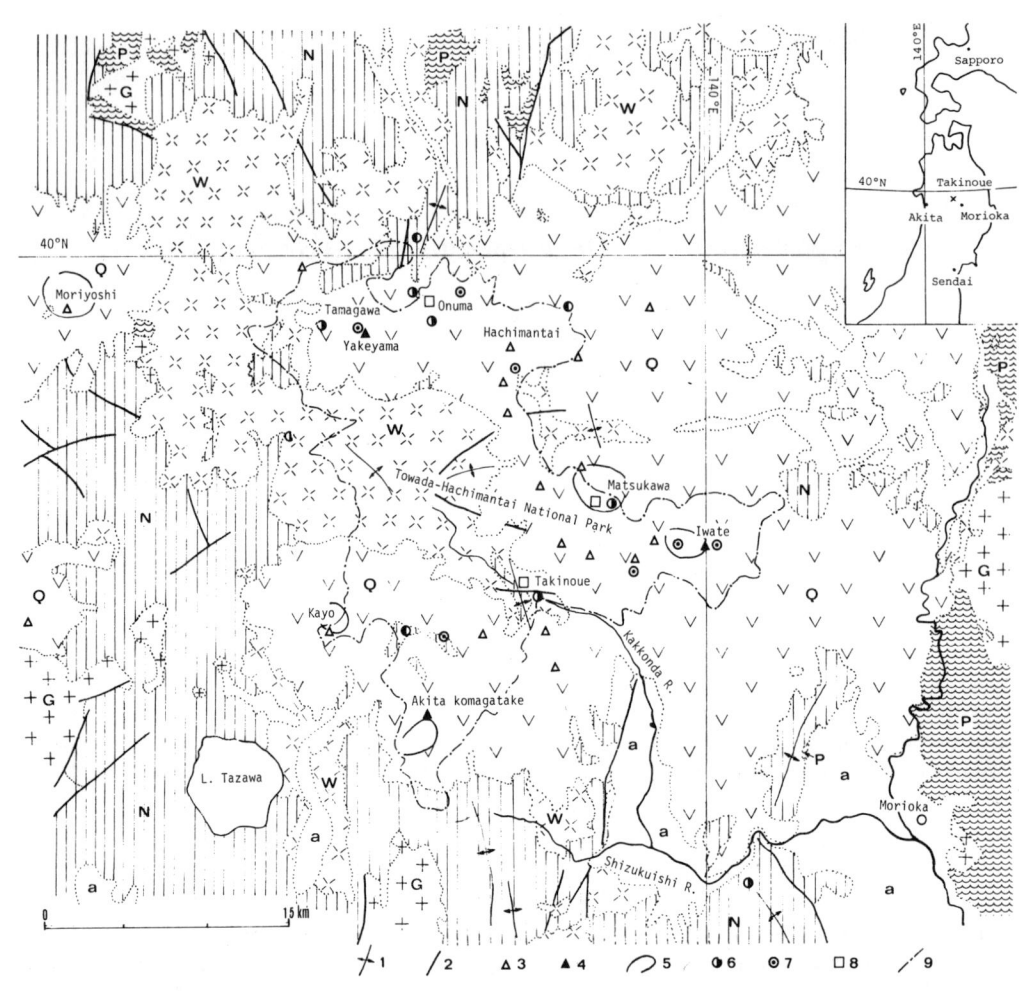

Figure 9.1. Geologic Map of the Hachimantai Volcanic region with distribution of geothermal areas. (P) Paleozoic sedimentary rocks; (G) Cretaceous granite, (N) Neogene rocks; (W) Tamagawa Welded Tuff; (Q) Quaternary volcanic rocks; (a) Quaternary sediments; (1) Folding axes; (2) Faults; (3) Extinct volcano; (4) Active volcano; (5) Caldera and crater; (6) Hot-spring area; (7) Steaming area; (8) Geothermal power plant; (9) National Park

Since the construction of the Kakkonda geothermal power plant (the name of the power plant constructed in the Takinoue area), the capacity of geothermal power plants in Japan has reached a total of 168 MWe (Matsukawa 22, Otake 11, Onuma 10, Onikobe 25, Hatchobaru 50, and Kakkonda 50 MWe). Besides these sites, a 50 MWe power development is being carried out in Nigorikawa on Hokkaido by the Donan Geothermal Energy Co., Ltd., with the assistance of JMC's engineers.

For the purpose of geothermal development in Takinoue, one billion yen (about five million U.S. dollars), which corresponds to about 10 per cent of the total construction cost of the steam supply system, was provided to JMC from the Research and Development Corporation of Japan (part of the Governmental Agency of Science and Technology) as funding for developing new technology related to exploration, drilling, and production.

In this chapter, the history of exploration (which is summarized in Table 9.1) and the production characteristics are described.

9. Exploration and Development at Takinoue, Japan

Table 9.1. History of geothermal exploration and development in the Takinoue area

	Activity	'58	'59	'60	'61	'62	'63	'64	'65	'66	'67	'68	'69	'70	'71	'72	'73	'74	'75	'76	'77	'78	'79	Remarks
(GS)	Geological and Geochemical Reconnaissance Survey	─																						GS-JMC joint survey
(GS)	Underground Temperature Survey at 30m Deep			───																				"
(GS)	Electric Resistivity Survey																							"
(JMC)	Drilling of AT-1					──																		"
(GS)	Drilling of GSR-2										─													"
(GS)	Temperature and Pressure Logging in GSR-2											─												"
(JMC)	Detailed Geological Survey												────────											Investigations by JMC for geothermal development
(JMC)	Rock Alteration Survey												────────											"
(JMC)	Chemical Analyses of Fumarolic Gases and Thermal Waters														────									"
(JMC)	Measurements of Soil Gases at 1m Deep													──										"
(JMC)	Electric Resistivity Survey														──									"
(JMC)	Underground Temperature Survey at 10m Deep													────										"
(JMC)	Measurements of Heat Flow														──									"
(JMC)	Seismic Survey																	──						"
(JMC)	Drilling of Exploration Wells														────────────									"
(JMC)	Drilling of Production Wells																	───────						Production of geothermal steam by JMC
(JMC)	Drilling of Re-injection Wells																	─────						"
(JMC)	Construction of Steam Gathering Pipelines																	────						"
(TEP)	Construction of Power Plant																	─────						Power generation by TEP
(TEP)	Test Run of Power Generation																					─		"
(TEP)	Operation of Power Plant																						──	"

GS: Geological Survey of Japan   JMC: Japan Metals & Chemicals Co., Ltd.   TEP: Tohoku Electric Power Co., Ltd.

9.2. Geological Framework

9.2.1. *General Geology of the Hachimantai Volcanic Region*

The Takinoue geothermal area is located in the central part of the Hachimantai volcanic region, which is one of the largest Quaternary volcanic fields in Japan with an area of more than 800 km². Geological formations in the region are classified into five stratigraphic units, from oldest to youngest: Paleozoic sedimentary rocks, Cretaceous granite, Neogene rocks, Pliocene–Pleistocene Tamagawa Welded Tuff, and Quaternary volcanoes. A schematic geologic map of the region is shown in Figure 9.1.

The sedimentary rocks and the Cretaceous granite are distributed randomly in the peripheral areas of the region. However, the central part of the region is covered with thick Cenozoic formations, the thickness of which reaches 500–1000 m even at marginal areas (Kitamura and Onishi, 1972). Judging from the xenoliths contained in the volcanic rocks of Quaternary age, the basement rocks beneath the Cenozoic formations are considered to be composed mainly of chert, slate, sandstone and limestone rather than granite.

The Neogene rocks are divided into the 'Green Tuff' of early to middle Miocene age and a sedimentary sequence of middle Miocene to Pliocene age.

The 'Green Tuff' is characterized by altered lavas and pyroclastic rocks with thin intercalated sedimentary beds and is distributed in the northern, western, and southern parts of the region. In Takinoue, cores of the 'Green Tuff' have been collected from near the bottom of wells 1600 m in depth.

The Neogene sedimentary sequence shows the same distribution as the conformably underlying 'Green Tuff,' although it is locally exposed in Takinoue as an inlier. The lower part of the Neogene sedimentary sequence is characterized by muddy facies with the fossils indicating a past open-sea environment. On the other hand, the upper part of the Neogene sedimentary sequence shows sandy and comglomeratic facies with the fossils of mollusca indicating a subcoastal environment. The Neogene sedimentary sequence can be correlated with the oil-bearing rocks found in the coastal region of the Japan Sea, about 50 km west of Takinoue.

The Tamagawa Welded Tuff, which unconformably covers the Neogene sequence, is composed of thick volcanic piles or pyroclastics of rhyolite intercalated with thin sedimentary layers deposited in shallow lakes. According to age determination by the fission-track method, the age of the welded tuff has a range from one to two million years

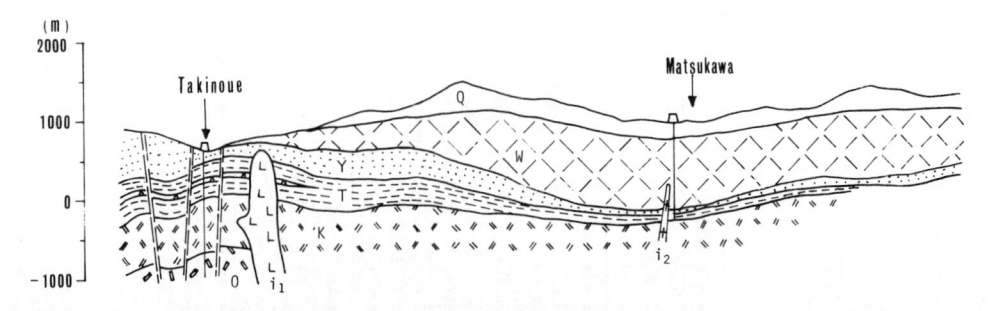

Figure 9.2. Geologic cross section through a line connecting Takinoue and Matsukawa. (O) Obonai Formation; (K) Kunimitoge Formation; (T) Takinoue-onsen Formation; (Y) Yamatsuda Formation, (W) Tamagawa Welded Tuff; (i_1) Dacite dike; (i_2) Andesite dike; Q: Quaternary volcanic rocks

(Tamanyu and Suto, 1978). The tuff has a thickness of several hundred meters, and in Matsukawa as much as 800 m, playing the role of a cap rock (Sato and Ide, 1975). The bottom of the tuff is situated at 200 m below sea level (Figure 9.2).

The Quaternary volcanoes are divided into six groups: Mounts Hachimantai, Yakiyama, Iwate, Akita-Komagatake, Moriyoshi and Kayo respectively. These are mostly andesitic or basaltic strato-volcanoes with the exceptions of Mt. Hachimantai (a shield volcano composed of andesitic lavas) and Mt. Yakeyama (having a small dacite dome on its summit) (Kawano and Aoki, 1958). Mounts Iwate and Akita-Komagatake have erupted lava flows and pyroclastics in historic time.

From the viewpoint of regional structure, the Hachimantai volcanic region is regarded as a large scale volcano-tectonic depression filled with the Tamagawa Welded Tuff and Quaternary volcanic rocks, with dimensions of 50 km in length (N–S) and 40 km in width (E–W). Peripheral areas of the region are composed of early and middle Miocene 'Green Tuff,' Cretaceous granite, and Paleozoic rocks.

Surface geothermal manifestations in the region are present throughout the volcano-tectonic depression, suggesting that the depression is favorable to hydrothermal systems. According to the results of a gravity survey carried out in the region, uplifted and basin structures are thought to be present in the basement, which is mantled with the younger sediments. Distribution of geothermal manifestations in the region seems to be related to structural directions of the basement rocks. Taking into account directions of the folding axes of the Neogene sedimentary rocks and distribution of the Quaternary volcanoes,

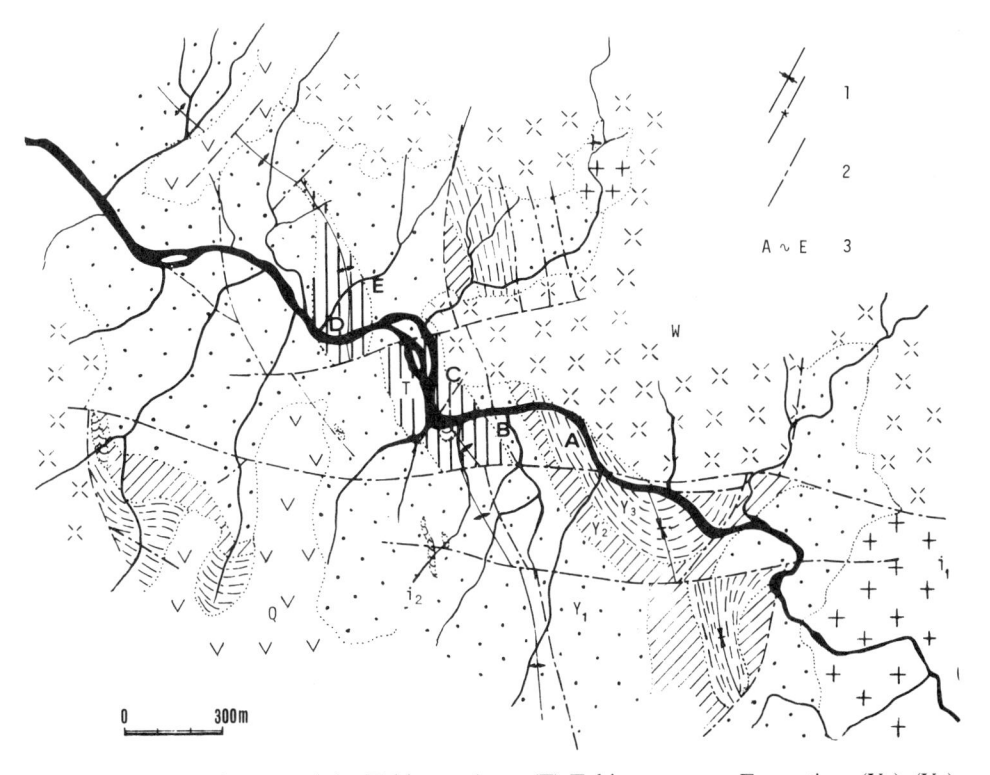

Figure 9.3. Geologic map of the Takinoue Area. (T) Takinoue-onsen Formation: (Y_1), (Y_2) and (Y_3) Yamatsuda Formation; i_1: Dacite dike; i_2: Andesite dike; (W) Tamagawa Welded Tuff; Q: Quaternary volcanic rocks; (1) Folding axes; (2) Faults; (3) Drilling site

fracture zones in the region are considered to be predominantly in NW–SE and E–W directions.

9.2.2. Geology in and around the Takinoue Geothermal Area

In Takinoue, the Neogene sedimentary sequence is exposed along the Kakkonda River as an inlier within the area covered with Tamagawa Welded Tuff and Quaternary volcanic rocks. The inlier was formed by a structural dome with NW–SE fold axes and by NW–SE and E–W faults (Figure 9.3).

Table 9.2 shows the stratigraphic sequence of the Takinoue area, compiled from both surface and drilling data. As seen in Table 9.2, the Neogene rocks are subdivided into Obonai, Kunimitoge, Takinoue-onsen and Yamatsuda Formations (from lower to upper). The former two and the latter two formations correspond respectively to the 'Green Tuff' and the 'Neogene sedimentary sequence' mentioned above.

The geothermal fluid is encountered in fractures developed in the sedimentary layers of the upper part of the Takinoue-onsen and Kunimitoge Formations and in fractures in the Obonai Formation. As seen in Figure 9.2, the geological structure of Takinoue contrasts remarkably with that of Matsukawa, where geothermal fluid is localized near the boundary between the thick Tamagawa Welded Tuff (which acts as a cap rock) and the Neogene sedimentary sequence. In Takinoue, a regional impermeable layer acting as a cap rock is not found. This difference in the geological structure is of considerable interest in connection with the hydrogeological relation between the vapor-dominated system of Matsukawa and the hot-water system of Takinoue.

9.3. History of Exploration

9.3.1. Cooperative Investigations with the Geological Survey (G.S.)

In 1958, in order to collect information on the Matsukawa area, geological and geochemical reconnaissance surveys were first undertaken in the Takinoue area (Nakamura et al., 1960). This reconnaissance confirmed that the Yamatsuda Formation (which underlies the Quaternary and Pliocene–Pleistocene volcanic rocks in the Matsukawa area) is exposed and also that geothermal manifestations such as fumaroles and hot springs with high water temperatures are emitted over a distance of about 3 km (Figure 9.4). These facts suggested that it would be necessary to conduct a more detailed survey to understand the relation between the occurrence of geothermal fluid and the geologic structure.

During the period from 1961 to 1964, a temperature survey at a depth of 30 m was carried out by the G.S. at about 50 different sites. The results showed that the high-temperature area is divided into two sections: one along the Kakkonda fault at the upper course of the Kakkonda River, the other along the periphery of the dacite intrusive body of the Takinoue hot-spring area (Figure 9.5).

In 1964, an electric resistivity survey was also carried out by the G.S. along survey line A parallel to the Kakkonda River (Figure 9.5). Resistivity values lower than 20 Ω-m were found not only around the high-temperature areas of Figure 9.5 but also at depths of more than 1000 m in areas lacking strong geothermal manifestations at the surface (Figure 9.6) (Baba et al., 1967). The resistivity survey suggested that geothermal fluid would be widespread in this area.

JMC, as a counterpart of the G.S., drilled a shallow hole to better understand the

Table 9.2. Stratigraphic sequence of the Takinoue area

Geological Age	Name of Formation		Thickness (m)	Rock Facies
Quaternary	Volcanic Rocks		—	Two pyroxene andesites
Pliocene-Pleistocene	Tamagawa Welded Tuff	Upper	600	Pumiceous tuff breccia
		Lower		Sandy tuff
Neogene Tertiary (N)	Yamatsuda Formation	Y₃	450	Tuffaceous sandstone and banded silt / Lapilli tuff
		Y₂	~	Tuffaceous siltstone
		Y₁	550	Tuffaceous sandstone / Tuff breccia and lapilli tuff
				Alternation of mud and sandstone
				Tuffaceous sand conglomerate / Volcanic breccia
	Takinoue-onsen Formation (T)		470	Tuffaceous shale, black shale and siltstone
			~	
			600	Tuffaceous sandstone and shale
	Kunimitoge Formation (K)		550	Altered andesite
				Tuff / Black Shale
				Alternation of shale and tuff
	Obonai Formation (O)		530+	Rhyolitic tuff

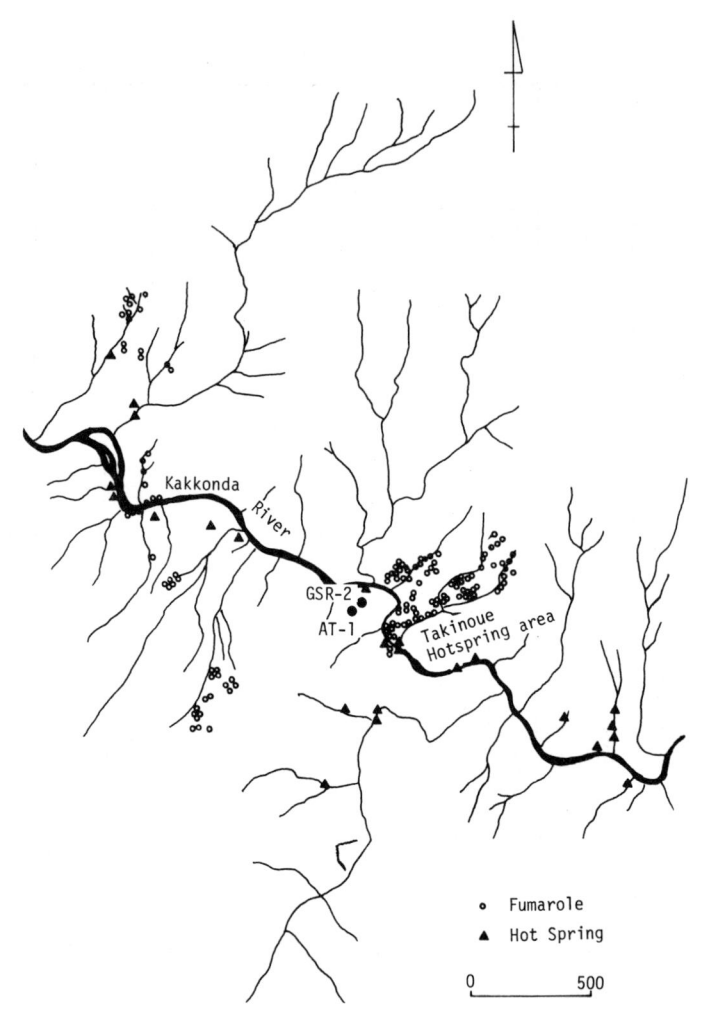

Figure 9.4. Map showing distribution of hot springs and fumaroles in Takinoue

geologic structure at the Takinoue hot spring area in 1963. This hole (named AT-1) encountered a big crack at a depth of 120 m, whereupon it became impossible to continue any deeper. Temperature logging thus was not attempted; however, emissions of steam and hot water were found at the head of the hole.

From the cooperative investigations described above, information about the geological structure relating to the occurrence of geothermal fluid was obtained; however, data on temperature of the deep-seated reservoir was not obtained.

A research well (GSR-2) was subsequently drilled near AT-1 to a depth of 404 m, and, in 1966, temperature and pressure logging of the hole were carried out by the G.S. (Figure 9.7) (Baba et al., 1969).

There were so many cracks found in the Yamatsuda Formation that upon reaching depths of 270 m steam and hot water spurted out from the well head at times. Dacite was encountered below 270 m and was regarded as a part of the intrusive body that crops out at the Takinoue hot spring area. In contrast to the Yamatsuda Formation, the intrusive

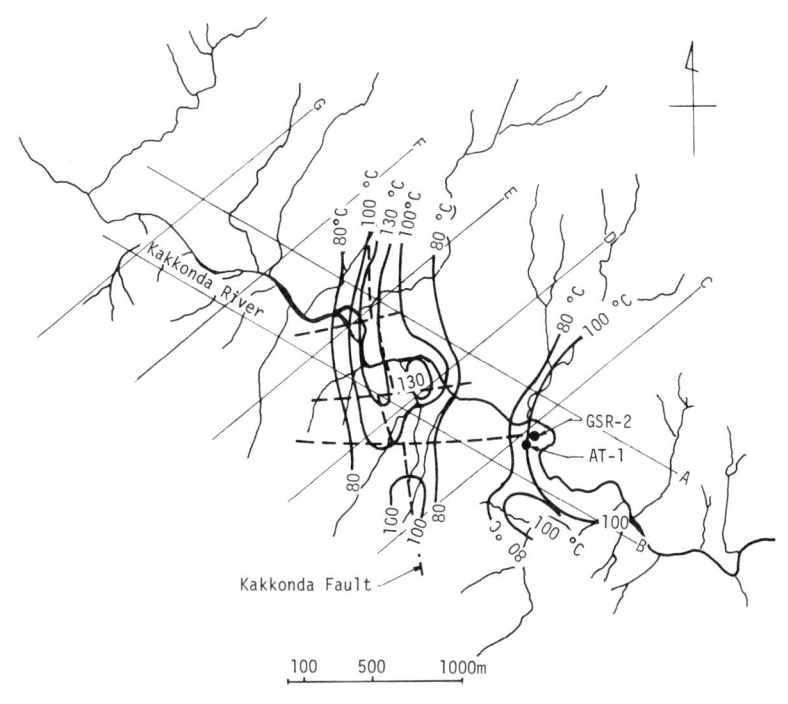

9. Exploration and Development at Takinoue, Japan

Figure 9.5. Isothermal contour map at 30 m depth in Takinoue. Geophysical survey lines are indicated by letters

body was lacking in fissures, although at the bottom of the hole temperatures reached as high as 200°C. After completion of drilling, a production test was attempted but resulted only in a periodic emission of steam and hot water, since cracks developed in the Yamatsuda Formation were sealed by iron casing pipes.

The reason for the prevailing cracks in the Yamatsuda Formation may be due to the intrusion and consolidation of dacite, which resulted in fractures along the boundary between these two differing rock types. Furthermore, judging from its occurrence at the

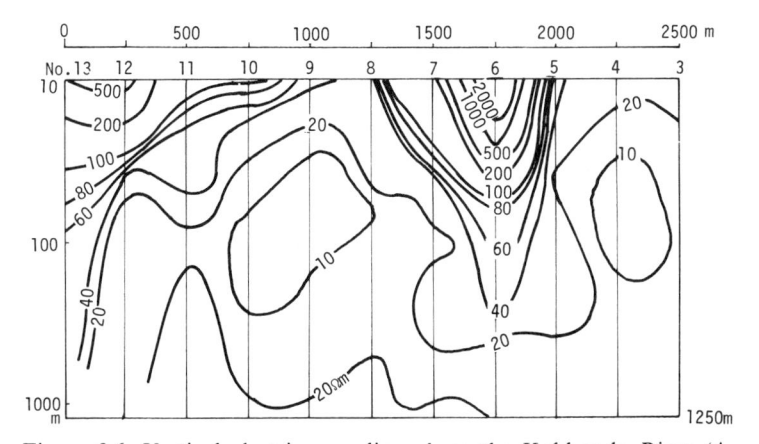

Figure 9.6. Vertical electric sounding along the Kakkonda River (A-line in Figure 9.5) in Takinoue

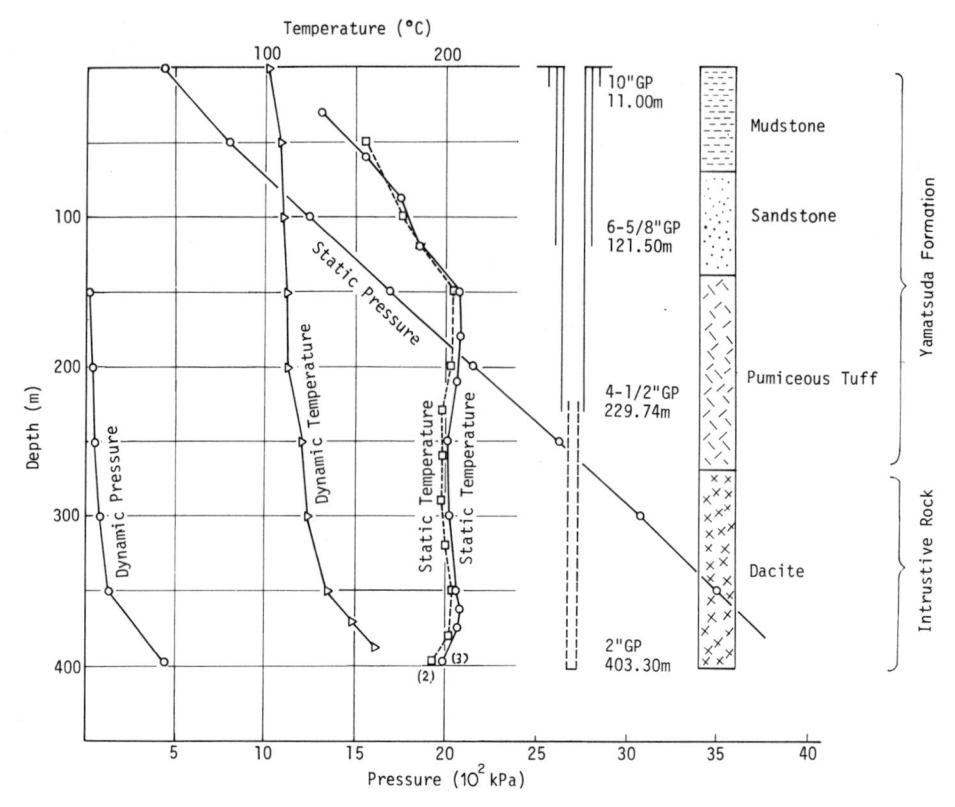

Figure 9.7. Temperature and Pressure log of GSR-2 Well; (2) Measured at five days after drilling; (3) Measured at about one year after drilling

ground surface, the intrusive dacite (which is lacking in cracks) appeared to be widely distributed throughout the underground regions of the Takinoue hot-spring area. Therefore, sufficient steam for power generation would not be expected at the hot-spring area, because the impermeable dacite would tend to obstruct supplies of geothermal fluid. For this reason, it was concluded from the GSR-2 survey that it would be better to avoid the hot-spring area for selection of a site for geothermal power development.

9.3.2 Exploration by JMC (Japan Metals & Chemicals Co.)

9.3.2.1. Surface exploration. The results of surface exploration conducted by JMC during the period from 1969 to 1972 are summarized as follows:

(1) *Geological survey.* Beginning in 1969, a detailed survey was conducted by JMC geologists in an area of about 50 km² using topographic maps at 1:10,000 scale. Figure 9.3 shows the geologic map around the developed area, and Figure 9.8 is a schematic cross section along the Kakkonda River. As shown in Figure 9.3 and Table 9.2, the formations which are exposed in the area of development are the Tamagawa Welded Tuff, the Yamatsuda Formation, and the upper part of the Takinoue-onsen Formation (from upper to lower respectively). Moreover, it has become apparent from the geothermal wells that there are two more formations in the substrata; namely, the Kunimitoge and Obonai

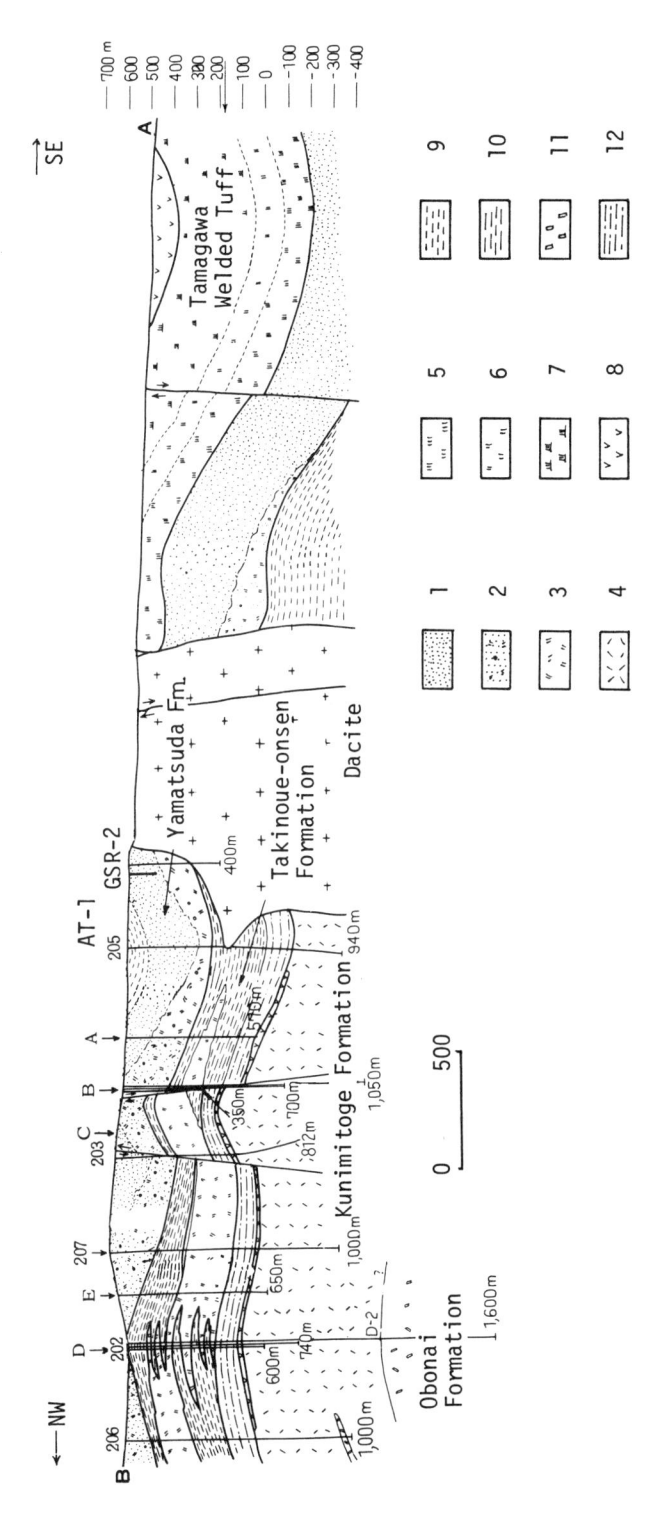

Figure 9.8. Schematic geologic section along the Kakkonda River in Takinoue. (1) Sandstone; (2) Tuffaceous conglomerate; (3) Dacitic tuff; (4) Andesitic tuff and shale; (5) Dacitic tuff; (6) Pumiceous tuff; (7) Pumiceous tuff breccia; (8) Andesite; (9) Siltstone; (10) Shale; (11) Rhyolitic tuff; (12) Sandy shale

258 *Geothermal Systems: Principles and Case Histories*

Formations. As basement rock for this area, the Paleozoic sedimentary rocks or the Mesozoic granite appears to underlie the Obonai Formation, but this has not yet been confirmed from drillings.

(2) *Rock alteration survey.* Four hundred and sixty samples were collected at the surface and analyzed by X-ray diffraction. The identified alteration minerals are distributed zonally as shown in Figure 9.9. The alteration can be classified into a regional one characterized by interstratified sericite/montmorillonite (S/M) minerals and a local one characterized by kaolin, alunite and/or pyrophyllite. The haloes formed by the local alteration are sporadically distributed around the present fumarolic areas, overlapping the zoning formed by the regional alteration.

The regionally altered area is divided from margin to center into five alteration zones based on the percentage of expandable layer in the S/M minerals: (1) free of interstratified S/M minerals, (2) 30–50 per cent expandable layer, (3) 15–30 per cent, (4) 10–15 per cent, (5) 0–10 per cent. The first zone is characterized by montmorillonite, quartz, and cristobalite. The latter four zones contain chlorite, quartz, and subordinate amounts of montmorillonite as well as S/M minerals.

Geologically, the main part of the regionally altered area characterized by the presence of S/M minerals corresponds roughly to the structural dome associated with NW–SE

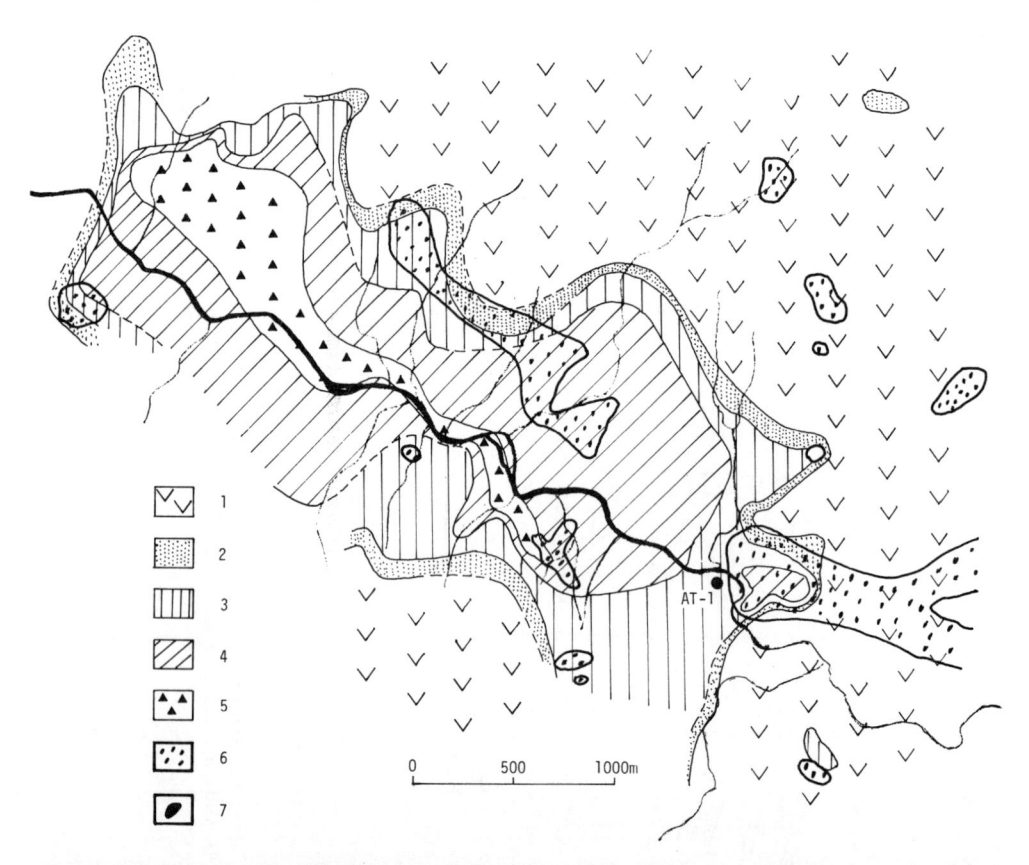

Figure 9.9. Map showing distribution of altered rocks in Takinoue. (1) Area free of sericite/montmorillonite interstratified (S/M) minerals, (2), (3), (4) and (5) 30–50, 15–30, 10–15, and 0–10% expandable layer in S/M minerals respectively, (6) Kaolin and alunite; (7) Pyrophyllite

trending anticlines where the Neogene sedimentary sequence is exposed. This fact suggests that the regional distribution of various S/M minerals is related to the crustal movement in addition to geothermal activity.

Recently, a mineralogical study on alteration minerals, especially on S/M minerals, was carried out by Kimbara and Sumi (1975) using core samples collected from AT-1 well, which is located in the mineral zone of 15–30 per cent expandable layer. According to the results of their study, the S/M minerals can be used as a geologic thermometer, and S/M minerals of about 30 per cent expandable layer at the shallowest part of AT-1 are considered to be formed at depths greater than 100 m under temperatures greater than 80–100°C. This conclusion implies the erosion of more than 100 m of rock.

(3) *Chemical analysis of thermal waters and fumarolic gases.* About 40 samples of thermal waters, 150 of fumarolic gases, and 70 of river waters were analyzed by the chemists of JMC during the period from 1969 to 1972. The fumaroles average 99·75 volume per cent H_2O and 0·25 per cent non-condensible gases, which consist of 75 per cent CO_2, 15 per cent H_2S, and 10 per cent residual gases including minor amounts of H_2, N_2, and CH_4. Thermal waters can be characterized as weakly alkaline NaCl type containing 400–570 mg/L of Cl^- and 240–375 mg/L of Na^+.

Hot springs found in the area along the Kakkonda River were distributed over 3 km. Fumaroles are distributed on the back slope of the Takinoue hot-spring area and at the upper streams of a few tributaries (Figure 9.4). As seen in Figure 9.5, thermal manifestations existing at the upper stream of the Kakkonda River are distributed along the high-temperature zone found by an underground temperature survey at 30 m in depth, in a NW–SE direction.

In order to estimate reservoir temperature of the area, a method to use silica in thermal water as a chemical geothermometer was attempted. In general, it was determined that the silica method could not be used directly on thermal waters diluted by surface or other underground waters. Thus, to estimate the original content of silica in the geothermal fluid, a ratio of dilution by fresh water had to first be determined.

Fortunately, since the content of chloride in the hot water emitted from the GSR-2 site was known, the silica content of hot water ($SiO_{2\,hw}$) can be calculated by the following equation:

$$SiO_{2\,hw} = (SiO_{2\,sp}) \times 610/Cl^-_{sp} - 20 \times (610/Cl^-_{sp} - 1) \tag{9.1}$$

where the original content of Cl^- in the hot water is 610 mg/kg, the content of Cl^- in the fresh water is 0 mg/kg, the content of SiO_2 in the fresh water is 20 mg/kg, Cl^-_{sp} is the content of Cl^- in a sample, and $SiO_{2\,sp}$ is the content of SiO_2 in a sample. From calculation by this equation, the temperature of the original hot water was estimated to be 180–255°C. Measurements in geothermal wells drilled subsequently showed that temperatures in the wells were within the range of 220–245°C. From these data, it can be concluded that the estimation of reservoir temperature by a modified silica method was relatively accurate.

(4) *Geochemical prospecting by CO_2 gas analysis.* Geochemical prospecting by CO_2 gas analysis attempts to discover the direction of fissures and cracks developed in the bed rocks concealed by the soil layer. The survey is conducted through a measurement of the concentration of acid gas (mainly carbon dioxide gas) encountered in small holes one meter in depth drilled in the soil. Measurements of acid gases were carried out at about 850 different points located along the electric resistivity survey lines at intervals of 25 m.

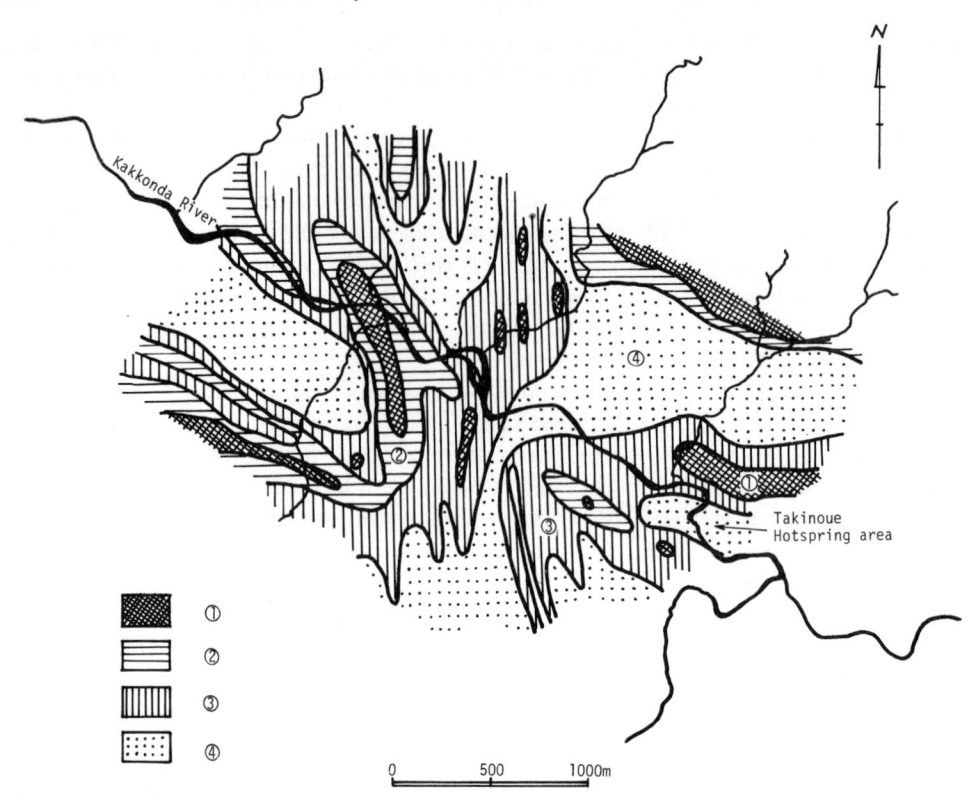

Figure 9.10. Map showing concentration of soil gas (mainly CO_2) at 1 m depth in Takinoue.
(1) >2·1 per cent; (2) 1·1–2·0 per cent (3) 0·6–1·0 per cent (4) 0·1–0·5 per cent

As seen in Fig. 9.10, zones of high gas concentration strike in N–S and E–W directions, while zones of lower gas concentration are distributed outside of the fracture zones and in the areas covered by volcanic rocks and mudstones where impermeable formations prevent gases from emanating.

(5) *Electric resistivity survey.* In 1970 and 1971, geophysical surveys were made by the JMC, among which a resistivity survey by the Schlumberger method to 300 m in depth was carried out along A, B, C, D, E and G survey lines (Figure 9.5), having a total length of 25 km. Judging from the distribution of the iso-resistivity lines, a low resistivity area seems to extend over the whole survey area, which indicates that the reservoir has an area larger than the entire survey surface area.

(6) *Measurements of heat discharge.* Measurements of heat discharge gave the following results:

from fumaroles	4.2 MWt
from hot springs	3.0 MWt
from river waters	31·9 MWt
by conduction	11·6 MWt
by evaporation	11·6 MWt
Total	62·3 MWt

(7) *Seismic survey.* A seismic survey was conducted in 1971 along A, B, C, E and G survey lines of 10 km in total length. Two stratigraphic boundaries were estimated, one being the L_2 horizon between the two formations (the Takinoue-onsen and Kunimitoge Formations), the other being the L_3 horizon between the Obonai Formation and the basement rock (cf. Table 9.2). The characteristics of the substrata obtained from the seismic survey appear similar to that concluded from the surface geological survey. Judging from the frequency of seismic waves, low frequency zones were recognized near L_2 and L_3 horizons and were regarded as fracture cavities filled with geothermal fluid.

9.3.2.2. Drilling of exploration wells. Surface exploration made from 1969 to 1972 showed that the Takinoue area is characterized by a folded structure with axes running in a NNW–SSE direction. Geothermal manifestations are distributed along faults accompanying the anticlinal axes and along structural lines traversing the axes in an E–W direction. These facts suggest that geothermal fluid would ascend to the ground surface through fractures accompanying faults. On the other hand, electric and seismic surveys indicate that low resistivity zones are widely distributed in the substrate below 300 m and that a low-frequency zone is located near the L_2 and L_3 horizons. These data suggest that a geothermal reservoir with a wide surface area extends under the survey area. Temperature of the reservoir was expected to reach more than 200°C at a depth of 400 m from the data obtained from the research well GSR-2 (Figure 9.7) and by the silica method as a chemical thermometer.

From these results, it was concluded that the survey area might be regarded as a potential field for geothermal development. However, as far as the geological background is concerned, the stratigraphic sequence below the Takinoue–onsen Formation was unknown at the time. Therefore, in order to collect information about the underground structure in the area, in addition to chemical and physical properties of the geothermal fluid, drilling of exploration wells was undertaken in 1972.

To select drilling sites for exploratory wells, the following matters were taken into account:

1. As pointed out before, there are two high-temperature zones in Takinoue: one is situated along the Kakkonda fault at the upper stream of the Kakkonda River; the other is on the back slope of the Takinoue hot-spring area.
2. An impermeable intrusive body of dacite is exposed around the hot-spring area and was encountered below a depth of 270 m in the GSR-2 well.

If the impermeable body of dacite is widely distributed underground in the hot-spring area, it may be impossible to tap much of the geothermal fluid due to this barrier. For this reason, the high temperature zone located at the upper stream of the Kakkonda River was selected for drilling.

Figure 9.11 shows the distribution map of the drilling sites where six exploration wells of 700–1000 m depth were drilled during the period from September 1972 to February 1974.

The results obtained from the exploration wells are summarized as follows:

1. The survey area consists of sandstone, mudstone, and tuffaceous rocks rich in cracks and fissures.
2. Cracks and fissures are developed preferentially along the upper and lower bedding planes of the mudstones and in fracture zones adjacent to faults.
3. From the measurements of down-hole temperature in the wells, it was noted that the

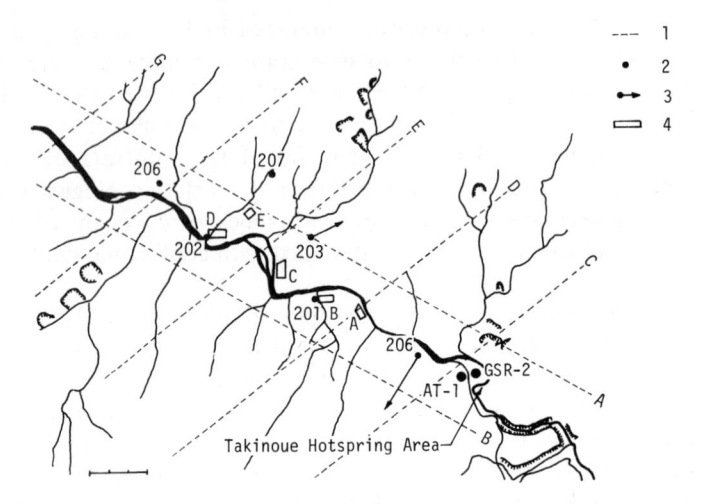

Figure 9.11. Map showing of drilling distribution sites of geothermal wells in Takinoue. (1) Survey line; (2) Exploration well (vertical); (3) Exploration well (directional); (4) Drilling site

reservoir temperature was 220–240°C. Temperatures of 233°C were recorded in a well in the flowing state (as shown in Figure 9.12).

From the data on distribution of cracks and fissures in the rock formations and from underground temperatures measured in wells, it was expected that geothermal fluid in the area would be restricted to the upper portion of the Takinoue-onsen Formation and to the Kunimitoge Formation; this was confirmed by the drilling of exploratory wells. However, the extent of the reservoir in the Takinoue-onsen Formation was found to be only 200 m and the depth from the ground surface was relatively shallow. Accordingly, the Kunimitoge Formation and the deeper-seated formations were regarded as the main reservoirs for this area. Also, from the measurements of the down-hole temperature, geothermal fluid was considered to be widely distributed in the area.

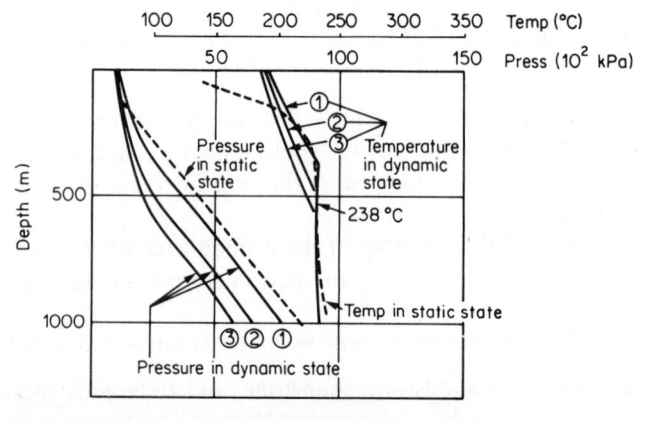

Figure 9.12. Temperature and pressure graph of static and dynamic states in the production well C-1. (1) Well head pressure, $120 \cdot 10^2$ kPa. Total discharge, 78 t/hr; (2) Well head pressure, $10.9 \cdot 10^2$ kPa; Total discharge, 148 t/hr (3) Well head pressure, $10.0 \cdot 10^2$ kPa; Total discharge, 200 t/hr

It was concluded that the potential to develop 50 MW of electrical power from Takinoue was indicated by the results of exploration during 1969 to 1974.

9.4. Production Characteristics

9.4.1. *Production of Geothermal Steam*

The amount of geothermal steam necessary for 50 MWe power generation is about 500 t/hr. To tap this amount of steam, 14–15 production wells are needed (for an average flow rate of 40–50 t/hr of steam per well).

At the planning stage of the geothermal development, the drilling of 14 wells (including two supplementary) was scheduled. However, the developing area was located within the boundaries of a National Park where land use for construction of roads, drilling sites, steam pipelines and power plants is normally prohibited by aesthetic and environmental concerns. Due to such restrictions, a method was adopted in which two to three wells were grouped in small base areas. This method also had other merits such as assembling several production wells in a small base area, thus allowing short-distance transportation of the piped steam and giving economic as well as environmental advantages to the system.

In Takinoue, five drilling base areas labeled A, B, C, D and E were outlined (Figure 9.11), on which individual wells were placed at intervals of only several meters. Thus, to tap the geothermal fluid effectively, directional drilling was applied to maximize the underground gathering area. Figure 9.13 shows the distribution of geothermal wells mostly constructed through directional drilling. So far, 13 production wells of 1000–1600 m in depth have been drilled, among which two were converted to reinjection wells with 11 remaining as production wells at present.

As mentioned above, geothermal fluid is produced from the reservoir developed in the Kunimitoge and Obonai Formations, whereupon the separated hot water is reinjected into the same reservoir rocks. Therefore, to avoid interference between production and injection wells, the casing in the production wells was inserted to a depth of 570–850 m so that geothermal fluid could be drawn only from the lower part of the reservoir without the effects of disposal water, which is injected into the upper part of the same reservoir.

9.4.2. *Reinjection of Hot Water*

Fortunately, it was known that an impermeable stratum existing near the boundary between the upper and lower parts of the Kunimitoge Formation could be expected to act as a cap rock. Consequently, it was determined to drill reinjection wells to a depth of 700–800 m, corresponding to the upper portion of the Kunimitoge Formation. Thirteen injection wells have been prepared thus far; moreover, two were added from production wells. At this point, 15 wells are being used for the disposal of hot water.

There were further problems related to the injection of hot water. One was precipitation of silica contained in the separated hot water, having a content of more than 500 mg/kg in Takinoue. From the result of experimentation on the deposition of silica scale in pipes carrying hot water from an exploratory well, it became clear that noticeable precipitation did not occur in the pipes even when the water temperature decreased, until the water was exposed to the air.

Another problem arose when reservoir pressure in Takinoue reached a level where well

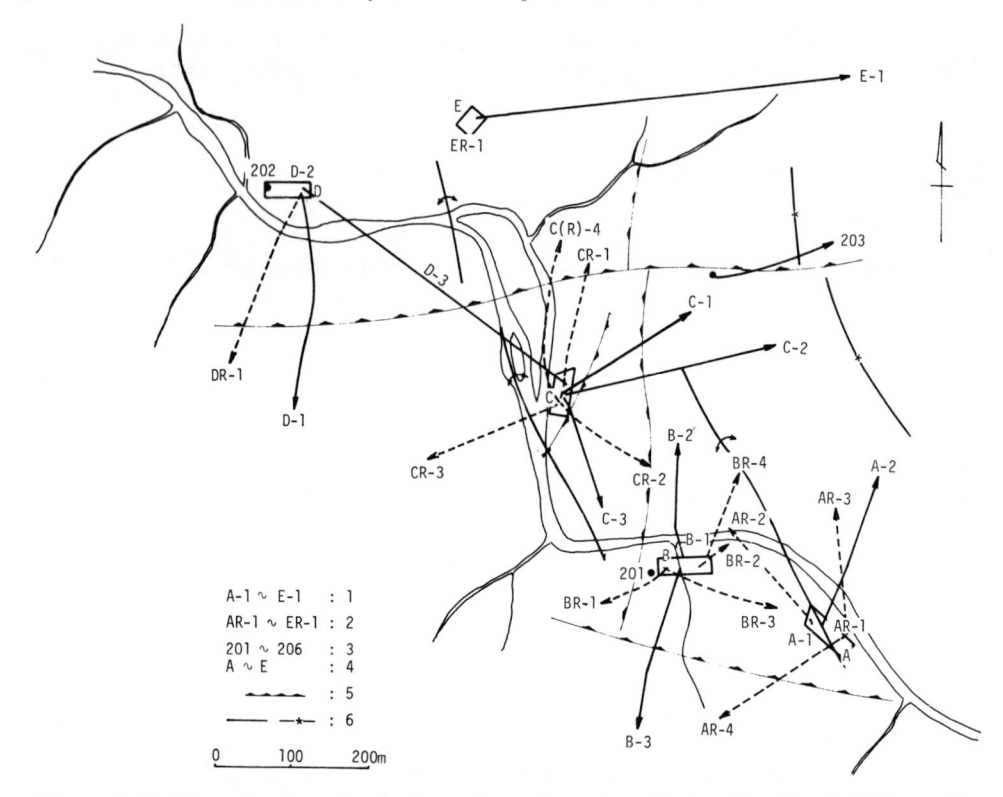

Figure 9.13. Map showing distribution of geothermal wells by directional drilling. (1) Production well; (2) Re-injection well; (3) Exploration well; (4) Drilling site; (5) Fault; (6) Folding

head pressure indicated several hundred kPa; thus pumping force was necessary to reinject the hot water into wells.

In order to solve these two problems, the direct injection method was tested, which meant that the separated hot water was to be injected into the wells directly through the pipes connected with a separator without coming in contact with the air. According to this method, the separated hot water was injected into the wells with a pressure and temperature of about 490 kPa and 160°C respectively.

Six months have passed since all of the geothermal wells were put into operation. To date, there are no apparent problems with precipitation of silica scale and the injection operation itself. This indicates that the direct injection method is useful for preventing silica in the hot water from precipitating.

9.4.3. Well and Reservoir Testing

To collect data on the permeability of the reservoir rocks and to estimate productivity and injectivity of the wells, the following tests were undertaken by the reservoir engineers of JMC: (1) static temperature and pressure readings, (2) production testings, dynamic temperature and pressure loggings, flow rate measurements at various well head pressures, and (3) injection testing.

(1) *Temperature and pressure logging (static and dynamic)*. This logging was carried out

after the well had been kept in a dormant state for such a period that the temperature and pressure in the well reached an approximate equilibrium with the geothermal environment. The results of the measurements could be regarded as the background data on temperature and pressure of the reservoir before development. Distribution of down-hole temperature and pressure in a production well, in both static and dynamic states, is shown in Figure 9.14. Figure 9.15 shows the vertical distribution of underground temperature measured in the production wells, and Figures 9.16 and 9.17 show a horizontal distribution map of reservoir temperature and pressure plotted on the surface of the water at 200 m below sea level (about 850 m beneath the ground surface).

(2) *Production testing.* The testing was carried out by measuring the flow rate of steam and hot water at various well head pressures simultaneously with temperature and pressure logging in the dynamic state. From such testing, data on the physical properties of the reservoir rocks and the geothermal fluid at a flowing state could be obtained.

In a dynamic state, where Pw is pressure at a given depth in a well and Pe is original reservoir pressure, the relation between flow rate (G) and pressure difference ($Pe - Pw$) is given as the following equation:

$$G = J(Pe - Pw) \tag{9.2}$$

J is called the production index which is related to the productivity of a well (namely the permeability of the well rock around the well). Table 9.3 shows the results of production testing, including the values of the production index of each well.

9.4.4. Reinjection Testing

As previously described, hot water is injected into a well with a pressure which corresponds to that in the separator. The injection well is thus filled with disposal water which is sent to the injection zone through cracks and fissures around the well, keeping a balance between the water column pressure in the well and the pressure of the injection

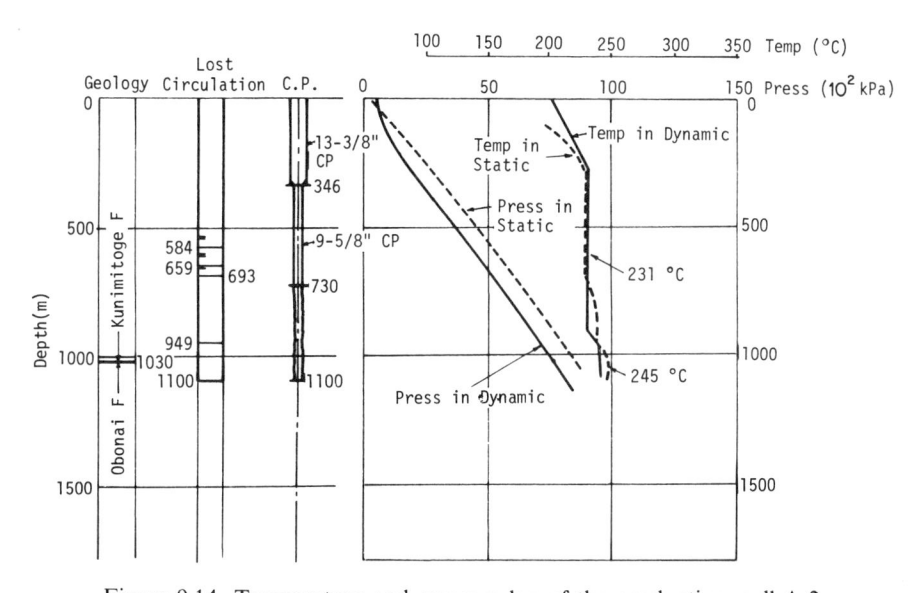

Figure 9.14. Temperature and pressure log of the production well A-2

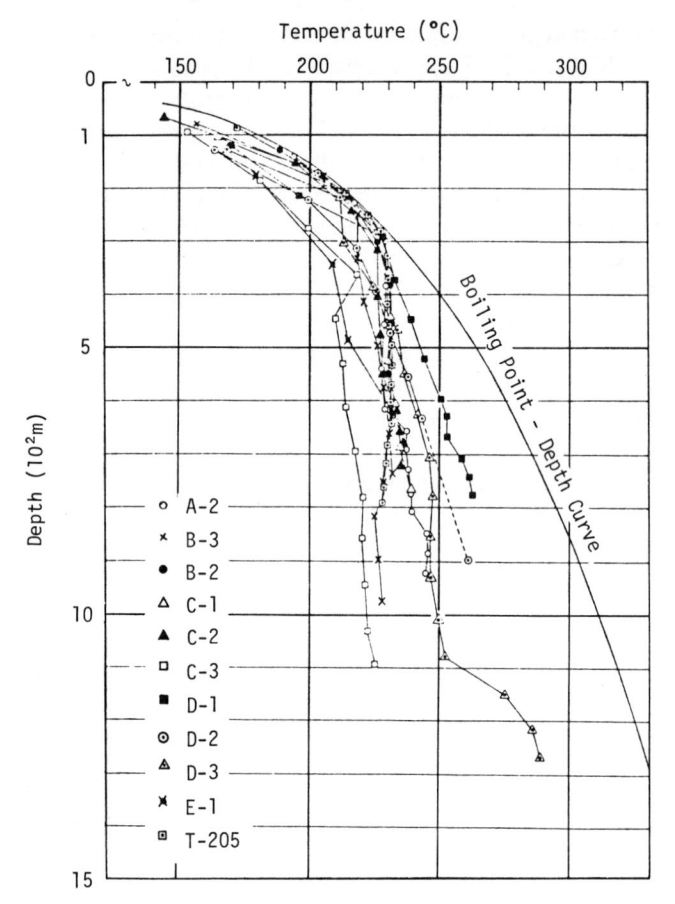

Figure 9.15. Graph of downhole temperature in production wells

zone (the reservoir). The equation of such a pressure balance, which is the same as that presented in the case of a production well (equation (9.2)), is:

$$G' = I(Pw - Pe) \qquad (9.3)$$

G' = injection flow rate (t/hr)
Pw = pressure at the selected depth of a well (kPa)
Pe = original reservoir pressure (kPa)
I = injection index

I is a factor related to the injection capacity of a well, namely the permeability of reservoir rocks, which corresponds to the production index (J) previously mentioned. Following that precedent, I can be defined as the injection index.

Figure 9.18 shows an injection flow rate matching the selected well head pressure in each well.

9.4.5. Steam Supply System

In Takinoue, reinjection wells have been drilled next to the production wells. However, because the injection flow rate differs from one base area to another, the following supply system (shown in Figure 9.19) was undertaken:

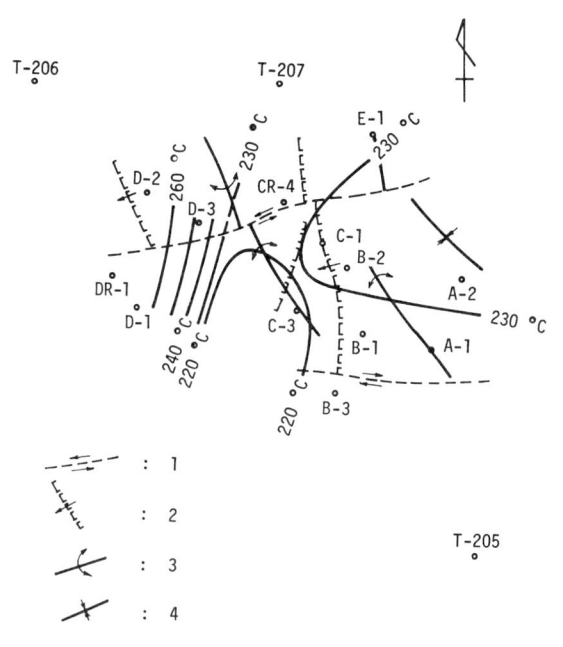

Figure 9.16. Map showing distribution of temperature in production wells at 200 m below sea level. (1) Strike-slip fault; (2) Dip-slip fault; (3) Anticline; (4) Syncline

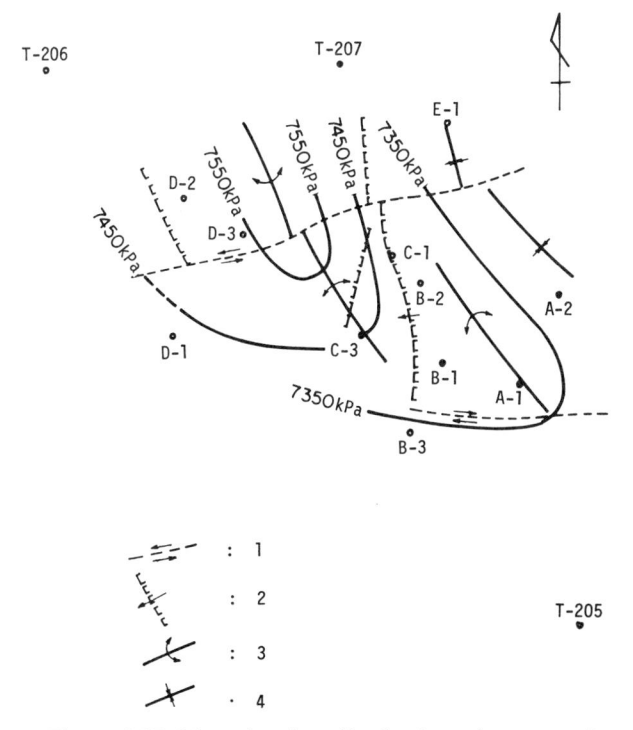

Figure 9.17. Map showing distribution of pressure in production wells at 200 m below sea level. (1) Strike-slip fault; (2) Dip-slip fault; (3) Anticline; (4) Syncline

Table 9.3. Results of production testing including values of production indexes

	Shut-in pressure at static state (in 10^2 kPa)	Maximum well head pressure at static state (in 10^2 kPa)	Fluid temperature (°C)	Saturated pressure (in 10^2 kPa)	Enthalpy (J/g)	Production index (t/hr)/(kPa)	Inflow depth (m)
A-1	3·6	9·8	231	27·4	992	0·09	800–900
A-2	4·1	16·0	231	27·4	992	0·44	900–950
B-2	3·4	14·3	234	28·9	1009	0·28	800–900
B-3	2·9	11·7	222	23·5	946	0·71	950,1200
C-1	1·5	12·7	233	28·4	1009	0·06	950–1000
C-2	1·0	14·7	235	29·9	1013	0·25	860–8800
C-3	0·0	11·7	222	23·5	938	1·02	850–1200
D-1	3·1	21·2	256	43·1	1109	0·12	950
D-2	3·5	24·3	265	50·0	1155	0·08	850–860
D-3	3·3	15·0	240	32·3	1038	0·03	1000–1100
E-1	0·0	15·4	244	34·8	1063	0·13	850–900
T-205	7·3	14·7	230	27·0	988	0·02	750–800

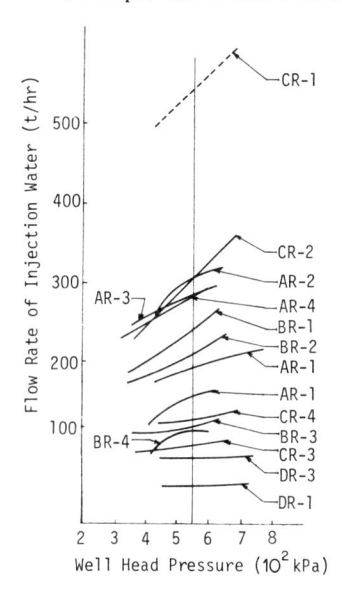

9. Exploration and Development at Takinoue, Japan

Figure 9.18. Relation of flow rate of injection water to well head pressure

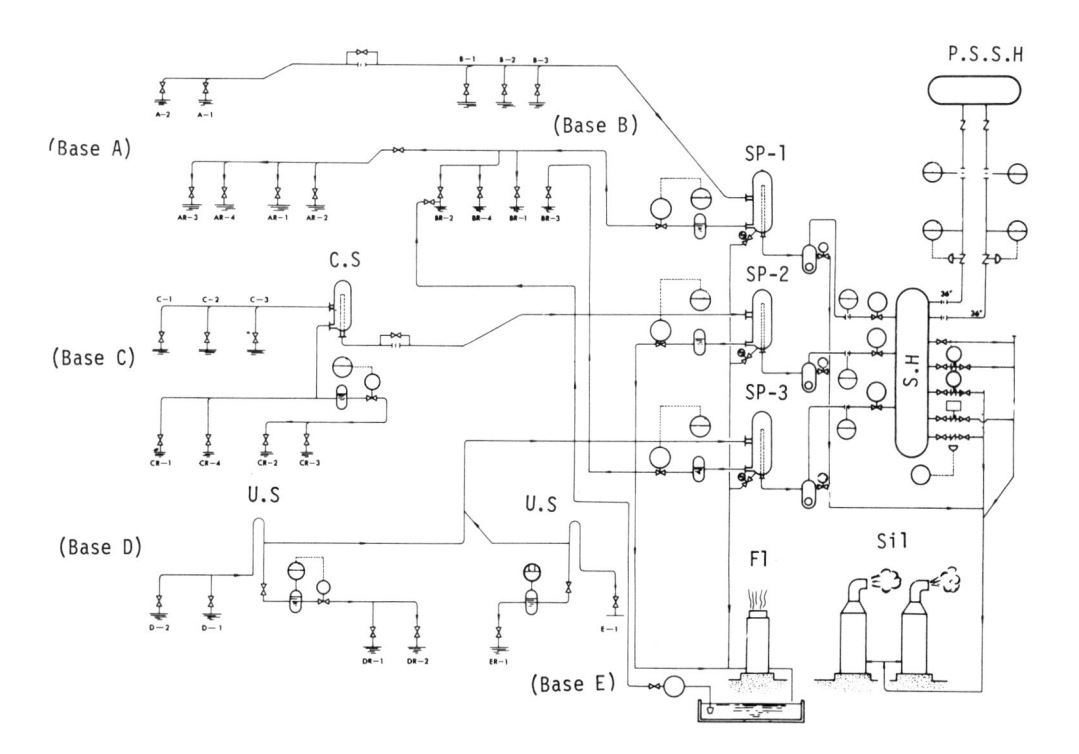

Figure 9.19. Schematic flow chart of the steam supply system in Takinoue. C.S.—Cyclone separator; U.S.—U-bend separator; S.H.—Steam header; P.S.S.H.—Power station steam header; SP-1—No. 1 Separator; Fl—Flusher; SP-2—No. 2 Separator; Sil—Silencer. SP-3—No. 3 Separator

(a) Geothermal fluid produced in Base A is sent to Base B (which is located near the power plant) in the state of two-phase flow. It is then separated by Separator No. 1 (SP-1) after being joined with the fluid produced from Base B, while the separated hot water is disposed in injection wells at Bases A and B.
(b) Geothermal fluid produced in Base C is separated into steam and hot water by a cyclone separator. The steam is sent to the steam header in Base B after being passed through SP-2 in Base B; the water is reinjected into the wells drilled at Base C.
(c) Geothermal fluid produced in Bases D and E is separated into wet steam and hot water by U-bend separators in both bases. Wet steam containing hot water of about 20 per cent is sent to SP-3 in Base B in the state of two-phase flow, and then steam which is separated by SP-3 is sent to the steam header. Hot water separated by SP-3 is disposed in wells at Bases C and B.

Dry steam contained in the steam header in Base B is kept under constant pressure of 440 kPa (gauge) and sent to the turbine whose entrance pressure is 340 kPa (gauge) through two transportation pipes.

Figure 9.20 shows characteristic flow rate curves of steam and hot water obtained prior to operation of the power plant. Table 9.4 shows the flow rate of steam and hot water and injection water measured at the time of the test run of the 50 MWe power generation. Steam from the production wells has water-gas ratio such as 99.86–99.96 volume percent of H_2O and 0.14–0.04 per cent of gases which consist of 61.6 per cent of CO_2, 23.5 per cent of H_2S and 14.9 per cent of residual gases as average values. The hotwater is characterized as weakly alkaline NaCl-type containing 430–460 mg/L of Cl^- and 60–90 mg/L of Na^+.

9.5. Effectiveness and Economics

Electric power produced by the Kakkonda geothermal power plant is consumed for domestic and industrial use in Iwate Prefecture. The amount of electric power produced

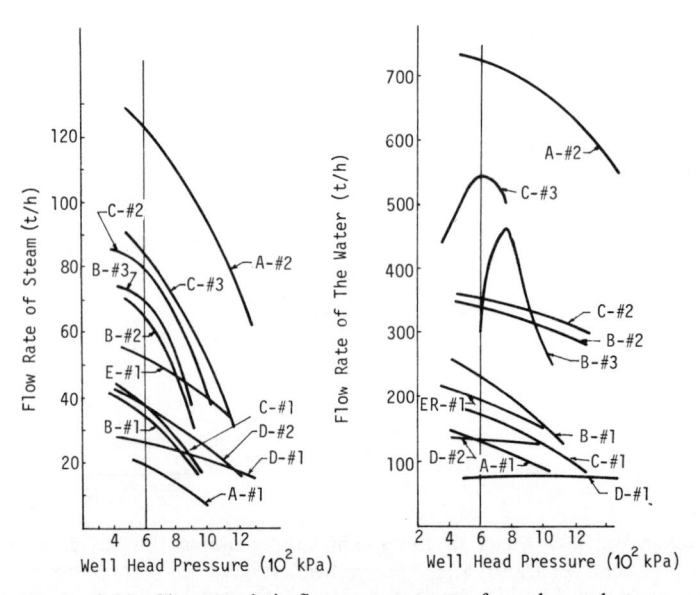

Figure 9.20. Characteristic flow rate curves of geothermal steam and hot water

Table 9.4. List of production of geothermal fluid and re-injection of hot water

Production wells					Re-injection wells			
		Discharge at $5.9 \cdot 10^2$ kPa (gauge)				Injection	
Name of well	Depth (m)	Steam (t/hr.)	Hot water (t/hr.)	Name of well	Depth (m)	Pressure 10^2 kPa (gauge)	Injection (t/hr.)
Base A							
A-1	1600	20	130	AR-1	570	5.4	162
A-2	1100	125	725	AR-2	700	5.4	312
				AR-3	700	5.4	289
				AR-4	700	5.5	287
Sub-Total		145	855				1050
Base B							
B-1	1160	35	230	BR-1	690	5.4	243
B-2	897	65	340	BR-2	615	5.4	212
B-3	1265	70	313	BR-3	700	5.4	122
				BR-4	700	5.4	118
Sub-Total		170	883				695
Base C							
C-1	1090	38	171	CR-1	724	5.4	546
C-2	887	80	354	CR-2	700	5.4	308
C-3	1298	85	546	CR-3	700	5.4	100
				CR-4	954	5.4	130
Sub-total		203	1071				1084
Base D							
D-1	954	27	77	DR-1	600	5.4	50
D-2	1565	38	135	DR-2	1600	5.4	85
Sub-Total		65	212				135
Base E							
E-1	1200	53	196	ER-1	650	5.4	196
Sub-Total		53	196				196
TOTAL		636	3217				3160

by the power plant during the period from June to October in 1978 was 130,000 MWh, which corresponds to 8.8 percent of the total consumption in the prefecture during the same period. Power plants (mainly hydroelectric) in the prefecture are few in number, and the total amount of electric power produced by those during the same period was 374,000 MWh. This means that the geothermal contribution of 130,000 MWh amounts to 34.7 per cent of the electric power produced in the prefecture. From these points, the Kakkonda geothermal power plant is regarded as an important power source in Iwate Prefecture.

The total construction costs from exploration to construction of the power plant amount to about 19 billion yen (about 95 million U.S. dollars). This consists of 11 billion yen (about 55 million U.S. dollars) for production and supply of steam and 8 billion yen (about 40 million U.S. dollars) for construction of the power plant. Cost of steam sold to the electric power company was settled at 7·4 yen per kWh (3·7 U.S. cents per kWh), based on the calculation that 10 Kg of steam generates 1 kWh of electricity. Moreover, as cost of construction of the power plant is added to that of steam supply system, the total operating costs for the Kakkonda geothermal power plant come to 12 yen per kWh (6 cents per kWh), which is comparable to costs in thermal power plants located elsewhere. In the case of Kakkonda, the cost was derived from calculation based on depreciation terms of five years for the geothermal wells, eight years for steam gathering pipelines, and fifteen years for the turbine and generator.

9.6. Future Development

Geothermal development in Takinoue has been completed with success in 50 MWe power generation and has been in active operation since the end of May, 1978. For the future development of this area, construction of another 50 MWe power plant is planned based on the data obtained from exploration and exploitation during the last decade. However, because the area is within the National Park boundaries, it becomes necessary to re-obtain permission from the Environment Agency despite being a second project on a previously approved site. Upon receiving approval from the Agency, construction work will commence immediately.

Concerning utilization of the separated hot water, the following plan was drafted by the prefectural government: first, hot water is to be transported by pipeline to the Shizukuishi basin (about 10 km SSE from Takinoue) to a heat exchanger to heat fresh water. After being utilized, the now-cool geothermal water is to be disposed in injection wells drilled in the basin or to be deposited in the Kakkonda River after As has been extracted by chemical treatment. The heated fresh water is then to be used as a heat source for space and greenhouses, for culture farming of fish, and for bathing water. This plan is now under consideration, but as yet has not been implemented.

References

Baba, K., Homma, I., and Takei, Y., 1967. 'Electric prospecting at Takinoue and Matsukawa geothermal areas', *Geol. Survey Japan Bull.*, **18**, 679–686.

Baba, K., 1969. 'Geothermal investigation wells at Takinoue and Takenoyu areas', *Geology News*, **176**, 10–15.

Kawano, Y., and Aoki, K., 1959. 'Petrology of Hachimantai and adjacent volcanoes', *Volcanological Soc. Japan Bull.*, 2nd Series, **4**(2), 61–76.

Kitamura, N., and Onishi, A., 1972. 'On the subsurface geological structure in the eastern foot-hill area of Mt. Iwate, Iwate Prefecture', *Prof. Junichi Iwai Memorial Volume*, 67–73.

Kimbara, K., and Sumi, K., 1975. 'Hydrothermal alteration minerals in the core sample at Takinoue geothermal area, Shizukuishi, Iwate Prefecture with special reference to the interstratified minerals of sericite and montmorillonite', *Chinetsu, Japan Geothermal Energy Assn.*, **12**(3) (Ser. No. 46), 59–68.

Nakamura, H., Ando, T., Sumi, K., and Suzuki, T., 1960. 'Geology and hot springs of Takinoue geothermal area, Iwate Prefecture', *Geol. Survey Japan Bull.*, **11**, 79–84.

Sato, K., and Ide, T., 1975. 'On structural characters and simulations of rock fracturing of geothermal areas in northeastern Japan', *Proc. 2nd United Nations Symposium on the Development and Use of Geothermal Resources*, **1**, 575–581.

Tamanyu, S., and Suto, S., 1979. 'Stratigraphy and geochronology of Tamagawa Welded Tuff in the western part of Hachimantai, Akita Prefecture', *Geol. Survey Japan Bull.*, **29**, 156–169.

Geothermal Systems: Principles and Case Histories
Edited by L. Rybach and L. J. P. Muffler
© 1981 John Wiley & Sons Ltd.

10. The Krafla Geothermal Field, Northeast Iceland

VALGARÐUR STEFÁNSSON

National Energy Authority, Geothermal Division, Reykjavík, Iceland

10.1. Introduction

Geothermal energy is extensively utilized in Iceland. The total installed geothermal capacity is now about 600 Megawatts thermal (MWt). The geothermal energy is mainly used for space heating, and the net geothermal consumption is about 500 MWt. As the total population of Iceland is 220,000, this is a large proportion of the energy consumed in the country (Pálmason et al., 1975). In fact more than 30 per cent of the net energy consumption in the country comes from geothermal energy sources (Stefánsson, 1975; Fridleifsson, 1978). Most of the geothermal energy used in Iceland comes from the low temperature fields, where geothermal water of the temperature 60–120°C is used directly for space heating and greenhouse cultivation. Figure 10.1 shows the volcanic zones in Iceland and the high-temperature geothermal fields as solid triangles, all located within active volcanic zones. The low-temperature fields are predominantly located on the flanks of the volcanic zones (Pálmason et al., 1975).

The utilization of the high temperature geothermal field at Námafjall started in 1967. The main purpose is processing of diatomite deposits from Lake Mývatn (Ragnars et al., 1970). The decision to build the diatomite plant was taken before major drilling for steam was started in the area. A 3 MWe power station was installed at Námafjall in 1969.

In 1973 a long standing plan for a hydroelectric power plant in NE Iceland was cancelled because of strong public opposition concerning the environmental consequences of the project. Time was short, and it was believed that the construction of a geothermal power plant would take less time than any hydroelectric alternative. In 1975 it was decided to build a geothermal power plant in Krafla, a high temperature field about 10 km north of Námafjall. In order to gain time, it was decided to build the power plant concomitant with the drilling for steam.

At the time, the development of the Svartsengi high temperature field is SW Iceland was initiated (Arnórsson et al., 1975), where a 30 MWt heat exchange plant is now in operation. A 235°C saline geothermal brine is used for the heating of fresh water of good quality for space heating and domestic use. A 1 MWe electric generation is used for the plant's internal power-supply. In addition a pilot plant for the production of sea chemicals was erected in 1978 at the Reykjanes thermal brine area. The exploration work has been described by Björnsson et al. (1970, 1972).

The exploration of the Krafla geothermal field started in 1970 and was initially carried

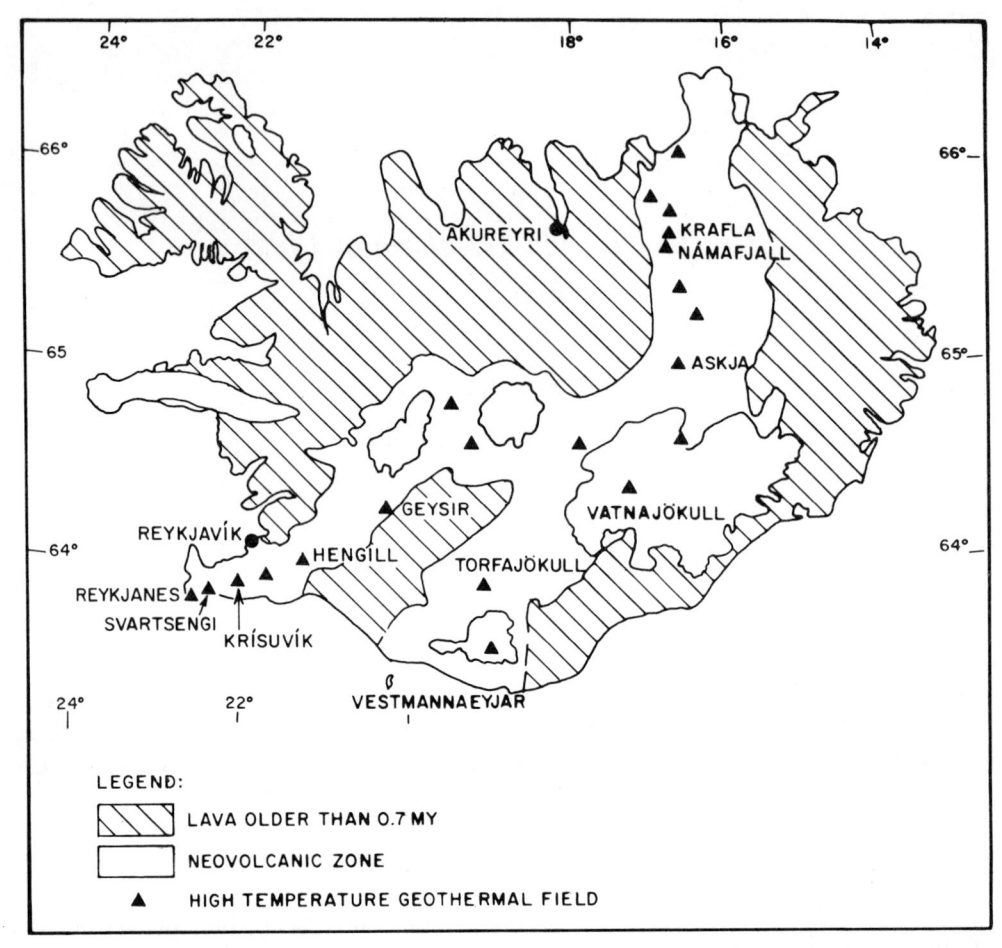

Figure 10.1. Schematic geological map of Iceland showing the high-temperature fields within the volcanic zones. Mapped by K. Saemundsson. Reproduced by permission of Orkustofnun

out according to the Program for the Exploration of High Temperature Areas in Iceland (Björnsson, 1970). A consequence of the decision to build the power plant concomitant with the drilling operation was that the investigation of the production characteristics of the field could not be made until during the production drilling phase. During drilling, it was found that the reservoir was partly boiling and that the production characteristics were quite different from those of a water-dominated field. Up to that time all high-temperature geothermal fields in Iceland had been found to be water-dominated fields, with base temperatures 200–300°C.

The reservoir in Krafla has been found to be complicated, consisting of two geothermal zones: an upper water-dominated zone with 205°C temperature, and a lower zone boiling at 300–350°C temperature. These unexpected circumstances greatly influenced the plans for the power plant. When the power plant was completed in 1977, the available steam was sufficient to produce 7 MW of electricity, whereas the power plant is designed for 60 MW.

In December 1975 a volcanic eruption occurred in Leirhnjúkur about 2 km from the Krafla power plant. This volcanic eruption was the beginning of a rifting episode in the fissure swarm intersecting the Krafla caldera. During the last three years this volcanic activity has proceeded with seven rifting episodes, three of which have resulted in volcanic eruptions. The magmatic activity has influenced the production characteristics of the Krafla field.

In the present paper the Krafla geothermal field and some of its production characteristics will be described. The utilization of the field has just started and the experience gained with this utilization is small.

10.2. Geological Framework

Iceland lies astride the Mid-Atlantic Ridge. The surface expressions of the ridge are the so called neovolcanic zones, which are divided into several branches. In general, the structure of the neovolcanic zones is dominated by fissure swarms and central volcanoes. The fissure swarms are usually about 10 km wide and 30 to 100 km long. Most of the fissure swarms pass through central volcanoes, which are the loci of highest lava production and are also defined by the presence of acid rock and high-temperature geothermal fields. The neovolcanic zones have been described in detail (Saemundsson, 1974; Walker, 1975; Jakobsson et al., 1978; Saemundsson, 1978), and the interaction between the central volcano in Krafla and the intersecting fissure swarm has recently been described (Björnsson et al., 1977; 1979).

The Krafla central volcano is situated on one of the five distinct fissure swarms in the northeast volcanic zone in Iceland. All fissure swarms in the northeast volcanic zone are associated with central volcanoes and have developed high-temperature geothermal fields (Figure 10.2). The Krafla central volcano developed a caldera during the last interglacial period, but since then the caldera has been almost filled with volcanic material. The caldera measures about 8×10 km. Figure 10.3 is a tectonic map of the area, showing the fissure swarm and the caldera.

A rifting episode, currently occurring in the Krafla fissure swarm (Björnsson et al., 1977), has shown that in the fine structure of continental drift discontinuous movements are present. During a rifting episode magma is stored temporarily in a magma trap under the Krafla central volcano from where it is expelled along the fissure swarm to form dykes. The geothermal fields are located in areas of preferred magma concentration at shallow levels in the swarm, such as in the Krafla caldera and on certain locations along the fissure swarm, e.g. in Námafjall. Figure 10.4 is a schematic picture relating the magma chamber, the fissure swarm and the geothermal fields (Björnsson et al., 1979). The longest distance of subterranean magma flow was recorded during the rifting event in December 1975, when earthquake activity progressed towards the north over a distance of about 50 km along the fissure swarm. A high-temperature geothermal field has been suggested to be present in the Axarfjördur area on the basis of surface geophysical and geochemical surveys (Stefánsson, 1979). The heat source for this field may be of a similar nature as described above for Námafjall.

The postglacial volcanism in the Krafla area has occurred in two main periods. The first was in early postglacial time and the second during the last 3000 years. In these two periods, there have been about 20 volcanic eruptions in the Krafla caldera and about 15 in the Námafjall area. The majority of the fissure eruptions are basaltic, but andesite and dacite flows have also occurred. Four subglacial silicic eruptions have produced large domes or ridges within and around the Krafla caldera. Several explosion craters are located within the caldera. The most recent one (Viti on Figure 10.5) was formed in 1724 at the beginning of the 1724–1729 volcanic and rifting episode, the Mývatn fires.

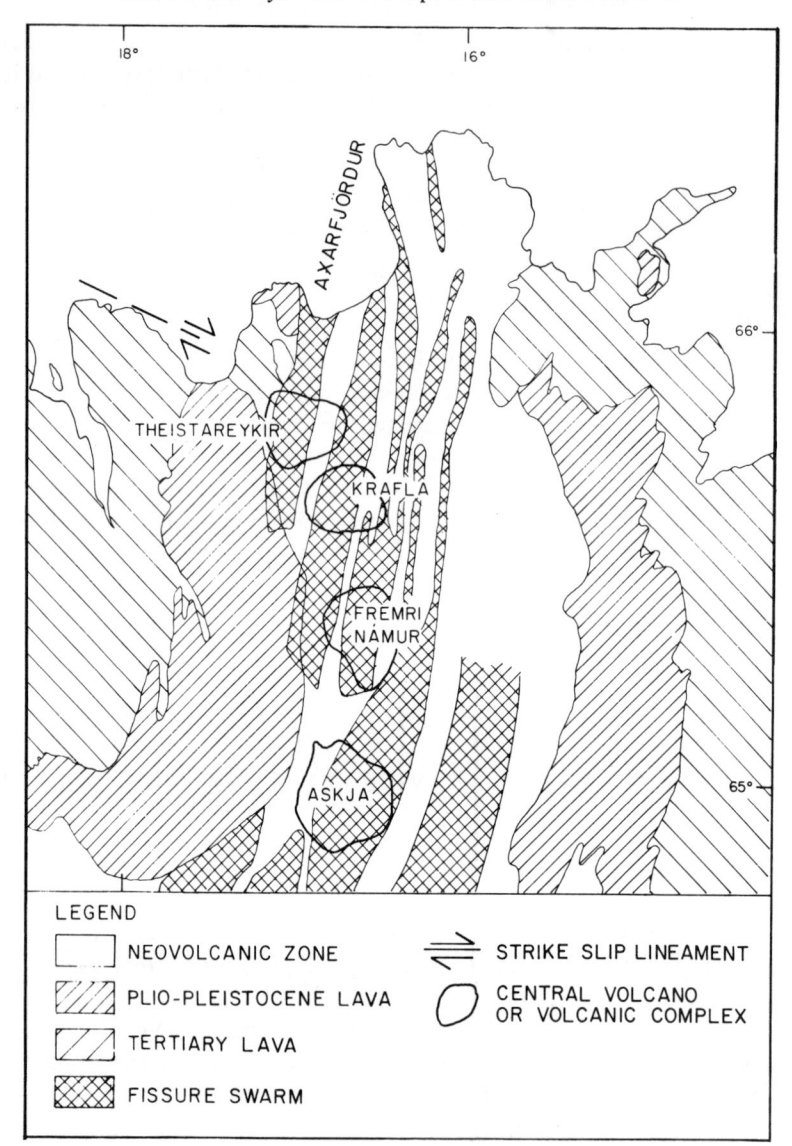

Figure 10.2. The fissure swarms and central volcanoes within the Northeast Volcanic zone in Iceland. Two volcanoes, Krafla and Askja have developed calderas. Mapped by K. Saemundsson. Reproduced by permission of Orkustofnun

The high-temperature geothermal field inside the Krafla caldera is elongated NW–SE. The main surface activity is in the center of the caldera (near Leirhnjúkur) and in the southeastern part of the caldera where a series of explosion craters have formed a gully called Hveragil. The surface area covered by thermal alteration and geothermal activity is about 35 km^2. Petrological study of the rocks in the Krafla area reveals composition ranging from olivine tholeiite to rhyolite. By drilling in the area it has been found that basalt as well as granophyre intrusions occur frequently below the 1100 m depth level. Figure 10.5 shows the relation of the wells to the underlying magma chamber.

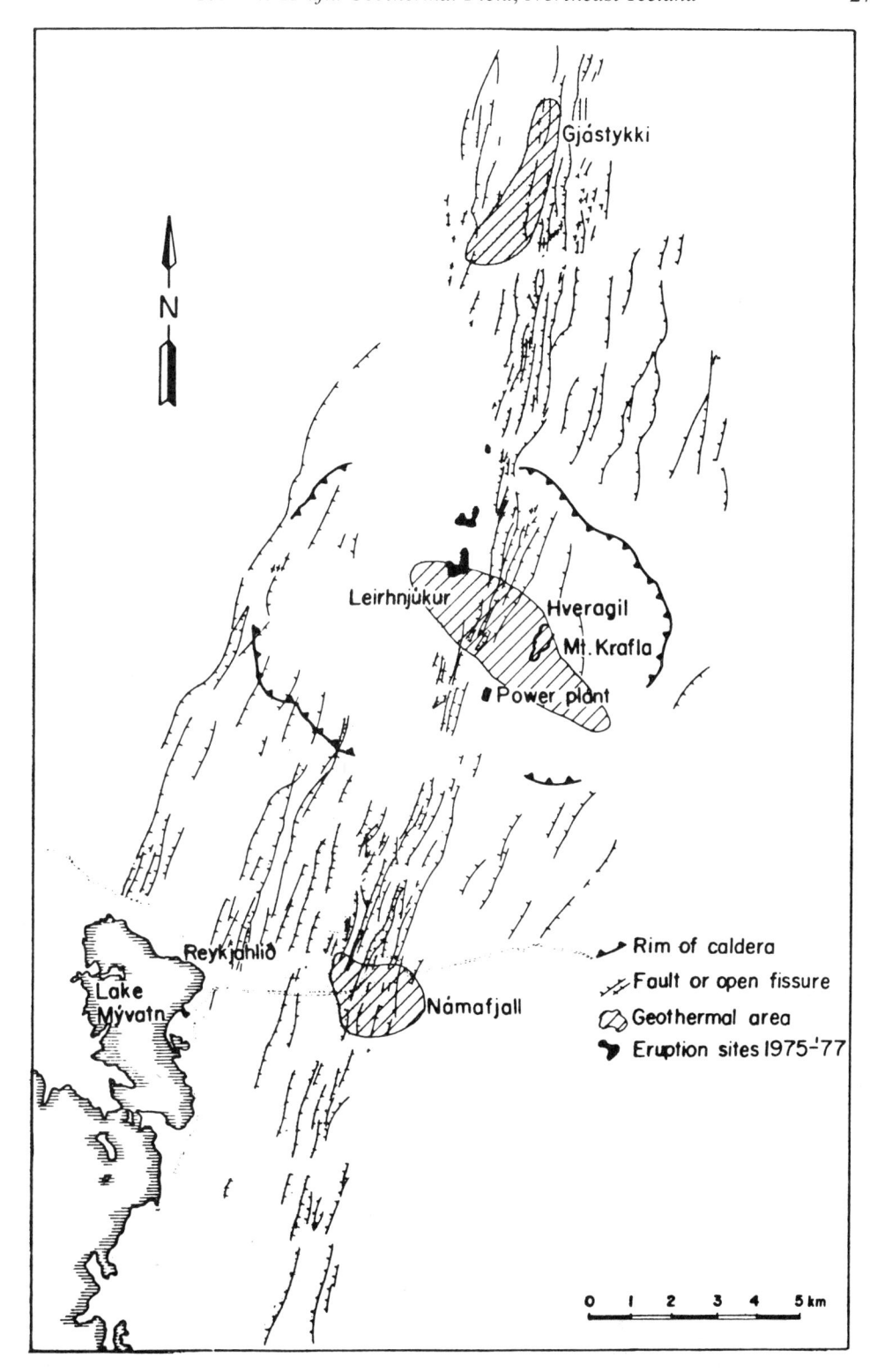

10. *The Krafla Geothermal Field, Northeast Iceland* 277

Figure 10.3. Tectonic map of the Krafla area showing the caldera and the active fissure swarm. Mapped by K. Saemundsson. Reproduced by permission of Orkustofnun

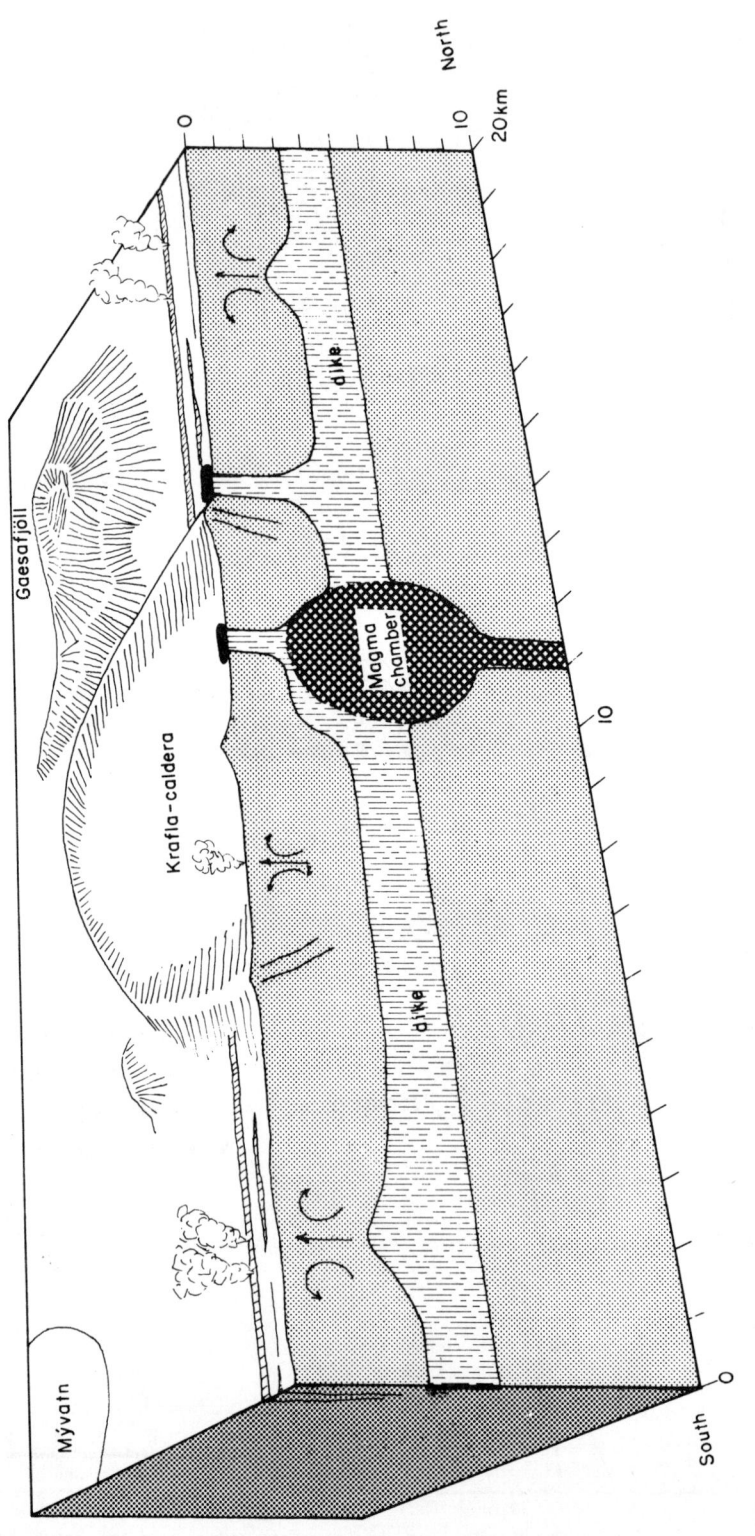

Figure 10.4. A block diagram showing schematically the magma body below the Krafla caldera and the dyke formed in the present tectonic episode. From Björnsson et al. (1979). Reproduced by permission of Orkustofnun

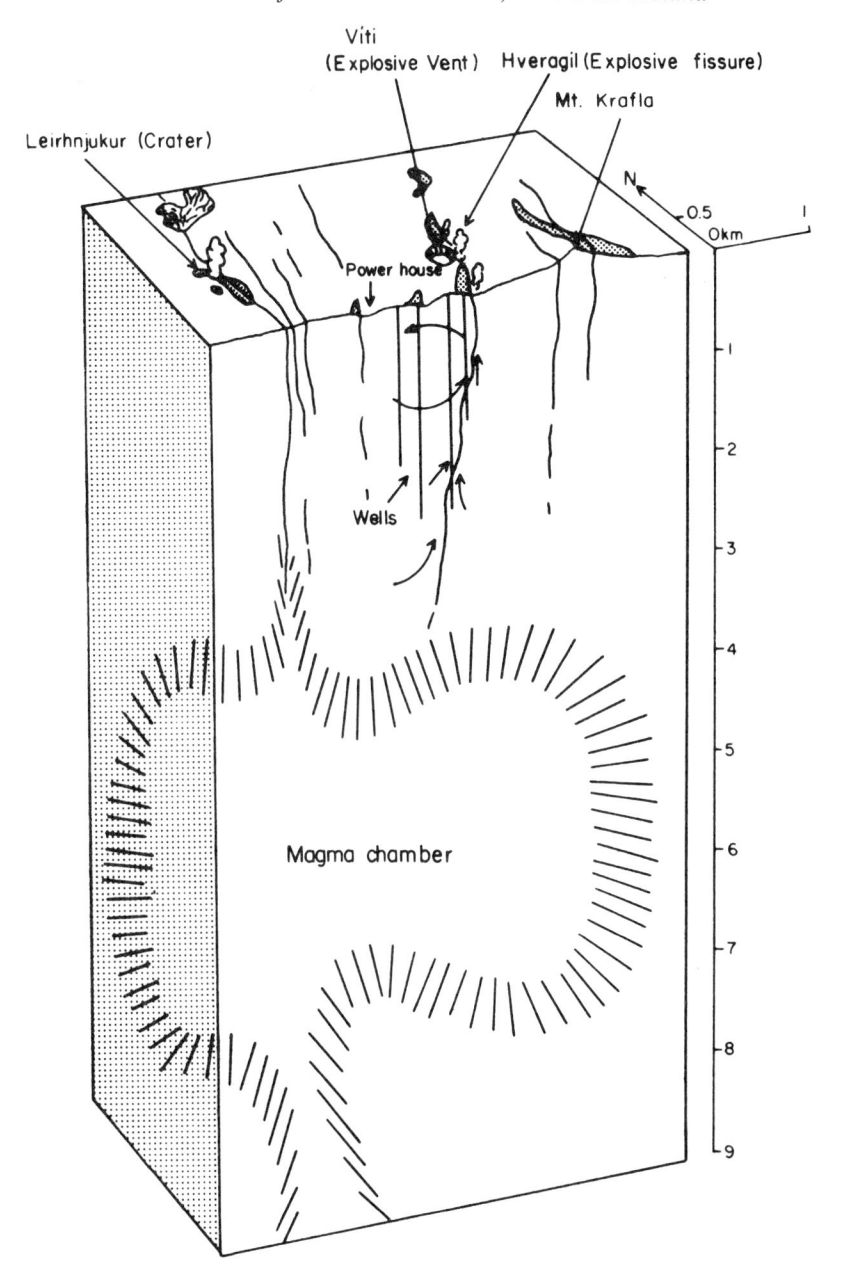

10. *The Krafla Geothermal Field, Northeast Iceland*   279

Figure 10.5. The Krafla Geothermal Field and the underlying magma body. From Gíslason et al. (1978). Reproduced by permission of Orkustofnun

10.3. Exploration History and the Model of the Field

10.3.1. *Surface Exploration*

Systematic exploration of the Krafla geothermal area started in 1970. The surface investigation included geological mapping, geochemical analysis of natural springs and fumaroles, aeromagnetic survey, resistivity survey, and seismic refraction measurements. After the start of the rifting episode in 1975, additional surface investigations (e.g. gravity measurements) have been carried out.

The geological investigation resulted in the description of the Krafla caldera and the history of volcanic activity in the area. Surface geothermal activity was mapped in detail, and the connection between tectonic manifestations and geothermal activity was found to be essential. A simplified tectonic map of the area is shown in Figure 10.3.

The investigation of the CO_2/H_2 ratio and the amount of H_2 in fumarole steam indicated that the hottest fluid (245–285°C) should be expected under the gully Hveragil.

The resistivity survey was performed in the years 1971 and 1972, and additional measurements were done in 1976 and 1977. The Schlumberger DC resistivity method has been mainly used, but several dipole soundings and magnetotelluric measurements have been done in the area as well. At shallow depth (less than 800 m) the resistivity measurements show a good correlation between surface alteration and the low resistivity region. For greater depths the resistivity measurements show a somewhat complicated picture. Inside the low resistivity region, the resistivity seems to increase beneath 800–1000 m depth. This observation could be a skin effect, but subsequent resistivity measurements in drillholes indicate that the increase in resistivity with depth is real. In Figure 10.6 the measured resistivity at 600 m depth is shown. The active geothermal area, mapped by surface alteration

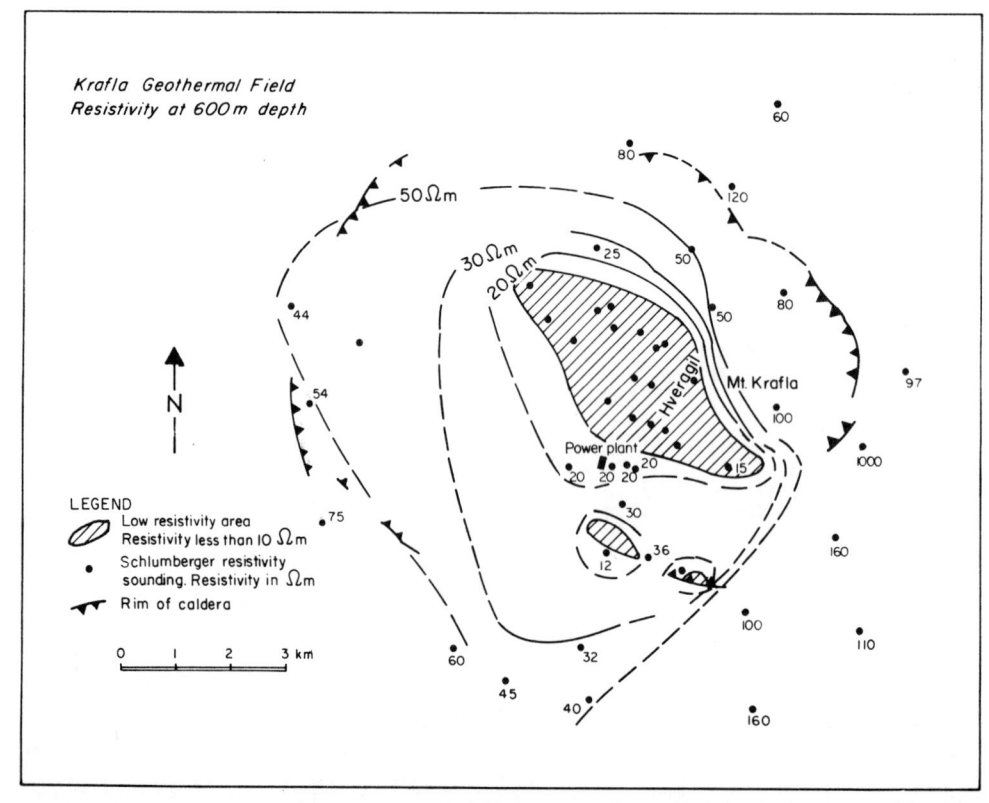

Figure 10.6. Resistivity at 600 m depth in the Krafla geothermal field. Mapped by R. Karlsdóttir. Reproduced by permission of Orkustofnun

and resistivity survey is approximately 30 km². This area is relatively large compared with other investigated high-temperature fields in Iceland. Therefore, it was initially concluded that the geothermal potential of the Krafla field was high.

The aeromagnetic survey showed a prominent magnetic low in the southern part of the caldera, and the strongest magnetic anomaly is found in the southern part of the geothermal field. The NW-SE elongation of the geothermal field is also prominent in the magnetic map shown in Figure 10.7. The magnetic low associated with geothermal activity is interpreted to result from alteration effects on magnetic minerals.

The seismic refraction measurements gave inconsistent results, and these have not been used further for the geothermal investigation of the field. In 1975 three seismometers were installed in the Krafla area. The seismic investigation revealed the existence of a magma body at a depth between 3 and 7 km inside the Krafla caldera (Einarsson, 1978).

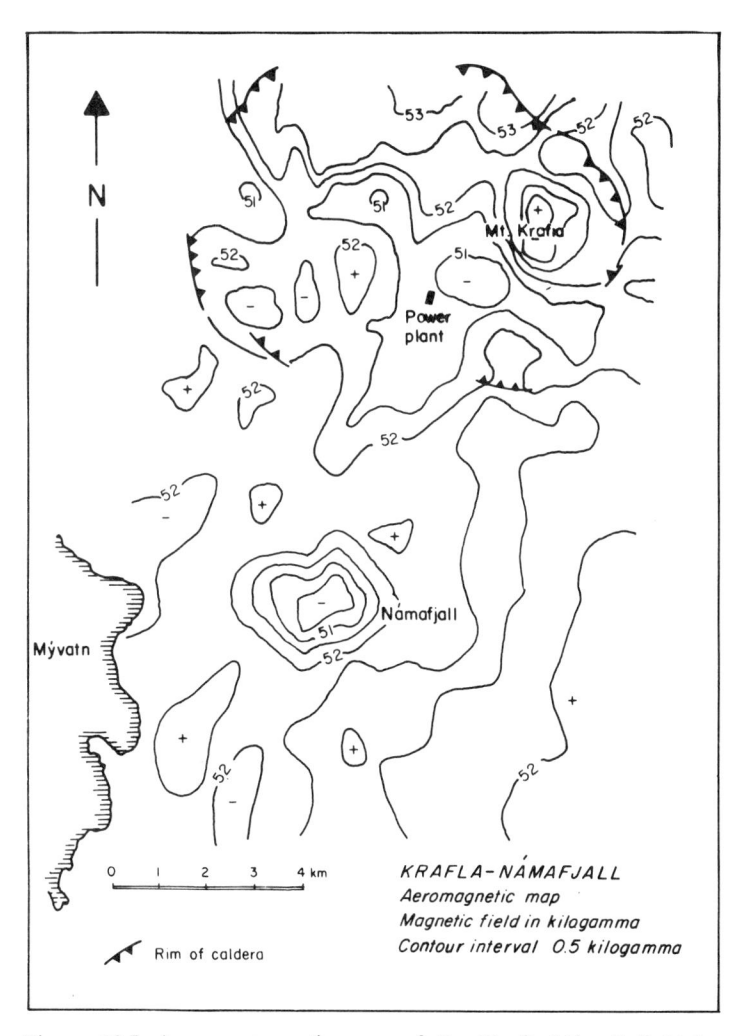

Figure 10.7. An aeromagnetic map of the Krafla-Námafjall high-temperature fields. From Pálmason (1975). Reproduced by permission of Orkustofnun

Gravity measurements have been repeated frequently in the area in connection with the rifting episode. A Bouguer map of the area is shown in Figure 10.8. The caldera effect, as well as the NW–SE elongation structure is prominent in the gravity map.

10.3.2. Subsurface Exploration

Two 1100 m deep exploratory wells were drilled in 1974. In one of the wells, a temperature of 300°C was found. At this stage the temperature conditions in the field were thought likely to be close to the boiling point curve. In the years 1975 and 1976 nine additional wells were drilled, and the twelfth well was drilled in 1978. The location of the wells is shown in Figure 10.9. The investigations carried out concomitant with the drilling and during flow tests were: petrological logging and mineralogical and geochemical study of the cuttings, temperature and pressure measurements in both static and flowing wells. resistivity and SP logs in wells, permeability tests both with injection and drawdown tests, measurements of enthalpy and chemical composition of the discharged fluid, production characteristics of wells, and analysis of deposits in the wells.

The subsurface rocks can be split into three main lithological units; the hyaloclastite formation, the lava formation, and the intrusive formation (Figure 10.10). The hyaloclastite formation in the uppermost 800–900 m is composed of primary and reworked products from subglacial eruptions. It is subdivided by a thick suite of subaerial lavas. Below 800–900 m depth, sequences of subaerial lava flows are dominant and hyaloclastites are rare. Dykes are common below 400 m depth. A multiple sill is located below the central part of the drilling area at 1100–1300 m depth. Below 1500–1600 m depth in the northern and southern part of the drilling area the rocks are entirely intrusive.

The compositions of the lavas and hyaloclastites are mostly near to saturated tholeite basalt, but range from olivine tholeiite to quartz tholeiite. Thin layers of acid tuff occur in the drilled section. The intrusive rocks are most frequently of basaltic composition, but granophyre intrusions appear also. The rocks in the hyaloclastite formation are completely recrystallized and the lavas are highly altered. The alteration pattern shows a fairly regular zoning (see Kristmannsdóttir, 1978). A retrograde transformation of the sheetsilicates above 1200 m depth level has been observed in some of the wells. The degree of alteration is comparable to that of zeolite to greenschist facies metamorphism. The transition between zeolite and greenschist facies alteration is at about 800 m depth.

The temperature measurements yielded values up to 345°C, which is the highest temperature recorded so far in geothermal wells in Iceland. The pressure profile of static wells indicated hydrostatic pressure in the field. The enthalpies of the wells were usually high, and the chemical composition of the fluid showed great variation within the field. In order to compile all the results and observations to a single solution, a new model of the geothermal field was necessary. The new model of the field was first presented in January 1977, but additional information confirmed its validity. All present observations are in agreement with this model.

10.3.3. The Model of the Field

The geothermal system in Krafla consists of two separate geothermal zones. The shallower one, extending down to ca. 1100 m depth, is a water-dominated system with a mean temperature of 205°C. The deeper zone ranges from about 1100–1300 m depth to at

Figure 10.8. Bouguer gravity map of the Krafla area. Mapped by G. Johnsen. Reproduced by permission of Orkustofnun

least 2200 m, which is the depth of the deepest well. The lower zone is boiling, i.e. the fluid in the formation is a mixture of steam, water and CO_2. The temperatures in this zone range from 300°C to 350°C, and both temperature and pressure are found to be close to saturation. The two zones are connected by an upflow channel near the gully Hveragil. A

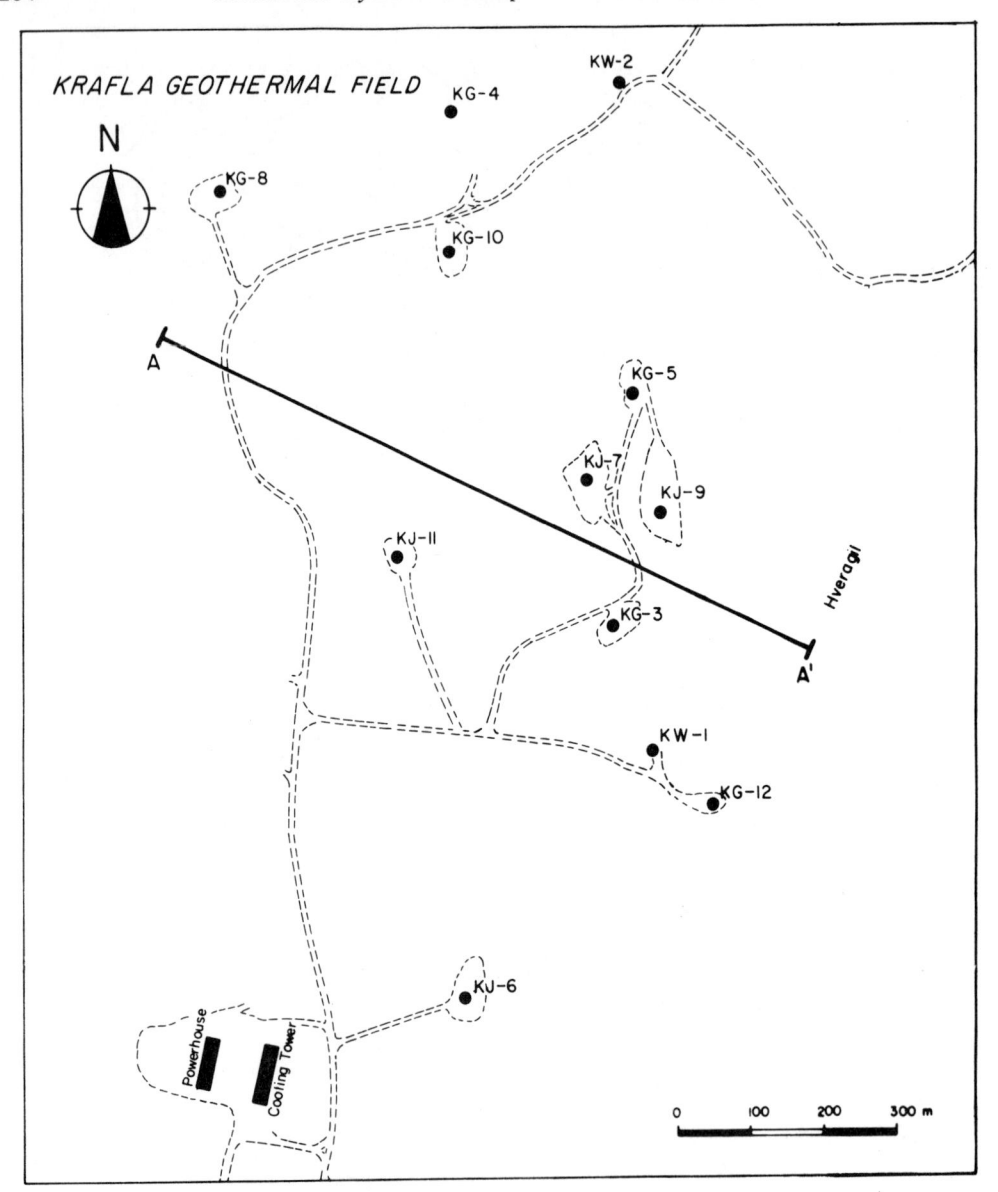

Figure 10.9. Location of the wells and of cross section in the Krafla geothermal field. A–A': trace of Figure 10.10. Reproduced by permission of Orkustofnun

simplified model of the Krafla geothermal field is shown in Figure 10.11. The facts that most of the wells have inflows from both zones and that the lower zone is a two-phase system, made the investigations more complicated. Most of the investigation methods had to be revised. As an example, all calculations of reservoir fluid chemistry had to be modified to include two-phase conditions in the initial stage. The complexity of the reservoir was realized at the same time as volcanic activity and spreading episodes began in the area. The influence of magmatic activity thus became an additional factor to be accounted for in the interpretation of the observations made in the geothermal field.

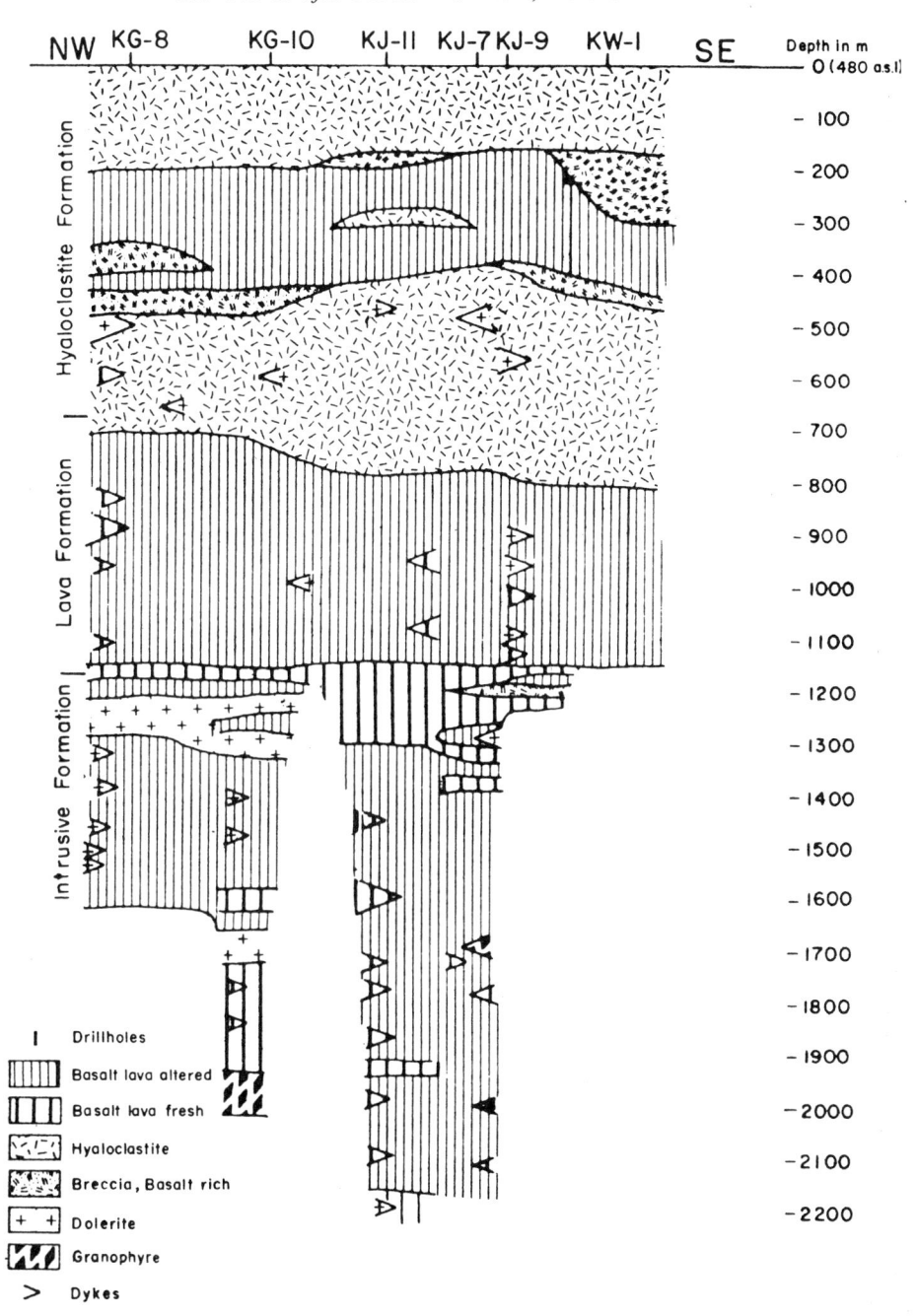

Figure 10.10. Geological cross section of the drilling area in Krafla. From Kristmannsdóttir (1978). Reproduced by permission of Orkustofnun

10.4. Production Characteristics of the Field

The production characteristics of the field cover a wide spectrum. The characteristics of the upper zone are rather uniform but quite different from the properties of the lower zone. In some of the wells an interaction between the two zones characterizes the

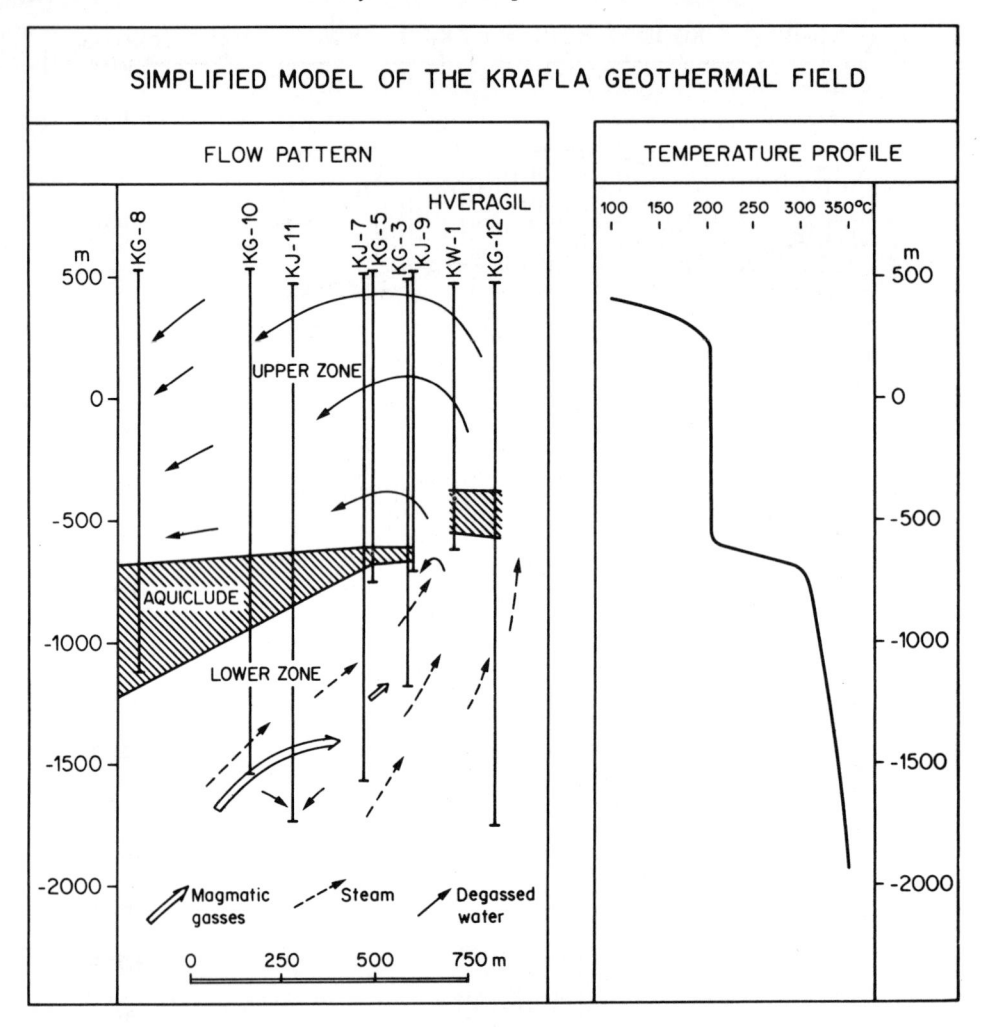

Figure 10.11. Simplified model of the Krafla geothermal field showing the main flow pattern and the average temperature profile of the field. Reproduced by permission of Orkustofnun

behaviour of the wells. Superimposed on these properties is the influence of the magmatic activity on the geothermal system.

10.4.1. *The Upper Zone*

The upper zone (200–1100 m depth) is a typical water-dominated geothermal reservoir. This zone is present in all the wells. The temperature is rather uniform, both horizontally and vertically over the 900 m thickness and is measured in the range 195–215°C, depending on the location of the wells. There is a good agreement between the measured temperature in the wells and the silica and Na–K–Ca temperatures. Pressure measurements in the discharging wells confirm the water phase nature of their inflow.

The distribution of the H_2/H_2S ratio of gases in the discharges of those wells which are fed only by the upper zone show (Figure 10.12) that the deep water in the more westerly

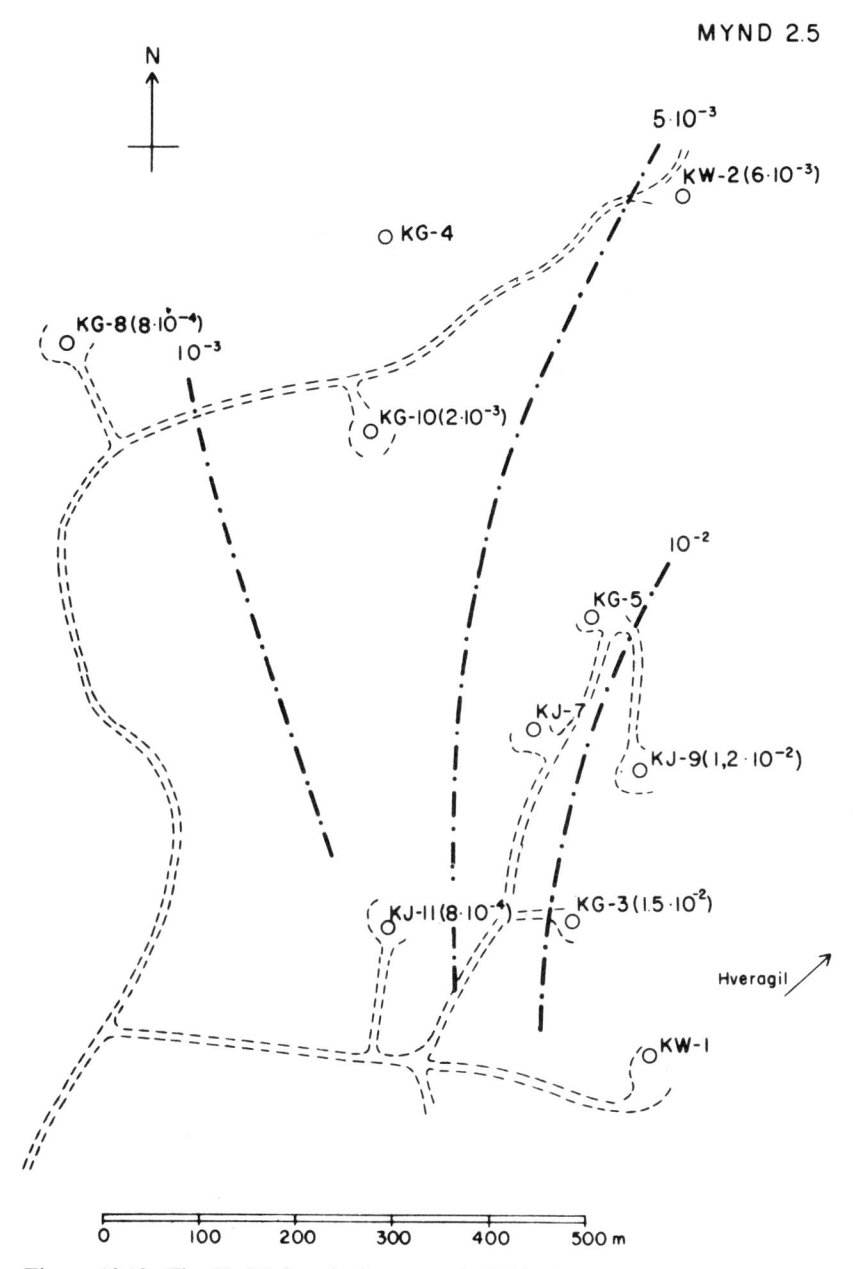

Figure 10.12. The H_2/H_2S ratio in reservoir fluid of the upper zone in Krafla. From Gíslason et al. (1978). Reproduced by permission of Orkustofnun

wells is more degassed than in the wells near Hveragil. A similar picture is found for the CO_2/H_2S ratio as well as for the total amount of gas in the discharge.

Pressure measurements in undisturbed wells indicate that pressure increases towards Hveragil. These observations indicate that the upper zone gets a through flow from the lower zone, rather than being an independent convection cell.

The chemical composition of the water from the upper zone is similar to other Icelandic water-dominated high-temperature fields. The major element composition of the unflashed reservoir fluid is shown in Table 10.1.

The water entering wells KW-2, KG-8 and KJ-9 is saturated with respect to calcium carbonate ($CaCO_3$). Degassing of flashing fluid in wells results in higher pH, which greatly increases the carbonate concentration. The calcium concentration increases upon boiling too. Due to boiling the activity product of calcium carbonate exceeds the saturation curve (Arnórsson, 1978a; 1978b). The calcium carbonate activity product for a representative upper zone well is shown in Figure 0.13 for a water decreasing in temperature as it boils. The supersaturation is greatest immediately after flashing, and the largest deposit should be formed at the level of boiling.

The rate of deposition in wells can be estimated by comparing the calcium concentration in deep water samples to that of the discharge at the well head. Such

Table 10.1. Average composition in mg/kg of the reservoir fluid in the upper zone.

pH	Na	K	Ca	Mg	SO_4	Cl	F	CO_2	H_2S	H_2	SiO_2/$TSiO_2$
7·5	183	18·2	2·2	0·04	186	27	0·7	228	54	0·24	300 210°C

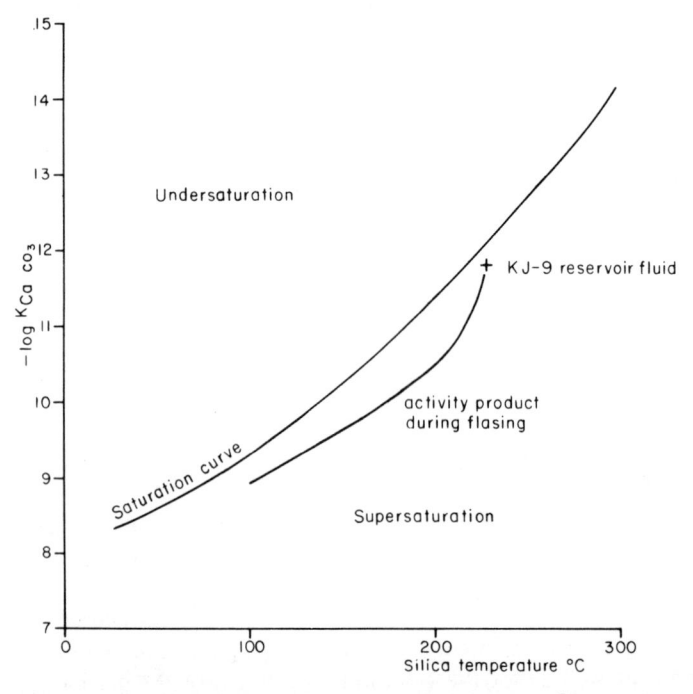

Figure 10.13. The computed activity product of Ca^{2+} and CO_3^{2-} in the geothermal water during one step adiabatic flashing in relation to the calcite solubility curve. Compiled by G. Gíslason and T. Hauksson. Reproduced by permission of Orkustofnun

10. The Krafla Geothermal Field, Northeast Iceland

investigation carried out in well KJ-9 revealed that there was an average 3·9 ppm decrease in the calcium concentration of the well fluid during boiling. For the actual flow of the well the deposition rate was 156 mg/s of $CaCO_3$, which corresponds to a deposition of 1·0 m^3 $CaCO_3$ during 180 days of flow. The well was actually worked over after a 180 days flowing period. The producing liner was pulled out of the well and the deposit measured. Its extent was greatest at the level of boiling (290 m depth), and the total volume was found to be 1·1 m^3. Chemical and X-ray analysis showed it to be almost pure calcite with traces of aragonite.

Permeability of the upper zone has been investigated in different ways. Injection tests, draw-down and recovery measurements have been made on individual wells. The mean permeability thickness is of the order of $10 \cdot 10^{-12}$ m^3.

The most spectacular method of permeability determination is the use of pressure pulses occurring in the upper zone coincident with rifting events and volcanic eruptions. The change in water level in well KG-5 during the event of September 8, 1977, is shown in Figure 10.14. By matching the decay part of the pressure curve with Theis log-log type curve (see Matthews and Russell, 1967) the response time of 8 hours is obtained. The distance from the well to the eruption site was 4·3 km, and by assuming the pressure transmission to be in 200°C water and the porosity to be 0·15 we obtain the permeability of $1·5 \cdot 10^{-12}$ m^2. Compared with the permeability·thickness of $10 \cdot 10^{-12}$ m^3, this indicates an aquifer thickness of the order of 10 m. As has been pointed out by Grant (1978), the pressure transient in Figure 10.14 is just the Green's function, and the late time response is decaying like time^{-1}. These circumstances are in agreement with the assumption of injection at a point source into a confined aquifer.

Due to the relatively low temperature of the upper zone and to the inconvenience of the calcite depositions associated with the water from the upper zone, an attempt is now made to avoid the upper zone water in the utilization of the field. This is done partly by casing off the upper zone, and partly by moving the drilling area towards the upflow channel, where the two zones are connected.

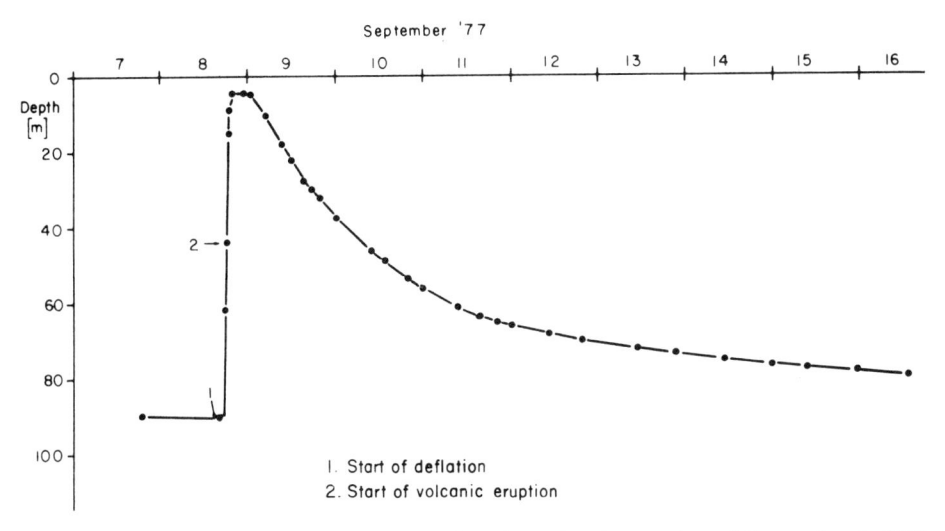

Figure 10.14. Water level in well KG-5 during the volcanic eruption of September 8, 1977. Compiled by B. Steingrímsson. Reproduced by permission of Orkustofnun

10.4.2. The Lower Zone

The undisturbed temperature and pressure in the lower zone are very close to saturation. The main characteristics of this zone are a high gas content and a high enthalpy of discharged fluid. Investigation of the chemical composition of the geothermal fluid has given very valuable information on the flow pattern within this zone. It has been demonstrated that volcanic gases (SO_2, H_2, CO_2) flow into the geothermal system from the underlying magma. The SO_2 is quickly reduced to H_2S during its passage through the formation. The gases CO_2, H_2S, and H_2 enter the investigated area in the neighbourhood of well KG-10 and propagate together with the steam up and eastward in the system. Other volatile elements like Hg and Rn follow this pattern closely. Radon is an interesting parameter in this respect as ^{222}Rn is radioactive with 3·8 days half life. By normalizing the Rn content to the CO_2 content in the fluid, the age of the fluid relative to the age of the fluid in well KG-10 can be determined. The following result was obtained.

	Age of fluid relative to KG-10
Well no. KG-10	0 days
Well no. KJ-7	1·9 days
Well no. KJ-6	5·4 days
Well no. KJ-9	9·9 days degassed water
Well no. KJ-11	11·2 days degassed water

For wells KJ-7 and KJ-6 this gives a flow velocity of 185 m/day.

Usual geochemical methods like silica and Na–K–Ca geothermometers cannot be used directly on a two-phase system. These methods assume the initial stage to be in pure water phase and the boiling to be isenthalpic. These conditions are not fulfilled for the lower zone in Krafla.

The main chemical characteristics of the fluid from the lower zone are shown in Table 10.2. Deposits consisting mainly of FeS, FeS_2, Fe_3O_4, and SiO_2 have been found in some of the wells tapping the lower zone fluid. The deposition rate has been most rapid in well KG-10, where the plug-in time has been recorded twice. Each time it was about three weeks. The deposition is believed to be at least partly due to magmatic influence on the geothermal fluid. The high concentration of SO_2 and other magmatic gases has a great influence on the pH-value, causing the fluid entering the wells to be highly supersaturated with respect to iron compounds.

The main characteristic of the boiling lower zone in Krafla is its response to production. Simple model calculations give that the steam fraction in the lower zone is of the order of 0·1–0·2 by volume. When utilization begins, the wells produce a mixture of steam and water, but the relative permeability of steam and water will influence the flowing history. The permeability of steam is higher than that of water, and when a certain steam fraction is reached, the water phase will be stagnant in the rock. The amount of water in the discharge will decrease while the amount of steam remains fairly constant. This means that the enthalpy of the discharge will increase. After a certain period the wells produce dry steam. The stagnant water then evaporates and contributes to the steam phase. During the evaporation, heat is withdrawn from the surrounding rock. When all the water close to the well has evaporated superheated steam will be produced.

One well, KG-12, produces exclusively from the lower zone. The flowing history of this well has followed the above description in detail.

Table 10.2. Chemical composition of fluid from wells, drawing predominantly from the Lower Zone

Well No.	KJ-6	KJ-7	KJ-9	KJ-11	KG-12
Sampling pressure MPa	0·98	0·78	1·04	0·22	0·83
Enthalpy of flow KJ/kg	1500	1900	1241	1483	2600
Steam fraction at P_s	0·37	0·58	0·23	0·44	0·92*
pH/°C of water phase	8·48/20	7·25/20	9·06/20	9·11/21	6·67/19
Ωm/°C of water phase	13·3/22	10·2/21	11·6/22	10·1/22	6·8/21
Silica† temperature (°C)	279	272	263	271	289
Chemical composition, expressed as mg/kg of total flow.					
CO_2	8436	17345	1357	5583	19048
H_2S	680	1046	75	199	801
H_2	8·1	26	1·4	1·1	49
CH_4	0·18	0.17	0·09	0·09	0·04
SiO_2	498	303	513	465	66
Na	95	82	158	132	19
K	16	12	18	22	4·8
Ca	0·79	0·98	1·3	0·95	2·1
Mg	0·006	0·008	0·008	0·006	0·01
SO_4	84	74	175	71	10
Cl	20	43	38	21	21
F	0·84	0·50	0·62	0·51	0·13
Fe	<0·06	0·21	0·13	<0·05	0·15
Total dissolved solids	831	564	962	852	140

*This flow developed into superheated steam.
†Calculated, taking enthalpy values into account.

10.4.3. Interaction Between the Zones

As demonstrated by the model, the two zones are connected near Hveragil, where a natural upflow channel keeps the lower zone in boiling condition. Other connections between the zones have not been discovered so far. If wells are open to both the upper and the lower zone, interference effects are observed in the behaviour of the wells. Effects like pressure oscillations during injection tests and flow from the upper aquifers down to the lower zone are common. One of the wells turned out to be very sensitive to load variation, as small pressure variation at wellhead could quench the lower aquifer.

As the discharge from wells taps fluids from both zones, the interpretation of chemical analysis becomes complicated. However, in the wells which produce from both zones it has been possible to estimate the amount of inflow from each zone by assuming the enthalpy and chemical composition of the upper zone to be fairly constant and using the total measured enthalpy of the discharge together with the measured concentration of the mixture at well head. The mass ratio of the lower zone contribution is found to be in the range 0·25 to 0·75 in the mixed wells.

10.4.4. Influence of Magmatic Activity on the Reservoir

As shown on Figure 10.14, the rifting events and magmatic activity cause a pressure impulse in the water-dominated upper zone. Similar pressure transients have not been detected in the lower zone. This is interpreted to be due to the two phase condition of the lower zone. The pressure transient is absorbed quickly in the two phase system. Quite noticeable magmatic influence has also been observed on the chemical composition of the lower zone fluid as shown in Figure 10.15.

The presence of magmatic gases in the reservoir has aided the mapping of the flowing directions as well as the flowing velocity in the reservoir as described in section 10.4.2. However, these same gases seem to cause serious depositions in wells. This has been strongly experienced in well KG-10, which is closest to the inflow of magmatic gases.

10.5. Experience with Utilization

The Krafla Power Plant is now operating with partial load and further drilling is needed to provide steam for the Power Plant. The experience gained with utilization is rather small.

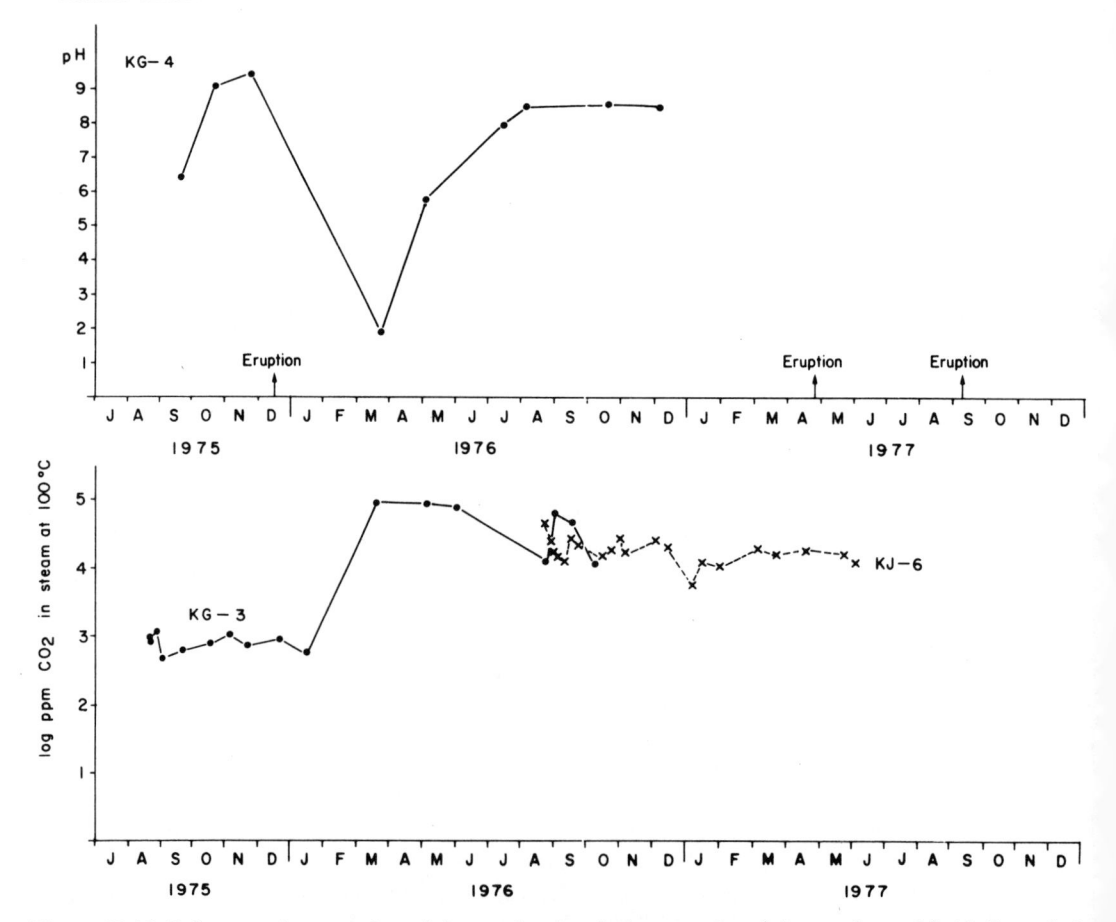

Figure 10.15. Influence of magmatic activity on the chemical properties of the geothermal fluid. Compiled by G. Gíslason. Reproduced by permission of Orkustofnun

The two turbine generators of the Krafla Power Plant consist of five stage double pressure, double flow steam turbines with 37·5 MVA brushless generators. The turbines are condensing with a direct contact condenser located under the turbine. The steam turbines have a maximum capacity of 35 MW each during the most favourable external conditions.

The steam separators are designed to separate the high pressure steam at 0·9 MPa pressure, and the HP inlet pressure of the turbine is assumed to be 0·77 MPa. The low-pressure separation is performed at 0·22 MPa pressure. The non-condensable gases (CO_2, H_2S) are removed by gas ejectors. The system can cope with 1–2 per cent of non-condensable gases in the steam. The pressure in the main condenser is estimated to be 0·012 MPa and the temperature of the condensed water 46°C. The condensate is cooled to about 22°C in cooling towers. Effluent water from the steam separators will be cooled down 10–20°C in cooling ponds and then dumped in a lava field south of the Krafla area. The first estimate of the amount of effluent water was 360 liters/s assuming the field to be water-dominated with the temperature of 270°C. The present knowledge of the field shows that utilization, restricted to the lower zone only, will give wells that produce more or less dry steam. Therefore, the environmental consequences of disposing effluent water in the area will be smaller than initially assumed.

The method of developing a geothermal field coincident with the construction phase of geothermal utilization has been experienced twice in Iceland. In the first case, the development of the Námafjall area (Ragnars et al. 1970), the risk taken did not have any consequences, as the production characteristics turned out to be more favorable than assumed. In the second case, the Krafla geothermal field, the production characteristics of the field turned out to be more complex than initially assumed, and the project became a very delicate political subject.

The experience gained in the development of the Krafla field has emphasized the necessity of the knowledge of the most essential production characteristics of a field prior to development, and that relatively higher cost for drilling should be accounted for when drilling is performed coincident with power plant construction.

The capital cost of the Krafla power plant inclusive of drilling and main transmission line is in November 1978 about 55 million U.S. dollars. The annual capital cost is estimated to be about 5 million US dollars. Additional finances will be needed for further drilling and transmission of steam.

From the scientific point of view, the experience gained in the investigation of the first proved boiling reservoir in Iceland has been very valuable. The coincidence that volcanic activity started to influence the geothermal field during the investigation has given rise to new ideas on the creation and nature of high-temperature geothermal fields in Iceland.

10.6. Current and Future Developments

The Krafla Power Plant is now operating at partial load and more drilling is needed for the production of steam. In the nearest future, additional drilling is planned east of the present drilling area in the neighbourhood of the upflow channel from the lower zone. The displacement of the drilling area in this direction is expected to result in better production characteristics of wells and less influence of magma on the reservoir fluid.

Volcanic activity is still going on in the Krafla area, and while this activity is continuing the investment in the field will probably be slow. However, assuming that volcanic activity will not cause a major damage to the Power Plant or the production wells, the aim of present and future operations in the area is to make the Power Plant operate at the rated capacity.

Acknowledgements

The present knowledge of the Krafla geothermal field is a result of an extensive team work of many people. Major contributions are due to: Dr. Halldór Ármannsson, Dr. Stefán Arnórsson, Dr. Axel Björnsson, Prof. Sveinbjörn Björnsson, Mr Gestur Gíslason,

Dr Jón S. Gudmundsson, Mr Gísli. K. Halldórsson, Mr Trausti Hauksson, Mr Gunnar Johnsen, Ms Ragna Karlsdóttir, Ms Hrefna Kristmannsdóttir, Dr Kristján Saemundsson, Mr Ómar Sigurdsson and Mr Benedikt Steingrímsson.

References

Arnórsson, S., 1978a. 'Precipitation of calcite from flashed geothermal waters in Iceland', *Contrib. Mineral. Petrol,* **66,** 21–28.

Arnórsson, S., 1978b. 'Framvinduskýrsla um efnafrae útfellinga í borholum við Kröflu', OS-JHD-7832, National Energy Authority, Reykjavik (Progress report on the chemistry of deposits in wells in Krafla–in Icelandic).

Arnórsson, S., Ragnars, K., Benediktsson, S., Gíslason, G., Thórhallsson, S., Björnsson, S., Grönvold, K., and Líndal, B., 1975. 'Exploitation of saline high-temperature water for space heating', *Proceedings Second United Nations Symposium on the Development and Use of Geothermal Resources,* 2077–2082.

Björnsson, A., Saemundsson, K., Einarsson, P., Tryggvason, E., and Grönvold, K., 1977. 'Current rifting episode in North Iceland', *Nature,* London, **266,** 318–323.

Björnsson, A., Johnsen, G., Sigurdsson, S. Thorbergsson, G., and Tryggvason, E., 1979. 'Rifting of the plate boundary in North Iceland 1975–1978', *J. Geophys. Res. 86,* 3029–3038.

Björnsson, S., 1970. 'A program for the exploration of high temperature areas in Iceland', *Geothermics,* Spec. Issue 2, **2**(2), 1050–1054.

Björnsson, S., Arnórsson, S., and Tómasson, J., 1970. 'Exploration of the Reykjanes thermal brine area', *Geothermics,* Spec. Issue 2, **2**(2), 1640–1650.

Björnsson, S., Arnórsson, S., and Tómasson, J., 1972. 'Economic evaluation of the Reykjanes thermal brine area', *Bull. Am. Ass. Petr. Geol.,* **56,** 2380–2391.

Einarsson, P., 1978. 'S-wave shadows in the Krafla Caldera in NE-Iceland, evidence for a magma chamber in the crust', *Bull. Volcanol. 41,* 1–9.

Fridleifsson, I. B., 1978. 'Applied Volcanology in Geothermal Exploration in Iceland', *Pageoph,* **117,** 242–252.

Gíslason, G., Ármannsson, H., and Hauksson, T., 1978. 'Krafla, Hitaástand og gastegundir í jarðhitakerfinu', OS-JHD-7846, National Energy Authority, Reykjavík (Krafla, *Temperature Conditions and Gases in the Geothermal Reservoir,* in Icelandic).

Grant, M., 1978. Personal communication.

Jakobsson, S. P., Jónsson, J., and Shido, F., 1978. 'Petrology of the Western Reykjanes Peninsula', *J. Petrol.* 19, 667–705.

Kristmannsdóttir, H., 1978. 'Alteration of basaltic rocks by hydrothermal activity of 100–300°C.' In *International Clay Conference 1978,* 359–367, M. M. Mortland and V. C. Former (Eds.). Elsevier, Amsterdam.

Matthews, C. S., and Russell D. G., 1967. *Pressure Build up and Flow Tests in Wells.* Monograph vol. 1, Soc. Pet Eng. AIME, 171pp.

Ragnars, K., Saemundsson, K., Benediktsson, S., and Einarsson, S. S., 1970. 'Development of the Námafjall Area, Northern Iceland', *Geothermics,* Special Issue 2, **2**(1), 925–935.

Pálmason, G., 1975. 'Geophysical Methods in Geothermal Exploration', *Proceedings Second United Nations Symposium on the Development and Use of Geothermal Resources,* 1175–1184.

Pálmason, G., Ragnars, K., and Zoega, J., 1975. 'Geothermal Energy Developments in Iceland 1970–1974', *Proceedings Second United Nations Symposium on the Development and Use of Geothermal Resources,* 213–217.

Saemundsson, K., 1974. 'Evolution of the Axial Rifting Zone in Northern Iceland and the Tjörnes Fracture Zone', *Geol. Soc. Am. Bull.* **85,** 495–504.

Saemundsson, K., 1978. 'Fissure swarms and central volcanoes of the neovolcanic zones of Iceland', *Geol. Journal,* Special Issue No. 10, 415–432.

Stefánsson, V., 1975. 'Jordvärme; Islands billigaste energy', *Forskning och Framsteg,* **5/75,** 15–17.

Stefánsson, V., 1979. Geothermal Activity in the Axarfjördur Area, *Jökull* (in press).

Walker, G. P. L., 1975. 'Excessive Spreading Axes and Spreading Rate in Iceland', *Nature,* London **255,** 468–471.

11. The Geothermal System of the Jemez Mountains, New Mexico and Its Exploration

A. WILLIAM LAUGHLIN

Geosciences Division, Los Alamos Scientific Laboratory, Los Alamos, New Mexico, U.S.A.

11.1 Introduction

The Jemez Mountains of north-central New Mexico, U.S.A. are volcanic in origin with the culminating activity occurring between 1·4 and 1·1 m.y. b.p. (Doell et al., 1968) when the Toledo and Valles Calderas were formed. This late-Cenozoic volcanic activity provided a large heat source causing the development of a complex geothermal system, which includes individual high and low temperature hydrothermal and hot dry rock systems. Various parts of the total system are currently under evaluation for electricity generation and for space heating by the U.S. Government, several large geothermal companies, and the small village of Jemez Springs, New Mexico. The variety of different exploration and development activities occurring within the area makes the Jemez Mountains unique among the geothermal provinces of the world.

A large portion of the central Jemez Mountains, surrounding the two calderas, has been designated a Known Geothermal Resource Area (KGRA) by the U.S. Geological Survey. The KGRA encloses a large privately owned tract of land (Baca Grant #1) on which the Union Oil Company of California has demonstrated the existence of a high-temperature hydrothermal system. The Baca Grant includes most of the area within the Valles Caldera, the younger of the two calderas. This hydrothermal system is presently under evaluation and development by Union Oil, Public Service Company of New Mexico, and the U.S. Department of Energy.

West of the Valles Caldera, the Los Alamos Scientific Laboratory (LASL) has for several years been testing a new method of extracting geothermal energy from rock that is hot but of very low permeability, Hot Dry Rock (HDR). A number of geological, geophysical, and geochemical investigations have been undertaken in support of this program.

Low-temperature hydrothermal systems are also present within the Jemez Mountains. These waters have long been used for balneological purposes, but recently their utility for space heating has been recognized, and several small scale projects are under development or are in the planning stage. On the west side of the mountains, the village of Jemez Springs has been funded by the State of New Mexico to develop some of the hot and warm springs for space heating of the village, and drilling began during the winter of 1979. On the east side of the Jemez Mountains, the Los Alamos Scientific Laboratory is also currently investigating the possibility of using low-temperature geothermal resources for heating the Laboratory.

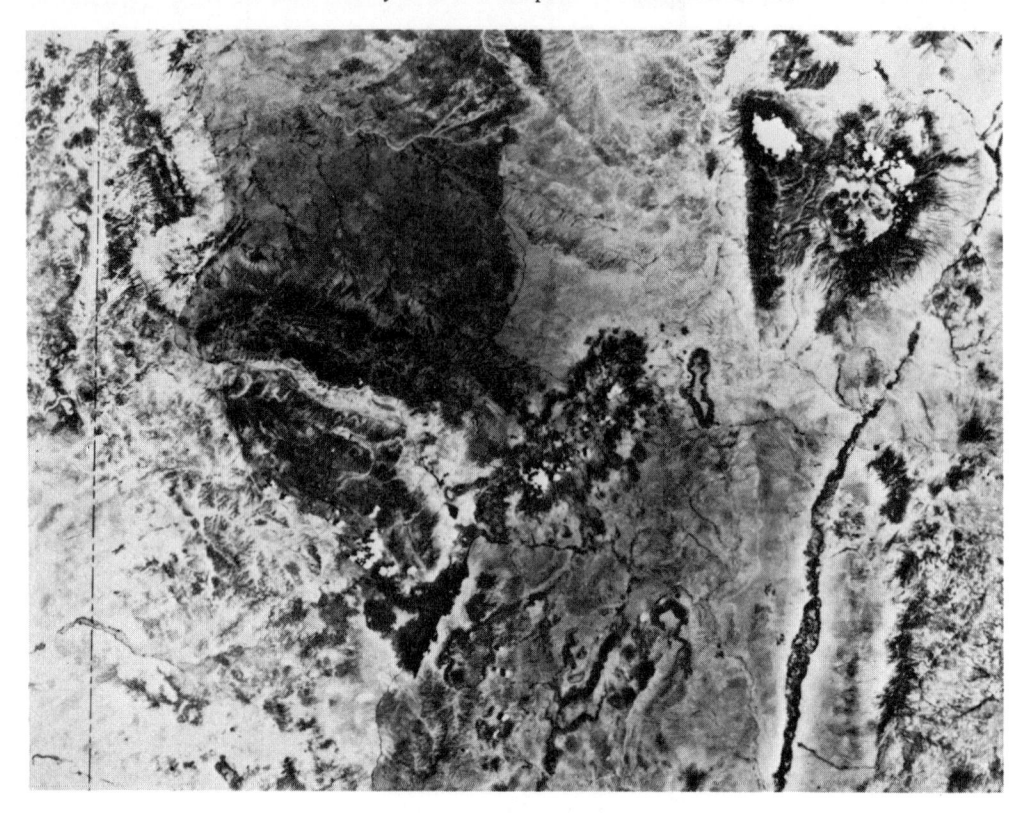

Figure 11.1A. LANDSAT photograph of North Central New Mexico showing the Jemez Mountains

11.2. Regional Geological and Geophysical Setting of the Jemez Mountains

The Jemez Mountains are located in northcentral New Mexico (Figures 11.1. and 11.2) on the boundary between the Colorado Plateau Province and the Rio Grande Rift. This latter feature may be considered a north-south trending extension of the Basin and Range Province, narrowing significantly to the north. The Jemez Mountains lie on the west side of the Espanola Basin, recently described by Manley (1978, 1979). The Jemez Mountains also can be thought of as lying at the intersection of the Rio Grande Rift and the Jemez Lineament or Zone (Mayo, 1958, Laughlin et al., 1972). As may be seen in Figure 11.1, the major Nacimiento Fault separates the Jemez Mountains from the San Juan Basin to the west. To the east, on the opposite side of the Rift, the Sangre de Cristo Mountains form a southern continuation of the Rocky Mountains.

Because the evolution of the Jemez Mountains is closely tied to the development of the Rio Grande Rift, considerable attention must be paid to the history of this feature. Several recent papers (Cordell, 1976, 1978, and Chapin and Seager, 1975) contain summaries and discussions of the geological and geophysical setting of the Rift and of its development and evolution in time. The results of a number of other geophysical investigations are also available (Reiter et al., 1975, Reiter et al., 1978; Sanford et al., 1973; Porath and Gough, 1971; Decker and Smithson, 1975; Pederson and Hermance, 1978; Bridwell, 1976, 1978),

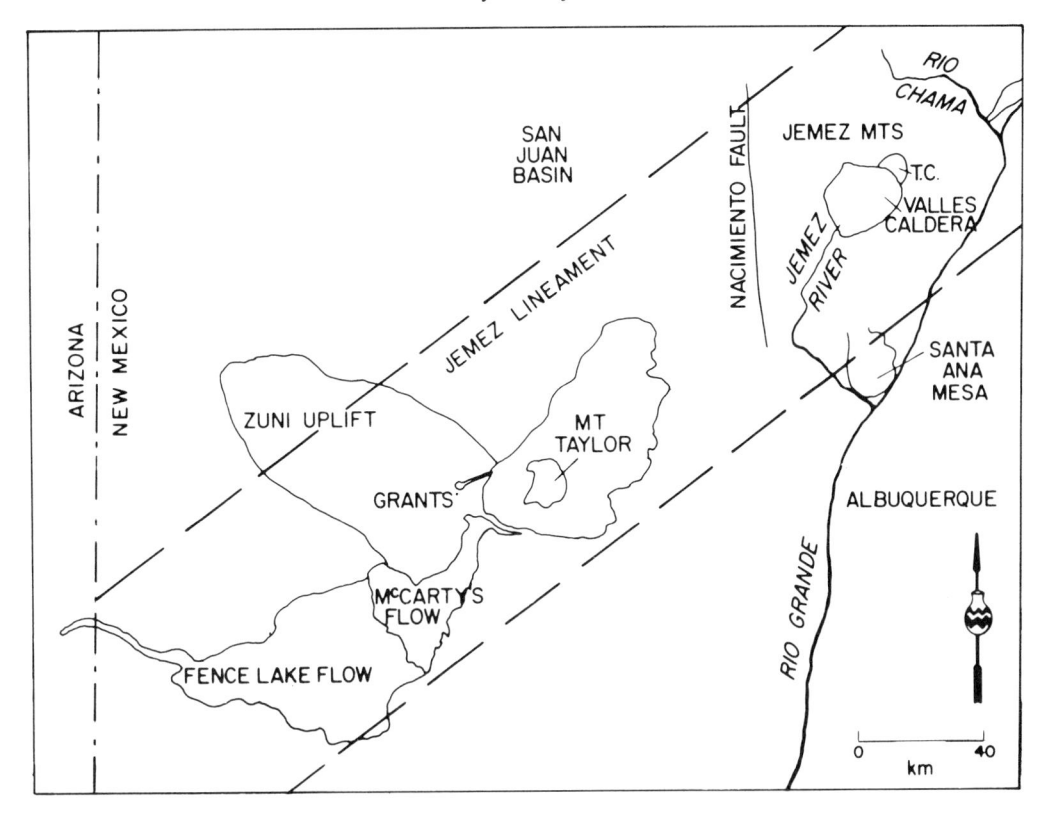

Figure 11.1B. Index map to accompany LANDSAT photograph 11.1A

clarifying the nature of deeper structures beneath the Rift and providing insight into its origin.

Although the cause is still in question, it is generally accepted that the Rio Grande Rift results from extensional tectonism. Cordell (1978) has suggested that the Rift may be coincident with a Precambrian basement suture where the Colorado Plateau was appended to the North American craton. On a smaller scale, the Precambrian basement structure, as recognized from aeromagnetic data, is also important in controlling the orientation of the graben faults which delineate the Rift. The dominant northwest-northeast Precambrian structural fabric results in a zig-zag pattern for the individual rift faults (Cordell, 1978) particularly on the west side of the Rift. On the east side, faults strike roughly N–S but are dextrally offset en echelon (Cordell, 1978). The locations of these graben faults are well defined by the abundant gravity data (Cordell, 1976, 1978; Aiken et al., 1977, 1978). At the site of the Jemez Mountains, the western graben boundary trends northwest beneath the Valles Caldera.

As has been discussed by Chapin and Seager (1975), the Rio Grande Rift began to form between 29 and 24 m.y. ago, and both faulting and volcanism occurred up until essentially the present. Volcanism has been common both spatially and temporally within the Rift, and this phenomena is well described by Lipman (1969), Aoki (1967), Ozima et al. (1967),

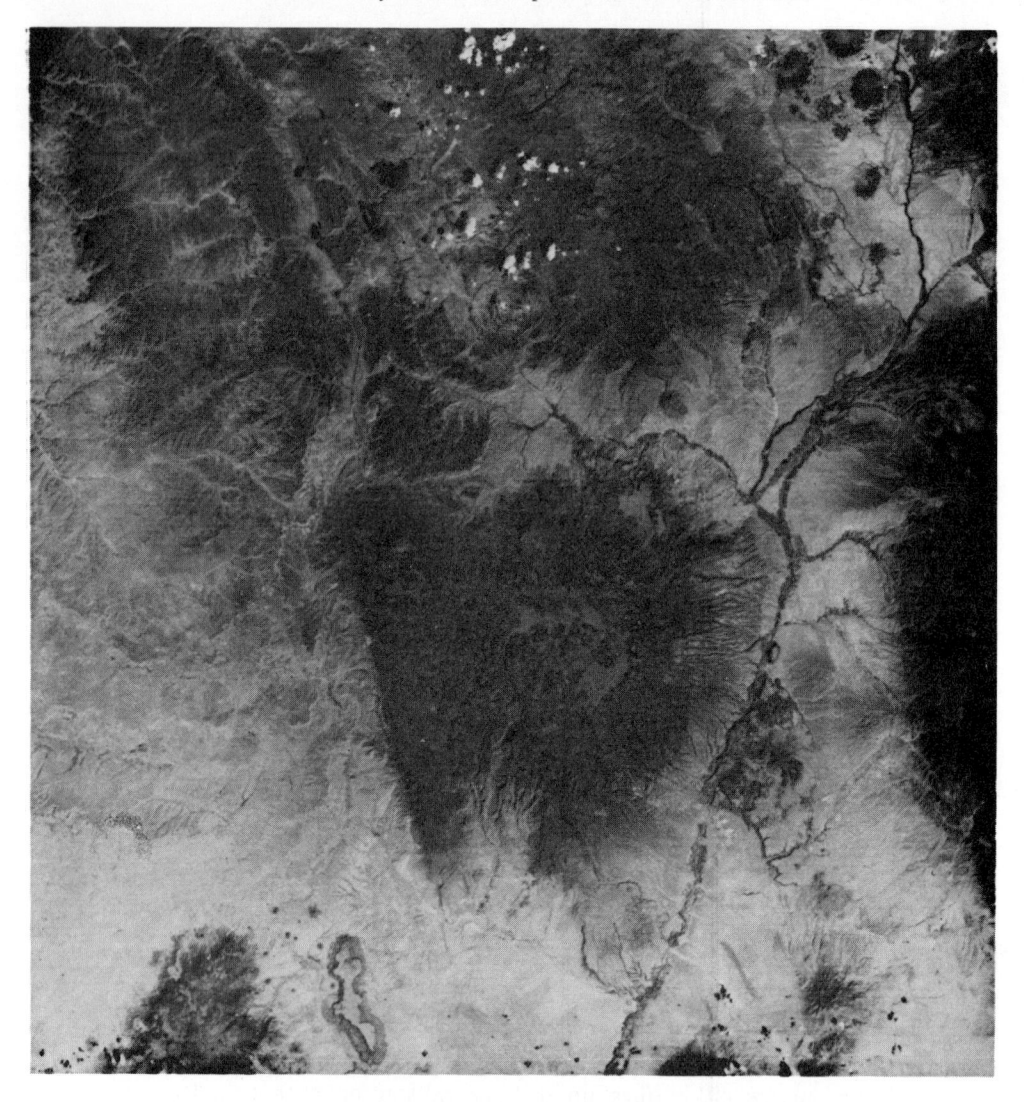

Figure 11.2A. LANDSAT photograph of Jemez Mountains

and Aoki and Kudo (1976). A general pattern has been noted of tholeiitic basalts erupting within the Rift and alkali basalts erupting on the flanks (Lipman, 1969). Lipman and Mehnert (1975) have postulated that the tholeiitic basalts fractionated at a depth of less than 35 km from a bulge in the upper mantle.

The Jemez Zone or Lineament, which intersects the Rio Grande Rift at the site of the Jemez Mountains, was first recognized by Mayo (1958). It is marked by an alignment of late-Cenozoic volcanic fields (Laughlin et al. 1972, 1976, 1979; Luedke and Smith, 1978) which extends from the Pinacate field of Sonora, Mexico to the NE corner of New Mexico. Although the majority of the fields are predominantly basaltic, intermediate to silicic magmas erupted at Mt. Taylor, in the Jemez Mountains, Taos Plateau, and the Raton/Clayton volcanic fields, NM.

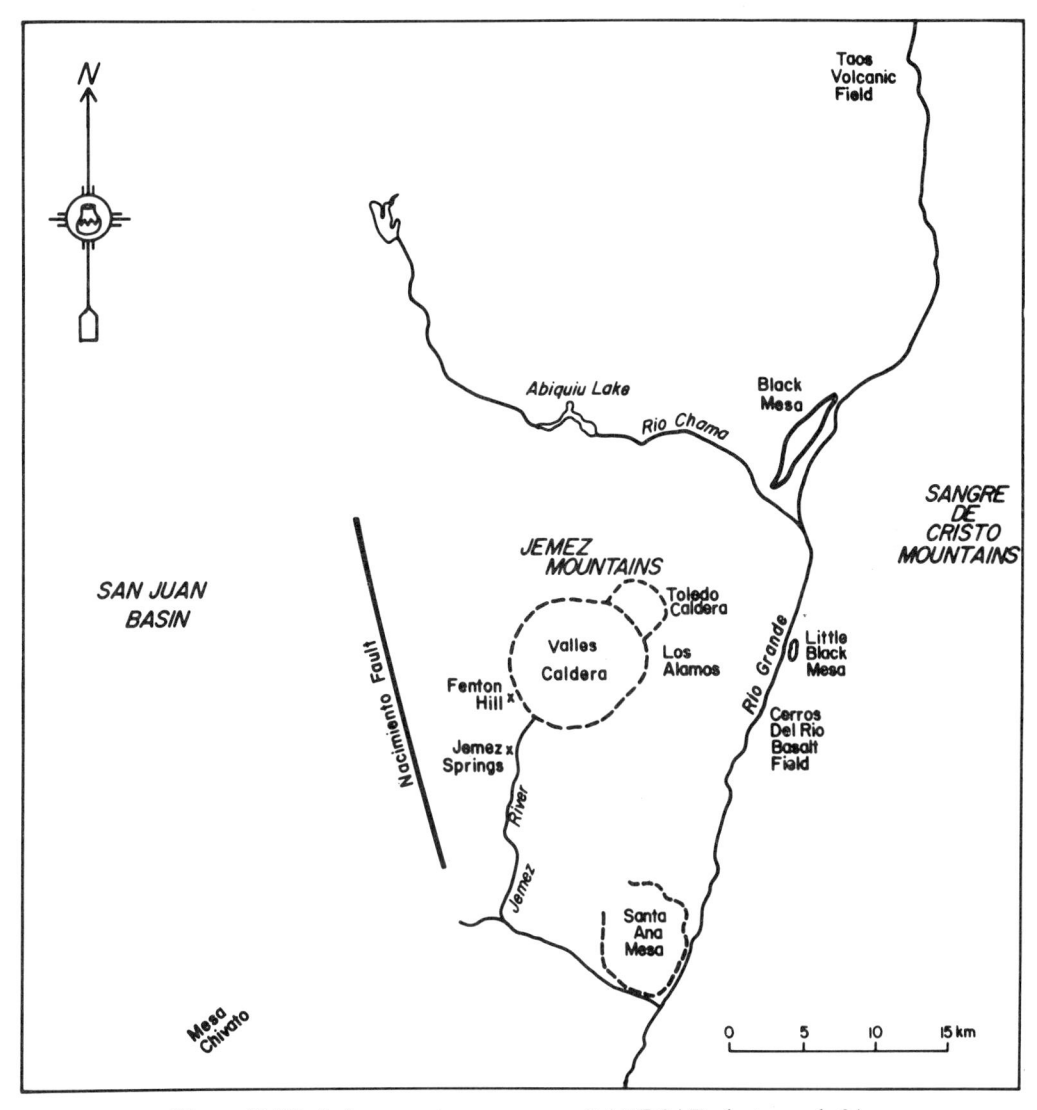

Figure 11.2B. Index map to accompany LANDSAT photograph 2A

Several investigations of heat flow have been made in New Mexico (Reiter et al. 1975, Decker et al. 1975, and Edwards et al. 1978), and more heat flow data are probably available for New Mexico than for any other state in the United States. The abundant data have been contoured most recently by Edwards et al. (1978) to present a reasonably complete picture of the regional heat flow within the state (Figure 11.3).

As can be seen in Figure 11.3, the state of New Mexico contains three general regions or zones of high heat flow. The most prominent of these is the belt of high heat flow along the Rio Grande Rift. Reiter et al. (1975) pointed out that the highest heat flow is not symmetrically located with respect to the Rift but is generally located near its western margin. To the west, a second zone abuts the Arizona–New Mexico state line and extends generally northeastward along the Jemez Lineament (Laughlin and West, 1976) roughly as far as Grants, NM. Unpublished data of Schearer and Reiter (oral communication,

300   *Geothermal Systems: Principles and Case Histories*

HEAT FLOW MAP of NEW MEXICO

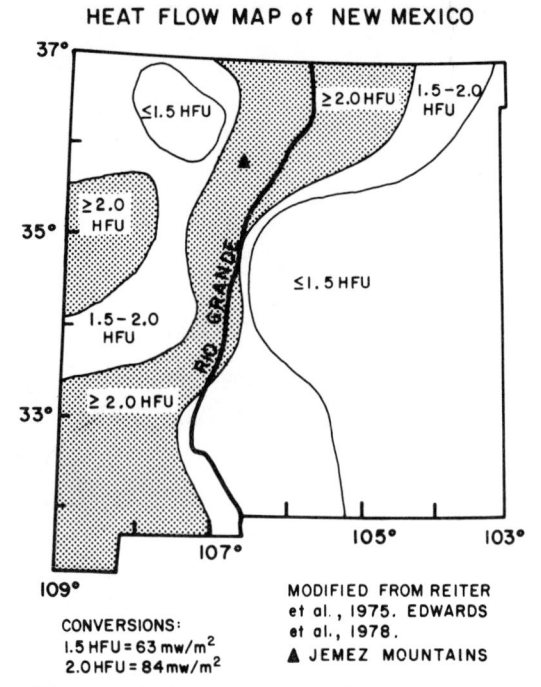

Figure 11.3. Generalized heat flow map of New Mexico, USA

1978) indicate that this zone extends southwestward into Arizona. The results of chemical geothermometry in Arizona by Swanberg (Hahman et al. 1978) support this conclusion. The third zone is in the northeast corner of the state along the extension of the Jemez Lineament. In the later interpretation by Edwards et al. (1978), this region is continuous with the high heat flow zone of the Rio Grande Rift. It is proposed here that the high heat flow in western New Mexico and in the NE corner of the state result from a northeastward trending zone of crustal weakness, the Jemez Lineament, which crosses the Rift in the vicinity of the Jemez Mountains. Basaltic magmas have risen along this structure for the last few million years (Laughlin et al., 1976, 1979) transporting both mass and energy from the mantle to the upper crust. All three zones are outlined by the 84 mW/m^2 contour and contain values >105 mW/m^2. The highest values are 250–670 mW/m^2 measured at four sites along the Rift (Reiter et al., 1978). These four sites are closely related to Holocene volcanic centers or to the intersections of caldera boundaries and large normal faults (Reiter et al., 1975). Edwards et al. (1978) conclude from heat generation data that the high heat flow values are not produced by high concentrations of the heat producing elements—uranium, thorium, and potassium—suggesting that tectonic or igneous processes are responsible for the positive anomalies.

Although the heat flow values along the Rift have a wide range as shown by Decker and Smithson (1975), Reiter et al. (1975), Reiter et al. (1978), and Edwards et al. (1978), the values are generally higher than those measured in the Great Plains Province to the east and the Colorado Plateau Province to the west.

11.3. Geology, Geophysics, and Hydrology of the Jemez Mountains

The Jemez Mountains are primarily volcanic, consisting of two central calderas, the Toledo and the Valles, surrounded by several plateaus capped by extrusive volcanic rocks.

11. The Geothermal System of the Jemez Mountains

The Pajarito Plateau on the east side of the mountains and the Jemez Plateau on the west are capped mainly by the Bandelier Tuff. South of the Valles Caldera, the flanks of the mountains are made up of the basalts of Santa Ana Mesa, the Cochiti Formation of Pliocene age, and basalts and andesites of the Pliocene Paliza Canyon Formation. Numerous small bodies of the intrusive Bearhead Rhyolite also occur. North of the Valles and Toledo Calderas, the main mountain mass consists of dacites, rhyodacites, and latites of the Tschicoma Formation, in places capped by the upper member of the Bandelier Tuff. A generalized geologic map of the mountains is presented in Figure 11.4.

Volcanism has been semi-continuous within the Jemez Mountains from approximately 10 m.y. ago until essentially the present. Bailey et al. (1969) recognize three main groups of volcanic rocks in the area: from oldest to youngest, the Keres Group, the Polvadera Group, and the Tewa Group. The Keres Group, which spans the interval from 10·4 to 6·5 m.y., is comprised of the basalt of Chamisa Mesa, the Canovas Canyon Rhyolite, the Paliza Canyon Formation (basalt and andesite), and the Bearhead Rhyolite. The Polvadera Group consists of three units which erupted over the interval of 7·4 to 2·0 m.y.; the Lobato Basalt, the Tschicoma Formation, and El Rechuelos Rhyolite. The Tewa Group is comprised of the Bandelier Tuff and the Valles Rhyolite. The Bandelier Tuff, in turn, is comprised of the older Otowi Member (1·4 m.y.) and the younger Tshirege

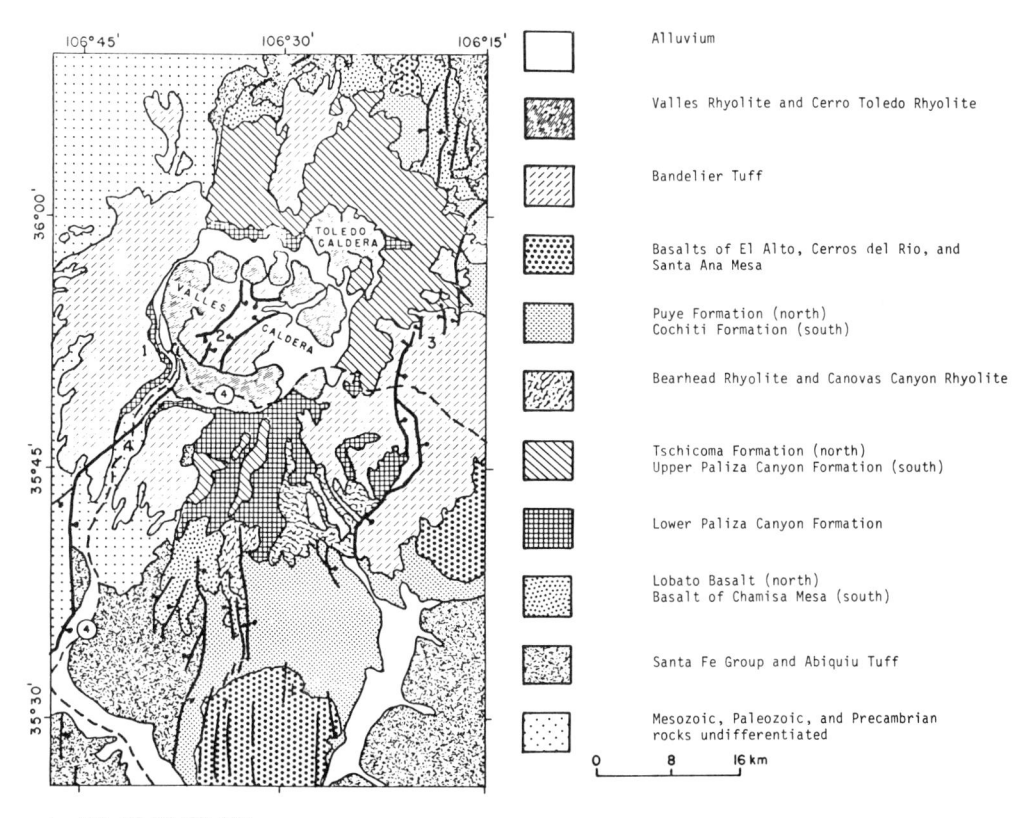

1. LASL HOT DRY ROCK SITE
2. UNION OIL SITE
3. LASL ALTERNATE ENERGY HOLE
4. JEMEZ SPRINGS SPACE HEATING HOLE

Figure 11.4. Generalized geologic map of Jemez Mountains, NM

Member (1·1 m.y.) separated by the Cerro Toledo Rhyolite. Both the Otowi and the Tshirege Members are made up of an older ash fall overlain by ash flows. The Valles Rhyolite is composed of several members, some of which may be as young as 0·1 m.y.

Two episodes of caldera collapse have occurred within the region (Smith and Bailey, 1966). The first took place approximately 1·4 m.y. ago after eruption of the Otowi Member of the Bandelier Tuff and formed the Toledo Caldera. Eruption of the Tshirege Member of the Bandelier Tuff 1·1 m.y. ago also led to caldera collapse, producing the Valles Caldera. Rhyolite domes were then extruded in a circular pattern within the Caldera. The vents for these domes were inferred by Smith et al. (1970) to lie along the main ring fracture of the caldera. Within this ring of rhyolite domes is a large resurgent dome, Redondo Peak, which rises about 1000 meters above the floor of the caldera. Landslide deposits, fan deposits, lacustrine sediments, and rhyolite domes fill the moat between the caldera rim and Redondo Peak.

The volcanic rocks of the Jemez Mountains are cut by faults with three general trends, NS, NW, and NE. North-south striking faults, downthrown to the east or towards the Rift, are common on the southern and eastern sides of the mountains. On the west side of the mountains and within the resurgent dome, NE and NW striking faults are most common while north-south striking faults are again generally most common on the north side of the mountains. The general picture of faulting within the mountains is probably best summarized by north-south striking faults parallel to the Rift and NE and NW striking faults reflecting Precambrian basement structures and possibly the influence of the Jemez Lineament.

A variety of geophysical methods have been utilized within the Jemez Mountains. These include seismic investigations of several types, gravity, aeromagnetic, heat flow, deep electrical resistivity, magnetotelluric (MT), and time-domain electromagnetic (TDEM). Because many of the more recent studies were directed toward geothermal exploration they will be summarized below within that section, and only the more general studies will be cited here.

Gravity surveys and compilations are available at several scales for the Jemez Mountains (Cordell, 1970, 1976; Aiken et al., 1977, 1978; Budding, 1978). The small-scale maps show a close relationship between the gravity patterns and such major structural features as the Rio Grande Rift, Nacimiento Uplift, and the Valles and Toledo Calderas. Steep gradients mark both the Nacimiento fault and the border faults of the Rio Grande Rift. A prominent gravity low lying beneath the Valles and Toledo Calderas is attributed by Dondanville (1978) to 1830 m of tuff filling the Caldera. The Precambrian structural contour map of Cordell (1976) suggests that the Rift boundary extends to the northwest beneath the calderas.

A large-scale gravity map of the Pajarito Plateau portion of the Jemez Mountains was recently published by Budding (1978). This map shows a NNE trending graben that lies essentially beneath the town of Los Alamos. The graben is marked by a 18 mgal gravity low. This structure may reflect underlying Precambrian basement structure since it is apparently a continuation of gravity lows to the northeast shown on the map of Cordell (1976).

Aeromagnetic data (Cordell, 1976) also indicate that the Rio Grande Rift trends northwestward beneath the Jemez Mountains. This trend continues into northwestern New Mexico where it appears as a magnetic gradient. An orthogonal (NW–NE) aeromagnetic pattern is apparent within the Jemez Mountains region.

No discussion of a geothermal system would be complete without an examination of the hydrogeology of the area. Such an examination is critical when evaluating reservoir recharge in hydrothermal systems and it is pertinent in predicting and assessing environmental problems related to any type of geothermal development. Several recent papers have presented the results of hydrologic investigations in the Jemez Mountains, (Trainer, 1974; Purtymun and Johansen, 1974). The results are summarized below.

The surface drainage of the Jemez Mountains is controlled by the Rio Grande and its several tributaries, including the Rio Chama, Rio Salado, Jemez River, and the Rio Guadalupe. These rivers essentially encircle the mountains; the Rio Grande drains the eastern flanks, the Rio Chama lies to the north, and the Jemez River and the Rio Salado close the circle to the west and north. The Jemez River and its tributaries drain the Valles Caldera while the Rio Guadalupe and its tributaries drain the area west of the caldera. Within the main mass of the mountains, modern alluvium covers at least the floor of all the canyons, and, as Trainer (1974) pointed out, this material has the highest permeability of any formation in the mountains. Tertiary valley fill generally has lower permeability because of a finer grain size and increased cementation (Trainer, 1974). These streams are recharged through alluvium, broken crustal rocks, and faulted and fractured exposures in the higher central portion of the mountains. Some recharge into these streams is due to influx from perched water tables in the near-surface volcanic and sedimentary rocks and from movement of deep seated mineralized waters upwards along faults. The relative fractions of the different recharge sources vary from one stream to another. In the Jemez River, deep seated geothermal waters are much more important than they are in the Rio Guadalupe, where recharge is mainly from perched aquifers (Trainer, 1974). Within the San Diego Canyon of the Jemez River, Trainer (1974) has shown that mineralized geothermal waters enter the river near Soda Dam and Jemez Springs. These waters carry relatively high concentrations of Li, B, F, and As.

Within the Valles Caldera, surface and ground water move towards the caldera moat from the mountains which form the topographic rim of the caldera. Infiltration into the alluvium covering the caldera floor is rapid. Water not penetrating to depth along faults in the caldera floor eventually flows out to the southwest-recharging the Jemez River.

Springs in the Jemez Mountains result from either the intersection of perched water tables with the rugged topography or from the upward movement of water along faults. Only the latter type is of significance in geothermal investigations. In San Diego Canyon, apparently fault-controlled warm to hot springs occur at Soda Dam and near the village of Jemez Springs. At Soda Dam, the fault control is more obvious, especially in the west wall of the canyon where mineralized, 50°C water escapes. Trainer (1974) has estimated that the total flow at Soda Dam is about 19 L/s of which 4–6 L/s comes from the springs in the west wall. These waters contain 50 mg/L SiO_2, 1500 mg/L Cl, and 900 mg/L HCO_3; abundant CO_2 gas and a trace of H_2S evolve with the waters, which have built a large travertine dam across the Jemez River at this point. The river has maintained its flow under the east abutment of the dam.

Near or within the village of Jemez Springs, ten springs flow out of the alluvium. Both the composition and a linear orientation parallel to the canyon suggest that the location of these springs is fault controlled (Trainer, 1974). The maximum temperature of the Jemez Springs is 75°C, and the total flow is 13 L/s (Trainer, 1974). The SiO_2 content is about 79 mg/L, Cl is 900 mg/L and HCO_3 is 700 mg/L.

East of the Valles Caldera, beneath the Pajarito Plateau, the main aquifer consists of the Puye and Tesuque Formations of the Santa Fe Group. Purtymun and Johansen (1974) state that the gradient on the aquifer in this area is about 11 m/km and that the depth to the aquifer decreases from about 365 m in the west to 183 m in the east near the

Rio Grande. Gradients are steeper near the river. The stretch of the Rio Grande east of the Jemez Mountains receives considerable recharge from this aquifer.

11.4. Geothermal Exploration in the Jemez Mountains

Because of surface expressions such as hot springs, fumaroles and rock alteration, and because of the youthfulness, magnitude, and duration of igneous activity, the Jemez Mountains have been the scene of geothermal exploration for over ten years. The U.S. government, several private geothermal companies, and the small village of Jemez Springs, NM have either conducted their own investigations or have subcontracted with academic or industrial workers for geothermal studies. Unfortunately, much of the industrial data are proprietary, and this report is based largely on data acquired by the USGS, by LASL, or by its subcontractors during development and testing of the HDR concept. The greatest part of these data, however, is pertinent to any type of geothermal exploration.

It should be kept in mind throughout this discussion that the different organizations had different goals in mind throughout their exploration programs. Union Oil and the other commercial geothermal companies were and are searching for high-temperature geothermal fluids that can be exploited in the conventional manner. This clearly requires a permeable reservoir as well as an adequate heat source. The LASL was also looking for a high-temperature heat source but desired an impermeable reservoir suitable for testing their energy extraction method. The village of Jemez Springs and more recently LASL have sought aquifers capable of producing large amounts of moderate-temperature water for space heating purposes. The different exploration goals have dictated different exploration strategies and have influenced the exploration techniques employed.

11.4.1. High-Temperature Hydrothermal Exploration

Industrial geothermal exploration of the Jemez Mountains was strongly influenced by three kinds of surface geothermal manifestations; (1) spatially and temporally widespread occurrence of silicic volcanism, (2) widespread recent rock alteration, suggestive of interaction of surface rocks and acidic hydrothermal fluids, and (3) numerous hot springs (Dondanville, 1978). As a result of the concentration of these surface 'shows' in the Sulphur Creek area, this area received early attention, and four wells were drilled in the vicinity. According to Dondanville (1978) the first of these, Westates Petroleum-Bond #1, was drilled as an oil test but flowed steam from a fracture at a depth of 1113 meters. Three hot but essentially dry holes were than drilled by the Baca Land and Cattle Company. Credit for the discovery steam well goes to Baca's fourth well in the nearby Redondo Creek area (Dondanville, 1978). Union Oil Company leased the area from the Baca Land and Cattle Company in 1971 and has continued exploration and drilling to the present time.

Although surface manifestations influenced the initial drilling within the Valles Caldera, Dondanville (1978) states that a number of geophysical techniques have been employed to evaluate the reservoir. These include gravity, aeromagnetics, thermal I.R., shallow temperature gradient drilling, microearthquake, seismic ground noise, dipole–dipole, dipole–bipole, expanding Schlumberger, electromagnetics, magnetotellurics, and telluric profiling. Except for those surveys conducted by the USGS and discussed elsewhere in this report, the results of these studies are proprietary and have not been released. Dondanville (1978) has published certain reservoir characteristics based on the results of the geophysical surveys and the drilling of 17 holes. He states, 'The principal geothermal resource

discovered to date, in Redondo Creek, is a liquid-dominated under
base temperature in excess of 260°C and salinity in the order of
total dissolved solids. A maximum temperature of 330°C has been
have encountered a vapor-dominated reservoir overlying the liqui
Production is principally from fractures in the pumicey basal
Bandelier Tuff, which is up to 6300 feet (1920 m) thick in the
Typical wells are 5000 feet (1525 m) to 9000 feet (2745 m) deep. Re___
fluid exist in all rocks underlying Redondo Creek, probably including the Precambrian granite.'

Dondanville (1978) attaches great importance to the graben faults within Redondo Peak for controlling the movement of hydrothermal fluids.

At the present time, negotiations are being concluded between the U.S. Department of Energy, Union Oil Company of California, and the Public Service Company of New Mexico to build a 50 MW electricity generation plant in the Redondo Creek area. Union Oil will provide steam to the plant, which will be built by the Public Service Company. The reserve potential of the field has been estimated to be at least 400 MWe.

11.4.2. The Hot Dry Rock Geothermal Energy Extraction Method

Because the Jemez Mountains of New Mexico are the site of the first Hot Dry Rock Demonstration Project, it is pertinent to review briefly here this energy extraction method and its present state of evaluation.

At the present time, all commercially developed geothermal systems rely on the presence of naturally occurring hydrothermal fluids to transport heat from a hot subsurface reservoir to the Earth's surface. The presence of these natural hydrothermal fluids implies that the reservoir is permeable, allowing principally meteoric water to move through, extracting heat from the reservoir rocks. Despite the best efforts of geothermal workers, however, numerous nonproductive wells are drilled even in very prolific fields, and it is clear that hot but impermeable rocks are common within the crust. In 1971, the Los Alamos Scientific Laboratory conceived the idea for a method of artificially creating a permeable reservoir and extracting the heat from within it (Robinson et al. 1971). The basic idea is simple, but actual development and testing have proved extremely difficult at times (Smith, 1975).

The concept calls for drilling a deep hole to some depth where the rock temperature is sufficiently high to be useful for either electrical generation or direct utilization. Assuming that the potential reservoir rocks are of low permeability, hydraulic fracturing is used to generate a large heat exchange surface at the bottom of the hole. A second well is drilled to intersect the heat exchange surface near its upper extent. Water can then be circulated down the deeper hole, along the fracture or fractures where it is heated and then back to the surface through the second hole. Depending upon the temperature, the water may be used for electrical generation and/or space heating before being returned down the first hole.

An initial system was created at Fenton Hill on the west side of the Valles Caldera in order to evaluate this extraction method. Two deep holes (GT-2 and EE-1) were drilled to a depth of about 3 km and a temperature of approximately 200°C. Hydraulic fracturing was used to connect the holes and to form a heat exchange surface. Circulation was established between the two holes, and to date, closed system circulation experiments have been conducted for approximately 2400 hours. A small, 10 MWt surface heat exchanger system was used to remove heat from the fluid before reinjection, and during the

inary evaluation period the heat was simply rejected to the atmosphere. Water loss
was very low during these experiments, decreasing to less than 2% at the end of the
experiments.

During the circulation experiments, the chemistry of the water was carefully monitored in all parts of the system: injection water, make-up water added to the system to replace the minor water losses, and the hot return water. The level of total dissolved solids in the returned water has remained low, i.e. about 2000 ppm, throughout the experiments.

With success demonstrated in the first evaluation of the extraction method, LASL is now beginning a second test wherein a deeper, hotter, and larger reservoir will be created. Drilling of EE-2 began in April, 1979 with an anticipated total depth of about 4 km and a bottomhole temperature of 250°C. Hydraulic fracturing will again be used to create the downhole heat exchange surface. Circulation will then be established by deepening one of the two present holes, GT-2 or EE-1, to intersect the fracture/s. Circulation experiments will again be conducted using the new and hotter reservoir.

11.4.3. *Hot Dry Rock Exploration*

The goal of the Hot Dry Rock exploration programs within the Jemez Mountains was to locate a site near Los Alamos where subsurface temperatures were high and where potential reservoir rocks could be expected to have low permeability. Several different exploration techniques were used prior to drilling, and several have been employed after drilling in order to evaluate exploration methods and to better characterize the site. The various methods utilized will be discussed below in order of their application to the exploration and evaluation program.

When it was decided to locate a drill site near Los Alamos for testing and developing the Hot Dry Rock method of geothermal energy extraction, reconnaissance work began in the nearby Jemez Mountains to select a suitable site. It was known from the work of Smith et al. (1970) that a youthful heat source probably existed beneath the Valles Caldera. Doell et al. (1968) had reported very young K-Ar ages on silicic volcanic rocks associated with formation of the two calderas, with the youngest post-caldera volcanism occurring 0·043 m.y. ago. The presence of the hot springs, fumaroles and very recent rock alteration discussed above also suggested a shallow still active heat source. Supporting this conclusion were newspaper accounts of the results of the early industrial drilling activity for hydrothermal fluids within the resurgent dome in the Valles Caldera.

Because the initial work suggested that a still active heat source was present beneath the Valles Caldera, attention was focused on locating a drill site where impermeable reservoir rocks might be expected. Emphasis was placed on areas where the impermeable reservoir rocks would lie at relatively modest depths. The map and cross-sections of Smith et al. (1970) indicated that the frequency of faults was high on the Pajarito Plateau on the east side of the caldera and within the caldera itself. It was anticipated that this faulting would result in fractured and permeable reservoir rocks. The relative scarcity of faulting in the Jemez Plateau and the shallow depths to the Precambrian basement suggested that exploration might profitably be concentrated in that area.

To further evaluate the heat source, shallow to intermediate depth gradient holes were drilled around the periphery of the Valles Caldera. Seven shallow holes, drilled to depths of about 30 m, indicated that the gradients were highest on the west side of the caldera (Potter, 1973; Reiter, 1976). It was suspected that the local hydrology might be perturbing the near-surface gradients, and four deeper gradient holes were drilled to confirm the earlier work (Figure 11.5). These holes (A–D) penetrated to depths of 152–229 m. Holes A,

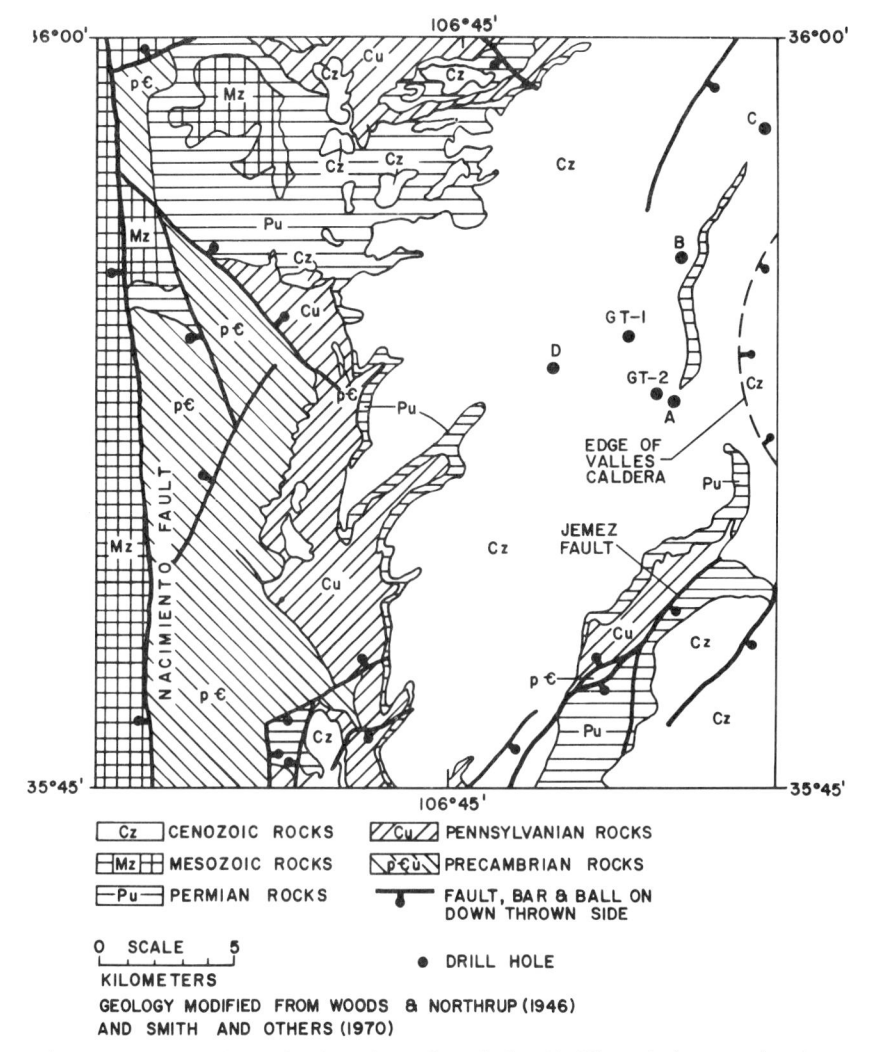

11. The Geothermal System of the Jemez Mountains

Figure 11.5. Location of LASL heat flow holes (A–D) and deep exploratory holes (GT-1, GT-2)

B, and C were drilled in an arcuate pattern 2·4 km outside the western portion of the caldera ring fault while 'D' hole was drilled at a distance of 6.4 km from the ring fault between and beyond holes B and A. Heat flow values of 213, 230, and 247 mW/m^2 were measured in holes A, B, and C respectively, indicating a slight increase in heat flow to the north. The heat flow in hole D was 92 mW/m^2, falling within the range of 63–105 mW/m^2 reported for the Basin and Range Province (Sass et al. 1971). These heat flow data suggest that the thermal anomaly associated with a possible Valles Caldera magma chamber falls off rapidly with radial distance from the caldera ring faults. The value of 92 mW/m^2 at a radial distance of 6·4 km may represent a regional background value since it corresponds closely to the average heat flow value of the Rio Grande Rift reported by Reiter et al. (1975).

Concurrently, deep slim hole drilling and an aerial-photographic fault investigation were employed to evaluate the Jemez Plateau as a Hot Dry Rock (HDR) site. A deep slim hole (GT-1) was drilled at a site in Barley Canyon, chosen primarily on the basis of the earlier heat flow measurements. Slim hole GT-1 reached the Precambrian basement at a depth of 642 m and had a total depth of 785 m. The bottom 47 m of the hole were continuously cored. Distinctly different gradient measurements were made in the Paleozoic and Precambrian sections of GT-1; 129°C/km and 45°C/km respectively. The change in gradient is thought to result from hydrologic disturbances produced by the flow of warm water through the permeable Madera Limestone near the Precambrian unconformity.

The drilling of GT-1 provided the opportunity for *in situ* measurements of the permeability of the Precambrian basement and for examination of core samples of these rocks. For overpressures of 1.3 to 17.7 MPa, the permeability measurements yielded values of 5×10^{-8} to 6×10^{-3} darcys (5×10^{-20} m^2 to 6×10^{-15} m^2). Hydraulic fracturing experiments were also performed in GT-1, and the results suggested very low-permeability rock. Core samples were principally granitic gneiss with small amounts of amphibolite (Perkins, 1973). The sealing of fractures by chlorite and calcite undoubtedly contributed to the low permeability of the basement rocks.

A preliminary version of the fault study by Slemmons (1975) indicated that the Barley Canyon area and adjacent parts of the Jemez Plateau were seismically quiet and lay within a large intact fault block. This block is bounded on the east by the caldera ring fault, on the north by the Calaveras Canyon fault, on the southeast by the Virgin Canyon and Jemez Springs faults and on the west by the Rio Cibolla fault.

After these results were evaluated, a new drill site was selected on Fenton Hill. Holes GT-2 and EE-1 were drilled at the Fenton Hill site. This site was logistically better than the Barley Canyon site since it was near the highway and power lines and was on the flat mesa top. The site is within a large burned area resulting from a 1971 forest fire. Drilling within the burn removed many environmental constraints which otherwise might have slowed testing of the HDR method. The work of Slemmons (1975) indicated that the nearest major fault was one of the caldera ring faults lying 1·5 km east of the site. The main caldera ring fault is 3 km east of Fenton Hill.

During and after the drilling of the deep holes GT-1, GT-2, and EE-1 at the Barley Canyon and Fenton Hill sites, a number of investigations were undertaken to either characterize the reservoir rocks or to evaluate techniques for exploration for hot dry rock.

The most directly applicable studies for reservoir characterization were the petrologic, geochemical, and geophysical logging investigations performed during drilling. During the drilling of GT-2, cuttings were collected at 1·5 m intervals, 27 cores were taken from the Precambrian section, and a number of geophysical logs were run. A detailed lithologic log of GT-2 was constructed from the results of the geophysical logging and from analysis of the cores and cuttings. This log is illustrated in Figure 11.6.

Drill hole GT-2 initially penetrated to a depth of 2·93 km. The upper 0·73 km consisted of Cenozoic and Paleozoic rocks; 2·20 km of Precambrian rocks were drilled. Laughlin and Eddy (1976, 1977) reported the results of petrographic and chemical analysis of the core samples obtained during this drilling. Summarizing briefly from their reports, the Precambrian section at Fenton Hill is largely made up of a metamorphic complex intruded by at least three igneous bodies. Roughly 75 per cent of the section consists of gneissic rocks, ranging in composition from syenogranitic to tonalitic. The majority of the sample were monzogranitic gneisses. A ferrohastingsite-biotite schist, which locally grades into amphibolite, is interlayered with the gneiss. In general, the metamorphic rocks are strongly foliated and compositional changes are very abrupt. At least two 15 m thick

Generalized Lithologic Logs of GT-1 & GT-2 -EE-1

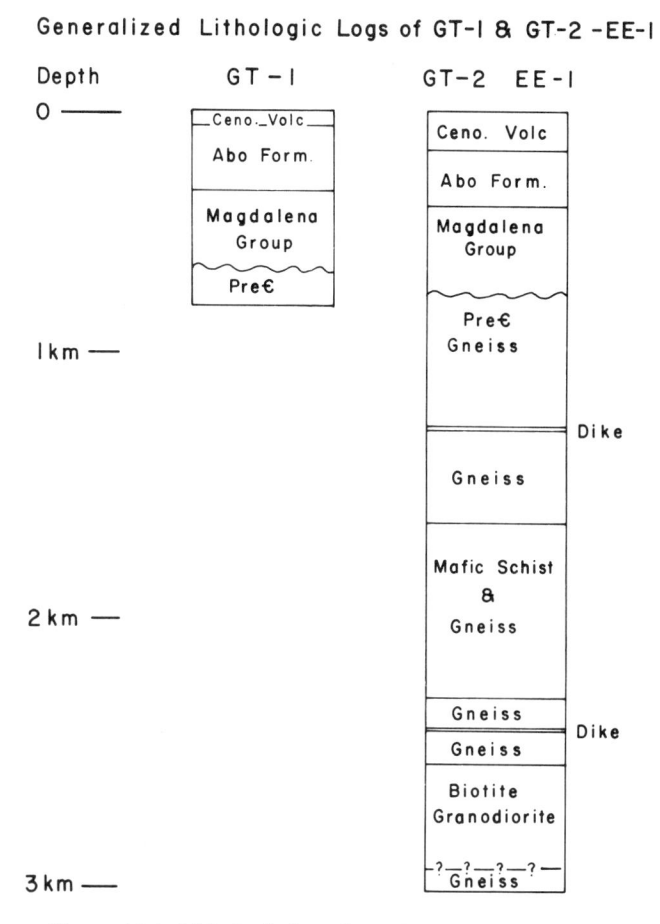

Figure 11.6. Lithologic logs from GT-1, GT-2, and EE-1

dikes of monzogranite intrude the metamorphic complex, as clearly indicated by the spectral-gamma logs (West and Laughlin, 1976) and one core. A large, biotite granodiorite body was penetrated by GT-2 at a depth of 2591 m. Based on cores, cuttings, and geophysical logs, the biotite granodiorite intrusive is approximately 385 m thick. Because the downhole heat exchange system was created within this unit, considerable attention was paid to this rock.

The biotite granodiorite is a very homogeneous, equigranular igneous rock characterized by high modal biotite and sphene contents and high concentrations of K_2O, TiO_2, and P_2O_5. Potassium feldspar contents range from 15 to 29 per cent, plagioclase (An_{31-36}) from 34 to 41 per cent and quartz from 23 to 35 per cent. Biotite makes up from 11 to 13 per cent of the rock and the sphene is relatively constant at about 2 per cent. The rock is only slightly altered with minor sericitization of the plagioclase and chloritization of the biotite.

Fractures were observed in all cores taken. The spacing between fractures commonly ranged from about 8 cm to 1 cm in the more intensely fractured regions. Horizontal, vertical, and steeply dipping fractures were present. In general, the fractures were tightly sealed with a variety of minerals. Calcite was the most common sealing mineral, but epidote, hematite, quartz, clays, and sulfides were also observed. Strontium isotopic studies by Brookins and

Laughlin (1976) indicated that the calcite was probably derived from the adjacent Precambrian rocks and not from the overlying Madera Limestone.

Several radiometric dating methods have been applied to the Precambrian rocks of the Fenton Hill site. The results of these investigations have been reported by Brookins et al. (1977), Brookins and Laughlin (1976), Naeser and Forbes (1976), Turner and Forbes (1976), and Zartman (1979). These results not only clarify age relations in the Precambrian reservoir rocks but aid in evaluating the thermal effects produced by formation of the Valles and Toledo Calderas. The Rb/Sr method yields data indicating that the metamorphic complex at Fenton Hill is approximately 1·66 b.y. old. This complex was intruded at roughly 1·3–1·4 b.y. by the biotite granodiorite magma. The age of the biotite granodiorite was confirmed by U–Th–Pb dating done by Zartman (1979). The results of the K/Ar dating (Turner and Forbes, 1976) indicate that the intrusion of the granodiorite was a major thermal event, since the K/Ar clock in the metamorphic complex was reset to agree with the age of the granodiorite, 1·3–1·4 b.y. Evidence for a second resetting of the K/Ar clock is apparent in the sample from the 2·90 km depth in GT-2 where the present temperature is 197°C. This temperature is sufficient to cause argon loss at the present time.

Because temperatures necessary for clock resetting (track annealing in this case) in the fission track method are relatively low, the results of Naeser and Forbes (1976) are particularly important in understanding the thermal effects produced by the latest igneous activity in the Jemez Mountains. An intermediate age tectonic event, uplift of Precambrian rocks, is also suggested by the fission-track results.

A large number of apatite and sphene fission-track ages have been measured on core samples of the Precambrian rocks from GT-1, GT-2, and EE-1. An apatite age was also determined on an amphibolite sample exposed at the surface in Guadalupe Box, 21 km southwest of Fenton Hill. Although the age of the Precambrian metamorphic rocks is well defined at 1·66 b.y. by the Rb/Sr results, the oldest apatite fission-track age is only 242 m.y. This age was obtained on the surface sample from Guadalupe Box. Naeser and Faul (1969) have shown that apatite completely anneals if held at temperatures above 170°C for times in excess of 1 m.y. Longer times will produce the same total annealing at even lower temperatures. These results suggest that the relatively young age of the Guadalupe Box sample resulted from uplift and consequent cooling of the Precambrian basement from a depth/temperature high enough to prevent track retention.

Fission-track ages on the core samples are still younger (Figure 11.7) reflecting the annealing caused by the present high geothermal gradients in the Barley Canyon-Fenton Hill areas. The apatite ages decrease from 69 m.y. at a depth of 0·79 km and a present temperature of 100°C to zero age at 1·88 km depth and 135°C present temperature. Sphene fission-track ages are concordant with the biotite granodiorite emplacement event down to a depth of 2·90 km where the temperature is now 197°C. At this depth and temperature, approximately 25 per cent track annealing has occurred in the sphene. Immediately up hole at 2·61 km and 177°C, zero track annealing has occurred in the sphene since the 1·3–1·4 b.y. event.

These data and the track loss curves of Naeser and Faul (1969) provide constraints on the geothermal gradient at Fenton Hill during the latest igneous event. At a depth of 1·88 km, apatite has undergone total annealing while the sphene shows zero annealing. If an age of 1 m.y. is assumed for the igneous event based on the age of the Bandelier Tuff, then the apatite results indicate that the temperature must have been in excess of 170°C at the 1·88 km depth.

Additional information on paleo-temperatures at Fenton Hill has been presented by

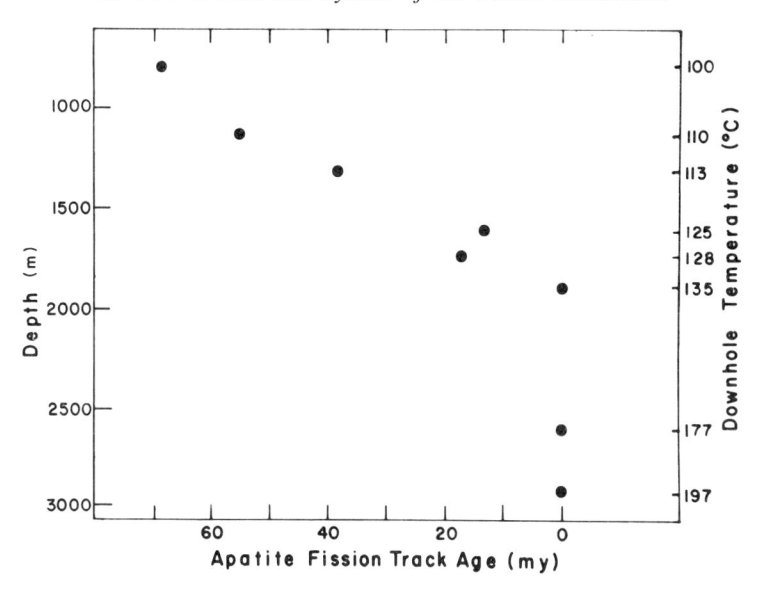

Figure 11.7. Variation of apatite fission track ages with depth in drill holes

Burruss and Hollister (1979). They measured homogenization and melting temperatures in over 300 fluid inclusions in samples from GT-2 and EE-1, finding that samples from depths greater than 1·5 km had reequilibrated in response to geothermal gradients greater than presently observed. They calculated a paleo-gradient of about 70°C/km and suggested that the maximum temperature at 3 km depth beneath Fenton Hill during the Valles event was approximately 230°C. This is 30°C higher than presently measured, indicating that the reservoir rocks are now cooling after this igneous event.

Knowledge of the nature of potential reservoir rocks is very important to a successful application of the HDR method. In the case of the Fenton Hill site in the Jemez Mountains, surface exposures of Precambrian basement rocks are very rare, and other ways of obtaining samples were examined. Eichelberger and Koch (1979) investigated Precambrian xenoliths within the Bandelier Tuff, assuming that these xenoliths represented basement rocks beneath the Jemez volcanic field.

Although the Bandelier Tuff contains an abundance of xenoliths, only trace amounts are of Precambrian rocks; a total of only 22 were found and examined in this study. Petrologic and geochemical work indicated considerable contrast between the xenoliths and the core samples from GT-2 and EE-1. Core samples were predominantly granitic in composition while xenoliths were roughly evenly divided between granitic and gabbroic compositions. The style of deformation also differed between the core samples and the xenoliths. The xenoliths commonly exhibited shear deformation affecting the biotite, quartz, and feldspars, and fractures were filled with the finely granulated material. Core samples showed evidence of more brittle fracturing, and the veins were sealed by minerals, such as calcite, deposited by fluids. The xenoliths were typically more altered and of a higher metamorphic grade than the core samples. High-fluorine biotite and actinolite were common in the xenoliths as alteration products of original hornblende.

The investigation of the xenoliths suggests that the Precambrian basement beneath the Valles Caldera is more mafic and more deformed and altered than is observed at Fenton Hill.

Jiracek (1975) has employed several electrical resistivity methods near the Fenton Hill site to characterize the manmade reservoir and the overlying volcanic and sedimentary rocks. Shallow Schlumberger soundings were used to evaluate the volcanic rocks down to depths of approximately 85 m. Both the surface Bandelier Tuff and the lower Bandelier Tuff and Paliza Canyon Formation at 85 m had very high resistivities. They were separated by a conductive zone within the Bandelier between 49 and 53 m, which Jiracek believes may be a clay zone. For deeper penetration, two lines of dipole-dipole resistivity soundings were laid out, intersecting at the Fenton Hill site. These survey lines were 3 km in length. A 50 kW resistivity transmitter was used to generate square-wave signals with peak amplitudes of 5 and 10 amperes. Although agreement was not good between models and the apparent resistivities, the preliminary work indicated that the resistant Precambrian basement could be probed using spacings greater than 2 km.

The roving dipole or bipole–dipole method was also used by Jiracek (1975) near Fenton Hill. In this case, 2 km long bipoles were maintained about 5 km west of Fenton Hill while over 100 receiver dipole stations were occupied. The bipole-dipole data suggest a relatively complex electrical structure at Fenton Hill. A NNE trending series of apparent resistivity lows passes through the drill site, while to the east, a resistivity high and then a low roughly parallel the caldera ring fault. Although Jiracek states that topography may produce some of the electrical effects, he believes that basement structures are also significant. In particular, he attributes the low resistivity zone near the ring fault to hot water moving up along the fault. He also suggests that the series of lows passing through the drill site may be caused by zones of weakness in the basement rocks.

Two seismic reflection surveys were run near Fenton Hill by LASL subcontractors. The results of the first of these surveys were reported by Kintzinger and West (1976). These results were combined with the results of the second survey and reinterpreted by Kintzinger et al. (1978) to help clarify the structural setting of the Fenton Hill drill site.

In the first survey, three sensor lines were laid out in close proximity to the Fenton Hill site. Two of these were essentially east-west lines; one along Lake Fork Canyon and one on top of Lake Fork Mesa. The third line ran northwest-southeast, through the GT-2 drill site. High explosive shots were used, and both colinear and broadside shooting were done.

Several faults were recognized by the reflection survey. A prominent north-northeast striking fault with a southeast dip lies immediately west of the GT-2 drill site. According to Kintzinger and West (1976), it appears to correlate with a surface lineament NE of the drill site, with the ring fault of Smith et al. (1970), and with the location of San Antonio hot springs. An east-west striking fault was detected with a surface trace north of GT-2. Its southward dip suggests that it may be the source of one of the fracture zones detected in GT-2 and EE-1.

A second reflection survey was performed to gain additional information on the structure of the Precambrian basement rocks. In particular, because a flow of hot water had been observed on top of the Precambrian surface during drilling, it was hoped that the top of this surface could be better defined. For the second survey, the dropping of a 150 kg bag of shot was used as an energy source. Recording was done along three lines, one along the Rio Cibolla, one up Barley Canyon past GT-1, and one along the ridge west of San Antonio Creek. Multiple shot drops were used where necessary, and the results were vertically stacked.

Three reflecting horizons were detected during the survey, the lowermost being within the top of the Precambrian basement. Structural contour maps were constructed for two reflectors, one within the volcanic rocks and one within the Precambrian. Four faults were observed to cut the reflector in the volcanic rocks. Three of these were related to the ring fault system while the fourth trends NE across the western part of the area. The Precambrian structural contour map indicates a gentle westward dip of about 5° for the reflector but with considerable superimposed structure. The most prominent structure is a very pronounced horst which strikes NE north of Barley Canyon. Kintzinger et al. (1978) called this the Bear Canyon High and stated that it was in part coincident with a gravity high mapped by Cordell (1976). The Bear Canyon High projects into the Valles Caldera to the NE, and to SW into an upthrown block of Precambrian rocks in the canyon of the Rio de Las Vacas. Blocks of Precambrian rocks were observed in the caldera fill within the Valles Caldera near the point of projection of the Bear Canyon high (Eichelberger and Laughlin, personal observation, 1978). The westward dip of the Precambrian rocks away from the Caldera agrees with the drilling observation that hot water is moving along this surface.

To assist in the characterization of the total geothermal system associated with the Valles Caldera, a telluric-magnetotelluric survey of the Jemez Mountains was performed by Hermance (1979) as part of a regional MT survey. With permission from Union Oil Company, 17 telluric sites were occupied on the Baca Grant within the Valles Caldera. Thirteen other stations were run in the Jemez Mountains. One of these was a base station recording over a period range of 10 to 10,000 seconds, while the 12 satellite stations recorded from 10 to 4000 seconds.

Hermance's work is pertinent to the understanding of regional setting of the Jemez Mountains as well as providing information on the more local geothermal system. His regional study indicates that west of the Jemez Mountains in the San Juan Basin, site F-2 is in an area that is relatively homogeneous, while sites east of the Sangre de Cristo Mountains near Las Vegas, New Mexico apparently see a heterogeneous crustal structure. Hermance (1979) notes that results from within the Jemez Mountains are very similar to those from within the Rio Grande Rift. This is not surprising when the evidence of the Rift extending under caldera is considered (Cordell, 1976). Within the Jemez Mountains, Hermance (1979) presents evidence that a partial melt accumulation may exist at a depth of about 15 km. He compares the Jemez results with data from near Santa Fe and El Paso, where Pedersen and Hermance (1976, 1978) have reported depths to melt accumulations of 15 and 27 km respectively. West of the Baca Grant, two adjacent sites (JE5 and JE6) have strikingly different polarizations indicating heterogeneous conditions in this area. Stations near Redondo Peak within the Valles Caldera show large telluric ellipses probably because of disturbances of the resurgent dome. Hermance (1979) states that four stations northwest of Redondo Peak have been affected by enhanced shallow crustal conductivity, perhaps from hydrothermal activity.

Drilling and coring at the Barley Canyon and Fenton Hill sites provided the opportunity for measuring geothermal gradients in shallow and very deep wells and for performing both thermal conductivity and heat generation measurements. Results of these measurements have been reported by Albright (1975), Sibbitt et al. (1979) and LASL HDR Project Staff (1978). Geothermal gradients measured in GT-1 and GT-2 indicate that both conductive and hydrologic heat transfer are significant at the Barley Canyon and Fenton Hill sites. As was discussed earlier, the geothermal gradient in the lower portion of the Madera Limestone, near its contact with the Sandia Formation, drops abruptly from 129°C/km to 45°C/km. This lower gradient continues into the relatively impermeable Precambrian section. Drill hole GT-2 penetrates a Precambrian basement with gradients of 55–60°C/km (Albright, 1975). The gradients of 129°C/km and 45–60°C/km are equivalent to heat flows of approximately 230 mW/m^2 and 155 mW/m^2 in the Paleozoic and Precambrian basement rocks respectively. The difference between these values is thought to result from hydrologic effects produced by warm water moving outward, away from the caldera, along the Precambrian unconformity. The difference of 75 mW/m² between the two measured heat flow values thus represents the hydrologic heat flow contribution to the total heat flow measured in the surficial rocks at Fenton Hill.

The conductive heat flow term of 155 mW/m² measured in the low-permeability Precambrian rocks can be examined in light of the reduced heat flow equation of Birch et al. (1968). In this equation, $Q = Q^* + Ad$, where Q = measured heat flow, Q^* = reduced that flow, i.e., heat flow from the lower crust and mantle, A = heat generation from the decay of U, Th, and K, and d = thickness in km of the radioactive layer. Birch et al. (1968) and Roy et al. (1968) found that this linear relationship between heat flow and heat generation held true for several geologic provinces (see also Rybach, this volume). For the Basin and Range Province, of which the Rio Grande Rift is a part, they calculated a value for Q^* of 59 mW/m² and for d, a value of 9·4 km. Balagna and Laughlin (unpub. data) measured U, Th, and K concentrations in the Precambrian rocks at Fenton Hill and, using a weighted average based on the abundance of different rock types in the drill hole, calculated a heat generation of $1·81 \times 10^{-6}$ W/m³. The 'regional' conductive heat flow at Fenton Hill should then be: $Q = 59$ mW/m² $+ 9·4$ km $(1·81 \times 10^{-6}$ W/m³$) = 75$ mW/m². The difference between the 155 mW/m² observed and 75 mW/m² calculated, 80 mW/m², is apparently the conductive heat flow contribution of the Valles Caldera event to the local heat flow. The conductive and hydrologic contributions, 80 and 75 mW/m² respectively, (see previous paragraph), of this event are approximately equal.

11.4.4. *Exploration for Low- to Moderate-Temperature Hydrothermal Systems*

The presence of numerous warm to hot springs within the Jemez Mountains indicates that the heat source beneath the Valles and Toledo Calderas has affected local aquifers, providing potential low- to moderate-temperature hydrothermal resources suitable for direct utilization in space heating. Exploration for such systems has begun in two parts of the mountains, and an initial hole was drilled in early 1979 by the village of Jemez Springs.

Exploration near Jemez Springs was relatively simple because a number of hot and warm springs occur along the length of the Canon de San Diego in which the village is situated. These springs have temperatures up to 76°C and are apparently localized by the Jemez Springs Fault, which extends southwestward from the breached southwest corner of the Valles Caldera. This is one of the largest faults in the area, with a throw of 244 to 305 m along the Jemez River (Purtymun et al. 1974). It is a normal fault, downthrown to the southeast.

Several hot springs occur in the immediate vicinity of the village. Purtymun et al. (1974) have estimated the total flow from the ten springs to be 13 L/s. They report temperatures of 34°C to 76°C and high concentrations of bicarbonate, chloride, and the cations calcium, magnesium, and sodium; sulfate concentrations are low. In part because of logistic and land ownership issues, a drill site was selected near the 'Bathhouse' spring, which has a surface water temperature of 55°C. A 17 cm diameter well was drilled to a total depth of 253 m, encountering several aquifers before bottoming in the Precambrian. A shallow 69°C aquifer was intersected at a depth of 24 m. Continued drilling led to intersection with a second aquifer at 160 m (Kleeman, oral communication, 1979) with a temperature of 61°C. Cuttings from the Jemez Springs drill hole were collected and are now being examined by LASL personnel. Kron (oral communication, 1979) states that about 21 m of alluvium overlies the Madera Limestone, which is approximately 186 m thick in this area. A 31 m thick section of Sandia Formation overlies the Precambrian basement.

The Los Alamos Scientific Laboratory, like other U.S. Government installations, has been directed to reduce its use of natural gas by 50% by 1985 and to be completely converted to an alternate energy source by the year 2000. Space heating, using low- to moderate-temperature geothermal fluids, is one of the alternatives under current investigation at LASL. Because of land ownership considerations it is almost essential that any geothermal resource developed must lie beneath the LASL-owned property, which encompasses an area on the east side of the Valles Caldera within the Rio Grande Rift. This general area was originally eliminated as a HDR site because of the large number of faults associated with the Rift, which, it was thought, would produce permeable reservoir rocks. It seems likely that any geothermal resource present will be hydrothermal with the heat originating either from the higher than normal gradients of the Rift or from magma beneath the calderas. Exploration techniques have therefore been biased towards a search for permeable reservoirs.

The geologic map of the Jemez Mountains by Smith et al. (1970) served as the starting point for the geothermal exploration of the area. From this map it can be seen that the entire LASL area is blanketed by Bandelier Tuff and is bounded on two sides by faults. On the west, the Pajarito fault, striking north-south, essentially marks the western boundary of LASL property. This fault has a displacement of 120 m, down to the east. Approximately two kilometers south of the southern boundary of the LASL-owned area, the Frijoles Canyon fault parallels the boundary. The Frijoles Canyon fault is also a normal fault, in this case with the northside down. The Pajarito fault parallels the Rio Grande Rift and is a major western boundary fault of the Rift, while the Frijoles Canyon fault is radial to the Valles Caldera and may result from the formation of the caldera.

Subsurface data for the area were provided by a detailed gravity survey (Budding, 1978), from the logs of deep water wells and from the results of a time-domain electromagnetic (TDEM) survey conducted specifically for geothermal exploration. A seismic refraction survey was also conducted, and the results are now being interpreted.

Budding's gravity map shows a NNE trending low of 18 milligals which passes through the area, 4·5 km southeast of the town of Los Alamos. This anomaly, underlying the Pajarito Plateau and interpreted as a graben, has as its western boundary the Pajarito Fault. It parallels and appears to be an extension of the Velarde graben within the Espanola Basin (Manley, 1978, 1979). Samples of the local stratigraphy were used for density measurement to aid in interpreting the results, and a gravity model was constructed which suggests that the anomaly results from a NNE trending graben, 4 to 6 km in width. Budding (1978) estimated that 2300 m of the Santa Fe Group sediments fill this graben.

Because the Santa Fe Group sediments have in part relatively high permeabilities, it was suspected that warm to hot water might be trapped within the graben. To test this hypothesis and to confirm the depth to the Precambrian basement, a TDEM survey was run in the area by an industrial subcontractor to LASL (Williston, McNeil and Associates, 1979). This group employed a single 200 kW dipole source and 50 receiver sites. Geologic complexities near the Pajarito Fault caused the results from some receiver stations to be uninterpretable. Williston, McNeil and Associates (1979) concluded that (1) a NE trending trough or graben lies beneath Los Alamos, (2) the graben is greater than 3000 m in depth, (3) graben-filling sediments are made up of two layers with distinctly different electrical properties, and (4) the lower layer is about 2000 m in thickness with resistivities as low as 2Ωm. They interpreted their data as suggesting that the graben-filling sediments must have a porosity of at least 20 per cent and could contain fluids with a temperature of 150°C.

The results of the TDEM survey apparently corroborate the gravity results of Budding (1978) with respect to the presence of a NE to NNE trending graben beneath the LASL. Consequently, a potential drill site has been selected for a hole so as to penetrate to the bottom of the graben. It is hoped that moderate temperature will be intersected in the lower portions of the Santa Fe Group. If not, the hole will be extended to penetrate the Precambrian basement—and the section will be evaluated as a possible HDR site.

11.5. The Geothermal System of the Jemez Mountains

Despite the fact that many of the recently acquired geothermal data from the Jemez Mountains are proprietary and unpublished, enough data are available to formulate a reasonable conceptual model of the total geothermal system. This model must explain the three different types of geothermal systems presently demonstrated to occur within the area. As time passes and new data are acquired, this model must be updated and quantified.

A variety of geological and geophysical evidence indicates that the Jemez Mountains volcanic complex developed on a major crustal structural discontinuity, the boundary between the Colorado Plateau and the Rio Grande Rift. Aeromagnetic and gravimetric data suggest that this boundary may be an older structure reactivated roughly 30 m.y. ago when the Rift began to form. The role of the Jemez Lineament in localizing the formation of the complex is unclear at the present time, but it does intersect the Rift at the site of the Jemez Mountains, and it has been volcanically active for at least the past 3 m.y. Volcanism along the Rift, however, began long before the Jemez Lineament became active, and the intersection may be fortuitous. In any case, volcanism began at the present site of the Jemez Mountains roughly 10 m.y. ago. From that time until essentially the present, basaltic through silicic magmatism has been almost continuous. Volcanism reached its peak roughly 1 m.y. ago when a shallow magma chamber or chambers erupted the two members of the Bandelier Tuff; this is after volcanism began along the Jemez Lineament. The effect of this long period of volcanic activity was to superimpose a local igneous heat source upon a large regional zone of high heat flow associated with the Rio Grande Rift.

During the peak of the Jemez Mountains volcanism about 1 m.y. ago, the magma chamber which supplied the Bandelier Tuff lay at relatively shallow depth beneath the Valles Caldera. The depth to the top of the chamber and its diameter are still unresolved, but both Smith et al. (1970) and Kolstad and McGetchin (1978) suggest that it extended beyond the present ring fault system of the Valles Caldera. From the cross-sections of Smith et al. (1970), the suggested depths at Fenton Hill and Redondo Peak are 4 km and 3 km respectively. Kolstad and McGetchin (1978) modeled heat loss from chambers at depths of 3 and 5 km. The deep drilling at the Fenton Hill site indicates that at this site the chamber must be deeper than 3 km. Metamorphic grade (low-grade amphibolite facies) and amount of Ar loss and fission track annealing in core samples from the deeper portion of GT-2 and EE-1 suggest that the chamber is considerably deeper than the 3 km depth. On the southeast side of the mountains, magnetotelluric data indicate that a high conductivity zone, which may represent a partial melt, lies at a depth of 15 km.

Despite the young igneous activity and hydrothermal manifestations, several workers have shown that the seismic activity within the area is low. This may, at least in part, be responsible for the very low permeability of basement rocks at Fenton Hill and the high percentage of dry holes on the Baca Grant.

Movement of ground water along faults and fractures obviously occurs throughout the mountains. At least some of this water encounters rock at elevated temperatures, being heated in the process. Hydrothermal wells drilled in the Redondo Peak and Sulfur Creek areas intersect either these zones of secondary permeability or permeable zones within the Bandelier Tuff. The Union Oil data suggest that, at least in some portions of the field, recharge is not sufficient to match discharge, leading to development of a vapor-dominated system above the liquid-dominated system.

At greater distance from the heat source, self-sealing has occurred within the reservoir rocks. This process plus the absence of rocks with primary permeability has produced large volumes of hot rock with very low permeability, the Hot Dry Rock system currently being investigated by LASL.

Major faults penetrating the Valles Caldera or interconnecting with faults that do so, permit escape of waters that have been heated by this igneous heat source. As these waters move down the hydraulic gradient they both cool and mix with colder ground waters. Despite the cooling and/or mixing, these waters still offer some potential as sources of low-grade geothermal energy, suitable for use in space heating.

Acknowledgements

This work was supported in part by the U.S. Department of Energy, Division of Geothermal Energy. Most of the LASL. geothermal work was performed by several groups within the Geosciences and Chemistry-Nuclear Chemistry Divisions. Special thanks go to F. G. West for many discussions during the writing of this manuscript and, when the original work was performed, to A. G. Blair (deceased), and M. C. Smith for support throughout.

References

Aiken, C. L. V., Laughlin, A. W., and West, F. G., 1977. 'Residual Bouguer gravity anomaly map of northern New Mexico', Los Alamos Scientific Laboratory Map (LA-6737-MAP).

Aiken, C. L. V., Laughlin, A. W., and West, F. G., 1978. 'Residual Bouguer gravity anomaly map of New Mexico', Los Alamos Scientific Laboratory map (LA-7466-MAP).

Albright, J. N., 1975 'Temperature measurements in the Precambrian section of geothermal test hole no. 2', Los Alamos Scientific Laboratory report LA-6022-MS.

Aoki, Ken-Ichiro, 1967. 'Petrography and petrochemistry of latest Pliocene olivine-tholeiites of Taos area, northern New Mexico' *U.S.A., Contr. Mineralogy and Petrology*, **14**, 191–203.

Aoki, Ken-Ichiro and Kudo, A. M., 1976. 'Major-element variations of late Cenozoic basalts of New Mexico', *New Mex. Geol. Soc. Spec. Pub. No. 5*, 82–88.

Bailey, R. A., Smith, R. L., and Ross, C. S., 1969. 'Stratigraphic nomenclature of the volcanic rocks in the Jemez Mountains, New Mexico', *U.S.G.S. Bull.* **1274-P**, 19 pp.

Birch, F., Roy, R. F., and Decker, E. R., 1968. 'Heat flow and thermal history in New England and New York.' In *Studies of Appalachian Geology: Northern and Maritime*, E-an Zen, W. S. White, J. B. Hadley, and, J. B. Thompson (Eds.), Interscience, New York, 437–451.

Bridwell, R. J., 1976. 'Lithospheric thinning and the late Cenozoic thermal and tectonic regime of the northern Rio Grande Rift.' In *New Mexico Geol. Soc. Guidebook*, 27th Field Conference, 283–292.

Bridwell, R. J., 1978. 'The Rio Grande rift and a diapiric mechanism for continental rifting. In *Tectonics and Geophysics of Continental Rifts*', I. B. Ramberg and E. R. Neumann (Eds.), D. Reidel Pub. Co., Dordrecht, Holland, 73–80.

Brookins, D. G., and Laughlin, A. W., 1976. 'Isotopic evidence for local derivation of strontium in deep-seated, fracture-filling calcite from granitic rocks in drill hole GT-2, Los Alamos Scientific Laboratory dry hot rock program', *Jour. Volc. and Geothermal Res.*, **1**, 193–196.

Brookins, D. G., Forbes, R. B., Turner, D. L., Laughlin, A. W., and Naeser, C. W., 1977. 'Rb–Sr, K–

Ar, and fission-track geochronolical studies of samples from L.A.S.L. drill holes GT-1, GT-2, and EE-1', Los Alamos Scientific Laboratory report LA-6829-MS.

Budding, A. J., 1978. 'Gravity survey of the Pajarito Plateau Los Alamos and Santa Fe Counties, New Mexico', Los Alamos Scientific Laboratory report LA-7419-MS.

Burruss, R. C. and Hollister, L. S., 1979. 'Evidence from fluid inclusions for a paleogeothermal gradient at the geothermal test well sites', Los Alamos, New Mexico,' *Jour. Vol. and Geothermal Res.*, **5**, 163–177.

Chapin, C. E. and Seager, W. R., 1975. 'Evolution of Rio Grande rift in the Socorro and Las Cruces areas', *New Mexico Geol. Soc. Guidebook 26th Field Conf., Las Cruces country*, 297–322.

Cordell, L., 1970. 'Gravity and aeromagnetic investigations of Rio Grande depression in northern New Mexico' (abstr.). In *Guidebook of the Tyrone–Big Hatchet Mountains—Florida mountains region*, N.-M. Geol. Soc., 158.

Cordell, L., 1976. 'Aeromagnetic and gravity studies of the Rio Grande graben in New Mexico between Belen and Pilar', *New Mexico Geol. Soc. Spec. Pub.* **6**, 62–70.

Cordell, L., 1978. 'Regional geophysical setting of the Rio Grande rift', *Geol. Soc. Amer. Bull.*, **89**, 1073–1090.

Decker, R. R. and Smithson, S. B., 1975. 'Heat flow and gravity interpretation across the Rio Grande rift in southern New Mexico and west Texas', *Jour. Geophys. Res.*, **80**, 2542–2552.

Doell, R. R., Dalrymple, G. B., Smith, R. H., and Bailey, R. A., 1968. 'Paleomagnetism, potassium-argon ages, and geology of rhyolites and associated rocks of Valles Caldera, New Mexico', *Geol. Soc. America Memoir*, **116**, 211–248.

Dondanville, R. F., 1978. 'Geologic characteristics of the Valles Caldera geothermal system, New Mexico' *Trans. Geothermal Res. Council.* **2**, 157–160.

Eddy, A. and Laughlin, A. W., 1978. 'Core studies applied to HDR geothermal energy extraction', A.A.A.S. Annual Meeting Abstract No. 613, Washington, DC.

Edwards, C. L., Reiter, M., Shearer, C., and Young, W., 1978. 'Terrestrial heat flow and crustal radioactivity in northeastern New Mexico and southeastern Colorado', *Geol. Soc. Amer. Bull.*, **89**, 1341–1350.

Eichelberger, J. C. and Koch, F. G., 1979. 'Lithic fragments in the Bandelier Tuff, Jemez Mountains, New Mexico', *Jour. of Volc. and Geothermal Res.*, **5**, 115–134.

Hahman, W. R., Sr., Stone, C., and Witcher, J. C., 1978. 'Preliminary map geothermal energy resources of Arizona', Geothermal Map No. 1, Bureau of Geology and Mineral Technology, Geological Survey Branch, University of Arizona, Tucson, Arizona.

Hermance, J. F., 1979. 'Toward assessing the geothermal potential of the Jemez Mountains volcanic complex: a telluric-magnetotelluric survey', Los Alamos Scientific Laboratory report LA-7656-MS.

Jiracek, G. R., 1975. 'Deep electrical resistivity investigations coupled with dry geothermal reservoir experiments in New Mexico', Technical Progress Report to Nat. Sci. Foundation/RANN, Contract No. NSF GI-42835, 47 pp.

Kintzinger, P. R., Reynolds, C. B., West, F. G., and Suhr, G., 1978. 'Seismic reflection surveys near LASL geothermal site', Los Alamos Scientific Laboratory report LA-7228-MS.

Kintzinger, P. R. and West, F. G., 1976. 'Seismic reconnaissance of the Los Alamos Scientific Laboratory's dry hot rock geothermal project area', Los Alamos Scientific Laboratory report LA-6435-MS.

Kolstad, C. D. and McGetchin, T. R., 1978. 'Thermal evolution models for the Valles Caldera with reference to a hot-dry-rock geothermal experiment', *Jour. Volc. and Geotherm. Res.*, **3**, 197–218.

LASL HDR Project Staff, 1978. 'Hot dry rock geothermal energy development project', Annual Report Fiscal Year 1977, Los Alamos Scientific Laboratory report LA-7109-PR.

Laughlin, A. W., Brookins, D. G., and Causey, J. D., 1972. 'Late Cenozoic basalts from the Bandera lava field, Valencia County, New Mexico', *Geol. Soc. Amer. Bull.*, **83**, 1543–1551.

Laughlin, A. W., Brookins, D. G., and Damon, P. E., 1976. 'Late-Cenozoic basaltic volcanism along the Jemez zone of New Mexico and Arizona', abs., *Geol. Soc. Amer. Abstracts with Programs*, **8**, 598.

Laughlin, A. W. and Eddy, A. C., 1976. 'Petrography and geochemistry of Precambrian core samples from GT-2 and EE-1', abs., *EOS Trans. Amer. Geophys. Union*, **57**, 350.

Laughlin, A. W. and Eddy, A. C., 1977. 'Petrography and geochemistry of Precambrian rocks from GT-2 and EE-1', Los Alamos Scientific Laboratory Report, LA-6930-MS.

Laughlin, A. W., Brookins, D. G., Damon, P. E., and Shafiqullah, M., 1979. 'Late Cenozoic volcanism of the central Jemez zone, Arizona-New Mexico', *Isochron West*, 5–8.

Laughlin, A. W. and West, F. G., 1976. 'The Zuni Mountains, New Mexico, as a potential dry hot rock geothermal energy site', Los Alamos Scientific Laboratory report, LA-6197-MS.

Lipman, P. W., 1969 'Alkalic and tholeiitic basaltic volcanism related to the Rio Grande depression', *Geol. Soc. Amer. Bull.*, **80**, 1343–1353.

Lipman, P. W., and Mehnert, H. H., 1975. 'Late Cenozoic basaltic volcanism and development of the Rio Grande depression in the southern Rocky Mountains', *Geol. Soc. Amer. Mem.* **144**, 119–154.

Luedke, R. G. and Smith, R. L., 1978. 'Map showing distribution, composition, and age of late Cenozoic volcanic centers in Arizona and New Mexico', U.S. Geol. Surv. Map I-1091-A.

Manley, K. 1978. 'Cenozoic geology of Espanola basin'. In *Guidebook to Rio Grande rift in New Mexico and Colorado*, J. W. Hawley (Compiler), New Mexico Bureau of Mines and Mineral Resources, Circ. **163**, 201–210.

Manley, K., 1979. 'Stratigraphy and structure of the Espanola basin, Rio Grande Rift, New Mexico', In *Rio Grande Rift: Tectonics and Magmatism*, R. E. Riecker, (Ed.), Amer. Geophys. Union, Washington, D.-C. 71–86.

Mayo, E. B., 1958. 'Lineament tectonics and some ore districts of the southwest', *AIME Trans.*, 1169–1175.

Naeser, C. W. and Faul, H., 1969. 'Fission track annealing in apatite and sphene', *Jour. Geophys. Res.*, **74**, 705–710.

Naeser, C. W., and Forbes, R. B., 1976. 'Variation of fission track ages with depth in two deep drill holes', abs., *EOS, Trans. Am. Geophys. Un.*, **57**, 352.

Ozima, M., Kono, M., Kaneoka, I., Kinoshita, H., Kobayaski, K., Nagata, T., Larson, E. E., and Strangway, D. W., 1967. 'Paleomagnetism and potassiumargon ages of some volcanic rocks from the Rio Grande Gorge, New Mexico', *Jour. Geophys. Res.*, **72**, 2615–2622.

Pedersen, J. and Hermance, J. F., 1976. 'Towards resolving the absence or presence of an active magma chamber under the southern Rio Grande rift zone', abs., *EOS, Trans. Am. Geophys. Un.*, **57**, 1014.

Pedersen, J. and Hermance, J. F., 1978. 'Evidence for molten material at shallow to intermediate crustal levels beneath the Rio Grande rift at Santa Fe', abs., *EOS, Trans. Am. Geophys. Un.*, **59**, 390.

Perkins, P. C., 1973. 'Petrology of some rock types of the Precambrian basement near the Los Alamos Scientific Laboratory Geothermal test site, Jemez Mountains, New Mexico', Los Alamos Scientific Laboratory report, LA-5129-MS.

Porath, H. and Gough, D. I., 1971. 'Mantle conductive structures in the western United States from magnetometer array studies', *Roy. Astron. Soc. Geophys. Jour.*, **22**, 261–275.

Potter, R. M., 1973. 'Heat flow of the Jemez Plateau' abs., *EOS, Trans. Amer. Geophys. Un.*, **54**, 1214.

Purtymun, W. D., and Johansen, 1974 'General geohydrology of the Pajarito Plateau', In *New Mexico Geological Society, Silver Anniversary Guidebook, Ghost Ranch, Central-Northern New Mexico*, C. T. Siemers, (Ed.), L. A. Woodward, and J. F. Callendar, (Assoc. Ed.), 347–350.

Purtymun, W. D., West, F. G., and Adams, W. H., 1974. 'Preliminary study of the quality of water in the drainage area of the Jemez River and Rio Guadalupe', Los Alamos Scientific Laboratory, report LA-5595-MS.

Reiter, M., Edwards, C. L., Hartman, H., and Weidman, C., 1975. 'Terrestrial heat flow along the Rio Grande rift, New Mexico and southern Colorado', *Geol. Soc. Amer. Bull.*, **86**, 811–818.

Reiter, M., Weidman, C., Edwards, C. L., and Hartman, H., 1976. 'Subsurface temperature data in Jemez Mountains, New Mexico', New Mexico Bureau of Mines and Mineral Resources Circular, **151**, 1–15.

Reiter, M., Shaerer, C., and Edwards, C. L., 1978. 'Geothermal anomalies along the Rio Grande rift in New Mexico', *Geology*, **6**, 85–88.

Robinson, E. S., Potter, R. M., McInteer, B. B., Rowley, J. C., Armstrong, D. E., and Mills, R. L., 1971. 'A preliminary study of the nuclear subterrene', Appendix F, Geothermal Energy, M. C. Smith, (Ed.), Los Alamos Scientific Laboratory report LA-4547.

Roy, R. F., Decker, E. R., Blackwell, D. D. and Birch, F., 1968. 'Heat flow in the United States', *Jour. Geophys. Res.*, **73**, 5207–5221.

Sanford, A. R., Budding, A. J., Hoffman, J. P., Alptekin, O. S., Rush, C. A., and Toppozada, T. R., 1973. 'Seismicity of the Rio Grande rift in New Mexico', New Mexico Bureau of Mines and Mineral Resources Circ. 120, 19 pp.

Sass, J. H., Lachenbruch, A. H., Monroe, R. J., Greene, G. W., and Moses, T. H., 1971. 'Heat flow in the western United States', *Jour. Geophys. Res.*, **76**, 6376–6413.

Sibbitt, W. L., Dodson, J. G., and Tester, J. W., 1979. 'Thermal conductivity of crystalline rocks associated with energy extraction from hot dry rock geothermal systems', *Jour. Geophys. Res.*, **84**, 1117–1124.

Slemmons, D. B., 1975. 'Fault activity and seismicity near the Los Alamos, Scientific Laboratory geothermal test site, Jemez Mountains, New Mexico', Los Alamos Scientific Laboratory report LA-5911-MS.

Smith, M. C., 1973. 'Geothermal energy', Los Alamos Scientific Laboratory report LA-5289-MS.

Smith, M. C., 1975. 'The Los Alamos Scientific Laboratory dry hot rock geothermal project', *Geothermics*, **4**, 27–39.

Smith, R. L. and Bailey, R. A., 1966. 'The Bandelier tuff: a study of ashflow eruption cycles from zoned magma chambers', *Bull. Volcanol.*, **29**, 83–104.

Smith, R. L., Baily, R. A., and Ross, C. S., 1970. 'Geologic map of the Jemez Mountains, New Mexico', USGS Misc. Geol. Inv. Map I-571.

Trainer, F. W., 1974. 'Ground water in the southwestern part of the Jemez Mountains volcanic region, New Mexico', In *New Mexico Geological Society Silver Anniversary Guidebook, Ghost Ranch, Central-Northern New Mexico*, C. T. Siemers, (Ed.), L. A. Woodward, and J. F. Callendar, (Assoc. Ed.), 337–346.

Turner, D. L. and Forbes, R. B., 1976. 'K–Ar studies in two deep basement drill holes: A new geologic estimate of argon blocking temperature for biotite', abs., *EOS, Trans. Amer. Geophys. Un.*, **57**, 352.

West, F. G. and Laughlin, A. W., 1976. 'Spectral gamma logging in crystalline basement rocks', *Geology*, **4**, 617–618.

Williston, McNeil and Associates, 1979. 'A time domain survey of the Los Alamos region, New Mexico', Los Alamos Scientific Laboratory report LA-7657-MS.

Zartman, R. E. (1979). 'U, Th, and Pb concentration and Pb isotopic composition of biotite granodiorite (sample 9527-2b) from LASL drill hole GT-2', Los Alamos Scientific Laboratory, LA-7923-MF, 18 pp.

Geothermal Systems: Principles and Case Histories
Edited by L. Rybach and L. J. P. Muffler
© 1981 John Wiley & Sons Ltd.

12. Extraction–Reinjection at Ahuachapán Geothermal Field, El Salvador

GUSTAVO CUÉLLAR, MARIO CHOUSSY, AND DAVID ESCOBAR

Comision Ejecutiva Hidroelectrica del Rio Lempa, El Salvador

12.1. Introduction

In order to increase the production of electric energy in El Salvador, a program of geothermal studies was started in 1965. This program led to the development of the Ahuachapán Geothermal Field, where the first of two 30 MWe medium-pressure units was commissioned in May 1975 and the second in June 1976. During 1977, these two units provided 32·3 per cent of the total electric energy generated in El Salvador. This statistic has greatly increased the interest and confidence in the use of underground steam as a source of energy. At the present time, a third 35 MWe unit is being installed at Ahuachapán and is expected to be in operation in January of 1980.

Several new geothermal areas in El Salvador are presently under investigation (figure 12.1), and recent drilling has proved the existence of extensive fractured zones with temperatures up to 300°C.

This chapter presents pressure, temperature, and chemical data from the Ahuachapán Geothermal Field in the production stage. Future efforts to better understand the reservoir characteristics will include a tracer study and the development of a mathematical model.

12.2. Geological Setting

The general geological structure of El Salvador has been interpreted as a major structural trough cutting approximately East–West across the southern part of the country. This trough has been largely filled by Quaternary cones that compose the main volcanic chain of the country (Figure 12.1). Structurally, El Salvador can thus be divided in to three units: (1) a tilted block bordering the Pacific Ocean and dipping towards it, (2) a median valley of variable width, and (3) an uplifted northern zone composed of a number of complex fault blocks. These simple features assumed by Williams and Meyer-Abich (1955), although probably true in general terms, are in fact more complex than suggested by this model. In particular, the form of the median valley is not a simple parallel-sided rift. The uplifted southern and northern sides are a series of blocks which have been faulted and tilted to give the present-day structure.

The Ahuachapán Geothermal Field is associated with the south flank of the central Salvadorean median trough and the northwest sector of the Cerro Laguna Verde volcanic group. This extrusive complex developed during Quaternary time near the Pliocene

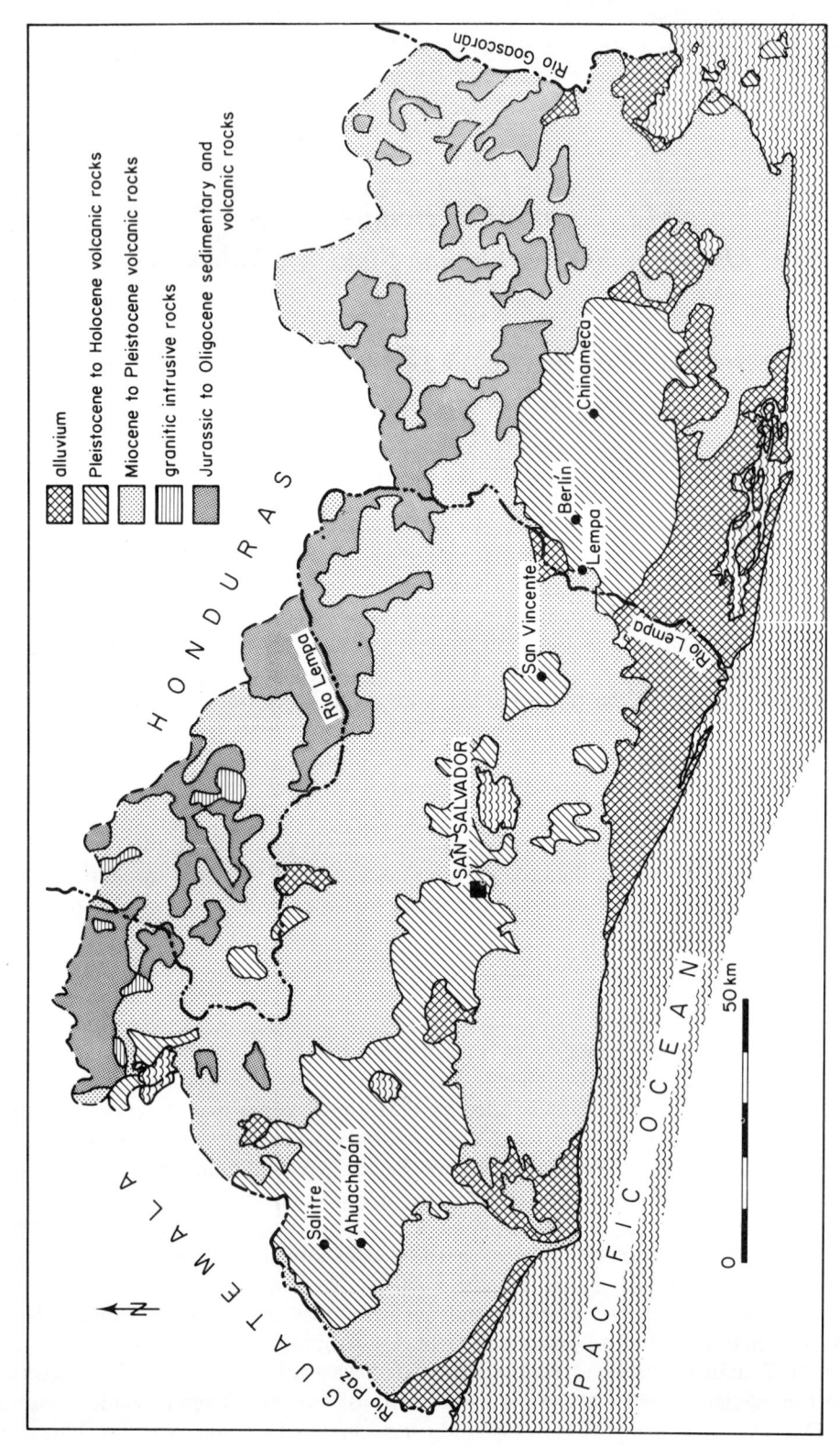

Figure 12.1. Geologic map of El Salvador (adapted from Wiesemann, 1973). Dots indicate geothermal areas under investigation

12. Extraction–Reinjection at Ahuachapán Geothermal Field

tectonic block of Tacuba-Apaneca, the regional faults of which have controlled first the sinking of the graben and subsequently the eruption of volcanic products.

The Ahuachapán Field lies in the southern part of a subelliptical basin that extends north and northwest, reflecting the subsidence of the graben.

Both the regional and the local structure are characterized by faults and fractures oriented along three main directions. An E–W system, approximately the trend of the main graben, consists of a series of step faults which limit the field to the north. To the W, the field is bounded by a second system of faults which strike NE. Finally, the most recent system of faulting and fracturing has a NNW trend and is associated with superficial hydrothermal activity. This youngest faulting probably has an important function in that it renders permeable the reservoir formations of the Ahuachapán Geothermal Field.

The stratigraphic sequence of the Ahuachapán area is described as follows (Figure 12.2):

- Laguna Verde volcanic complex: andesitic lava flows and some pyroclastics (Holocene). Thickness up to 200 m. Not shown on Figure 12.2.
- Tuff and lava: Tuffs prevail in the upper part and lava intercalactions in the lower part (Pleistocene). Thickness up to 500 m.
- Young agglomerate: Volcanic agglomerate with occasional lava intercalations (Pleistocene). Thickness up to 400 m. The unit is essentially impermeable and forms the caprock of the geothermal reservoir.
- Ahuachapán andesites: Lavas with pyroclastic intercalations (Pliocene–Pleistocene). Thickness up to 300 m. The Ahuachapán andesites constitute the reservoir formation and typically have a secondary permeability caused in part by columnar jointing related to cooling, in part by the contact surfaces of the different formations, but mainly by tectonic fracturing. In Figure 12.2, permeable zones are indicated at the top of the Ahuachapán andesites. Structure contours on the top of the Ahuachapán andesites are shown in Figure 12.3.
- Old agglomerate: Agglomerate with breccia intercalations in the lower part. Thickness in excess of 400 m.

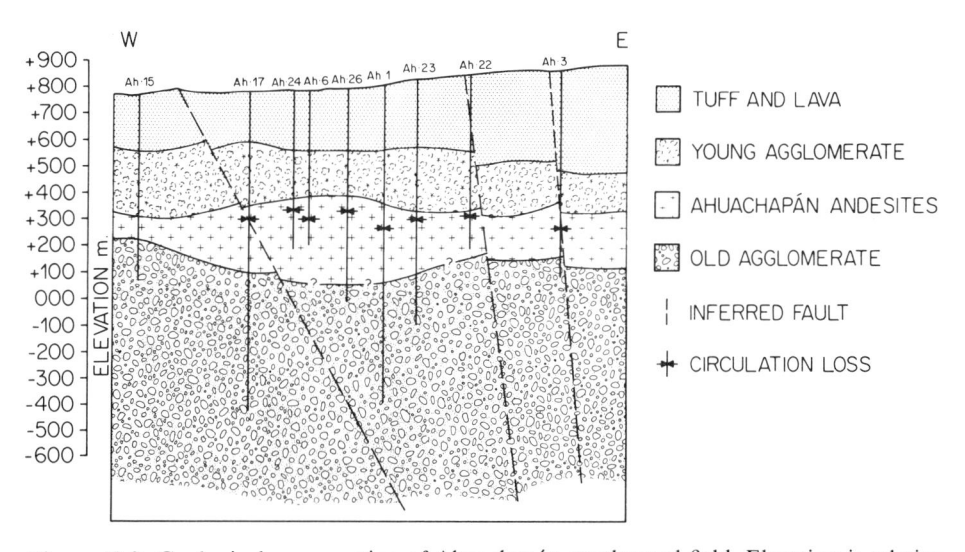

Figure 12.2. Geological cross section of Ahuachapán geothermal field. Elevation is relative to sea level

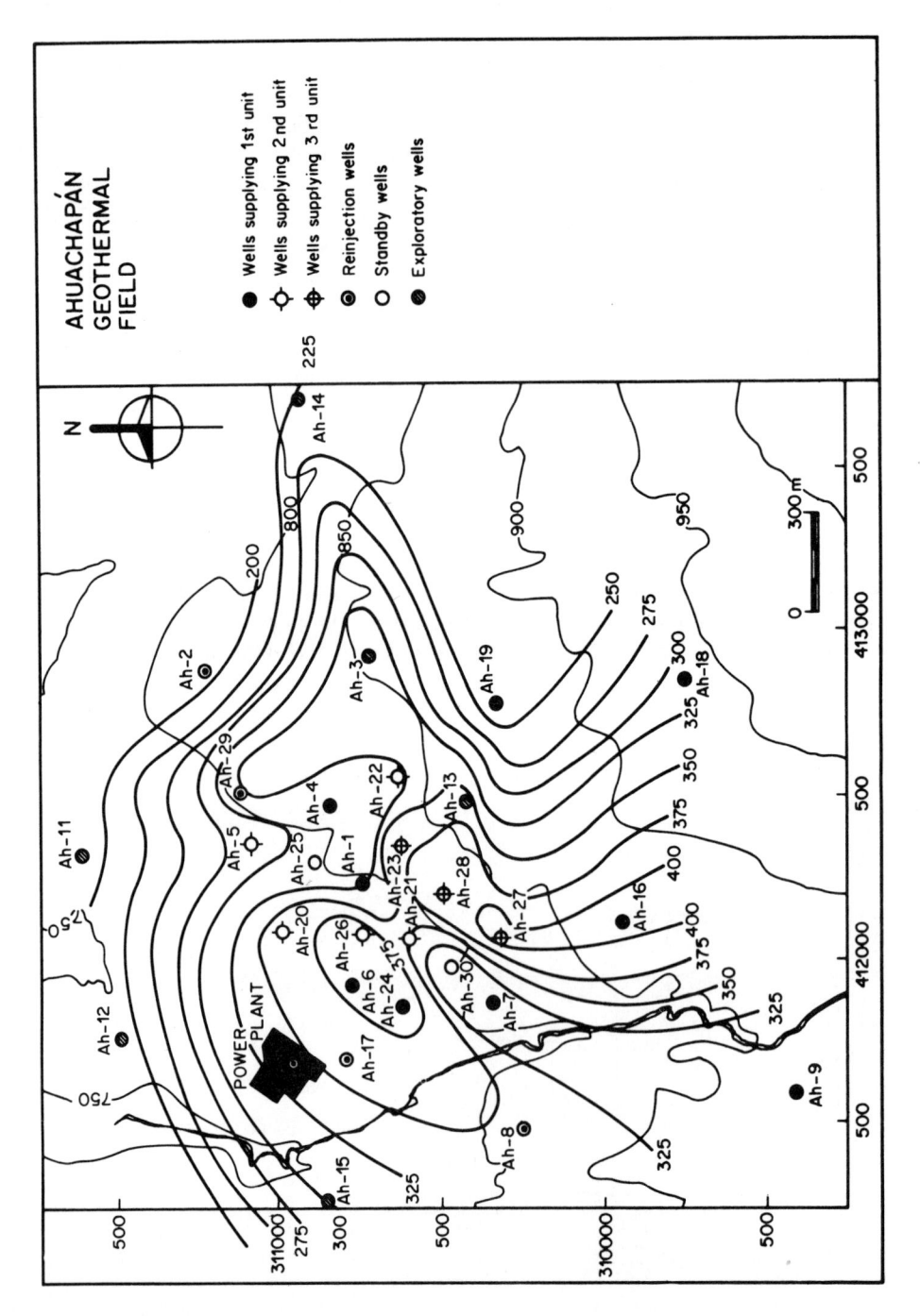

Figure 12.3. Map showing wells in the Ahuachapán geothermal field and structural contours (heavy lines, in meters above sea level) on the top of the Ahuachapán andesites. Light lines are topographic contours in meters above sea level

12.3. Hydrogeology

In the region around Ahuachapán, there are three aquifers, here termed the shallow aquifer, the saturated aquifer, and the saline aquifer. The saline aquifer is the deepest and constitutes the geothermal reservoir of the Ahuachapán field. The shallow aquifer appears to be distinct, whereas there is some communication between the saturated aquifer and the saline aquifer.

12.3.1. Shallow Aquifer

The shallow aquifer consists of tuffs, pumice detritus, and talus covering the lavas of the Laguna Verde complex. This unconfined aquifer is recharged by infiltrating rain water and feeds several springs located on the slopes of the Laguna Verde and Laguna de Las Ninfas volcanoes at the contact with the underlying lavas that constitute the aquiclude of this system.

The variations in flow rate are controlled by precipitation, showing very fast response. The waters are generally of calcium carbonate type, locally sulfatic, with residues below 500 mg/L. This very shallow aquifer is of interest only locally in the uphill area of the geothermal field.

12.3.2. Saturated Aquifer

The saturated aquifer consists of fractured lavas and pyroclastic deposits of the tuff and lava formation, while the young agglomerate, of low or no permeability, represents the impermeable basal stratum. Recharge takes place by direct infiltration, which produces a shallow free surface, tapped by several wells for domestic purposes and surfacing at several springs on the plain north of the geothermal area.

The piezometric surface in the area of the plain exhibits a concave shape that is open to the north, having a gradient (and therefore a principal flow component) in a northerly direction. The response of the piezometric level to the variations in rainfall is much slower than in the case of the shallow aquifer.

The water of the saturated aquifer is of calcium-sodium carbonate type, with residues generally below 400 mg/L. An exception to this rule is a group of springs, of which the most important one is the Salitre spring, with a flow rate of 1000 L/s and a temperature of 70°C. Water from this spring differs from the usual water of the saturated aquifer in its chemistry (sodium-chloride type) and in its much higher residues (600 to 1700 mg/L). These differences are attributed to admixture with waters that migrate upward along fractures from the underlying saline aquifer.

12.3.3. Saline Aquifer

The saline aquifer corresponds to the geothermal reservoir of the Ahuachapán Field and consists of the Ahuachapán andesites. The permeability of the Ahuachapán andesites is predominantly secondary, due to fractures, thus explaining the circulation losses observed during drilling. The permeability of the aquifer is therefore extremely anisotropic and variable; however, it is logical to assume that the permeability is highest along the previously mentioned principal directions of faulting and fracturing.

12.4. Characteristics of the Ahuachapán Wells

By early 1979, 29 wells had been drilled in the Ahuachapán Field, with depths between 591 and 1524 m. All the wells are located in an approximate area of 4 km². The zone of production, however, covers only 2 km², with a minimum well spacing of about 150 m. No significant interference has been detected among the wells.

The wells are distributed as follows (Figure 12.3): ten production wells provide steam for the first and second unit, three wells will provide steam for the third unit, two production wells are kept as standby, four wells are used for reinjection purposes, and there are ten exploratory wells.

Typical completions of production and reinjection wells are shown in Figure 12.4. These completions are based on a good knowledge of the geological conditions and have been standardized for the whole field.

The characteristics of the production wells are presented in Table 12.1, including depths and elevations of the top of Ahuachapán andesites. From the production characteristics it can be observed that there exists no relation between the well depths and mass discharge. The production characteristics, however, do correlate with the elevation of the top of the formation. The wells on the structural high have discharge enthalpies that correspond very closely to the enthalpy calculated from the adiabatic expansion of water at the initial reservoir temperature (230°C) and pressure. However, wells off the structural high have lower enthalpies, with the greatest deviations observed in the wells that showed, during drilling, circulation loss at relatively shallow levels within the reservoir.

The wells with higher steam percentage, Ah-6 and Ah-26, correspond exactly to the structural high of the reservoir (Figure 12.3). In both wells, 78 per cent of the produced steam fraction comes from steam in the reservoir water. However, well Ah-24, which is located in the same structural high, has a much lower enthalpy and steam content. This discrepancy is believed to be due to the sealing of the shallower fractures during the drilling of Ah-24.

12.5. Reinjection Program

The search for a better way for disposing of the hot, saline water and for maintaining the reservoir pressure at stable conditions led to a reinjection program. This program started in August 1975 when Ah-2 (a production well with low permeability) was converted to reinjection by using residual water from well Ah-4 at atmospheric pressure. Under these conditions a flow of 126 ton/h was injected into well Ah-2. In January 1976, injection tests under pressure were started between the producing well Ah-7 and the injection well Ah-8. The Ah-7 separator pressure was used as a driving force between the wells, with the water being thus kept free of any contact with the atmosphere. This process avoided silica scaling, allowed the reinjection at higher temperatures ($\pm 160°C$), and increased the reinjection capacity. In April 1976, this injection system was put into full operation, and in October 1976 fluid from the system Ah-6 and Ah-21 began to be reinjected into Ah-17. It is important to point out that at the beginning Ah-17 received only water from Ah-6. However, Ah-17 showed great absorption capacity, and Ah-21 water was subsequently added to this reinjection system.

12.5.1. Completion of the Reinjection Wells

The reinjection wells have, at the present time, two kinds of completion:

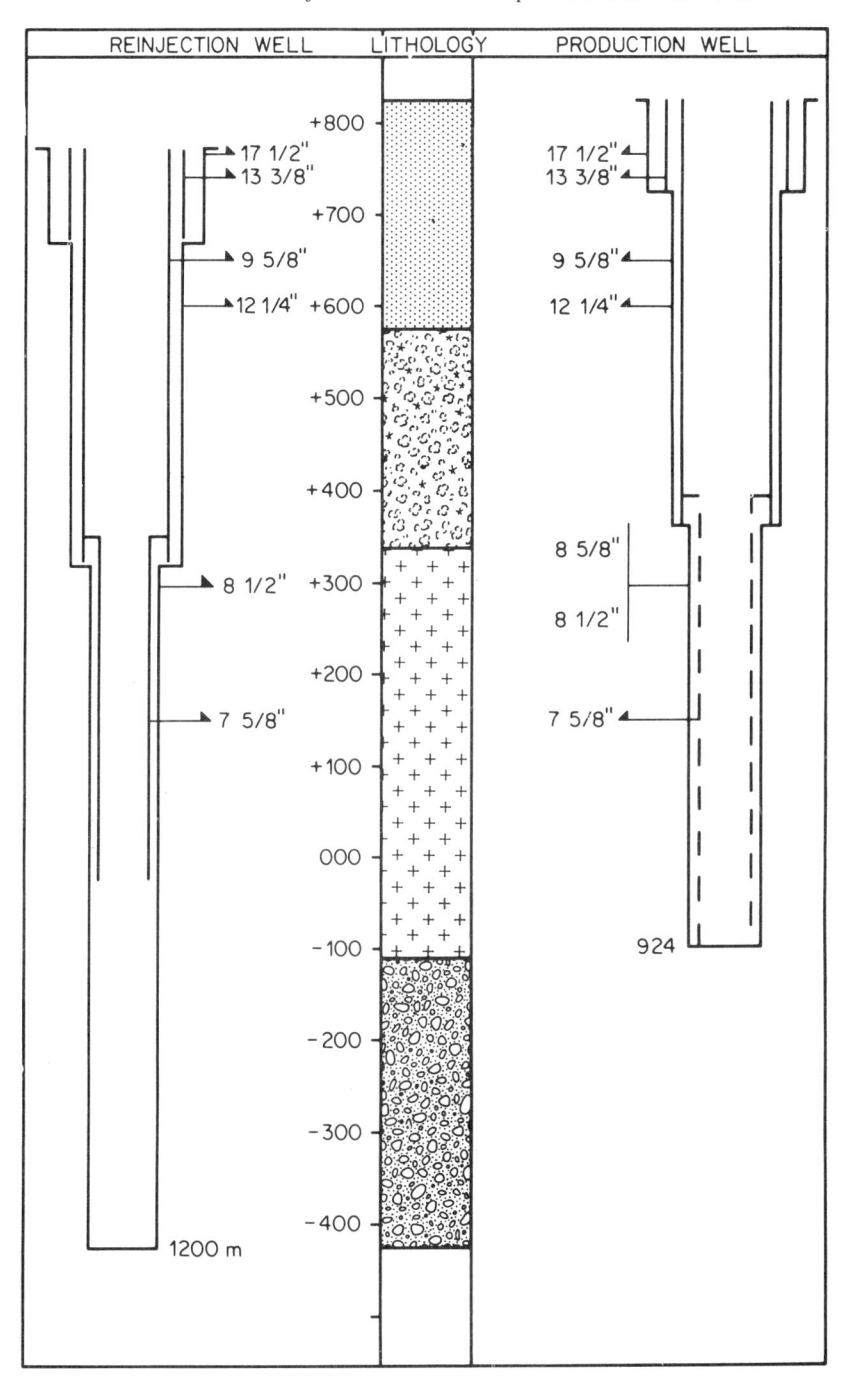

Figure 12.4. Typical completions of production and reinjection wells at Ahuachapán. Depth scale in meters relative to sea level. Stratigraphic units as in Figure 12.2.

Table 12.1. Characteristics of Ahuachapán production wells

Well	Separator pressure (10^2 kPa)	Mass rate (kg/s) total (TM)	Mass rate (kg/s) steam (SM)	Mass rate (kg/s) water	Enthalpy (J/g)	Steam ratio SM/TM	Total depth (m)	Elevation of top of Ahuachapán andesites (m)
Ah-1	5·4	85·08	12·50	72·58	984	0·15	1205	325
Ah-4	5·9	99·46	23·33	76·13	1176	0·23	640	315
Ah-5	5·4	49·97	5·56	44·41	908	0·11	952	284
Ah-6	5·5	32·96	13·61	19·35	1536	0·41	591	383
Ah-7	5·3	44·98	7·64	37·34	1026	0·17	950	285
Ah-20	5·3	47·28	11·94	35·34	1197	0·25	600	370
Ah-21	5·6	82·22	11·94	70·28	984	0·15	849	350
Ah-22	5·4	70·68	16·34	54·34	1160	0·23	659·5	315
Ah-24	5·3	51·93	7·50	44·43	975	0·14	850	380
Ah-26	5·3	21·44	8·89	12·55	1536	0·41	804	391
Totals		586·00	119·25	466·75				

1. Wells which were designed for production purposes and recently transformed to reinjection wells (Ah-2 and Ah-8). These wells still keep the production completion (Figure 12.4), having production casing cemented down to the top of the reservoir and open hole to the bottom.
2. Wells completed with a double purpose (production-reinjection). These wells have production casing cemented down to the top of the reservoir, with uncemented casing subsequently hung down to the bottom of the Ahuachapán andesites.

Reinjection wells Ah-17 and Ah-29 (double purpose) are very close to the production wells, and the lithologic columns show a considerable reservoir thickness (400 and 325 m, respectively). However, the wells Ah-2 and Ah-8 are farther away from the production zone and show a lesser reservoir thickness (105 m and 75 m, respectively).

12.5.2. Capacity of Reinjection Wells

The capacity of the reinjection wells is closely related to the formation permeability and to the reinjection pressure. For example, at atmospheric pressure well Ah-2 absorbed only 126 ton/h, but, using the separation pressure of the well Ah-4, increased to 245 ton/h. The quantities of the reinjected fluids are illustrated in Table 12.2.

12.5.3. Reinjected and Extracted Mass

In order to have a clearer knowledge of the reservoir behavior, the extraction in the Ahuachapán Field can be divided into two periods: a development period from August 1968 to July 1975, and a production period from July 1975 to the present.

The total mass extracted and reinjected during these two periods is shown in Table 12.3. As can be seen from the table, the reinjected mass has been, up to the present time, 29·5 per cent of the total mass extracted and 39·9 per cent during the production period alone.

Table 12.2. Summary of the Ahuachapán reinjection program

Well	Date	Reinjection pressure (10^2 kPa gauge)	Mass capacity (ton/hr)	Cumulative mass capacity (ton/hr)
Ah-2	Aug. 1976	atmospheric	130	130
Ah-8	Jan. 1976	4·9	200·4	330·4
Ah-2	Mar. 1976	6·0	114.9*	445.3
Ah-29	Apr. 1976	5·1	306·4	751·7
Ah-17	Oct. 1976	5·9	167·1	918·8
Ah-17	Dec. 1977	6.1	306.1*	1224·8

*Increment with respect to its initial capacity.

Table 12.3. Fluid extracted and reinjected at Ahuachapán

	1st Period (tons)	2nd Period (tons)	Total
Extracted mass	23,317,800	48,228,933	71,546,733
Reinjected mass	1,850,060	19,218,384	21,068,444
Net mass extracted	21,467,740	29,010,549	50,478,289

12.5.4. Reinjection Control

Several questions arising from the reinjection program led to a study of the effects of injection on the field condition. As a part of the study, measurements of temperature and pressure in the production and non-production wells were taken. Figure 12.5 shows that there is an immediate pressure response to variations in reinjection or extraction. However, temperature variations probably are masked by other kinds of changes. In order to detect small pressure changes, a continual recording pressure gauge has been installed in well Ah-25. These data have not yet been processed.

In order to detect any change in permeability, spinner equipment has been used in the reinjection wells to detect absorption zones (Figure 12.6). Well Ah-2 has permeability in the zone between 20 and 250 m above sea level and none in the bottom. Well Ah-17 absorbs all the injected water immediately after the bottom of the hanger liner. This can be interpreted to indicate that water circulates upward through the annulus ring to the total lost circulation zone at +300 m. In the open hole at depth, the formation has a low permeability. Well Ah-29 shows permeability throughout the open hole; water may flow to the annulus, but to a lesser extent than in well Ah-17.

The paths taken by the water injected into wells Ah-17 and Ah-29 can be determined by repetitive analyses for chloride in production wells. The water injected into Ah-29 goes partly to the central field and partly to the east, whereas it appears that most of the water injected into Ah-17 goes toward the center of the field.

12.6. Effects of the Reinjection-Extraction Rate

The various changes from the original reservoir conditions are due to the different rates of reinjection and extraction used in field operation. The reservoir behavior during

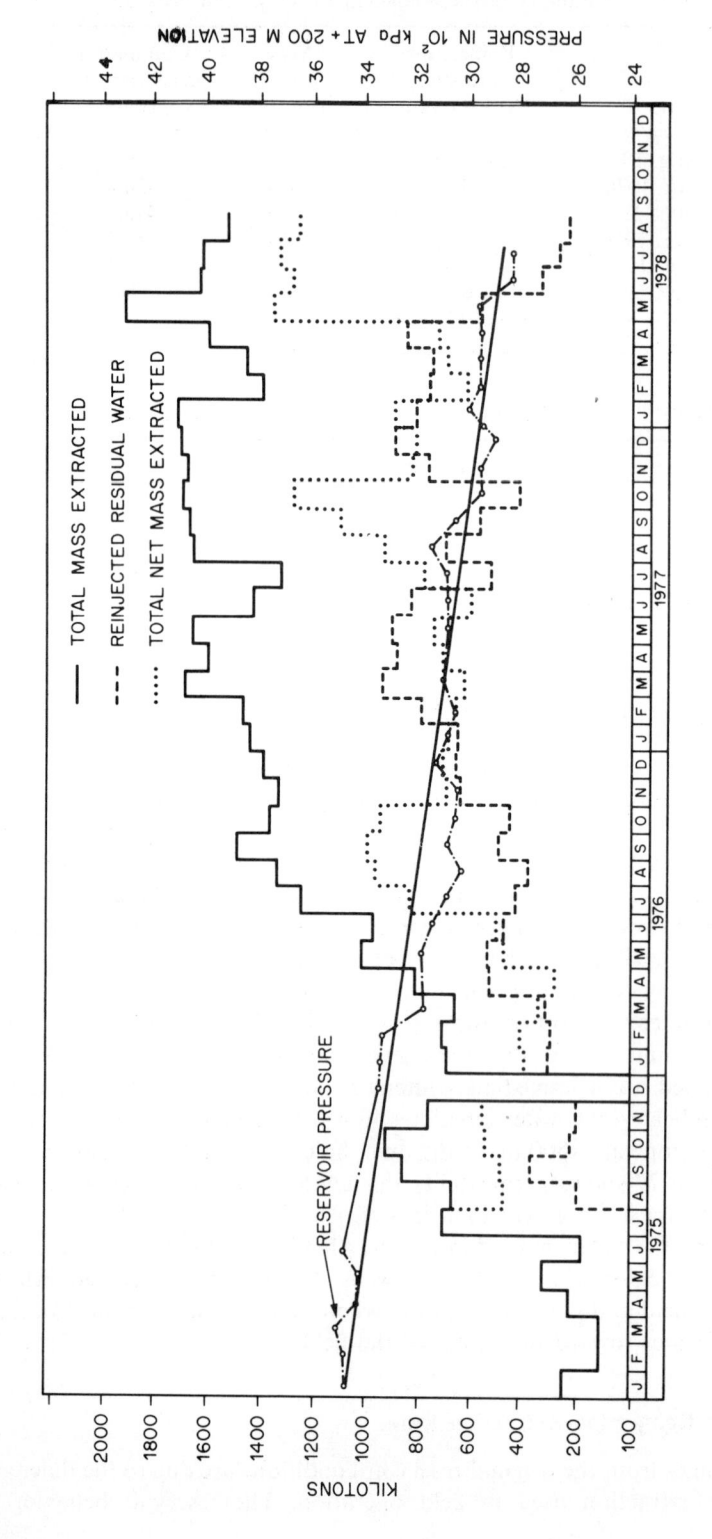

Figure 12.5. Graph showing monthly measurements of fluid production, fluid reinjection, and reservoir pressure at Ahuachapán from January 1975 to August 1978

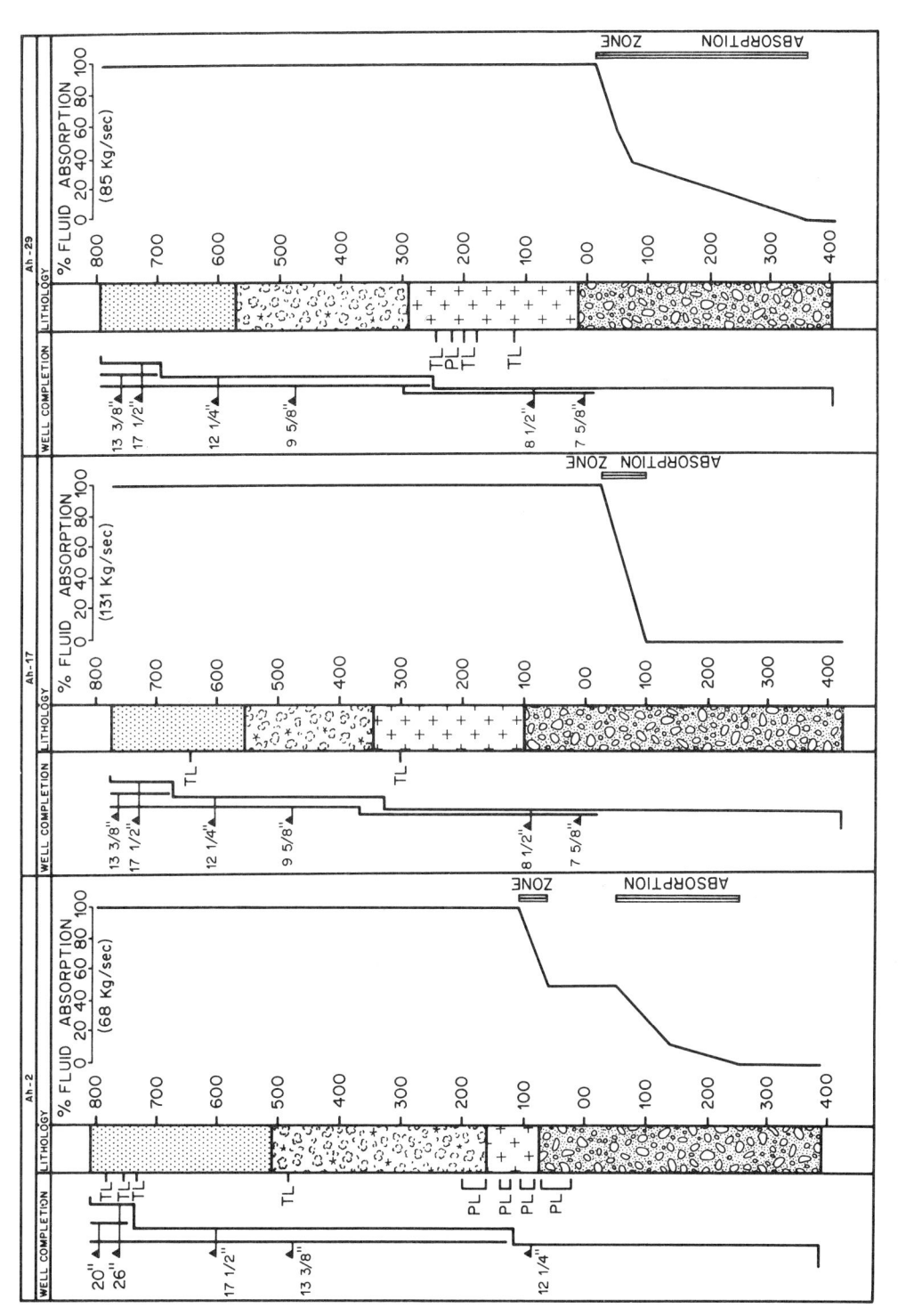

12. Extraction–Reinjection at Ahuachapán Geothermal Field 331

Figure 12.6. Graphs showing absorption of fluid as determined from spinner logs during injection, for wells Ah-2, Ah-17, and Ah-29. TL indicates total loss of circulation during drilling; PL indicates partial loss of circulation

production has been, according to experience in other places, typical of a water-dominated field. However, the general trend depends on the reinjection rates. In this section, we shall analyze how varying extraction-injection rates effect the pressure and temperature of discharged fluids.

12.6.1. *Pressures*

Figure 12.7 illustrates the reservoir pressure change as a function of the cumulative net mass extraction. At the beginning, the reservoir had a pressure (at 200 m above sea level) close to 3600 kPa, which is greater than the saturation pressure at the measured temperature (2960 kPa at 232°C). Subsequently, as a result of extraction, the pressure fell to new equilbrium values dependent on the extraction rate, reaching at the end of the development period in July 1975 a value of 3560 kPa. The existence of recharge is suggested by the fact that the pressure change was only 40 kPa in spite of the large amount of mass extraction. The anomalously high pressure values observed in 1971 may be due to the first injection tests.

During the production period subsequent to July 1975, the pressure decreased rapidly, reaching values close to the saturation pressure (2960 kPa). Then followed a stabilization period controlled by the reinjection effects and the developing of a steam zone in that reservoir.

Figure 12.5 presents in more detail the changes in the pressure during the production period, showing also the different extraction-reinjection rates. It can be seen clearly that

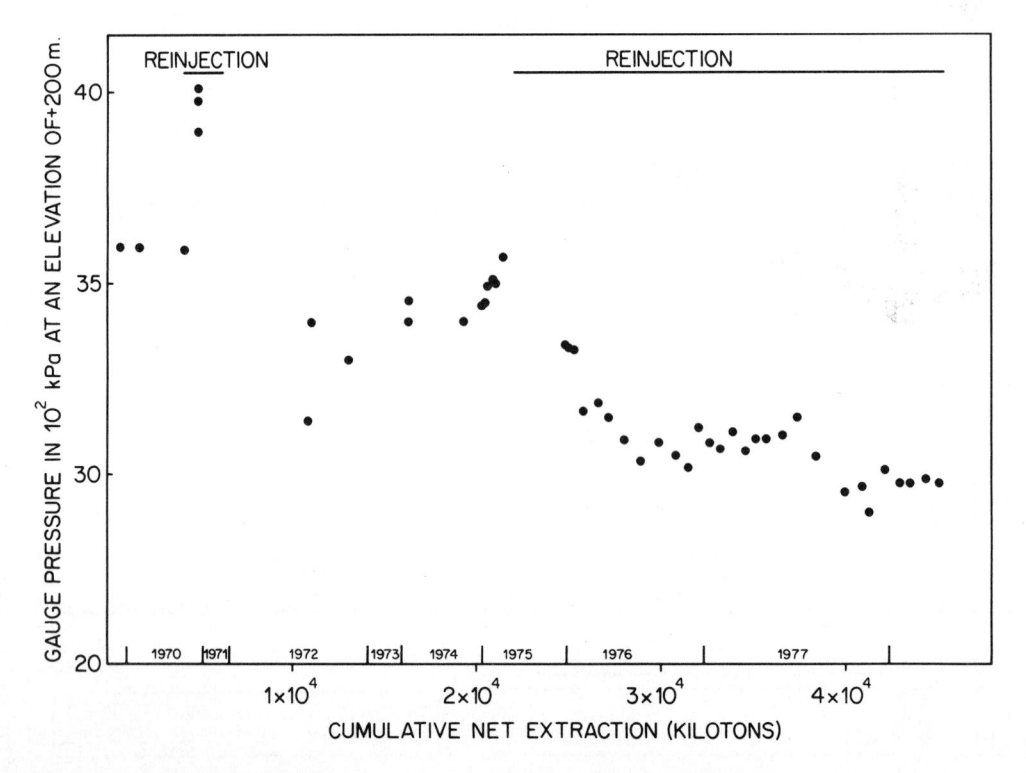

Figure 12.7. Graph showing reservoir pressure change at Ahuachapán as a function of cumulative net extraction of fluid. Note that time scale is not linear

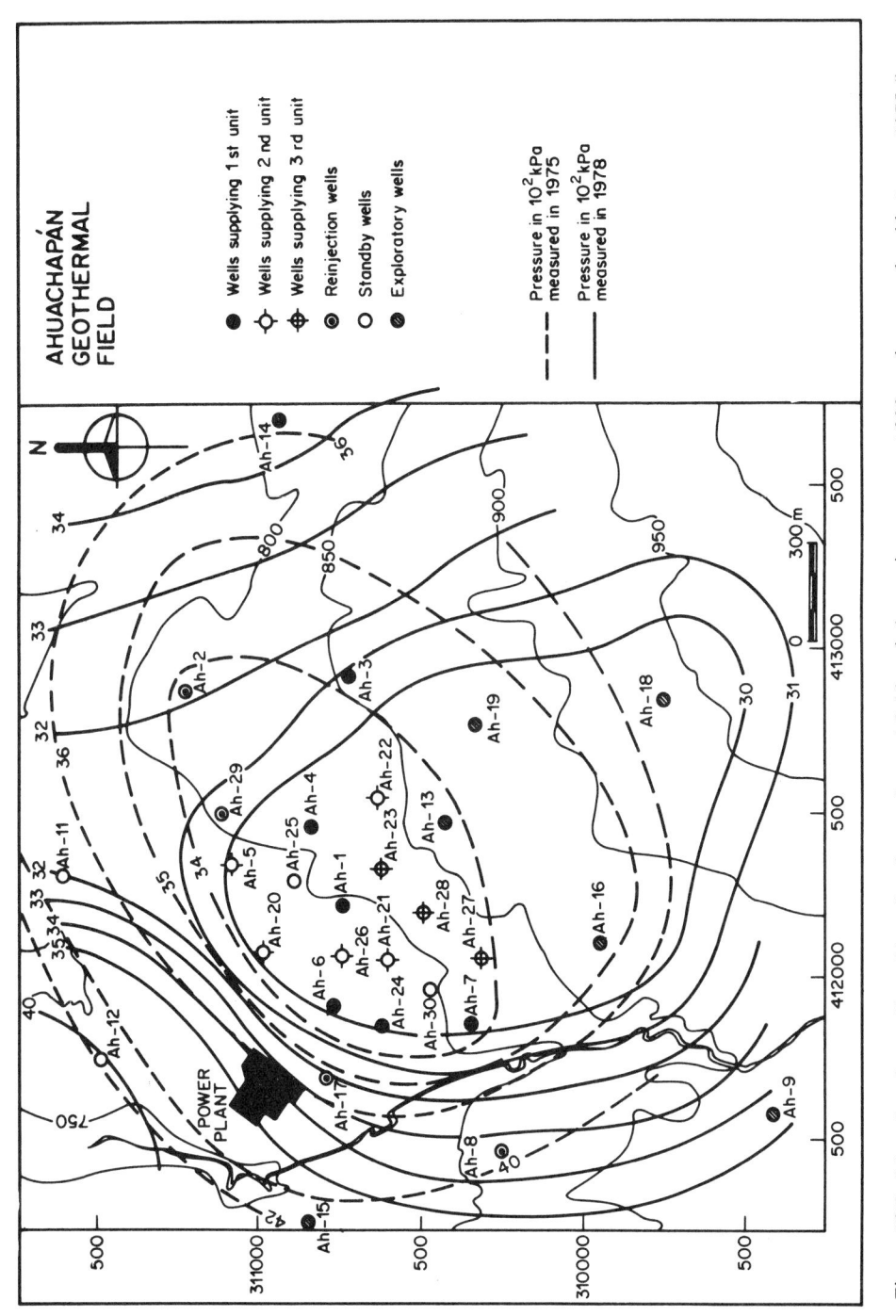

Figure 12.8. Map showing wells at Ahuachapán geothermal field and change in pressure at 200 m above sea level between 1975 (heavy dashed lines) and 1978 (heavy solid lines). Light lines are topographic contours in m above sea level

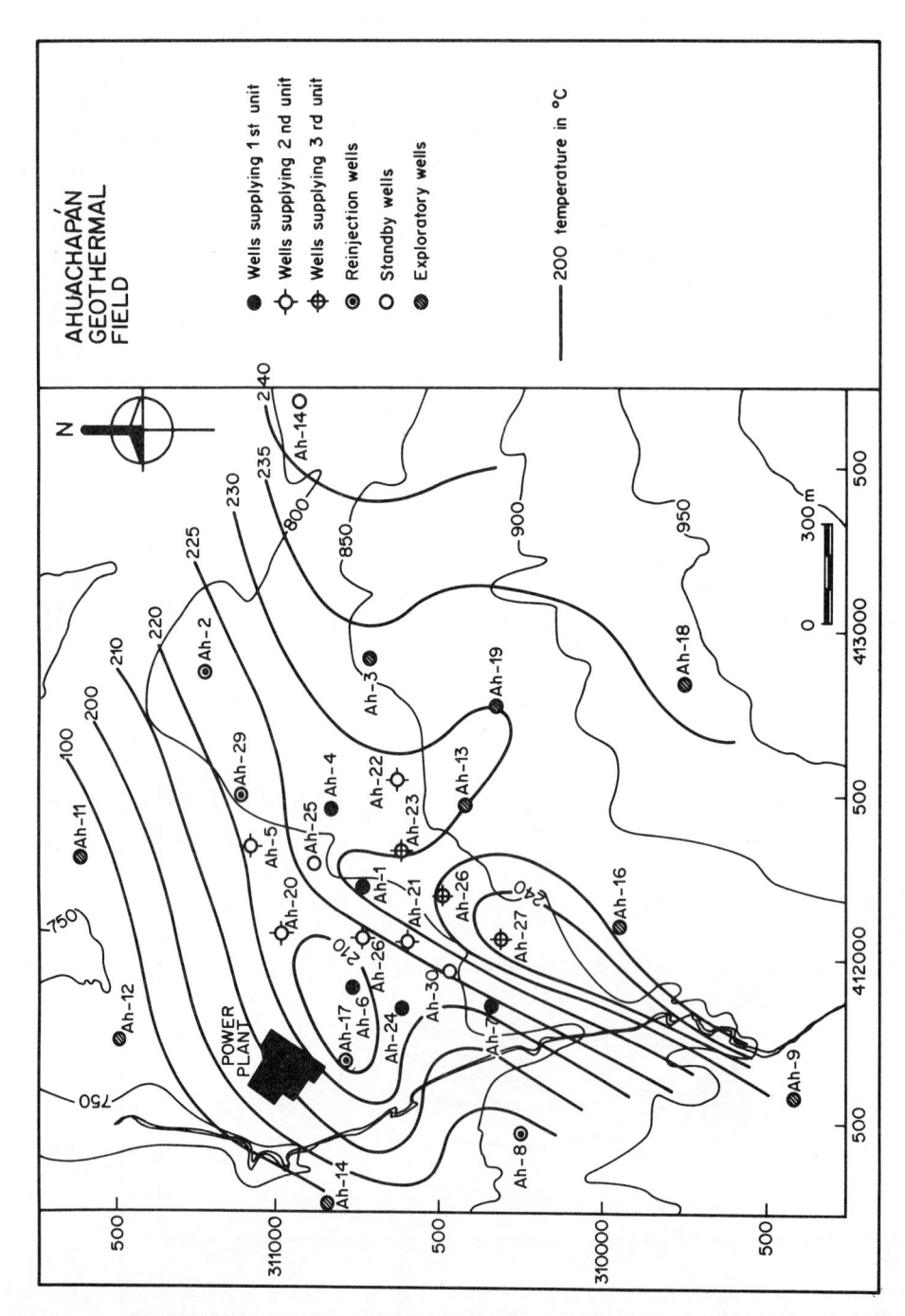

Figure 12.9. Map showing geothermal wells and reservoir temperatures (heavy lines) at Ahuachapán based on experimental quartz-water equilibria and on chemical analyses of silica in reservoir water in 1978. Light lines are topographic contours in m above sea level

pressure depends on the reinjection and extraction rates and therefore on the net mass extraction. With the starting of intensive production, the pressure decreases but tends to stabilize once the new equilibrium state is reached as a consequence of the reinjection and the development of a steam zone in the structural high of the reservoir. It seems difficult to determine which of the two effects is dominant.

The pressure distributions in the reservoir before and after intensive extraction are shown in Figure 12.8. The pressure sink is spreading toward the south, where more permeability has been found. In the east and west directions, in which the reservoir is limited, the pressure response has been minimum.

12.6.2. *Temperatures*

Reservoir temperatures have been calculated from chemical analyses for silica in reservoir water in 1975 and 1978, using the experimental equilibrium between quartz and water. In 1975 the 230°C isothermal line included all the production wells, but Figure 12.9 shows that in 1978 the same isotherm had contracted to include only production wells Ah-27, Ah-26, and Ah-1.

The distribution of reservoir temperatures (Figure 12.9) is influenced by well depth, by reinjection, and by the development of a steam zone in the structural high of the reservoir. The low-temperature zone in the region of well Ah-6 corresponds to this structural high and is near to the reinjection well, Ah-17. It is believed, however, that the predominant effect is due to boiling, since the temperature decrease in the zone occurred prior to operation of the reinjection well.

The distribution of temperatures also shows two additional things: (1) temperatures increase to the southwest, which seems to be the direction from which recharge occurs, and (2) temperatures increase towards the southeast, where the rocks have lower permeability and conductive heat transfer plays a bigger role.

12.7. Conclusions

The results of the intensive extraction of Ahuachapán can be summarized as follows:

1. The reservoir pressure has decreased below the saturation pressure corresponding to the reservoir temperature, thus causing boiling in the formation and the development of a steam zone in the structural high.
2. The temperature has reached an equilibrium state in response to the new pressure.
3. The extraction rate for total mass has decreased as a consequence of the decreasing pressure. However, this does not indicate a remarkable decrease in the amount of available steam.
4. Stabilization of the pressure appears to result both from control of the reinjection rate and from the development of a steam zone.

References

Wiesemann, G., 1973. 'Arbeiten der Bundesanstalt für Bodenforschung in der Republik El Salvador in den Jahren 1967–1971', *Münster. Forsch. Geol. Paläont.*, **31/32**, 277–285.
Williams, Howel, and Meyer-Abich, Helmut, 1955, 'Volcanism in the southern part of El Salvador, with particular reference to the collapse basins of Lakes Coatequepue and Ilopango', *Univ. Calif. Pubs. Geol. Sci.*, **32**, 1–64.

Bibliography

Baldo, L., Cozzini, M., and Carrena, E., 1977. 'Aprovechamiento del fluido geotermico del campo de Ahuachapán (El Salvador)', *Proc. Internat. Symposium on Geothermal Energy, Guatemala City, October 1976*, Rome, Istituto Italo–Latino Americano, 411–431.

Choussy, M. E., and Penate, T. S., 1977. 'Campo geotermico de Ahuachapán despues de un ano de exploitacion', *Proc. Internat. Symposium on Geothermal Energy, Guatemala City, October 1976*, Rome. Istituto Italo–Latino Americano, 527–552.

Cuéllar, Gustavo, 1976. 'Comportiamento del la silice en aguas geotermicas de desecho', *Proc. 2nd United Nations Symposium on the Development and Use of Geothermal Resources, San Francisco, May 1975*, 1337–1343 (English translation on pp. 1343–1347).

Cuéllar, Gustavo, Rentana, M. E., and Diaz, Oscar, 1978. 'Recursos geotermicos en El Salvador', *Simposio sobre Exploration*, OLADE, Quito, Equador, March 1978, 15 pp.

DiPippo, Ronald, 1978. 'The geothermal power station at Ahuachapán, El Salvador', *Geothermal Energy Magazine*, **6**, 11–22.

Retana, M. E., and A. Maltez, M., 1977. 'Estado actual de las investigaciones geotermicas en la zona oriente de El Salvador', *Proc. International Symposium on Geothermal Energy, Guatemala City, October 1976*, Rome, Istituto Italo-Latino Americano, 273–313.

Romagnoli, P., Cuéllar, G., Jimenez, M., and Ghezzi, G. (1976), 'Aspectos hidrogeologicos del camp geotermico de Ahuachapán, El Salvador', *Proc. 2nd United Nations Symposium on the Development and Use of Geothermal Resources, San Francisco, May 1875*, 563–570 (with English translation 541–574).

Sigvaldason, G. E., and Cuéllar, G., 1970. 'Geochemistry of the Ahuachapán thermal area, El Salvador, Central America', *Geothermics*, Special Issue 2, **2**, 1392–1399.

Vides, Alberto, 1976. 'Investigaciones recientes en el campo geotermico de Ahuachapán', *Proc. 2nd United Nations Symposium on the Development and Use of Geothermal Resources, San Francisco, May 1975*, 1835–1850 (with English translation, 1851–1854).

Geothermal Systems: Principles and Case Histories
Edited by L. Rybach and L. J. P. Muffler
© 1981 John Wiley & Sons Ltd.

Index

abatement procedures, *201f*
absorption zones, 329
accessible fluid resource base, 192, 194
accessible resource base, 184
   residual, 185
   undiscovered, 192
acid-alteration of silicate minerals, 117
acid-altered hot ground, 128
acid-sulfate springs, 128
active faults, 208
active geothermal systems, 109
activity, earthquake, 275
   diagrams, 111
   diagram method, 113
   hydrogen ion, 110
   microseismic, 210
   of the sulfate ion, 86
   product, calcium carbonate, 228
   ratio, 113
   ratios, in solutions, 111
      of aqueous species, 111
adiabatic cooling, 115, 126, 127
   maximum, 139
adiabatic expansion, of the fluid, 137
   of water, 326
adiabatic flow, 60
aerial photography, 82
aeromagnetic survey, 82, 100, 280
Afar Depression, East Africa, 29
African rift, 29
age determination by the fission-track method, 250
agricultural use of geothermal energy, 242
Ahuachapán geothermal field, El Salvador, 97, 129, 130, 135, 200, 210, *321f*
air lifting, 243
air quality, *201f*
Alaska, 190
   Aleutian volcanic chain, 192
alkali geothermometer, 129
alkaline hot spring water, 118
alkaline springs, 128
alluvial geothermal systems, 49
Alps, Northern Foreland of the, 19, 25
alteration, effects on magnetic minerals, 281
   hydrothermal, 55, 57, 82, 83, 99

local, 258
minerals, 258
of magnetic minerals, hydrothermal, 100
of silicate minerals, acid, 117
recent rock, 306
regional, 258
rock, 304
thermal, 276, 307
aluminium resources, 184
America, Central, 209
amorphous silica, 115
   polymerization of, 116
   precipitation of, 116
   solubility of, 116
anisotropic permeability, 17, 39
anisotropic variation of resistivity, 92
anisotropy, 92
   electrical resistivity, 93
anomalies, Bouguer gravity, 6
   detection of thermal, 80
   geothermal, *25f*, 29, 31, 38, 150
   heat flow, 31
   intraplate, melting, 4, 26
      thermal, *31f*
   low electrical resistivity, 91
   magnetic, 281
   negative, Bouguer, 99
      magnetic, 100
   positive, geothermal, 3
      thermal, 233
   radioactive, 204
   residual gravity, 99
   seismic, 97
   sublithospheric thermal, 32
   temperature, 11
   thermal, 307
aqueous silica, 116, 118
aquifer, fractured-fissured, 224
   high-salinity, 8
   recharge of, 38
   multiple, 137
   near-surface, 57
      cold, 193
   permeability, 39, 57
   pressure, curves of, 162
   saline, 325

saturated, 325
sedimentary, 5
shallow, 325
temperatures, 5, 7, 138
   in wells, 136
arc, Carpathian, 223
arc basin, back-, 223
area, estimation of fracture, 186
area of maximum pressure reduction, 210
area of maximum reinjection, 170
area of maximum subsidence, 212
areal model of the Wairakei geothermal system, 72, 73
areas, continental, 8
   discharge, 5
   recharge, 5
arsenic, 203
   effluents, 200
artesian overpressure, 59
artesian pressure, 5
   gradient, 47
artificial fluid circulation, 8
artificial fracturing, 175
artificial recharge, 146
assessment, exploration well, *101f*
   geothermal resource, *181f*
   methodology for geothermal resource, 185
   recent geothermal resource, *186f*
   systematic geothermal resource, *186f*
asthenosphere, 26
   partial melting of the, 31
   rheological properties of the, 31
   updoming of the, 233
atmospheric pressure, 326
   flashing at, 138
atomic ratio of chloride, 87
attenuation of seismic waves, 6, 97, 98
   in geothermal areas, 99
available work, 191
Azores, 31

background flow, horizontal, 48
   impressed, 45
   temperature field, 48
   vertical, 47
Bagnore, Italy, 171
Baikal rift, 29
bailing, 139
balneology, 242
basaltic magma chambers, 4
basaltic magmas, 26
basaltic terrain of Iceland, 117
basalts of Iceland, unfolded flood, 186
base inflow, *150f*
basement, fracture system, 59
   fracture zone, 7
   surface, 190
   suture, Precambrian, 297
basin, back-arc, 223

Carpathian, 223
Northern Gulf of Mexico, 5, *188f*, 192, 194
Pannonian, 7, 233, *221f*, 236
Paris, France, 7, 195
Basin and Range, province, U.S.A., 17, 26, 29, 193, 296
   topography, 228
bath area, hot spring, 247
batholith, 16
Benioff zone, 30
bipole–dipole method, 93
bipole source array, multiple, 93
Black Sand-type waters, 126
boiling, 335
   pools, 90
   springs, 5, 83, 84, 115
   underground, 127
   water interface, 168
   zone, two-phase, 80
boron, 129, 203
bottom-surface heat flux, 56
Bouguer anomalies, negative, 99
Bouguer gravity anomalies, 6
boundaries, field, 153, 156
   zone, 149
boundary, irregularities, 149
   locations, field, 95
   regions, side, 170
   resistivity, 69
   sealing, 67
   thermal, 160
   zone, 93
Boussinesq approximation, *42f*, 63
Brawley fault zone, 65, 97
Brazoria geothermal fairway, Texas Gulf Coast, U.S.A., 173
brine, hot, 66, 67
   layer, deep, 172
Broadlands geothermal field, New Zealand, 4, *37f*, 53, 85, *91f*, *97f*, 131, 137, 138, 148, 149, 156, 157, *168f*, 175, 203
   water composition, 111
Bruneau-Grand View, 192
buffering of hydrogen ion, 117
buoyancy, 37, 63
   effect, 157
buoyant convection, 47
buoyant flow, 60

calcium carbonate activity product, 288
calcium-rich waters, 119
caldera, Long Valley, 6, 62
   collapse, 302
California, Coso Hot Springs, 128, 136, 138, 139, 192
   East Mesa field, Imperial Valley, 37, 59, 60, 97, 99, 200, 210
   Heber field, 37, 147, 159, 160
   Imperial Valley, 4, *37f*, 98, 99, 148, 206, 213

Long Valley, 38, 62, 97, 122, 192
Sierra Nevada, east-central, 62
The Geysers, 4, 6, 21, *37f*, 68, 168, 171, 176, 201, 205, 206, 210, 212
cap rock, 65, 80, 251, 263
　impervious, 6
capping structures, 171
carbon dioxide, 86
　gas analysis, geochemical prospecting by, 259
Carlsberg ridge, 29
Carpathian arc, 223
Carpathian basin, 150, 223
Cascade Range, 193
cation exchange, 119
　reactions, 121
cation geothermometer temperature, 139
cations, ratios of, 109
cellular convection, 53
Central America, 209
Central Europe, Upper Rhine Graben, 29
Cerro Prieto geothermal field, Northern Mexico, 29, 39, 131, 135, 137, 138, 140, 200
　springs, 129
Cesano, Italy, 99
chalcedony, 117, 124
charge zone, 54
charging of a geothermal reservoir, fault zone controlled, 59
chemical contaminants, 148
chemical re-equilibration, complete or partial, 123
chemical sampling, 84
chemistry, isotope, 83
　of geothermal systems, 84
Chile, 90, 95
　El Tatio, 31, 129, 130, 135, 147
Chingshu, Taiwan, 132
chloride, atomic ratios of, 87
　concentration, 109
　　variation of, 86, 123, 137
　content, 86
　evaporative concentration of, 126
　waters, 88
chlorite, 258
circulation, artificial fluid, 8
　closed-loop, 214
　convective, 5
　deep water, 83
　hydrothermal, 3, 29
　path for injected fluid, 173
　system, *4f*
　　heat exchange, 9
　　with meteoric water, *6f*
　water, 5
clay dehydration or shale dewatering, 173
clays, 111
Clear Lake area, The Geysers–, 6, 100
climate, effects of, 79
cold front, rate of advance of the local, 176
　speed of advance of the, 170

collinear dipole–dipole traversing method, 95
Colorado, Denver, 210
　Plateau, 316
　　province, 296
　Rangely oil field, 211
　River, 213
colorimetric methods of analysis, 116
columnar jointing, 323
compaction, uniaxial, 41
complex ions, 111
composition, of discharging hot springs, 6
　of geothermal fluids in hot-water systems, 200
　of natural waters in drill holes, 116
　oxygen isotopic, 7
compressibility, 153
　of dissolved silica, 109
　water, 173
condensate injection, reinjection of collapsed pores by, 39
condensate layer, liquid-dominated, 145
condensation surface, 56
conduction, 9, 10, 28, 55, 57
　dominated environments, regional, 188
　temperature profile, steady-state, 47
conductive and hydrologic heat transfer, 313
conductive cooling, 28, 115, 122, 124, 233
　models, 188, 194
conductive heat, flow, *16f*, 89, 314
　component, 7
　flux, 53
　transfer, 3, 6, *9f*, 90, 186, 335
conductive geothermal gradient, 12
conductive geothermal system, *4f*
conductive regional heat flow, 5
conductive surface heat flux, 60
　maximum, 56
conductivity, electrical, 7
　hydraulic, 3, 235, 237
　thermal, *10f*, 18, 39, 41, 89, 230
　uniform thermal, 188
connate pore water, 5
connate water, 8
contaminants, chemical, 148
　gaseous, 148
continental areas, 8
continental drift, 275
continental lithosphere, 29
　cratonic parts of, 32
continental rifts, 4, *29f*
convection, 10, 52, 90, 233
　buoyant, 47
　cell, 37, 287
　cell models, large scale, 54
　cellular, 53
　cooling of an igneous body by hydrothermal, 190, 194
　energy transport by, 59
　hydrodynamics of, 43
　hydrothermal, 26, 28

in a rectangular reservoir, 50
in a saturated porous media, *45f*
natural, 160
of heat and mass in porous media, 38
oscillatory, 50
patterns, asymmetric steady, 48
process, natural, 57, 59
processes, oscillatory, 55
roll, 53
rolls, 58
systems,
  hydrologic characteristics of hydrothermal, 194
  hydrothermal, 15, 17, 186, 188
  identified hydrothermal, 192
  recovery of thermal energy from hydrothermal, 189, *193f*
  undiscovered hydrothermal, *193f*
velocity, characteristic, 44
  oscillatory, 55
  vertical, 45
convective charging, 56
convective flow, 67
  process, 56
convective geothermal systems, *4f*
convective heat-flow component, 7
convective heat transfer, 12, 55
convective heat transport, 230
convective hydrothermal systems, *5f*, 38
convective mass and heat transfer, 57
convective plume, 55
convective–radiative heat loss, 57
convergent plate-margins, 4
conversion cycle, 191
Cook–Austral Islands, 31
cooling, adiabatic, 115, 126, 127
  conductive, 28, 115, 122, 124, 233
  effect, 60
  maximum adiabatic, 139
  of an igneous body by hydrothermal convection, 190, 194
  of intrusions, hydrothermal, 194
  time, post-solidification, 23
coring, 101
Coso Hot Springs, California, 128, 138, 139, 136, 192
counterflowing steam, 148
craton, North American, 297
cristobalite, 258
criteria of Fournier and Truesdell, 120
crust, oceanic, 28
crustal fracture permeability, 55
crustal structural discontinuity, 116
crustal volcanic belts, 193
crystallization, latent heat of, 23
Curie isotherm, 7
cyclic pumping tests, 211

Darcy's law, 39, 154

dead core, 157
dead field, 152
decompression, 150, 152, 153, 173
degree of economic feasibility, 183
degree of geologic assurance, 183
dehydration, clay, 173
density, example of a triangular probability, 191
  logging, 102
  triangular probability, 191, 193
Denver, Colorado, 210
deposition rate, 288, 290
deposits, high-temperature, 199
  low-temperature, 199
Desert Peak, Nevada, 7
dewatering, shale, 173
diagram, activity, 111
  enthalpy chloride, *125f*, 129
  enthalpy composition, 123
  for geothermal energy, McKelvey, 185
  McKelvey, 183, 185
  method, activity, 113
  phase, 112
diapirism, mantle, 244
  thermal mantle, 223
diffusion, front, 152, 155
  pressure front, 156
  rate of pressure, 162
diffusivity, 10
  hydraulic, 154, 168, 170
  thermal, 18
digital stacking techniques, 98
dike system, 150
dilution, of thermal water by cold water, subsurface, 118
  on mixing, 87
dipole, configuration, 92
  –dipole soundings, 95
  –dipole traversing method, collinear, 95
  method, bipole–, 93
  technique, roving, 92
directional drilling, 263
discharge, enthalpy, mean, 175
  well, 167, 170
  features, natural, 89
  groundwater, 239
  heat, 82, 171, 260
  initial gas content of the, 171
  natural, 221
  of thermal fluid, 186
  rates, optimum, 103
  salinity of the surface, 81
  steam, 171
  surface, 78, 151, 164, 204
  temperature, spring, 7
  well, *151f*
  wells, 152
discontinuity, crustal structural, 316
dispersed water phase, 172
distribution network, 152

Dixie Valley, Nevada, 193
dome, rifted, 32
    thermal, 32
    unrifted, 32
down-flow of meteoric water, 230
down-hole, flow measurements, 102
    heat exchange system, 309
    pressures, 102
    pressure distribution, 265
    samples, 102
    water samples, 138
drainage, surface, 303
drainage model, simple, 164
drawdown, 137, 150, 176, 289
drift, continental, 275

earthquake, activity, 275
    man-made, 9
    San Fernando, 208
East Africa, Afar Depression, 29
East African Rift, 29
    Kavirondo, 32
East Mesa geothermal field, Imperial Valley, California, 37, 59, 60, 97, 99, 200, 210
East Pacific Rise, 29, 208
East Puna, Hawaii, 97
Easter Island, 31
Eastern type of intermediate heat flow, 188
economic feasibility, degree of, 183
effects, 83
    geochemical evidence of, 139
    pattern of uniform, 165
    pressure, 102, 145
    profile of, 157
    pulse, 170
    zone of, 157
effects, of climate, 79
    of injection, potential, 151
    of magmatic pre-heating, 188, 194
    of pH, *117f*
    of the reinjection–extraction rate, *329f*
electrical and electromagnetic methods, *91f*
El Tatio, Chile, 31, 129, 130, 135, 147
El Salvador, 95
    Ahuachapán geothermal field, 97, 130, 200, 210, *321f*
    geological structure of, *321f*
electrical conductivity, 7
electrical logging, 102
electrical methods, 6
electrical resistivity, 80, 91
    anomaly, low-, 91
    measurements, 8, 87
    survey, 247
electrolyte, 91
electromagnetic methods, controlled-source, 96
electromagnetic sounding, 96
energy, amount of recoverable thermal, 180
    content of the field, 83

dissolved methane, 192
economic reservoirs of geothermal, 77
    from a geothermal reservoir, utilization of, 148
    from geopressured-geothermal fluids, recovery of, *194f*
    igneous-related thermal, *181f*
    loss, rate of thermal, 186
    low-temperature, geothermal, 195
    of methane, 189
    quality, 175
    recoverable geothermal, 183
        resource, available, 151
        estimation, 182
    source of extracted, 153
    thermal, 181, *189f*
    transport by convection, 59
    utilization of geothermal, 289
    yield of, 156, 157
enthalpy, 123, 150, 153
    –chloride diagrams, *125f*, 129
    –composition diagrams, 123
    discharge, 151, 163, 326
    fluid, high-, 78
    history of the discharged fluid, 164
    initial, 125
    mean discharge, 175
    measurements, wellhead, 137
    of discharged fluid, 164
    of fluid, 78, 175
    of the warm-water reservoir, 151
    output, 146
    recharge, 170
    reservoir, 150
    rise, transient, 170
    temperature, 138
    well discharge, 167, 170
environment, open-sea, 250
    regional conduction-dominated, 188
    subcoastal, 250
environmental aspects of geothermal development, *199f*
environmental concern, evaluation of, *199f*
environmental problems, *201f*, 204, 303
erosion, subcrustal, 223
estimation, heat content, 146
    of fracture area, 186
    of reservoir temperature, 6, *113f*
Europe, Upper Rhine Graben, Central, 29
eutrophication, 214
evaporative concentration, 109
    of chloride, 126
exchange reaction, cation, 121
expellation, zone of, 236
exploitation, flow, natural pre-, 146
    geological effects of, 82
exploration, drilling, 101
    geophysical, *87f*
    geothermal, 79, 82, 86, 87, *109f*
    high-temperature hydrothermal, *304f*

programme, 78, 79, 81
  geothermal, 89
  Hot Dry Rock, *306f*
  techniques, 80
  well, siting of an, 101
    assessment, *101f*
extensional tectonics, 6, 29, 297
extracted energy, source of, *153f*
extraction, fluid, 146
  from geothermal reservoirs, heat-, *145f*
  heat, 146
  rate, fluid, 39
  system, man-made heat-, 4

falling water-level model, 171
fault/fracture zone, 7
  hot spring discharges in, 6
fault movement, 211
fault zone, 57, 58, 60
  Brawley, 65, 97
  controlled charging of a geothermal reservoir, 59
  heat and mass transfer in a, 58
faults, active, 208
  graben, 297, 305
  injection of fluid into active, 208
  San Andreas, 210
  step, 323
feed point of a well, *147f*
feed zone, 164
  permeable, 167
  temperature, 148
feeding features, 152
field boundaries, 153, 156
field boundary locations, 95
field response, change in, 157
finite element method, 95
fission-track method, 310
  age determination by the, 250
fissure, and porous bed permeability, 102
  eruptions, 275
  near-vertical, 80
  swarms, 275
flashing, 72, *137f*, 288
  at atmospheric pressure, 138
  front, 137, 138
flow, resistance, 53
  testing, 136
  velocity, 290
flow distribution of water and steam, 172
flow measurements, down-hole, 102
flow number, mass, 53
flow rate, injection, 266
  curves of geothermal steam and hot water, characteristic, 270
  measurements, 264
  rate, mass, 54, 57, 116, 123, 150
flowing, directions, 292
  history, 290

velocity, 292
circulation, 64
fluid,
  composition, 109
  enthalpy, 78
    history of the, 78, 175
    of discharged, 138, 290
  extraction, 146
    rate, 39, 176
  flow, 65
    in fissured rocks, 38
    through fractured systems, 39
    through porous media, *39f*
    two-phase, 42
  geopressured, 181
  geothermal, 5, 99, 109
  hot, 80
  intrusions, 311
  pore, 5
  pressure, 41, 127, 211
    local, 39
  residence time, 123, 140
  supply from mantle/deep crustal sources, 6
  temperature, 127, 137
  underground, 80
  virtual ore-depositing, 204
  yield, pattern of, 135
    rate of, 155
fossil-fuel sources, 214
fossil groundwater, 80
Fournier and Truesdell, criteria of, 120
fractionation, of oxygen isotopes, 113
  of the isotopes, temperature dependence of the, 85
fracture, area, estimation of, 186
  cavities, 261
  control, 146
  -fissured aquifer, 224
  -induced permeability, 193
  method, planar, 186
  permeability, 56, 147
    crustal, 55
  planar, 185
  porosity, 38
  spacing, 186
  structure, high-density, 174
  system, 152
    basement, 59
    geometry of, 38
  zone, basement, 7
fractured reservoir, 167
fractured rock system, water-saturated, 55
fractured system, 149, 175
  fluid flow through, 39
  recovery factor for geothermal, 167
fractured zones, 321
fracturing, artificial, 175
  hydraulic, 9, 88, 98, *305f*, 308
  interwell, 174

natural hydraulic, 83, 98
  tectonic, 323
France, Paris basin, 7
friction, intergranular, 211
frictional heating, 13, 30
fumaroles, 5, *82f*, 90, 128, 186, 252, 259, 304, 306

gaseous contaminants, 148
gas content of the discharge, initial, 171
gases, noncondensable, 136, 201
  in geothermal areas, noncondensable, 202
geochemical analysis of natural springs, 279
geochemical evidence of drawdown, 139
geochemical prospecting by carbon-dioxide gas analysis, 259
geochemical survey, 247
geochemical thermometer, 113, 140, 244
geochemistry, 83
  isotope, 7, 83
  water, *109f*
geographic distribution of springs, 136
geohydrologic model, integrated, 62
geohydrology, 302
geologic assurance, degree of, 183
geologic effects of exploitation, 82
geologic mapping, 7, 279
geologic provinces of the U.S.A., major, 192
geological studies, 82
geology, regional, 79, 82
geophysical exploration, *87f*
geophysical methods, *87f*
geopressured basins of the U.S.A., 192
geopressured fluids, 181
geopressured-geothermal fluids, recovery of energy from, *194f*
geopressured-geothermal resource, 195
geopressured system, 188, 189
geopressured waters, dissolved methane in, 195
  thermal energy in, 189
geopressured zone, 4, 80
geotherm, 19
geothermal activity, 26, 29, 79, 81, 276
  in urban areas, 91
  surface, 82, 280
  surface expression, 90
geothermal anomaly, *25f*, 29, 31, 38, 150
  positive, 3
geothermal aquifer production, 152
geothermal area, 52, 57, 82, 97, 199
  active, 54
geothermal areas, attenuation of seismic waves in, 99
  non-condensible gases in, 202
geothermal brine, saline, 273
geothermal development, environmental aspects of, *199f*
geothermal effluents, gaseous components of, *201f*
geothermal electrical capacity, world-wide installed, 182

geothermal energy, 109, 183, 199, 214, *273f*, 295
  economic reservoirs of, 77
  evaluation of igneous-related, *194f*
  extraction method, Hot Dry Rock, *305f*
  low-temperature, 195
  McKelvey diagram for, 185
  prospecting for, 101
  recoverable, 183
  resource, *77f*, 97, *181f*, 184, 239
    information, basic, 82
  utilization, low-enthalpy, 7
    and recovery of, 239
geothermal exploration, 79, 82, 86, 87, *109f*
  programme, 89
  in the Takinoue area, history of, 249
geothermal fields, 77, 79, *80f*, *86f*, 96, 98, 103
  delineation of, 87
  high-temperature, 273, 275
  Hot Dry Rock, 80
  lifetime of a, 206
geothermal fluid, 5, 99, 109
  flow, 152
    in hot-water systems, composition of, 200
  temperature, 101
  ratios of the cations in, 140
  saline, 181
geothermal gradient, 6, 11, 32, 56, 88, 109, 310, 313
  conductive, 12
geothermal mapping, regional, 7
geothermal models, 16
  calculation, 233
geothermal phenomena, preliminary location of, 78
geothermal potential, map of, 82
geothermal production field, 78
geothermal prospecting, 77, 88, 96
geothermal provinces, 295
geothermal recovery factors for fractured systems, 167
geothermal region, 77
geothermal reserve, 185
geothermal reservoir, 3, 5, 46, 48, 60, 110, 261
  beneath irrigated fields, 206
  drilled, 110, 111
  heat extraction from, *145f*
  fault zone controlled charging of a, 59
  in volcanic environments, 6
  island, 54
  utilization of energy from a, 148
  vapour-dominated, 172
  volume of a, 186
geothermal resource, 56, *77f*, 97, *181f*, 184, 304, 315
  assessment, *181f*
    methodology for, *185f*
    recent, *186f*
    systematic, *186f*
  base, *25f*, 184

estimation, 181
low-enthalpy, *221f*
prospecting for, *77f*
terminology, *184f*
utilization of, 9
geothermal steam, 263
geothermal systems, *3f*, 38, 43, 47, 57, 61, 90, 111, 116, 146, 282, 302, 313
  active, 109
  alluvial, 49
  areal model of the Wairakei, 72, 73
  blind high-temperature, 7
  chemistry of, 84
  conductive, *4f*
  convective, *4f*
  geopressured, 188, 189
  horizontal and vertical flows in, *147f*
  identified hot-water, 191
  liquid-dominated, 68, 152
  man-made, 8
  sedimentary and alluvial, 49
  thermally driven flow in single-phase, 42
geothermal water, *273f*
  deep, 121
  high-temperature, 119
  in sedimentary rocks, low-temperature, 195
  low-temperature, 190
  temperature of deep, 84
geothermal wells in New Zealand, 114
geothermometer, 103, 113, 140, 192
  alkali, 129
  application of chemical, 129
  chemical, 6, 7, 85, 188, 259
  effect of boiling upon, *127f*
  effect of dilution on the Na–K–Ca, 119
  effects of mixing on the Na–K–Ca, 119
  magnesium corrected Na–K–Ca, 120
  Na–K, *118f*, 127, 136, 138
  Na–K–Ca, 85, *119f*, 127, 136
  ^{18}Oxygen (SO$_4$–H$_2$O), *121f*, 127
  quartz, 115
  silica, 8, *113f*, 126, 127, 138
  silica and Na–K–Ca, 290
  sulfate oxygen isotope, *121f*
  temperature, dependence of selected, 114
    cation, 139
geothermometric techniques, 84
geothermometry, 188
  chemical, 81, 84, 300
Germany, Urach anomaly, South, 23
Geyser Hill-type waters, 125, 126
geysers, 37
graben fault, 297, 305
graben formation, 29
granite terrains, 117
Grass Valley, Nevada, 95
gravity, and magnetic methods, *99f*
  anomalies, residual, 99
  instability, 151

  mapping, 6
  residuals, negative, 99
    positive, 99
  survey, 100
  vector, 44
Green's function, 289
groundnoise, 97, 98
  survey, 97
groundwater, 117
  discharge, 239
  flow, 63, 67
  fossil, 80
  inflow of meteoric, 37
  near-surface, 102
  shallow, 127
Gulf of California, 29
Gulf of Mexico Basin, Northern, 5, *188f*, 192, 194

Hachimantai volcanic region, 250
Hawaii, 79, 97
  East Puna, 97
  Oahu, 95
  Pahoa geothermal field, 32
Hawaiian Islands, 31
heat and mass, in porous media, convection of, 38
  transfer, convective, 57
    in a fault zone, 58
    in a water-saturated porous medium, 48
    processes, 52
    transport in geothermal reservoir systems, 61
  budget, magmatic, 186, 185
  capacity, 153
    volumetric, 9, 25
  content, 9, 188
    estimation, 146
    of rocks, *25f*
  discharge, 82, 171, 260
  exchange, circulation system, 9
    surface, 9, *305f*
    system, down-hole, 309
  extraction, 146
    from geothermal reservoirs, *145f*
    system, man-made, 4
heat, radiogenic, 13
  resupply of, 186
  specific, 10
  volumetric specific, 186
heat flow, *10f*, 28, 88, 156, 228, 299, 307
  anomaly, 31
  Basin and Range type of high, 188
  component, conductive, 7
  conductive, *16f*, 89, 314
    regional, 5
  determination, methods of, 10
  distribution, regional, 22
    patterns, 7
  Eastern type of intermediate, 188
  hydrologic, 314

mantle, 14
map, of Hungary, 233
  of New Mexico, 300
mapping, 10
measurements, *89f*
natural, 70
paleo, 8
pre-production fluid and, 62
profile, 21, 30
provinces, *17f*, 188
  regional, 5
rate, 89
regional, *9f*, 26, 188
Sierra Nevada type of heat flow, 188
surface, 13, 14, *16f*, 26, 56, 57, 89, 102
survey, 90, 101
vector, 10
heat flux, bottom-surface, 56
  conductive, 53
  conductive surface, 60
  distribution, near-surface, 48
  maximum conductive surface, 56
  maximum surface, 60
  overall, 54
  surface, 56, 57
heat generation, 13, 17
  by radioactive decay, 10
  radioactive, 9, 188
heat influx, 150
heat loss, convective-radiative, 57
  natural, 81
heat of crystallisation, latent, 23
heat production, 18
  by radioactivity, *13f*
  constants, 13
  in igneous rocks, 14
  in metamorphic rocks, 14
  in sedimentary rocks, 15
  radiogenic, 17
heat recovery factor, 152
heat removal, 167
heat source, *13f*, 72, 295
  deep, 96
  long-lived radiogenic, 12
  magmatic, 3
  transient, *15f*
heat transfer, 137
  coefficient, 57
  conductive, 3, 6, *9f*, 90, 186, 355
    and hydrologic, 313
  convective, 12, 55, 57
  rate of, 176
  surface, 52
  transport, convective, 230
heat traps, 5
heat withdrawal front, mass and, 156
heating, district and space, 241
  domestic and agricultural, 195
  frictional, 13, 30

Heber geothermal field, California, 37, 147, 160, 159
history, of geothermal exploration in the Takinoue area, 249
  solidification, 15
  stress, 41
horizons, hydrostratigraphic, 241
  paleokarst, 241
hot brine, 66, 67
Hot Dry Rock, 4, *8f*, 90, 295
  demonstration project, 305
  exploration program, *306f*
  geothermal energy extraction method, *305f*
  geothermal fields, 80
  system, 5, *173f*
  technology, extraction of energy, 194
Hot ground, 82
  acid-altered, 128
Hot hydrostatic head, 52
Hot pools, 90
Hot spot, 54
Hot spring, and well water, 115
  bath area, 247
  water, alkaline, 118
Hot water, dominated field, 88
  geothermal system, identified, 191
  reinjection of, 263
  system, 80, 86, *127f*, 139, 252
    composition of geothermal fluids in, 200
Hubbert–Rubey principle, 211
Hungary, 195
  heat flow map of, 233
  Pannonian basin, 7, *221f*, 233, 236
hydraulic conductivity, 3, 235, 237
hydraulic diffusivity, 154, 168, 170
hydraulic fracturing, 9, 88, 98, *305f*, 308
  natural, 83, 98
hydraulic gradient, 236
hydraulic pressure, 236
hydrochemical zonality, 239
hydrochemistry, *238f*
hydrochloric acid, 244
hydrodynamic gradient, 38
  of convection, 43
hydrodynamic transients, 50
hydrogen ion, buffering of, 117
  activity, 110
hydrogen isotope, 86
hydrogen sulphide, 200
hydrogeology, *233f*, *325f*
hydrologic characteristics of hydrothermal convection systems, 194
hydrologic heat flow, 314
hydrologic heat transfer, conductive and, 313
hydrolysis reactions, 110
  silicate, 117
hydrostatic gradient, 170
  of pressure, 167
hydrostatic head, 124, 126, 128, 205

hot, 52
hydrostatic pressure, 5, 45, 128, 236, 282
  distribution, reference, 44
hydrostatic steady-temperature field, reference, 44
hydrostratigraphic horizons, 241
hydrothermal activity, 313
  superficial, 323
hydrothermal alteration, 55, 57, 82, 83, 99
  of magnetic minerals, 100
hydrothermal circulation, 3, 29
hydrothermal convection, 26, 28
  cooling of an igneous body by, 190, 194
  systems, 15, 17, 186, 188
    hydrologic characteristics of, 194
    identified, 192
    recovery of thermal energy from, *193f*
    undiscovered, *193f*
hydrothermal cooling of intrusions, 194
hydrothermal exploration, high-temperature, *304f*
hydrothermal fluid, acidic, 304
  reserves of, 305
hydrothermal reactions, 109
hydrothermal reserves, 62
hydrothermal reservoir performance, prediction of, 60
hydrothermal resources, low to moderate, 314
hydrothermal systems, *4f*, 23, 58, 62, 63, 109, 122, 136, 251, 295, 303
  associated, 190
  border region, 9
  convective, *5f*, 38
  mathematical modelling of, *37f*
  model of, 113
  pore fluid in, 39
  pre-production models of, 50, 60
hydrothermal waters, 90
hypersaline brines, reinjection of, 205

Iceland, 101, 113, 195, 28
  basaltic terrain of, 117
  Krafla geothermal field, North-east, 100, 168, *273f*
  Reykjanes, 130, 135
  unfolded flood basalts of, 186
Icelandic field, 150
Icelandic low-temperature system, 159
Idaho, Island Park area of, 192
  Raft River, 95, 101
igneous activity, 304
igneous body by hydrothermal convection, cooling of an, 190, 194
igneous-related geothermal energy, evaluation of, *194f*
igneous-related systems, 188
igneous-related thermal energy, *191f*
igneous rock, solidified, 190
igneous system, young silicic, 190
ignimbrites, 70, 72

Imperial Valley, 4, *37f*, 98, 99, 148, 206, 213
  –Mexicali area, 208
impermeable boundaries, 69
inaccessible resource base, 185
India, Parbati Valley of the Himalayas, north-west, 31
Indonesia, Kawah Kamodjang, 31, 168, 171, 172, 191
induced microearthquakes, 210
induced seismicity, 210
infiltration, direct, 325
  secondary, 236
  zone of, 6, 236
inflow, 164
  base, *150f*
  cold water, 164, 165
  natural, 173
  of meteoric groundwater, 37
  side, *150f*
influx, heat, 150
  rate, total, 66
infrared imagery survey, 82
infrared scanners, 90
infrared surveying, aerial, 91
injected fluid, circulation path for, 173
  capacity of a well, 266
injection, effects of, 329
  flow rate, 266
  method, direct, 264
  of fluids into active faults, 208
  of liquid waste, 210
  rate, 157
  reinjection of collapsed pores by condensate, 39
  tests, 210
    pressure oscillations during, 291
  well, 174
interface, boiling water, 168
  level, 156
  water/two-phase, 162
intergranular flow, 187
intergranular friction, 211
intergranular pore spaces, 127
intergranular vapourization, 187, 173
interstitial fluids, 91
inter-well fracturing, 174
intraplate melting anomalies, 4, 26
intraplate thermal anomalies, *31f*
intraplate volcanism, 31
intrusions, cooling, 23
  hydrothermal cooling of, 194
  igneous, 10
  magmatic, 13
  of hot mafic magma, 6
  shallow, 15
  silicic, 188
  young magmatic, 4
intrusive activity, 55
  volcanic or, 83
inventory, low-temperature, *195f*
investigation, of microearthquakes, 89

Index

preliminary, 81
program, 82
ions, complex, 111
irrigation waters, trace-element tolerances for, 205
Island Park area of Idaho, 192
isobutane, 208
isotherm, Curie, 7
isotope, chemistry, 83
   exchange reaction, sulfate oxygen, 121
   fractionation of oxygen, 113
   geochemistry, 7, 83
   hydrogen, 86
   oxygen, 86
   temperature dependence of the fractionation of, 85
   variations in oxygen and hydrogen, 123
isotopic reactions, 113
Italy, Bagnore, 171
   Cesano, 99
   Lardarello, 4, *37f*, 68, 140, 168, 171, 203, 214
   Preappenine belt of central, 186
   Travale, 171
   Viterbo region of, 210

Japan, 208
   Kurikoma field, 99
   Matsukawa geothermal area, 98, 170
   Otaki area, 210
   Takinoue area, *247f*
Jemez Mountains, New Mexico, 17, 95, 173, *295f*, *316f*
jointing, columnar, 323

Kakkonda geothermal power plant, 248
K/Ar dating, 310
karstic water system, 235
Kavirondo, East African Rift, 32
Kawah Kamodjang, Indonesia, 31, 91, 168, 171, 172
Kawerau geothermal field, New Zealand, 39, 99, 149, 165, 175, 176
Kenya, 29, 90, 95
Kizildere, Turkey, 99, 133, 148
Klamath Falls, 192
Krafla caldera, 275
Krafla geothermal field, north-east Iceland, 100, 167, *273f*
Kurikoma field, Japan, 99
Kuril Islands, 30
Kyle Hot Springs, Northern Nevada, 204
Kyushu, north-central, 99

lake water, closed basin saline, 123
Landau area, Upper Rhine Graben, Western Germany, 8
landslide movement, casing failure as result of, 207
landslides, *206f*

Lardarello geothermal field, 4, *37f*, 68, 140, 168, 171, 172, 203, 214
lifetime, of a field, 165
   of a geothermal field, 206
   lifetime, reservoir, 152
line source method, differentiated, 230
liquid and two-phase zones, 169
liquid-dominated condensate layer, 145
   field, 168, 170
   geothermal system, 68, 152
   systems, 4, 37, 145, 156
   two-phase system, 156
   underpressured system, 305
liquid waste, injection of, 210
literature survey, 81
lithosphere, 26
   continental, 29
   cratonic parts of continental, 32
lithostatic pressure, 5, 29, 80
location of productive zone, 78
logging, density, 102
   electrical, 102
   neutron-gamma, 102
   sonic, 102
Long Valley, caldera, 6, 62
   hydrothermal system, California, 38, 62, 97, 122, 192
Los Alamos Hot Dry Rock concept, *8f*
Los Alturas Estates, New Mexico, 93
Louisiana, 189
low-temperature system, 195

mafic magma, intrusions of hot, 6
magma, basaltic, 26
   chamber, 25, 64, 99, *174f*, 188, 275, 307
      basaltic, 4
   flow, subterranean, 275
   silicic, 26
magmatic activity, 26, 29, 284, 291
magmatic differentiation, 13
magmatic heat, budget, 185, 186
   sources, 3
magmatic intrusions, 13
   young, 4
magmatic pre-heating, 190
   effects of, 188, 194
Magnesium-corrected Na–K–Ca geothermometer, 120
   -rich waters, 121
   temperature correction, 121, 136
magnetic anomaly, 281
   negative, 100
magnetic gradient, 302
magnetic methods, gravity and, *99f*
magnetic minerals, alteration effects on, 281
   hydrothermal alteration of, 100
magnetic properties, changes in the, 88
magnetotelluric methods, 96
mantle, deep-crustal sources, fluid supply from, 6
   diapirism, 244

thermal, 223
heat flow, 14
plume, 31
mapping, geologic, 7, 279
gravity, 6
heat flow, 10
regional geothermal, 7
resistivity, 6
Marysville, Montana, 7, 90
mass and heat transfer, convective, 57
pre-production, 71
in a fault zone, 58
and heat withdrawal front, 156
discharge, 70f, 326
at Wairakei, mean, 163
surface, 73
flow number, 53
flow rate, 54, 57, 115, 123, 150
flux, surface, 56
leakage, 73
loss, net, 100
transfer, in a water-saturated porous
medium, 48
processes, heat and, 52
spatial distribution of, 54
Matsukawa geothermal area, Japan, 98, 170
Matsukawa geothermal power plant, 247
McKelvey diagram, 183, 185
for geothermal energy, 185
melting anomalies, intraplate, 4
melting of the astenosphere, partial, 31
melting pressure-release, 28
Momotombo reservoir, Nicaragua, 165
mercury, 203
contamination, 200
metamorphic fluids, melting/migration of, 15
metamorphic reactions, 31
metamorphism, 13, 99
of sediments, low-grade, 99
thermal, 55
zeolite to greenschist facies, 282
meteoric groundwater, inflow of, 37
meteoric water, 3, 6, 205, 236
circulation system with, 6f
cold, 150
downflow of, 230
methane, 86, 181
energy, dissolved, 192
energy of, 189
gas, evolution of, 173
in geopressured waters, dissolved, 195
method of analogy, 187
Mexico, 95
Mexico, Cerro Prieto geothermal field,
Northern, 29, 38, 131, 137, 138, 140, 200
micas, 111
micro-earthquake, 97, 98, 304
induced, 210
investigation of, 89

studies, 97
survey, 88
microseismic activity, 210
microseismicity, 97
Mid-Atlantic Ridge, 275
mid-oceanic spreading ridge, 28, 29
migration velocity, 237
mineral prospecting, petroleum or, 79
mineral stability fields, 111
mineralization, 238
mineralized water, 236, 303
minerals, alteration-, 258
effects on magnetic, 281
mining operation, efficiency of the, 146, 176
mining process, margin, 175
mixed waters, recognition of, 122f
mixing model, 7
silica, 123
mixture, two-phase, 153, 154
steam–water, 153
model; calculation, geothermal, 233
mixing, 7
of conduction cooling, 194
of hydrothermal system, 113
pipe, 25f
pre-production, 38, 67
simple drainage, 164
monomeric silica, 116
Montana, Marysville, 90
Monte Carlo method, 191
montmorillonite, 258
mud flow rate, 101
mud pot, 128
multiple aquifers, 137
multiple bipole source array, 93
multiple reflections, 98

Na–K, geothermometer, 118f, 127, 136, 138
method, 85
ratio, 140
variation in the, 137
temperature, 129, 138
Na–K–Ca geothermometer, 85, 119f, 127, 136
effects of mixing on the, 119
magnesium-corrected, 120
silica and, 290
Na–K–Ca temperature, 85, 136
nano-earthquakes, 98
natural geothermal waters, 118
near-surface, aquifer, 57
cold aquifer, 193
gradient, 53
groundwater, 102
thermal properties, 57
water, 43
zone, 148
near-vertical fissures, 80
neovolcanic zones, 275
neutron gamma logging, 102

Nevada, 97
  Desert Peak, 7
  Dixie Valley, 193
  Grass Valey, 95
  Kyle Hot Springs, Northern, 204
  Steamboat Springs, 58
New Mexico, heat flow map of, 300
  Jemez Mountains, 17, 95, 174, *295f, 316f*
  Los Alturas Estates, 93
  Radium Springs, 98
  Rio Grande Rift, 29
  U.S.A., 8
  Valles Caldera, 9
New Zealand, 30, 69, 87, 91, 92, 101, 102, 148, *187f*, 214
  Broadlands geothermal fields, 4, *37f*, 85, 97, 98, 131, 137, 138, *168f*, 175, 203
  geothermal wells in, 114
  Kawerau geothermal system, 39, 99, 149, 165, 175, 176
  Ngawha, 48, 131, 170, 175
  Orakeikorako, 129, 132, 135
  Rotokawa, 132, 168, 170
  Tauhara field, 89, 165
  Taupo volcanic zone, 47
  thermal belt of, 165
  Wairakei geothermal field, 4, *37f*, 137, 138, 158, 160, 199, 200, 211, 214
Nicaragua, Momotombo reservoir, 165
North American craton, 297
Northern Gulf of Mexico basin, 5, *188f*, 192, 194
Nusselt number, *45f*, 48, 51, 53

Oahu, Hawaii, 95
ocean water, 123
oceanic crust, 28
oceanic spreading ridges, 4, 26, 28, 29
open-sea environment, 250
Orakeikorako, New Zealand, 129, 132, 135
ore-depositing fluids, virtual, 204
Otaki area, Japan, 210
outflow, surface, 150
overheat ratio, 44
overpressure, 44
  artesian, 59
oxidation, of H_2S, formation of sulfate by, 122
  of sulfides, 117
  processes, 86
oxygen, isotope, 86
    fractionation of, 113
    geothermometer, sulfate, *121f*
    of sulfate, re-equilibration of, 122
    isotopic composition, 7
    18 oxygen (SO_4–H_2O) geothermometer, *121f*, 127

Pahoa geothermal field, Hawaii, 32
paleo heat flow, 8
paleokarst horizons, 241

Pannonian basin, 7, *221f*, 233, 236
  high-enthalpy reservoir possibilities in the, 244
  Hungary, 7, *221f*,
  reservoir system, Upper, 239
  thermal activity, 228
  thermal water resources of the, 228
paragenetic sequences of silica minerals, 116
Parathetys sea, 223
Parbati Valley of the Himalayas, north-west India, 31
Paris basin, France, 7, 195
permeability, 5, 41, 88, 100, 109, 146, 154, 157, 160, 186, 289, 303
  anisotropic, 39, 47
  aquifer, 39, 57
  barriers, low, 149
  channels, higher, 176
  crustal fracture, 55
  distribution, 3
  effects, relative, 173
  effects of, 56, 57
  fissure and porous bed, 102
  formation, 328
  fracture, 56, 147
  fracture-induced, 38
  horizontal, 80, 165
  horizontal and vertical, 47
  local, 170
  natural, 8
  natural and induced, 193
  of steam and water, 290
  of the pluton, 56
  overall, 102
  profile, 168
  reservoir, 63, 55
  reservoir, low, 175
  reservoir of ideal, 193
  secondary, 83, 323
  system, fracture, 58
  variation, 169
  variations in, 80
    of fracture, 8
  vertical, 11, 84, 86, 147, 159
  zones, high, 80
permeable feed zone, 167
permeable reservoir, 172
petroleum, resource calculations, 183
  or mineral prospecting, 79
pH correction, 118
phase changes, 13
  diagram, 112
  dominant mobile, 145
  flow, single, 152
  fluid flow, two-, 152
  steam, 126
phase system, liquid-dominated two-, 156
  two-, *167f*
    phases, solubility of silica, 116
Philippine Sea, 30

photography, aerial, 82
piezometric pressure, 238
piezometric surface, 325
pipe models, 52f
pipe system, 52
Pitot tube, 90
planar fracture, 185
plate-boundaries, 26, 208
plate-margins, convergent, 4
plate tectonics, 26, 208
   model for the evolution of the Pannonian basin, 223
plume, convective, 55
plumes of hot water, 99
pluton, 55
P_n velocity, 6
polymerization of amorphous silica, 116
polymerized silica, 116
pools, boiling, 90
pore, collapse, 39, 173
   fluid, 5
      in hydrothermal systems, 39
         pressure, 9, 83
   pressure, 41, 80, 98
   spaces, intergranular, 127
   water, connate, 5
porosity, 5, 41, 65, 91, 139, 153, 186, 212
   effective, 187
   fracture, 38
   joint, 223
   matrix 38
   rock, 39
   secondary, 233
   total, 187
   vug, 233
porous-bed permeability, fissure- and, 102
   porous media, convection in saturated, 45f
   convection of heat and mass in, 38
   fluid flow through, 39f
porous medium, heat and mass transfer in a water-saturated, 48
   two-phase fluid flow through, 42
post-glacial volcanism, 275
potassium, 13
power generating capacity, potential, 103
Preapennine belt of central Italy, 186
Precambrian basement suture, 297
Precambrian shields, 17
precipitation of amorphous silica, 116
   of silica, 115f, 263
   rate of quartz, 115
prediction, of hydrothermal reservoir performance, 60
   of reservoir behaviour, 73
pre-exploitation, conditions, 71
   flow, natural, 146
   surface manifestations, 165
   values of environmentally sensitive parameter, determination of, 78

preheating, effects of magmatic, 188, 194
   magmatic, 190
preliminary investigations, 81
preliminary location of
   geothermal phenomena, 78
pre-production, flow, 61
   fluid and heat flow, 62
   mass and heat transfer, 71
   model, 38, 67
   model of hydrothermal systems, 50, 60
   model temperature contours, 67
   model velocity field, 67
   temperature, 66, 73, 60
      and flow field, 60
pressure, artesian, 5
   at Wairakei, 163
   atmospheric, 326
   change, reservoir, 332
   control, 244
   curves of aquifer, 162
   decay curve, 165
   defect, 162
   diffusion, rate of, 162
   distribution, initial, 165
   distribution of down-hole, 265
   disturbances, 168
   down-hole, 102
   drawdown, 102, 145
      profile of, 155
   drop, 72, 153, 156, 162, 164, 175, 213
      internal, 150
      during production, 73
      history, 72
   field, 56
   fluid, 41, 127, 211
   flashing at atmospheric, 138
   front, 154
      diffusive, 156
   gradient, 162, 164
      artesian, 47
      horizontal, 52
      natural, 167
   horizontal transmission of, 165
   hydraulic, 236
   hydrostatic, 5, 45, 128, 236, 282
   hydrostatic gradient of, 167
   injection, 9
   lithostatic, 5, 29, 80
   local fluid, 39
   natural gradient of, 159
   of the production zone, 171
   oscillations during injection tests, 291
   piezometric, 238
   pore, 41, 80, 98
   pore fluid, 9, 83
   pulse, 152, 170, 289
   reduction, area of maximum, 210
   reinjection, 326
   release, 29

release melting, 28
reservoir, 147, 263, 326
saturation, 332
separation, 326
transmission, 146, *152f*, 156
water column, 265
wave, sideways propagation of the, 156
well head, 71, 189, 263, 264
production, characteristics, 326
field, geothermal, 78
index, 265
testing, 265
wells, 326
  temperature changes in, 136
zone, 139, 169
  location of, 78
  pressure of the, 171
productive capacity, potential, 101
propagation, two-phase, 156
propagation effects, vertical, 156
propagation of the pressure wave, sideways, 156
propane, 208
prospecting, for geothermal energy, 101
  for geothermal resources, *77f*
  geothermal, 77, 88, 96,
  petroleum or mineral, 79
  resistivity, 81
  strategies, 78, 79
provinces, geothermal, 295
  heat flow, *17f*, 188
  regional heat-flow, 5
pumping tests, cyclic, 211
P-wave delay, teleseismic, 6, 98

quartz, geothermometer, 115
  precipitation, rate of, 115
  solubility curve, 123, 124
  solubility of, 114, 115, 117
quasi-steady field condition, 169
quasi-steady flow, 156
quasi-steady state, 152

radiation, 10
radiative heat loss, convective, 57
radioactive anomalies, 204
radioactive components in springs, 204
radioactive decay, heat generation by, 10
radioactive heat generation, 9, 118, 188
radioactive thermal waters, 204
radioactivity, heat production by, *13f*
radioelements, distribution of natural, 13
radiogenic heat, 13
  production, 17
  sources, long-lived, 12
radiometric dating methods, 310
Radium Springs, New Mexico, 98
radon, 203, 290
Raft River, Idaho, 95, 101
Rangely oil field, Colorado, 211

rate, fluid extraction, 39, 176
  heat flow, 89
  injection, 157
    flow, 266
  mass flow, 54, 57, 115, 123, 150
  mud flow, 101
  of advance of the local cold front, 176
  of deposition, 288, 290
  of equilibration, 121
  fluid yield, 155
  of heat transfer, 176
  of pressure diffusion, 162
  of quartz precipitation, 115
  of reinjection and extraction, 151
  of scaling, 244
  of upflow, 124
  of withdrawal, 151
  optimum discharge, 103
  sedimentation, 8
  total influx, 66
ratio, Na–K, 140
  of cations in solution, 109
  of cations in the geothermal fluid, 140
  overheat, 44
Rayleigh number, *44f*, 46, 48–60
recent rock alteration, 306
recent silicic volcanism, 208
recharge, areas of, 5
  artificial, 146
  enthalpy, 170
  external, 172
  fluid, 170
  of aquifer, 38
  of deep carstic water, 233
  reservoir, 302, 303
  sources, 303
recirculation of fluid, 173
reconnaissance survey, 81
recoverability, 167, 172, 183
recoverable geothermal energy, 183
recoverable thermal energy, amount of, 186
recovery, analysis, 192
  factor, 165, 173, 182, 183, 186, 187, 191, 193
    for fractured systems, geothermal, 167
    heat, 152
    ideal, 176
    reserve, 183
    resource, 183, 187
  of energy from geopressured-geothermal fluids, *194f*
  of geothermal energy, 239
  of thermal energy from hydrothermal convection systems, 189, *193f*
Red Sea, 29
re-equilibration, chemical, 123, 136
  of the oxygen isotope of sulfate, 122
  water-rock, 111, 120, 122
reflections, multiple, 98
  seismic survey, refraction and, *98f*

regional alteration, 258
regional geology, 79, 82
regional heat flow, 9f, 26, 188
  conductive, 5
  distribution, 22
    patterns, 7
  provinces, 5
regional seismicity, 208
regional tectonic stresses, 55
regions, side boundary, 170
reinfiltration, zone of, 6, 236
reinjection, 205
  and extraction, rates of, 151
  area of maximum, 170
  capacity, 326
  –extraction rate, effects of the, 329
  for subsidence control, 206
  of collapsed pores by condensate injection, 39
  of fluid, 152
  of hot water, 263
  of hypersaline brines, 205
  potential effects of, 151
  pressure, 326
  program, 326f
  test, 157, 265
  well, 152, 157, 326
  wells, capacity of, 328f
  water, 157
remote sensing methods, 90
reserve, 25, 182, 187
  geothermal, 185
  hydrothermal, 62
reserve calculations, 183
reserves, of hydrothermal fluid, 305
  of thermal waters, 221
  recovery factor, 183
reservoir, behaviour, 326, 329
    prediction of, 73
  beneath irrigated fields, geothermal, 206
  boundaries, 67
  carbonate thermal water, 241
  characteristics, 103, 148
  engineering, 109, 136f
  enthalpy, 150
  enthalpy of the warm-water, 151
  fluids, 110
    unflashed, 288
    chemistry, 284
  formation, 323
  fractured, 167
  geothermal, 3, 5, 46, 48, 60, 110, 261
  heat extraction from geothermal, 145f
  high porosity/permeability, 4
  high-temperature, 118
  hot subsurface, 305
  hot-water, 91, 145, 150, 157, 160
  lifetime, 152
  low-permeability, 175
  minimum temperature of the, 125

of geothermal energy, economic, 77
of ideal permeability, 193
performance, prediction of hydrothermal, 60
permeable, 172
permeability, 55, 63
possibilities in the Pannonian basin, high-
    enthalpy, 244
pressure, 147, 205, 263, 326
  change, 332
  drop, 42
  recharge, 302, 303
regional thermal water, 226
rock, 137, 211
  high-porosity, 6
rocks, thermomechanical properties of, 63, 67
single-phase, 150, 164
  water-saturated, 156
steam, 171
steam-saturated, 99
system, Upper Pannonian, 239
temperature, 57, 72, 103, 115, 129, 136, 138,
    139, 140, 175, 199, 259, 261, 262, 335
  estimation of, 6, 133f
  initial, 326
  maximum, 9
two-phase, 145, 148
vapour, 172
vapour-dominated, 148, 305
volume, 192, 195
warm-water, 145, 157f
water-saturated, 150
residence time, 5, 121, 140
  fluid, 123, 140
residual brine, 128
residual gravity anomalies, 99
residual liquid water, 124
residual saturation, 172
resistivity, anisotropic variation of, 92
  anisotropy, electrical, 93
  anomaly, low electrical, 91
  boundary, 69
  electrical, 80, 91
  mapping, 6
  measurements, 93
    electrical, 8, 87
  of formation water, 8
  prospecting, 81
  rock, 6
  survey, 166, 280
    electric, 247
    Schlumberger or Wenner, 79
  zone, low, 95, 261
resource, 182, 188
resource, aluminium, 184
  assessment, geothermal, 181f
    methodology for geothermal, 185f
    recent geothermal, 186f
    systematic geothermal, 186f
  available energy, 151

available geopressured-geothermal, 195
base, 182
    accessible, 184
    accessible fluid, 192, 194
    fluid, 188
    geothermal, 25f, 184
    inacessible, 185
    residual accessible, 185
    undiscovered accessible, 192
calculations, petroleum, 183
estimation, energy, 182
    geothermal, 181
geothermal, 56, 77f, 97, 181f, 183, 184, 304, 315
hydrothermal, low to moderate temperature, 314
information, basic geothermal energy, 82
recovery factor, 183, 187
prospecting for geothermal, 77f
terminology, 181f
    geothermal, 184f
resupply of heat, 186
Reykjanes, Iceland, 135, 130
rheological properties of the astenosphere, 31
Ridge, Carlsberg, 29
Ridge, Mid-Atlantic, 275
ridge, mid-oceanic spreading, 28, 29
ridges, spreading, 4, 26
rift, African, 29
    Baikal, 29
    continental, 4, 29f
    East African, Kavirondo, 32
    Rio Grande, 29, 296, 316
rifting episode, 274
rifting event, 291
Rio Grande rift, New Mexico, 29, 296, 316
Risk's electrode configuration, 93
rock, cap, 65, 80, 251, 263
    fluid flow in fissured, 38
rock alteration, 304
    recent, 306
rock and water, specific heats of, 186
rock porosity, 39
rock properties, 67
rock re-equilibration, water-, 111, 120, 122
rock resistivity, 6
rock system, water-saturated fractured, 54
rock temperatures, 137
    initial, 186
rock–water system, 55
rocks, heat content of, 25f
    heat production, in igneous, 14
        in metamorphic, 14
        in sedimentary, 15
    hot country, 190
    impermeable, 12
    low-permeability, 190
    porous permeable, 37
    reservoir, 137, 211

    solidified igneous, 190
    thermomechanical properties of reservoir, 63, 67
    water-saturated, 171
    water-storing, 235
Rotokauwa, New Zealand, 132, 168, 170
roving dipole technique, 92
Rubidium/Strontium method, 310
run-down characteristics, 103
rupture, surface, 209

saline aquifer, 325
saline aquifer,
    geothermal fluids, 181
salinity, 6, 85, 109, 115, 123, 140, 204, 305
    depth distribution, 8
    effect, 8, 56
    of formation water, 8
    of the surface discharge, 81
Salton Sea geothermal reservoir, California, 29, 37, 64, 181
sampling, chemical, 84
San Andreas fault, 210
San Fernando earthquake, 208
saturation, residual, 172
saturation pressure, 332
scaling, rate of, 244
scaling control, 244
Schlumberger, electrode configuration, 92
    method, 95
    resistivity survey, 79
sealing boundary, 67
    sealing, self-, 317
seamounts, 31
seasonal variation in a single spring, 123
sedimentary aquifer, 5
sedimentary geothermal system, 49
sedimentary rocks, low-temperature geothermal waters in, 195
sedimentation, 13
    rate, 8
seepages, hot, 82, 90
seismic activity, 88
seismic groundnoise, 97, 304
seismic noise anomaly, 97
seismic methods, passive, 6, 89
seismic refraction measurements, 280
seismic regions, active, 38
seismic signals, high-frequency, 89
seismic source, location of the, 89
seismic surveys, refraction and reflection, 98f
seismic techniques, active, 7
seismic waves, attenuation of, 6, 97, 98
seismic waves in geothermal areas, attenuation of, 99
seismicity, 208f
    induced, 210
    regional, 208
self-potential method, 96
sensing techniques, remote, 7
separation pressure, 328

serpentine, 117
shale dewatering, clay deydration or, 173
side inflow, *150f*
Sierra Nevada, east-central California, 62
   type of low heat flow, 188
silica, 113
   amorphous, 115
   aqueous, 116, 118
   concentration, 137
      in flashed well waters, 138
      initial, 124
      of dissolved, 109
   content of hot water, 259
   deposition, 124
   enthalpy method, 138
   geothermometer, 8, *113f*, 126, 127, 138
   minerals, paragenetic sequences of, 116
   mixing models, 123
   monomeric, 116
   phases, solubility of, 116
   polymerization of amorphous, 116
   polymerized, 116
   precipitation of amorphous silica, 116
   precipitation of, *115f*, 263
   solubility, 84, 121, 138
      of amorphous, 116
   temperature, uncorrected, 85
silicate hydrolysis reactions, 117
silicate minerals, acid alteration of, 117
silicic igneous system, young, 190
silicic intrusions, 188
silicic magmas, 26
single-phase flow, 152
single-phase geothermal systems, thermally driven flow in, 42
siting of an initial exploration well, 101
skin effect, 280
Snake River Plain, Eastern, 193
solidification history, 15
solubility, curve, 115
      quartz, 123, 124
   of amorphous silica, 116
   of quartz, 114, 115, 117
   of silica phases, 116
   retrograde, 84
   silica, 84, 121
sonic logging, 102
sources, fluid supply from mantle/deep crustal, 6
   low-temperature, 204
   long-lived radiogenic heat, 12
   magmatic heat, 3
   of extracted energy, 153
   short-lived (transient), 13
   stationary, 10
   steady-state, 13
   transient, 10, 13
spatial distribution of mass transfer, 54
specific heat, of rock and water, 186
   volumetric, 186

speed of advance of the cold front, 170
spreading rate, 29
spreading ridges, 4, 26
   mid-oceanic, 28
   oceanic, 29
spreading velocity, 28
springs, acid-sulfate, 128
   alkaline, 128
   areas, hot, 116
   boiling, 5, 83, 84, 115
   Cerro Prieto, 129
   composition of discharging hot, 6
   discharges in fault/fracture zones, hot, 6
   discharge temperature, 7
   geochemical analysis of natural, 279
   geographic distribution of, 136
   hot, 37, 56, 57, 62, 82, 84, 86, 90, 120, 186, 252, 259, 304, 306
   location of thermal, 83
   radioactive components in, 204
   seasonal variations in a single, 123
   thermal, 5, 204, 228, 233
   water, alkaline hot, 118
      hot, 122, 125, 118, 140
      sodium bicarbonate, 128
stability fields, mineral, 111
stabilization period, 332
steady-state conduction temperature profile, 47
steady-state initial conditions, 150
steady-state temperature distribution, 71
steam, cap, 137
   counterflowing, 148
   discharge, 171
   dry, 91
   excess, 137
   field, Lardarello, Italy, 38
      The Geysers, California, 38
   filled zone, 100
   flow distribution of water and, 172
   geothermal, 263
   loss, continuous, 122, 124
      maximum, 124, 126
      one step, 122
      single-stage, 124
   mobility, 170
   phase, 126
   reservoir, 171
   saturated reservoir, 99
   segregation, 138
   separation, 119, 122, 125
      effects of, 115
   superheated, 147, 172, 290
   underground, 321
      –water, mixture, two-phase, 153
      ratio, 156
   systems, 41
   zone, 148
   well, 156

zone, 127, 332
  dry, 157
Steamboat Springs, Nevada, 58
steaming ground, 83, 90, 186
step faults, 323
storage coefficient, 3
stream function, 45
stress, natural regional, 210
stress history, 41
stresses, regional tectonic, 55
Stretford process, 201
subcoastal environment, 250
subcrustal erosion, 223
subduction velocity, 30
subduction zone, 26, *30f*
subsidence, 83, 206, *211f*
  area of maximum, 212
  control, reinjection for, 202
subsurface dilution of thermal water by cold water, 118
subsurface reservoir, hot, 305
subsurface temperature, 84, 86, 93, 115, 188
  estimated, 83
subsurface water movements, 236
subsurface waters, 236
subterranean magma flow, 275
sulfate, by oxidation of H_2S, formation of, 122
  ion, activity of the, 86
  low-temperature, 122
  re-equilibration of the oxygen isotopes of, 122
sulfate oxygen isotope, 113
  exchange reaction, 121
  geothermometer, *121f*
sulfate springs, acid, 128
sulfides, hydrogen, 200
  oxidation of, 117
sulfur, 214
surface, basement, 190
  condensation, 56
  heat exchange, 9, *305f*
  piezometric, 325
  conductive heat flux, 53
  discharge, 78, 151f, 164, 204
    salinity of the, 81
  drainage, 303
  expressions of geothermal activity, 90
  flux, 60
  geothermal activity, 82, 280
  heat flow, 13, 14, *16f*, 26, 56, 57, 89, 102
  heat flux, 56, 57
    conductive, 60
    maximum, 60
      conductive, 56
    distribution, 48
  heat transfer, 52
  manifestations, pre-exploitation, 165
  mass discharge, 73
  mass flux, 56
  outflow, 150

rupture, 209
spring temperature, 57
temperature, 57, 59
thermal flux, 185
  method of, 186
water, 58, 81, 83, 214
suture, Precambrian basement, 297

Taiwan, Chingshu, 132
  Tuchang, 133
Takinoue area, history of geothermal exploration in the, 249
  Japan, *247f, 252f*
  stratigraphic sequence, 253
Tauhara field, New Zealand, 89, 165
Taupo volcanic zone, New Zealand, 47
tectonic fracturing, 323
tectonic setting, 82
tectonic stresses, regional, 55
tectonism, extensional, 6, 29, 297
teleseismic P-wave delays, 6, 98
telluric current measurements, 96
temperature, anomaly, 11
  aquifer, 5, 7, 138
  aquifers in high-porosity environments, low, *7f*
  bottom-hole, 91, 230
  cation geothermometer, 139
  change in underground, 138
  changes in production wells, 136
  contours, pre-production model, 67
  decline, progressive, 167
  dependence of the fractionation of the isotopes, 85
  dependence of selected geothermometers, 114
  deposits, high-, 199
    low-, 199
  -depth curve, 13, 9
  distribution, 67, 160
    steady-state, 71
    subsurface, 18
  distribution of down-hole, 265
  down-hole, 102, 261
  drop, 153
  enthalpy, 138
  equilibrium reservoir, 230
  equilibrium rock, 230
  estimated subsurface, 83
  estimation of reservoir, *113f*, 6
  feed zone, 148
  field, background, 48
    reference hydrostatic steady-, 44
  fluid, 127, 137
  geothermal fluid, 101
  gradient, 10, 45, 53, 71, 89, 102
    survey, 89
  homogenisation and melting, 311
  in wells, aquifer, 136
  initial reservoir, 326
  initial rock, 186

isotherm, distortion of high, 52
local mean annual, 184
magnesium-corrected, 136
maximum reservoir, 9
mean, 186
mean annual, 181, 183
  surface, 25
measurement of, 88
minimum acceptable outflow, 186
Na–K, 129
Na–K–Ca, 85, 136
of deep geothermal waters, 84
of the reservoir, minimum, 125
outflow, 233
paleo, 310
pre-production, 60, 66, 73
profile, 168
  at Kawerau, New Zealand, 166
  steady-state conduction, 47
quasi-equilibrium, 230
reservoir, 57, 72, 103, 115, 121, 129, 136, 138, 139, 140, 175, 199, 259, 335, 118
rock, 137
silica, 138
spring discharge, 7
subsurface, 84, 86, 93, 115, 188
sulfate/oxygen isotope geothermometer, 121
surface, 57, 59, 90
surface spring, 57
survey, underground, 247
uncorrected silica, 85
underground, 109, 228
variations, diurnal or seasonal, 89
wellhead, 191
Texas, 189
  Gulf Coast, U.S.A., Brazoria geothermal fairway, 173
The Geysers, California, U.S.A., 4, 6, 21, *37f*, 68, 168, 171, 172, 176, 201, 205, 206, 210, 212
The Geysers–Clear Lake area, 6, 100
The Philippines, 208
thermal activity of the Pannonian basin, 228
thermal alteration, 276, 307
thermal anomaly, 307
  detection of, 80
  intraplate, *31f*
  positive, 233
  sublithospheric, 32
thermal belt, New Zealand, 165
thermal boundary, 160
  layer, 57
thermal conductivity, *10f*, 18, 41, 89, 39, 230
  uniform, 188
thermal diffusivity, 18
thermal dome, 32
thermal energy, 181, *189f*,
  amount of recoverable, 186
  from hydrothermal convection systems, recovery of, *193f*

igneous-related, *191f*
in geopressured waters, 189
loss, rate of, 186
thermal expansion, 6
  effects, 42
thermal fluids, 98
  discharge of, 186
thermal flux, method of surface, 186
  surface, 185
thermal front, 151
  velocity of the, 151
thermal history of sedimentary formations, 8
thermal loading, 214
thermal mantle diapirism, 223
thermal metamorphism, 55
thermal properties, near-surface, 57
thermal springs, 5, 204, 228
  location of, 83
thermal stressing, 55
thermal waters, 80, 83, 122
  bearing formations, 235
  multipurpose utilization of, *243f*
  natural, 113
  radioactive, 204
  reserves of, 221
  reservoir, carbonate, 241
    regional, 226
  resources of the Pannonian basin, 228
  utilization, 241
thermally driven flow in single-phase geothermal systems, 42
thermistors, array of thermocouples or, 90
thermomechanical properties of reservoir rocks, 63, 67
thermometer, geochemical, 113, 140, 244,
  geologic, 259
  mercury, 230
thorium, 13, 195
Tibesti, 32
time, fluid-residence, 123, 140
  post-solidification cooling, 23
  residence, 121
  water-residence, 110
time-domain technique, transient, 96
time of emplacement, 188, 190, 194
trace-element tolerances for irrigation waters, *205f*
tracer-tests, 67
transient flow, 154
transmission, pressure, 146, 152
transmissivity, 5, 147
Travale, Italy, 171
triangular probability density, 193
  example of a, 191
  maximum estimates of a, 191
triple point, 111
tritium, 123
Tuchang, Taiwan, 133
Turkey, Kizildere, 99, 133, 148
Tuscany, central and southern, *187f*

two-phase, flow, 59
  interface, water, 169
  /liquid interface, 175
  mixture, 153, 154
  reservoir, 145, 148
  system, 152f, 167

Uganda, 29
underground, boiling, 127
    effects of, *124f*
  flow, 90
  fluid, 80
  hot water movement, direction of, 136
  mixing, *122f*
  mixing and boiling, 118
  steam, 321
  temperature, change in, 138
  water, flow of, 97
  water movements, 83
unflashed reservoir fluid, 288
uniaxial compaction, 41
updoming of the astenosphere, 233
upflow, rate of, 124
upflow region, 54
Upper Rhine Graben, Central Europe, 29
  Western Germany, 8
Urach, Western Germany, 32
Urach anomaly, South Germany, 23
Uranium, 13, 195
U.S.A., 189, 195, 188f, *190f*
  Basin and Range Province, 17, 26, 29
  Brazoria geothermal fairway, Texas Gulf Coast, 173
  Eastern, 17
  geopressured basins of the, 192
  Heber, California, 147, 159,
  Imperial Valley, California, 148, 206
  major geologic provinces of, 192
  New Mexico, 8
  Rio Grande Rift, New Mexico, 29, 296, 316
  Roosevelt Hot Springs, Utah, 211
  Salton Sea, 29, 37, 64, 181
  Sierra Nevada, 17
  The Geysers, 4, 6, *37f*, 21, 68, 168, 171, 172, 176, 201, 210, 212, 205, 206
  Valles Caldera, New Mexico, 9
  Wilmington oil-field, 210
  Western, 32, 190
  Yellowstone Park, 134
Utah, Roosevelt Hot Springs, U.S.A., 211
U–Th–Pb dating, 310
utilization energy from a geothermal reservoir, 148
utilization factor, 191
utilization of geothermal energy, 239
utilization of geothermal resources, 9
utilization of thermal water, multipurpose, *243f*

Valles Caldera, New Mexico, U.S.A., 9

vapour, mobility, 170
  reservoir, 172
vapour-dominated geothermal reservoirs, 172
vapour-dominated geothermal system, 191
vapour-dominated reservoir, 148, 176, 305
vapour-dominated systems, 4, 23, 37, 90, 91, *127f*, 139, 145, 152, 156, 157, 168, *171f*, 252
vapour flow, natural, 170
vapour or water dominated field, 78
vapourization, intergranular, 173, 187
velocity plot, 67
velocity of the thermal front, 151
venturi meter, 90
vibroseis, 99
viscosity, kinematic, 42, 154
  system, 45
  total dynamic, 154
Viterbo region, Italy, 210
volcanic activity, 275, 284, 295
volcanic belts, crustal, 193
volcanic environments, geothermal reservoirs in, 6
volcanic or intrusive activity, 83
volcanism, 28, 297
  intraplate, 31
  postglacial, 275
  recent silicic, 208, 304

Waiora aquifer, 70
Waiora formation, 70, 164, 212
Wairakei, and Broadlands, 187
  and Waiotapu, 97
  breccia, 70
  geothermal field, geological section across the, 161
  geothermal field, New Zealand, 4, *37f*, 56, *86f*, 85, 86, 90, 100, 137, 147, 148, 149, 150, 152, 160, 162, 164, 165, 166, 168, 176, 199, 200, 211, 214
  geothermal system, a real model of, 72, 73
  ignimbrites, 70
  reservoir, 162
water, adiabatic expansion of, 326
  alkaline hot spring, 118
  and steam, flow distribution of, 172
  and steam, permeability of, 290
  bearing formations, thermal, 235
  bearing strata, 172
  bicarbonate, 238
  Black Sand-type, 126,
  calcium-rich, 119
  chloride, 88
  circulation, 5
  circulation, deep, 83
  circulation systems with meteoric, *6f*
  closed basin saline lake, 123
  cold meteoric, 150
  column, pressure, 265
  composition, Broadlands, 111
  compressibility, 173

connate, 8
  pore, 5
deep geothermal, 121
dissolved methane in geopressured, 195
domestic, 243
-dominated field, hot, 88
-dominated geothermal reservoir, 286
  identified hot, 191
-dominated system, 282
downflow of meteoric, 230
fed wells, 147
flow of, underground, 97
  Geyser Hill type, 125, 126
front, cold, 173
geochemistry, *109f*
high-temperature geothermal water, 119
hot-spring, 115, 122, 125, 140
hydrothermal, 90
immobile, 156
in drillholes, composition of natural, 116
in sedimentary rocks, low-temperature geothermal, 195
inflow, cold, 164, 165
injection, 306
interface boiling, 168
level, equilibrium, 238
  model, falling, 171
low-enthalpy, 239
low-temperature, 204
low-temperature geothermal, 190
magnesium-rich, 121
meteoric, 3, 6, 205, 236
migration, 230
  directions, 238
  effects of, 11
  systems, 236
mineralized, 236, 303
mixture, two-phase steam-, 153
mobility, 170
movement of recharge, 139
movements, subsurface, 236
  underground, 83
    direction of hot, 136
natural, 111
natural geothermal, 118
natural hot-spring, 118
natural thermal, 113
near-surface, 43
ocean, 123
phase, dispersed, 172
plumes of hot, 99
pore, 189
radioactive thermal, 204
quality, *204f*
recharge, 38
recharge of deep carstic, 233
recognition of mixed, *122f*
reinjection, 157
reinjection of hot, 263

reserve of thermal, 221
reservoir, hot, 91, 145, 150, 160
  warm, 145, *157f*
residence time, 110
residual liquid, 124
resistivity of formation, 8
resources of the Pannonian basin, thermal, 228
rock, equilibration, 111, 120, 122
  equilibrium, 113
  reactions, 120
  re-equilibration, 122
  systems, natural, 110
salinity of formation, 8
samples, down-hole, 138
  wellhead, 138
saturated fractured rock system, 55
saturated porous medium, heat and mass transfer in a, 48
saturated reservoir, 150
  single-phase, 156
saturated rocks, 171
saturated zone, hot, 148
saturation, uniform, 154
silica concentration in flashed wells, 138
silica content of, 259
single-phase, 102
sodium bicarbonate spring, 128
specific heats of rock and, 186
storing rocks, 235
subsurface, 236
surface, 7, 38, 58, 81, 83, 148, 214
system, composition of geothermal fluids in hot-, 200
  hot-, 80, 86, *127f*, 139, 252
  karstic, 235
  rock, 55
steam-, 41
warm-, 150
tables, perched, 303
temperature of deep geothermal, 84
thermal, 80, 83, 122
thermal energy in geopressured, 184
trace-element tolerances for irrigation, 205
two-phase interface, 169
viscous properties of, 47
well, 115, 119
well, behaviour, 137
  capacity of reinjection, *328f*
  discharge, 152
  enthalpy, 167, 170
  feed channel, 152
  level, 150
  feed point of a, *147f*
  injection, 174
  capacity of a, 266
  logging method, 102
  production, 326
  reinjection, 152, 157, 326
  siting of an exploration, 101

response analysis, 147
steam, 156
temperature changes in production, 136
two-phase, 147
water-fed, 147
waters, 115, 119
   silica concentration in, 138
wellhead, enthalpy measurements, 137
  pressure, 71, 189, 263, 264
  temperature, 191
  water samples, 138
Wenner electrode configuration, 92
Wenner resistivity survey, 79
Western Germany, Landau area, Upper Rhine Graben, 8
  Urach, 32
Wilmington oil field, U.S.A., 210
withdrawal front, mass and heat, 156
work, available, 191
World Health Organisation, 200

xenoliths, 250, 311
X-ray analysis, 289
X-ray diffraction, 258

Yellowstone caldera, 6
Yellowstone National Park, U.S.A., 91, 92, 122, 125, 126, 134
yield, of energy, 156, 157
  of mass, 156, 157
  pattern of fluid, 155

rate of fluid, 155

zeolite to greenschist facies metamorphism, 282
zeolites, 111
zonality, hydrochemical, 239
zone, absorption, 329
  Benioff, 30
  boundaries, 149
  boundary, 93
  charge, 54
  dry steam, 157
  feed, 164
  geopressured, 4, 80
  hot-water saturated, 148
  liquid and two-phase, 169
  location of the production, 78
  low resistivity, 95, 261
  mass and heat transfer in a fault, 58
  near-surface, 148
  neovolcanic, 275
  of drawdown, 157
  of expellation, 236
  of high gas concentration, 260
  of infiltration, 6, 236
  permeable feed, 167
  pressure of the production, 171
  production, 139, 169
  steam, 127, 332
    water, 148
  temperature, feed, 148
  two-phase, 157